COMMUNICATION

SYSTEMS

by
Charles D. Johnson
Professor, Communication Technology
Department of Industrial Technology
University of Northern Iowa
Cedar Falls, Iowa

Publisher
The Goodheart-Willcox Company, Inc.
Tinley Park, Illinois

ABOUT THE AUTHOR

Charles D. Johnson's professional background includes over 20 years of teaching communication courses at the junior high, high school, and college levels. Dr. Johnson is active in Technology Education at both state and national levels. He has served on a variety of state and national committees and has authored articles for national publications as well as serving on the Editorial Review Board for *The Technology Teacher*. He has also co-authored a guide on integrating special needs students into technology classes. He received his undergraduate degree from Florida State University, a master's degree from Western Carolina University, and a doctorate from North Carolina State University.

Library of Congress Catalog Number 99-10560
International Standard Book Number 1-56637-615-7

1 2 3 4 5 6 7 8 9 10 00 03 02 01 00 99

Cover Image: Lonnie Duka / Tony Stone Images

Library of Congress Cataloging-in-Publication Data

Johnson, Charles D. (Charles David),
Communication Systems / by Charles D. Johnson

p. cm.
Includes index.
ISBN 1-56637-615-7
1. Communication and Technology. I. Johnson, Charles D.
II. Title.
P96.T42J64 2000
302.2--dc21 99-10560
CIP

INTRODUCTION

COMMUNICATION SYSTEMS introduces you to the variety of ways people use technology to communicate. The communications field is a dynamic one that is growing and changing rapidly. This text gives you an overview of technological advancements. It also describes conventional methods of communication.

COMMUNICATION SYSTEMS is divided into seven sections, including:

- An introduction to communication technology.
- Basic communication skills, including design, problem solving, sketching, measurement systems, and safety and health concerns.
- Technical Drawing and Illustration.
- Photography.
- Graphic arts and desktop publishing.
- Electronic communication.
- Business, career exploration, and the impact of technology.

The 41 chapters of this text provide knowledge about the various communication technologies. You can apply much of this information in your everyday life, in a laboratory setting, and in the future.

COMMUNICATION SYSTEMS is heavily illustrated with colorful photographs, drawings, diagrams, and original artwork. References to all illustrations are included in the copy to help you associate the visual images with the written material.

Each chapter begins with a list of instructional objectives. New vocabulary terms appear in bold. Following each chapter, a list of words to know is given along with review questions covering the material presented in each chapter. Several end-of-chapter activities are also given that will allow you to apply your knowledge. Following each section, basic, advanced, and above and beyond activities will challenge you to apply concepts you have learned throughout the section.

This text includes an expanded table of contents to give you an overview of the wide variety of topics you will be studying in this book. An index/glossary reference is included to help you locate the terms and information you want quickly and easily.

A Student Activity Manual is available for use with COMMUNICATION SYSTEMS. A study guide will help you to review concepts presented in the text. Many of the suggested activities will provide you with hands-on experiences.

You will find that communications is an exciting area of study. COMMUNICATION SYSTEMS is a comprehensive text for anyone interested in the career opportunities available in the field of communications and communications-related jobs. It will enable you to obtain a working knowledge of what people involved in communication do. It will also show you how communication technology affects and will continue to affect many aspects of your life now and in the future.

Charles D. Johnson

CONTENTS

section one:
Introduction to Communication

section two
Basic Communication Skills

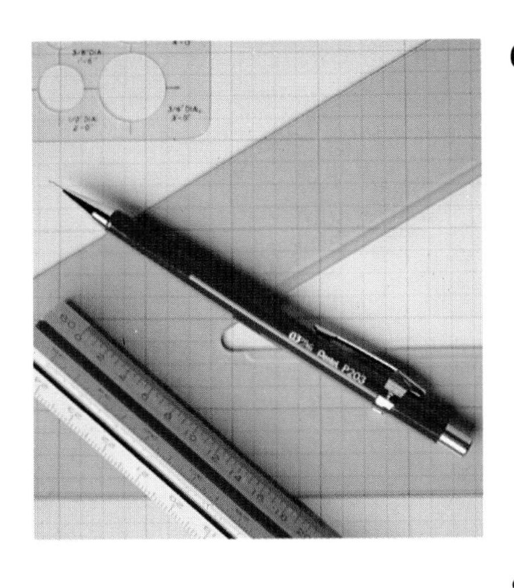

section three

Technical Drawing and Illustration

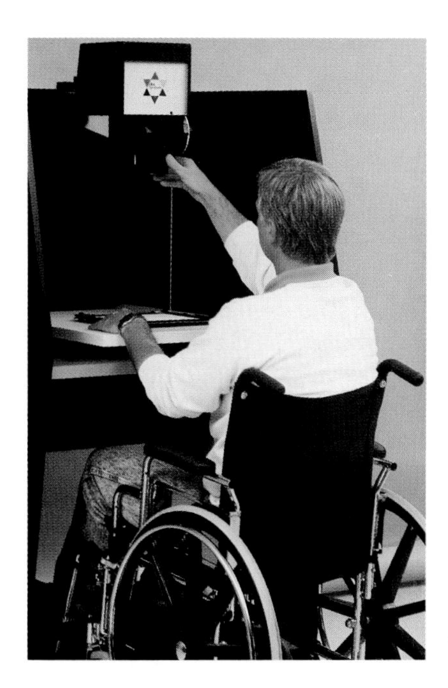

section four

Graphic Communication-Photography

section five

Graphic Communication-Graphic Arts and Desktop Publishing

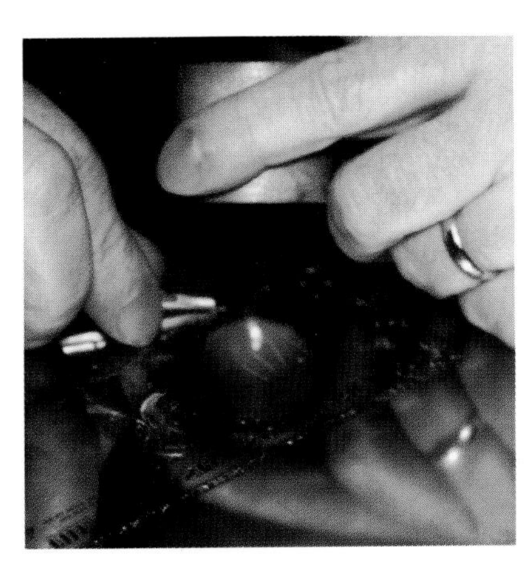

section six

Electronic Communication

section seven

Communication, Jobs, and You

NORAND CORP.

NASA

BOWLING GREEN STATE UNIVERSITY, VISUAL COMMUNICATION TECHNOLOGY PROGRAM

section 1

Introduction to Communication

IBM CORP.

IBM CORP.

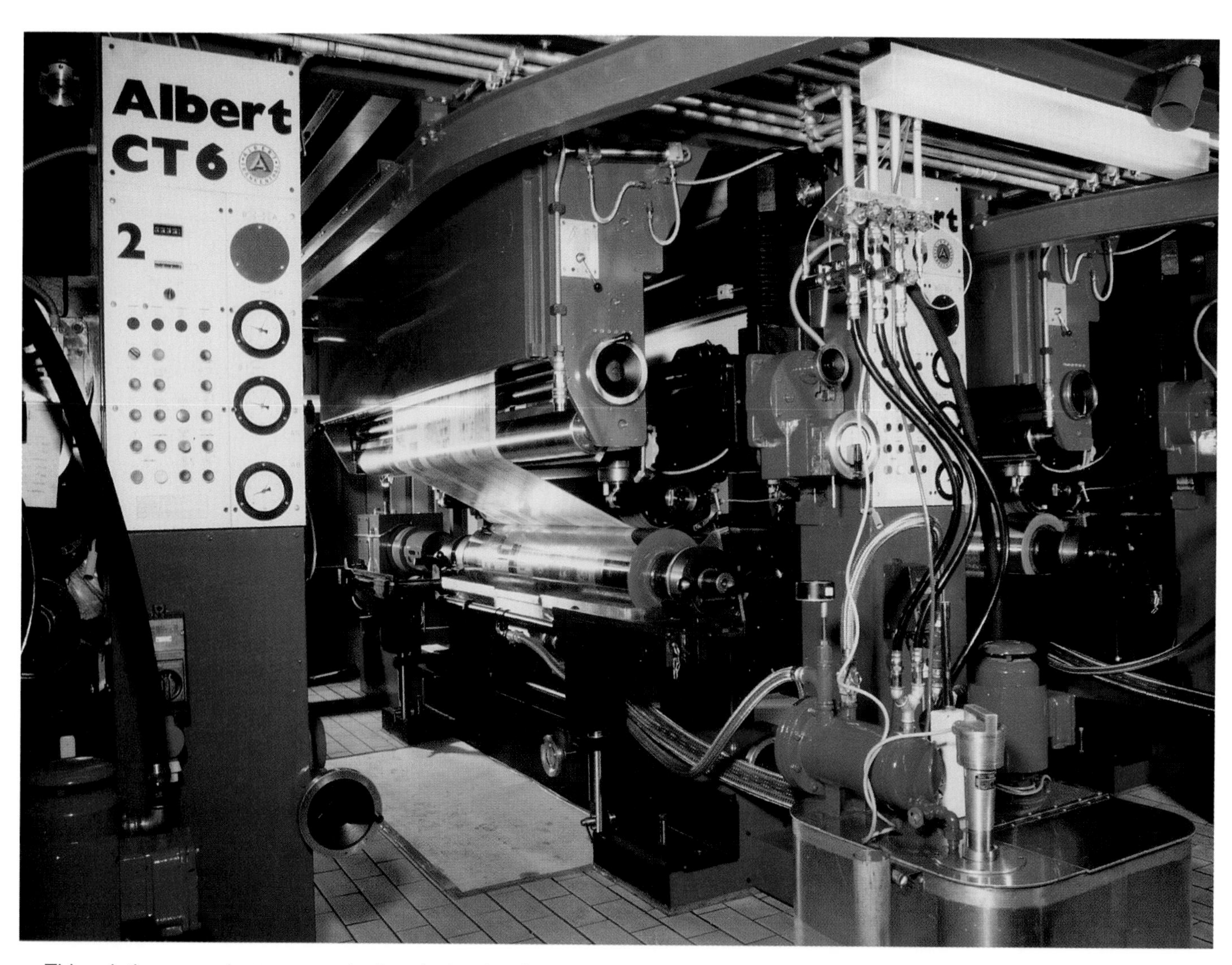

This printing press is a communication device. It prints magazines and other products that are read by millions of people. (Koenig & Bauer)

1

CHAPTER

INTRODUCTION TO COMMUNICATION

After studying this chapter, you will be able to:
- *Define communication.*
- *Describe the components of a communication model.*
- *List five major purposes of communication.*
- *Describe various types of communication involving humans and/or machines.*
- *Explain how human sensory systems can aid in communication.*
- *Distinguish between verbal and nonverbal methods of communication.*
- *Give examples of how perception affects human communication.*
- *Outline the history of human communication.*
- *Cite some predictions of future communication technologies.*

What is communication? **Communication** is the sending and receiving of information or messages. For instance, this textbook is a message. If you read and understand the words, you have received the message, and communication has occurred. However, suppose this textbook were written in Arabic or a language in which you were not fluent. Suppose that you were tired or distracted while trying to read. Would you say that communication took place? Probably not, since you most likely did not receive the intended message. Therefore, it is possible to send or transmit information without it reaching its destination. In order for communication to occur, a message must be both sent and received.

A SIMPLE COMMUNICATION MODEL

At the most basic level, complete communication includes a sender, message, and receiver. However, we can be more precise by adding a few more elements.

The message needs to go through a channel, or medium, of some sort. The channel holds the message as it is being sent. This might be a wire in the case of a telephone, or even the air as in the case of a cellular phone. In some cases, the message may be stored in the channel until used, as with a videotape or book. In addition, there are often several channels the message takes before reaching the receiver. A television program signal might be sent through the air, then a wire, and last, the television itself. The term medium is often used to describe the last channel in the system, such as radio, television, books, and newspapers. An example would be the statement radio is the best medium for advertising this product.

Marshall McLuhan wrote a book in 1967 entitled *The Medium is the Message*, and this title is now a famous quote. He believed that the medium has an effect on the message received. For example, a book or radio show requires you to imagine a visual image, while a television provides this image for you. Sending the same message by e-mail and phone is another example. The phone message will provide the listener information on the sender s voice and mood, while the e-mail message contains other information, such as the sender s ability to organize thoughts in letter form.

Another helpful addition to our model is feedback. How do we know that communication has occurred unless we get this information back from the receiver? If you tell a joke and get a laugh from the other person,

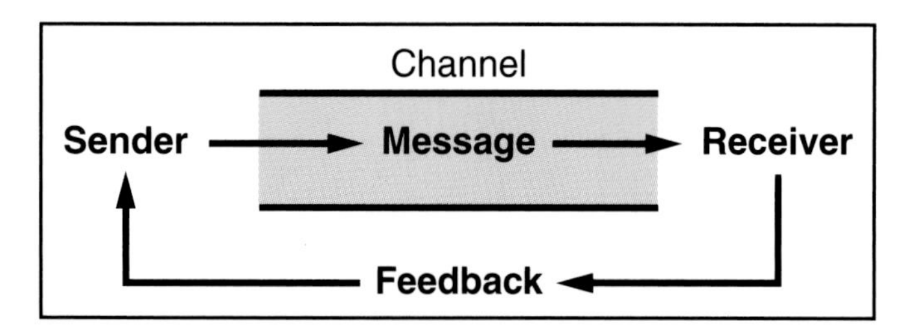

Fig. 1-1. The basic communication model has five elements.

this is feedback that they think the joke is funny. Our more complete model, shown in Fig. 1-1, now has the five elements of sender, message, channel (or medium), receiver, and feedback. Can you think of another communication example using these five elements?

WHY DO WE COMMUNICATE?

Humans have certain basic needs. A baby may cry (communicate) because he or she is hungry, cold, or scared. Food is a basic need for survival, along with shelter and security. Without them, neither the baby nor anyone else could survive. Communication enables us to meet our basic needs.

Communication has five major purposes, Fig. 1-2. These are to:

- Inform.
- Educate.
- Persuade.
- Entertain.
- Control.

In the past, humans survived by informing others that danger was near. Children were educated on how to make shelters and weapons. If people had disagreements about which trail to travel, they used communication to try to persuade each other. When free time was available, people learned to use communication to entertain one another through art and music. Communication was used, as it is today, to control people. For example, societies create rules for their members to follow.

Although communication is used for the same basic reasons today as in the past, technology has made it much more complex. You can receive information via satellites from all over the world. Textbooks, computers, and videos aid teachers in educating students. Advertisements on billboards, magazines, television, and radio constantly persuade you to buy products and services. Videos, books, and CDs keep you entertained. Communication can be used to control people and machines. For example, an alarm clock often helps to control how long you sleep. A light switch controls whether the light is on or off. More complex control devices function automatically without human intervention. For instance, an automatic camera senses the amount of light needed for an exposure and closes the shutter at the correct time. Computers can be used to control other machines such as printers and robots.

TYPES OF COMMUNICATION

As discussed earlier, both senders and receivers are needed for communication to take place. Senders and receivers can be people, animals, or machines. See Fig. 1-3. Communication, as covered in this text, primarily occurs in the following ways:

- *Human-to-human communication:* When people speak to one another, write notes, or gesture, this type of communication takes place. A teacher giving a lecture and using the chalkboard is an example of this.
- *Human-to-machine communication:* People often operate machines to communicate. Examples of these machines are computers, telephones, and musical instruments.
- *Machine-to-machine communication:* When one machine signals or controls another, this type of communication takes place. This process is often called **cybernetics**. Examples include computers that control machine tools, thermostats, and timers. Another example would be automatic switches that can sense an unsafe condition and stop tool motion.
- *Machine-to-human communication:* In this type of communication, machines transfer messages to humans. Examples include smoke and fire alarms, alarm clocks, gauges, and buzzers. This type of communication serves several purposes. Machines are used to entertain, give information, and save time and energy. In vehicles, these machines can warn of necessary service or breakdown.
- *Supplementary types of communication:* These types of communication are other systems of exchanging information that can involve humans, animals or machines. Examples of supplemental systems include ESP (extrasensory perception) and animal communication (both with humans and with other animals). Other types include mineral communications (using a compass, magnet, or seismograph) or extraterrestrial communication (messages received from outer space).

HUMAN COMMUNICATION

Humans have developed a highly complex spoken and written language. They also have the ability to change their world. Humans have extended their

MAJOR PURPOSES OF COMMUNICATION

TO INFORM

TO EDUCATE

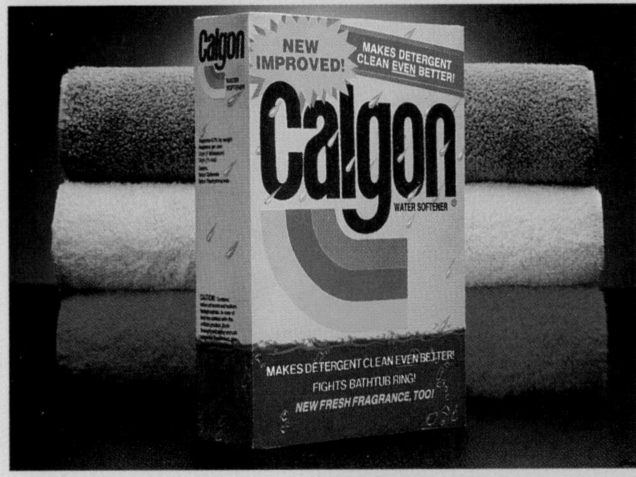

TO PERSUADE

TO ENTERTAIN

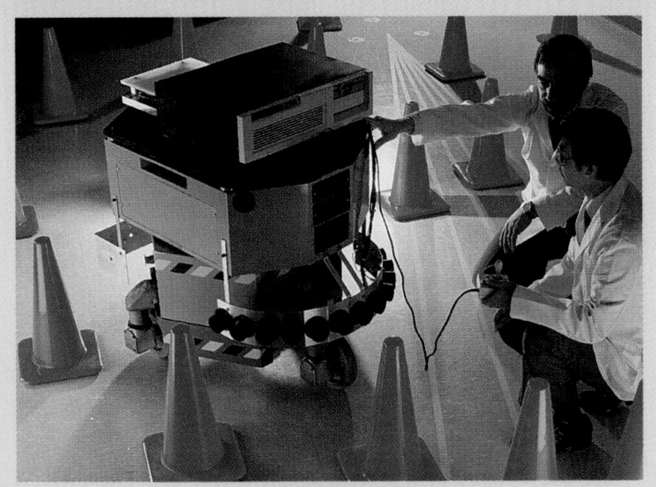

TO CONTROL

Fig. 1-2. Communication serves five major purposes: **To inform**—This faucet part identification system uses a computer to give customers information. (Galaxy Chemical Co., Inc.) **To educate**—Teachers use communication skills in educating students. The use of videos, the chalkboard, and projectors aids in this process. (Chrysler Corp.) **To persuade**—Packaging and advertisements can be used to persuade you to buy products. (Beecham Products) **To entertain**—Comedy shows are often televised, providing entertainment. (WGN Television.) **To control**—A computer program would be needed to control the movements of this robot. (Siemens AG)

Fig. 1-3. These are various types of communication: **Human-to-human**—Talking with another person is a good way to share thoughts and ideas. (Nancy Wehlage) **Human-to-machine**—A cellular phone allows people to conduct business while on the road. (Motorola, Inc.) **Machine-to-machine**—This bar code scanner and hand-held computer send inventory information to a host computer. (Norand Corp.) **Machine-to-human**—This computer display screen provides "real time" information to a system operator. (National Instruments) **Supplementary**—Animals often communicate with humans, and humans can communicate with animals. (Colgate-Palmolive Co.)

ability to communicate by using machines. As technology has evolved, machines have played a larger role in communication.

Most humans have five senses. These are *seeing, hearing, touching, tasting,* and *smelling.* In each of these sensory systems, nerves send electrical signals to the brain. The brain then interprets or decodes these signals. Your most important senses for communicating are seeing and hearing. Seeing is the sense that allows you to receive visual messages. Hearing allows you to receive sound waves. Touch allows you to detect contact, temperatures, textures, and shapes. Taste allows you to distinguish various flavors in foods. The sense of smell enables you to detect various scents and odors.

The five senses can be used by themselves, in combination, or all at once. See Fig. 1-4. When you talk on the phone (hearing) or read a book (seeing), you are using only one sensory system at a time. When you watch and listen to a TV program, you are using a combination of senses (seeing and hearing). Think about which senses would be involved in a visit to a restaurant. Probably all five of your senses would become involved in the experience.

Some people are not able to communicate using one or more of their senses. People unable to hear or speak often use a language called **signing**. In signing, hand signals are used to spell out words. See Fig. 1-5. **Braille** enables blind people to read. Braille words are printed as raised dots. A person can "feel" the words by moving his or her fingers across a page.

Fig. 1-4. Listening to music requires one sense, hearing. Playing musical instruments may require several senses, including touching and hearing.

Verbal and nonverbal communication

We often send both verbal and nonverbal messages when we communicate. Human communication can be described as either **verbal communication** (using words) or **nonverbal communication** (using factors other than words). See Fig. 1-6.

If you were asked to identify how people normally communicate, how would you respond? Most of us would think of verbal communication. This is the process of exchanging information using a language that people have in common. People use language to solve problems and to think about past and future events. In fact, language is thought to be one of the primary reasons

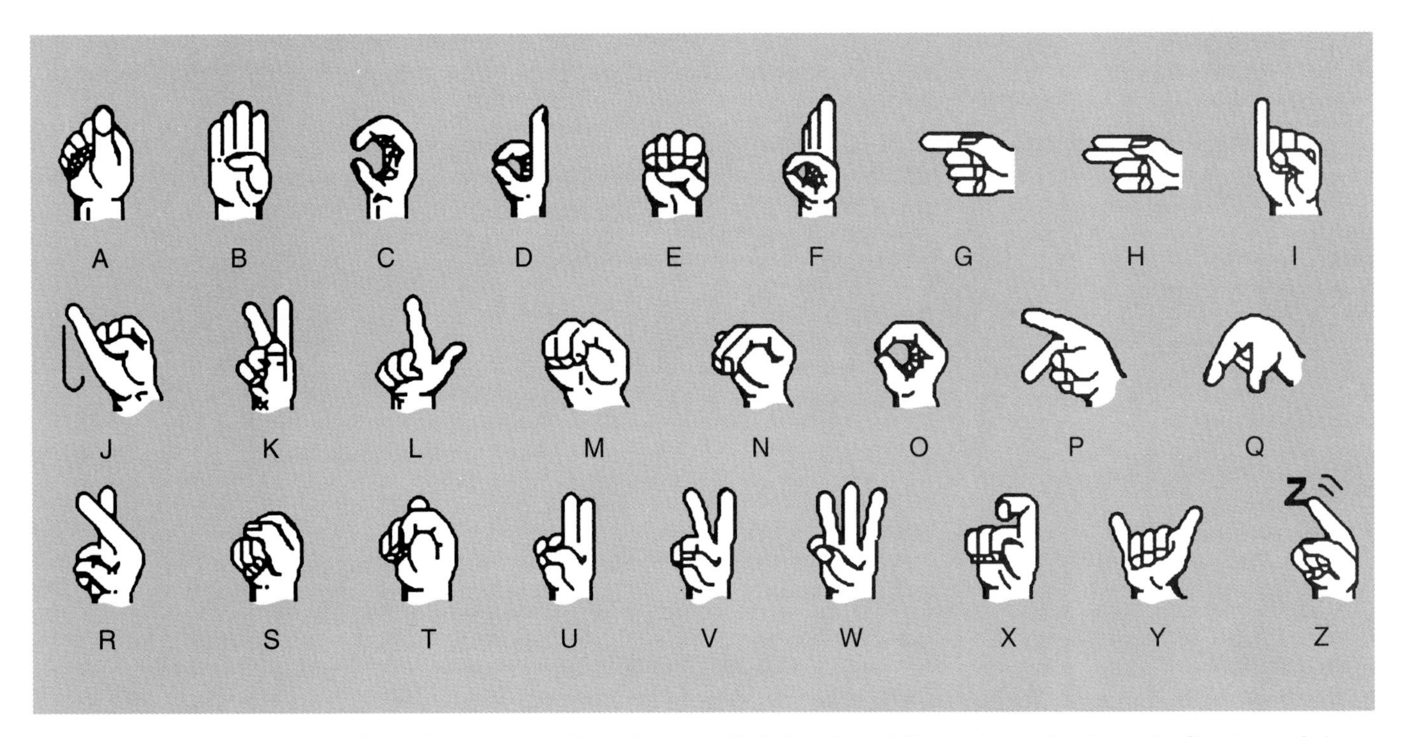

Fig. 1-5. Signing enables deaf people to communicate by using their hands and fingers to spell out words. One type of sign language is finger spelling. The signs for the letters A through Z are shown.

A

B

Fig. 1-6. Communication can be verbal or nonverbal. A—These two people are communicating verbally. (Sonoco) B—These people are using nonverbal communication. (Nancy Wehlage)

humans have evolved so quickly. Our methods of communication give us the power to think about and change our world.

Nonverbal communication can be as important as verbal communication in sending a message. Nonverbal communication is how we communicate without words. For example, think of two people interviewing for a job. The first person is unkempt, offers a weak handshake, doesn t look the interviewer in the eye, and appears to be bored. The second person interviewed is neatly dressed, offers a firm handshake, looks the interviewer in the eye, and is pleasant. Now, let s assume that the verbal communication in both situations is exactly the same. Can you make your judgment on verbal communication alone? No, how the message is presented is also important, which, in this case, is nonverbal communication. The second person would most likely get the job.

Nonverbal signals are learned in a culture just as verbal communication is learned. For example, in some societies people stand very close together when talking, while in other societies, people prefer more distance. This distance, often called *personal space*, may also vary depending on how well people know one another. Another example of nonverbal communication is personal appearance. This is often used for judging another person, as we saw in the job interview example. Gestures, eye contact, and the way the hands are used also play a role in how a message is received.

Perception

The process of understanding a message is known as **perception**. When you receive verbal and nonverbal messages from someone, these have to be integrated into a message that is meaningful to you. This process takes place in the brain after you receive the sensory information. You have to think about the sensory information in order to give it meaning.

Many of our perceptions are learned. For example, a baby learns that the world is three-dimensional by exploring it. Soon, the baby can recognize the shape of an object and determine how far away it is, not by touching it, but by visual clues such as the lines of the object. We continually improve our perception through childhood, so that line and colors in the world have meaning.

Upon first meeting someone, have you ever gotten the wrong impression? Have you ever said or written something that was misinterpreted? Since people are not always aware of the signals they are sending, and since the same signals can mean different things to two people, it is important to continually check your perceptions to be sure they are accurate. Your opinions may change about the same topic as your senses provide you with additional information.

Although people have many perceptions in common, they perceive differently as well. This is immediately evident if an audience is asked about a speech or movie. Although everyone heard and saw the same thing, they will give different answers because their perceptions were different.

Optical illusions can often trick your perception. Many perceptions are formed out of habit, so you may not think about them. For instance, when looking at a row of telephone poles, they seem to become smaller, but you understand that they are not really smaller. They are just farther away. Optical illusions have a way of playing tricks with these normal perceptions. An example is shown in Fig. 1-7.

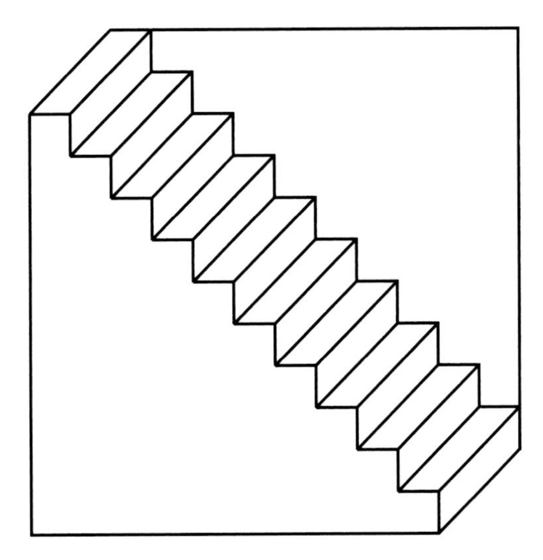

Fig. 1-7. Depending on how you first perceive this optical illusion, you will see the topside or the underside of the cellar stairs. As you look at it, it will change from one to the other.

A BRIEF HISTORY OF HUMAN COMMUNICATION

We normally take communication methods, such as writing and speaking on the telephone, for granted. However, these methods have evolved over thousands of years. A brief review of this history will give you a better appreciation of the communication systems you encounter now and in the future.

Primitive people probably first communicated with gestures, such as pointing, and guttural sounds (sounds in the throat). This was the beginning of our verbal (spoken) language. As this language developed, it became necessary to extend the capabilities of humans, such as communicating over long distances. This was the beginning of communication technology. For example, smoke signals and drum beats were used to send coded information, Fig. 1-8. These signals were probably first used to warn of danger.

Another way of transmitting information is to record it in symbolic (picture-like) form. In the past, pictures or crude drawings were made on the walls of caves to tell stories. Pictures or drawings are still used today to communicate ideas. A **pictogram** is an easily recognized symbol. For example, a school crosswalk sign might have a pictogram of children walking. Pictograms evolved into ideograms. See Fig. 1-9. **Ideograms** are symbols that have a meaning that must be learned. For example, the meaning of traffic sign shapes must be learned. Egyptian hieroglyphics were an ideographic way of writing.

A true **alphabet** was formed when symbols came to represent sounds. The first alphabet was invented in the Middle East. It was eventually used by different countries for their languages.

As writing developed, the surface for writing improved. Stone, clay tablets, and animal skins were first used. Around 2500 B.C., the Egyptians discovered a process of making paper from a reed known as papyrus. The Chinese were the first to make paper from wood pulp.

Until the 15th century, most books in Europe were handwritten by scribes, Fig. 1-10. These books were very expensive, and only the rich could afford them. When Gutenberg invented the printing press and movable metal type around 1450, producing books became less costly. This was the beginning of *mass communication*. Written materials were available to the "masses," both rich and poor.

While written communication was evolving, picture communication was also developing. Artists developed

Fig. 1-8. Smoke signals were used by primitive people to transmit coded information.

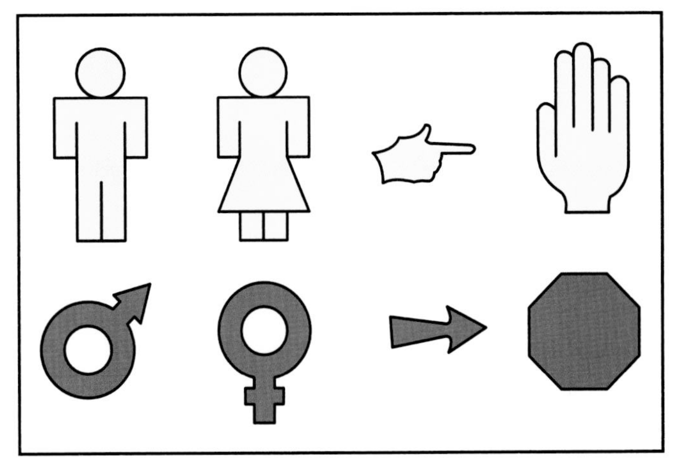

Fig. 1-9. The top row shows examples of pictograms. The corresponding ideograms are on the bottom row.

Fig. 1-10. Before printing presses were invented, books were produced by scribes.

freehand drawing and painting techniques that were used for painting and book illustrations. As instruments were designed, drafting or mechanical drawing techniques were being developed. Drafting was used to convey accurate information on building houses and machines.

In the early 1800s, Joseph Niepce devised a way to record the world, known as *photography*. Thomas Edison expanded on this idea by developing a method to show motion pictures. At the same time, developments in printing made it possible to print photographs. This was known as the halftone process. Printed materials of all kinds, such as books and newspapers, could now use illustrations and photographs produced by this method.

Long distance communication has always been important to humans. Smoke signals were used to warn of danger. Later, transportation such as the Pony Express, was used for sending written information across America. After electricity was discovered, it was possible to communicate much faster. The telegraph sent information in coded form across electrical wires. The telephone provided voice communication over long distances.

After electromagnetic waves were discovered, inventions allowed wireless communication. The radio was used to communicate with ships at sea. Later, these waves were used to send pictures and sounds to television sets. These waves are bounced off communication satellites today to extend the signal range, Fig. 1-11. In addition, high frequency electromagnetic waves in the form of a laser beam are sent through glass fibers to communicate messages, Fig. 1-12. This is known as *fiber optics*.

When digital computers became less expensive in the 1980s, they changed virtually every technology used to communicate. Older analog technology used continually varying levels of electricity (analog signal) to send a message over a distance. Early telephones are a good example. Newer digital technology uses on-off states of electricity to send a sample of the message at equal intervals in time (digital signal), such as 5 times per second. So, a voice gradually getting louder would be represented in analog as a smooth rising curve and in digital as a stairstep. However, the steps can be so close together that it is hard to distinguish between the analog and digital signals. In some cases, a message is sent in both digital and analog form. For example, analog-to-digital converters are used with computers to

Fig. 1-11. Satellite dishes like this one are used to send and receive electromagnetic waves such as radio and television signals.

Fig. 1-12. Optical fibers can be used for transmitting messages with light. (U.S. Sprint)

change analog signals such as a sound into a digital signal.

The computer and peripherals (e.g. a scanner), provided a means for converting any information into digital form, such as text, graphics, sounds, or moving images. This digital information, actually electrical signals, could then be processed (changed) in some way or transmitted (sent). This made it possible to rapidly send information long distances in digital form, whether over wire or air. See Fig. 1-13. When these signals are transmitted, a computer at a distant site can decode them so that they can be understood. This process of communicating between computers is known as **data communication**.

The *Internet* began in 1969 as a Department of Defense project that would allow computer communication among scientists even in wartime when part of the network might be destroyed. It has continued to grow as a computer network since that time, and is used worldwide for a variety of communication purposes today.

As can be seen by this brief history, communication is much more complex now as compared with earlier methods. Machines have allowed us to extend our communication capabilities. Still, all these methods are the same in that they allow us to extend human senses and to exchange information. Communication technology will continue to evolve because it is important in the survival and evolution of humans.

EMERGING TECHNOLOGIES

James Baldwin, an American author, once said, No one can possibly know what is about to happen: it is happening, each time, for the first time, for the only time. He was correct in that it is impossible to predict with 100% accuracy. However, we can make some educated guesses. In fact, there are organizations devoted to making these predictions, like the World Future Society.

Fig. 1-13. This computer and printer are carried in the truck on delivery routes and can communicate with a host computer at the home office. (Norand Corp.)

The future will certainly bring more powerful and less expensive computing. This was predicted in 1965 by Gordon Moore, who said that microchips double in performance, while staying about the same size, about every 18 months. This is now known as *Moore s Law.* In the next twenty years, tiny and powerful computers are predicted to serve a variety of purposes. They will be part of smart cards, like a credit card, that can contain our medical history or be our currency (electronic cash). They will also be used to plug into computers away from home to give us access to our own software, e-mail, etc. Small, voice-based computers will be able to fit on our wrist and provide Internet connections, and even be able to tell us our exact location on the planet. Books will be completely electronic.

Small and powerful computers will also allow sensors to be embedded wherever they are needed. Cars will be able to drive themselves, adjust to road conditions and avoid accidents without help from a human. Sensors in homes will be able to determine not only simple tasks such as lighting and heating, but also what groceries are needed for the refrigerator, and even a person s health conditions.

Virtual reality, which provides a computer-simulated world, will become much more realistic and commonplace. We will be able to use this technology to explore a vacation spot or a home that we cannot personally visit before making a purchase. Virtual reality will also be used more frequently for entertainment.

The *Internet*, in some form, is predicted to continue to expand, so that we will all be interconnected by voice, video, and data whenever and wherever wanted. Along with this, more businesses will have operations around the world, to bring together the necessary resources for products and services.

SUMMARY

This chapter has provided a brief overview of communication, beginning with an explanation of a simple communication model. The five major purposes of communication were explained which are to inform, educate, persuade, entertain, and control. There are various types of communication. These include: human-to-human communication, human-to-machine communication, machine-to-machine communication, machine-to-human communication, and supplementary types of communication.

Humans use their senses when communicating. Human communication can be described as either verbal communication (using words) or nonverbal communication (using gestures other than words). As humans receive communication, they form perceptions about these messages.

Throughout history, humans have attempted to communicate with one another. From the drums used by primitive people, to the complex telecommunication used today, technology has played a vital role in the evolution of communication. Futurists provide us a glimpse at the future of communication technology, as described in the last part of this chapter.

WORDS TO KNOW

All of the following words have been used in this chapter. Do you know their meanings?

alphabet
analog technology
Braille
communication
communication model
cybernetics
data communication
digital technology
ideogram
nonverbal communication
perception
pictogram
signing
verbal communication
virtual reality

REVIEWING YOUR KNOWLEDGE

Please do not write in this text. Write your answers on a separate sheet.

1. In order for communication to occur, a message must be:
   a. Sent.
   b. Sent and received.
   c. Received.
   d. None of the above.
2. List and describe the five major purposes of communication.
3. Cybernetics is an example of which type of communication?
   a. Human-to-human communication.
   b. Human-to-machine communication.
   c. Machine-to-machine communication.
   d. Machine-to-human communication.
4. Which of the following senses are most important for human communication?
   a. Seeing and hearing.
   b. Touching and tasting.
   c. Smelling and hearing.
   d. Seeing and touching.
5. An example of nonverbal communication is:
   a. A handshake.
   b. A nod.
   c. A speech.
   d. Both a and b.
6. True or false? Personal appearance is often an example of nonverbal communication.
7. Perception is the process of:
   a. Sending a message.
   b. Receiving a message.
   c. Understanding a message.
   d. None of the above.
8. True or false? Pictograms are symbols that have a meaning that must be learned.
9. How did the invention of the printing press and movable metal type affect communication?
10. True or false? Data communication involves sending and receiving electrical signals by computer.

APPLYING YOUR KNOWLEDGE

1. Discuss the common ways humans communicate each day. Analyze these methods using the communication model.
2. Divide into small groups. Demonstrate examples of both verbal and nonverbal communication. Share your examples with the other groups.
3. Write a report on the historical roots of a form of communication technology that is used today. Examples can include computers, telephones, satellites, etc.
4. Research how people with hearing, speaking, or sight impairments communicate. Learn several words using one of these methods. Share them with the class.

2

CHAPTER

TECHNOLOGY AND COMMUNICATION

After studying this chapter, you will be able to:
- *Explain how humans use technology to change the world to meet their needs and wants.*
- *Describe the relationship between technology and change.*
- *Describe the impact of the information age upon communication technology.*
- *Give examples of how technological literacy can affect your life.*
- *List the major technology systems.*
- *Define the four processes of technology.*
- *Name and describe the major categories of communication technology.*
- *Describe the relationship between communication technology and other subject areas.*

Technology plays an ever increasing role in the world in which we live. That is why it is important to understand what this term means and its importance in your life. In this chapter, the broad topic of technology will be explored as well as communication technology.

TECHNOLOGY

What is technology? **Technology** is a constantly evolving body of knowledge that deals with the technical way in which we change the world to meet our needs and wants. Technology allows us to extend our human potential through the use of tools. Technology can also be described as the practical application of scientific knowledge.

For example, when radio waves were discovered, scientific knowledge was broadened. Radio technology was then developed to use these waves. Use of the radio has saved many lives by providing a way for people aboard ships, airplanes, and other vehicles to signal for help when in danger. Radio is used to warn people of approaching hurricanes or tornadoes. This is just one example of how technology has helped people to meet their needs and extend their human potential. See Fig. 2-1.

Knowledge of tools, materials, and processes allows people to adapt to their environment. Suppose this knowledge did not exist? You would have to learn to build a fire without matches. You would have to find ways to gather enough food to live. Survival would be your primary interest, and most of your time would be spent finding shelter and food. If your eyesight was poor, your world would always be blurred, and you might have difficulty surviving. Without antibiotics, a cut might threaten your life.

How does technology differ from industry? **Industry** is an organization that uses technology to produce goods and provide services for a profit. The goal of industry is to make money by selling goods and

Fig. 2-1. Radio is an important communication tool for police and fire departments in emergency situations.

services. So, industry, which is an economic institution, utilizes technology, which is a body of knowledge. See Fig. 2-2.

TECHNOLOGY AND CHANGE

Change has always been a part of human existence. However, technology has had a large impact on the rate of change. In the book, *Future Shock*, Alvin Toffler describes human existence in terms of lifetimes. In the past 50,000 years, people have had over 800 lifetimes of 62 years each. Of these, 650 have been spent in caves. Writing has only been possible in the past 70 lifetimes and printing in the last 6 lifetimes. Electric motors have only been around for 2 lifetimes. Most of the technology that surrounds us today has been invented in this, our 800th lifetime.

Today, the time lapse between new discoveries and innovations, and practical application of these discoveries, is decreasing. In other words, the rate of change has actually accelerated. Think about devices such as computers, satellites, cable television, and copying machines. These are all technological innovations we have witnessed in recent years. See Fig. 2-3. New technological devices actually increase the rate of change, by providing ways in which faster change can occur. For example, computers and telecommunication make it possible to share and combine ideas from around the world.

As change accelerates, the amount of knowledge in the world increases as well. Presently, the amount of knowledge in the world doubles in about five years. Therefore, rapid change is something with which we must all learn to live.

Throughout most generations of the human race, most technological innovations have been for basic survival needs, such as food, clothing, and shelter. Prior to the invention of complex machines, most people worked in agriculture. Farming families grew their own food, built their own homes, and made their own clothes. However, as technology evolved, fewer people were needed in agriculture to produce the amount of food needed.

The shift from an agricultural economy to an industrial economy began in the early 1800s and is known as the **industrial revolution**. New inventions, such as the steam engine and spinning jenny, provided a way to mass-produce goods quickly and inexpensively. Factories were built to produce the goods, and as people were hired to work in the factories, cities were formed. Transportation was needed to send the goods to consumers, so railways and steamships were used for this purpose.

As the number of industries increased during this period, the greatest number of jobs in the nation were in industry. The industrial revolution lasted until the 1950s, when information-related jobs became more prevalent than industry jobs. Today, there are still many

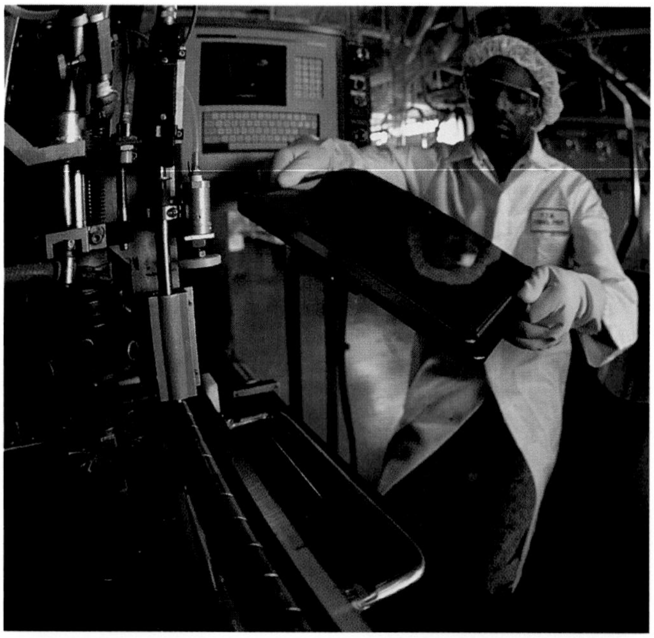

Fig. 2-2. Industry utilizes the technology that was used to develop this laser welder that assures precision alignment of television parts. (Zenith)

Fig. 2-3. Barcoding technology has made important contributions to both retail business and industry in only a few years. (Intermec Corp.)

industries and industry jobs. However, there are now more service-related industries than goods-producing industries since fewer people are needed to produce the goods. Many of the services provided by industry are involved with information technology. See Fig. 2-4.

During the industrial revolution, there were many innovations that provided the basis for the information age of today. Electricity and electromagnetism were discovered during this period. This resulted in communication devices such as the telegraph, telephone, and radar. Communication devices, such as cameras, radios, televisions, and video recorders were invented during this era as well. In addition, many technological innovations were being used in factories that allowed more goods to be produced by fewer people.

The information age

Today, we are in the midst of an **information age**. An information age can be defined as an era when a majority of people are involved in jobs related to information. Developments in electronics have resulted in computers and telecommunication that allow us to communicate around the world. Most jobs are in some way related to the transfer of information and utilize information technology. These jobs include those of clerks, secretaries, teachers, lawyers, printers, etc. See Fig. 2-5.

Fig. 2-5. These people, working in a military computer center, have jobs that are primarily associated with the transfer of information. (Official U.S. Navy Photograph by PH1 Paul Pappas)

TECHNOLOGICAL LITERACY

It is obvious, based on the impact that technology has on our world, that knowledge of technology is extremely important. Technology is accelerating the rate of change, and understanding it is important if we are to play a role in deciding its future use. It is important to be technologically literate.

What is **technological literacy?** In the *Technology for All Americans* project, it is defined as being able to *use, manage,* and *understand* technology. What exactly does this mean?

Using technology involves being able to utilize current technology to accomplish a task. It also involves using the technology safely and efficiently. Since technology

MAJOR ERAS IN AMERICAN HISTORY

ERA	Agricultural	Industrial	Information
TIME PERIOD	PAST ——— 1800	——— 1950	——— PRESENT
TYPICAL JOB	Farmer	Laborer	Clerk

Fig. 2-4. This time line shows how technology has evolved.

is continually evolving, use of technology also requires continual updating if one is to remain current.

Managing means directing or controlling the use of something. So, in addition to using technology, technological literacy involves managing the technology with which we work. This includes being sure that the technology used is appropriate for the task, used efficiently, and maintained. For example, if you are taking a large number of photos with a digital camera, you need to ensure that everything is working properly before each photo session. You also need to maintain the camera as well as computer software used to edit the pictures. When the pictures are completed, you must devise a system for organizing the pictures. These are just a few examples of how technology is managed.

Most technologies, even those that we use every day to get our work done, are developing faster than we can keep pace. However, being technologically literate involves continual effort to understand the technology and how it affects us as workers, citizens, and even innovators. Technology is important in many jobs, so understanding technology is certainly important in getting and keeping these jobs. As citizens, we need to understand how technology affects our lives. Only then can we make informed decisions about its use, both personally and politically. In addition, understanding past and present technology is an important factor in the development of new technology. In fact, new technology is an important element of economic growth and continued prosperity.

TYPES OF TECHNOLOGY SYSTEMS

There are various ways to classify technology. One way is to divide technology based on the purpose of the system. In this case, common divisions include:

- Communication.
- Construction.
- Energy, power, and transportation.
- Manufacturing.
- Bio-related.

The purpose of **communication technology** is to transfer information by technical means. Typical technologies in this category include telecommunications, printing, radio, and television.

Construction involves building structures such as houses, office buildings, factories, malls, and roads. This system provides us shelter for family, work, and leisure. It also provides us with bridges, railroad tracks, and runways needed for transportation.

Energy, power, and transportation are interrelated. **Energy** is the capacity to do work. Sources of energy can be unlimited (the sun or a river) or limited (fossil fuels such as oil, gas, and coal). This energy can then be harnessed to produce power. An electric motor in a washing machine, which uses electricity to run, is an example. Power systems are often used for **transportation**. This is the movement of people or products. Typical transportation technology includes cars, buses, airplanes, trains, and ships.

The purpose of **manufacturing** is to convert raw materials into goods for society. These goods include furniture, cars, televisions, homes, and food.

Biotechnology is the use of living things to serve our needs, such as growing food. Animal cloning and using bacteria to clean up an oil spill are examples of bio-related technologies.

Most types of technological systems are interrelated, Fig. 2-6. For instance, in constructing a house, communication is used as the plans are drafted. These plans communicate to the builder what is to be built and how. Energy, power, and transportation systems are used to transport materials to the building site as well as for providing electricity. Can you think of other ways in which the various types of technologies might be interrelated?

TECHNOLOGY PROCESSES

Technology processes can be considered the action part of technology. Developing and using products and services requires effort on the part of humans. These

Fig. 2-6. Manufactured housing utilizes many technology systems. For instance, communication occurs as the house is planned and as it is being constructed. Energy, power, and transportation are utilized as the house is transported to its final destination. (Manufactured Housing Institute)

actions are those that allow us to *develop, produce, use,* and *assess* technology.

Developing technology products and services involves design and problemsolving. A variety of design ideas are developed and one is chosen for use. A problemsolving model, with specific steps, can be used for this process. When done with the development stage production can begin.

The production stage is when products are made or services are provided. In the case of communication, the production stage involves information transfer. Examples include notepads, pictures, radio announcements, etc.

Technology products and services are created based on the needs and wants of individuals, and are meant to be used. If they are not used, there is no need to continue production. Normally, there are competing products and services, so careful decisions need to be made when making a selection.

Using technology involves knowing how to use devices and services correctly and safely. For example, if the directions for using a video camera are not clear, or if they are not read, this can cause it to be used incorrectly, resulting in poor video and dissatisfaction. If a device requiring a grounded outlet is not plugged into this type of outlet, there is a risk of electrical shock. Using technology also includes being responsible. Recycling paper and drink cans is a good example. Getting maximum use out of a product before throwing it away is another example.

All technology has desired and undesired results. People may enjoy the roominess and power of bigger cars, but they may create more pollution and affect our health. This is an undesired result. Assessment involves carefully considering the consequences of technologies and determining if the benefits outweigh the problems. On a larger scale, these decisions are made by local, state, and federal government. On a smaller scale, we can make responsible decisions almost every time we make a purchase.

COMMUNICATION TECHNOLOGY

Communication technology extends the capabilities of our human senses. For instance, a telephone allows us to hear someone thousands of miles away. A photo and article in a newspaper allow us to see and understand stories that take place without our actually being there.

It is important to remember that communication technology extends our ability to communicate, Fig. 2-7. The two primary methods of human communication involved in this process are *visual communication* and *acoustical (sound) communication*. In other words, these technologies extend our ability to communicate using symbols across long distances visually, acoustically, or both.

Fig. 2-7. Signs like this warning to drivers are an example of visual communication.

Categories of communication technology

There are various ways we can classify communication technology. One way is to divide into graphic communication and electronic communication, Fig. 2-8. **Graphic communication** involves transmitting information with visual symbols, such as words and pictures. Examples include technical drawing and illustration (also known as technical graphics), photography, and printed graphics (printing). **Electronic communication** includes ways of communicating that require sending messages over a distance. The message must be first changed into an electronic signal in order to be transmitted. Examples include the telegraph, telephone, radio, television, and telecommunication. Major topics in this text related to graphic and electronic communication include technical graphics (technical drawing and illustration), photography and image editing, printed graphics, and electronic communication. Each will be described in more detail.

Technical drawing and illustration involves communicating information in graphic form, usually with drawings that require special skills and tools to produce. In the case of technical drawing, tools such as a T-square and triangle, or a computer and drafting software, are

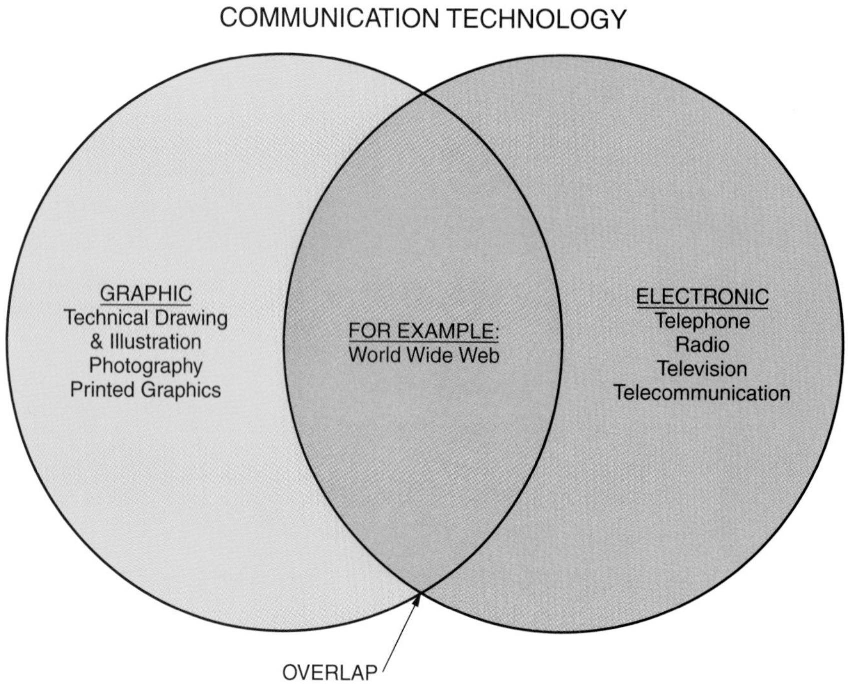

COMMUNICATION TECHNOLOGY

GRAPHIC
Technical Drawing
& Illustration
Photography
Printed Graphics

FOR EXAMPLE:
World Wide Web

ELECTRONIC
Telephone
Radio
Television
Telecommunication

OVERLAP

Fig. 2-8. Examples are shown for graphic communication, electronic communication, and technologies that overlap the two categories.

used to create the drawing. These drawings are most often used as a plan for creating something, such as a set of blueprints for a house. Technical illustration involves making technical drawings for publications. Many of these are 3D drawings and in color. Some illustrations are created using manual or computer drafting techniques while others may require the artistic and freehand skills of a graphic artist.

Photography is a system that uses reflected light to produce pictures. The image is stored on light-sensitive film or in digital format on a magnetic medium such as a disk. Once the image is captured, it may require further change, which is known as *image editing*. In still photography, a single scene is recorded. In motion picture photography, a series of still pictures are taken in rapid succession. When played back, this gives the illusion of a moving picture.

Printed graphics is the communication system involved with mass production of graphic images. This is often simply called **printing**. Examples include newspapers, books, magazines, shirts, and even printing on cans and bottles.

Electronic communication involves putting information in coded form and then transmitting the message with electrical signals. An early example of this was Morse code, a means of sending messages as dots and dashes. Today, many messages are converted into digital format before being transmitted. A digital signal created by computer consists of binary digits (bits), with each bit being either on or off, like a light switch. A combination of these bits can be used as code for text, pictures, and sound.

Electronic communication over long distances is known as **telecommunications.** An example is television. Television signals are sent by electromagnetic waves through the air and through cables into homes. The signal is in coded form until it reaches the television where it is converted into a picture and sound. Other common telecommunication devices include telephones and radios

Although two separate categories are being presented for communication technology (graphic and electronic communication), it is important to remember that overlap exists. Most images can be created in digital form on computer or can be converted to digital form, which involves electronics. These images are also transmitted for various purposes. For example, a graphic image may be created on a computer, Fig. 2-9, and then sent by computer modem and telephone lines to another location for printing.

Most communication technology, except for machine-to-machine communication, is related to the human senses of seeing and hearing. This is logical since these are the senses humans use most often for communication purposes.

Importance of communication technology

As mentioned earlier in this chapter, we are in the midst of an information age. Communication technology has an impact on most jobs. Many businesses, such

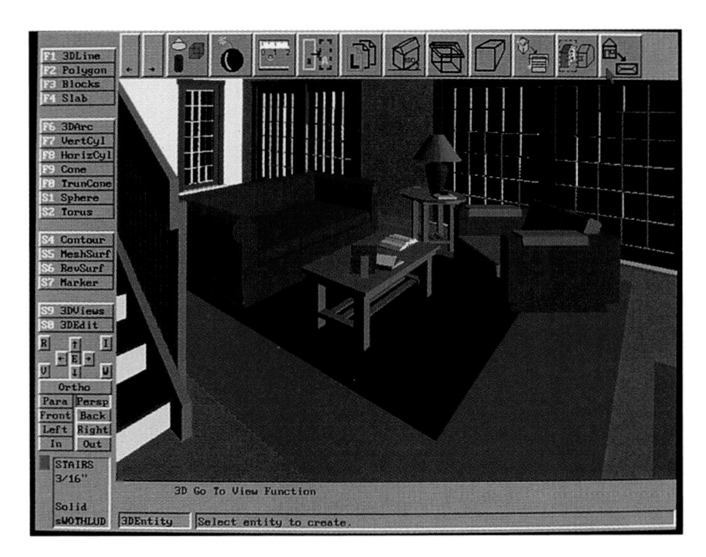

Fig. 2-9. Technical illustrations, such as this 3D view of a living room, can be created on a computer and printed out. The information, in digital form, can also be transmitted by telephone lines. (DataCAD)

Fig. 2-10. Communication technology involves an understanding of reading, writing, mathematics, and science, such as computer science.

as radio, television, printing, advertising, music, and telecommunications, are established solely to distribute information. Many of the goods produced by industry, such as telephones and computers, are related to information jobs. In addition, almost all businesses utilize communication technology in order to work more efficiently and effectively.

In today s society we are virtually surrounded by information technology. Understanding communication technology is an important part of becoming technologically literate. It affects all aspects of your life from work to leisure.

Communication technology and other subject areas

Although communication technology is studied as a separate field, it overlaps other subject areas, Fig. 2-10. Academic skills such as reading and writing are very important in effective communication. Math skills play a role in communication as well. For example, technical drawing requires an understanding of accurate measurement, geometry, and proportions. Printed graphics require the use of formulas for page calculations. In photography, photos are typically scaled before they are printed. This involves the use of percentages. These are just a few examples that show the importance of basic academic skills in communication technology.

Science is also important in communication technology. Scientific discoveries are made in order to understand the world around us. Technology is the application of that knowledge. Science has played a role in virtually all communication technologies. For example, an understanding of the physics of light was necessary in order to have fiber optic communication systems. The science of color was important for developing devices such as color television sets.

Since technology has played a key role in the evolution of people and their culture, this area is important in humanities as well. Communication technology has provided the way to transmit information around the world. Other countries and people are seen daily on television and in the newspaper. This worldwide communication network is playing an important role in understanding other cultures and the shaping of individual cultures.

Just as basic academic skills, science, and humanities play an important role in communication technology, communication technology can also help in understanding other subjects. When academic skills are used as part of technology activities, we realize their usefulness, and it makes them more meaningful and fun. For example, geometry can be much more interesting when making designs with geometric shapes. The physics of electromagnetic waves is more relevant when actual messages can be sent by these waves.

SUMMARY

In this chapter, you have explored technology and communication technology. Technology plays a key role in the changes in the world so it is important that we all continually strive for technological literacy, including being able to use, manage, and understand the technology around us. It is also important to know the processes involved in using technology to provide products and services, as well as to create new technology.

These include developing, producing, using, and assessing technology.

Communication technology can be divided into graphic and electronic communication although some overlap exists. This subject area requires the use of other academic skills such as math and science, but it provides a practical application of these skills. Learning about communication technology can aid in the understanding of other subjects and vice versa.

WORDS TO KNOW

All of the following words have been used in this chapter. Do you know their meanings?

bio-related technology
construction technology
communication technology
drafting
electronic communication
energy, power, and transportation technology
graphic arts
graphic communication
industrial revolution
industry
information age
manufacturing technology
photography
printed graphics
printing
production technology
technical drawing and illustration
technological literacy
technology
telecommunications

REVIEWING YOUR KNOWLEDGE

Please do not write in this text. Write your answers on a separate sheet.

1. Technology:
   a. Is an evolving body of knowledge that deals with the technical way in which we change the world to meet our needs and wants.
   b. Allows us to extend our human potential through the use of tools.
   c. Is the practical application of scientific knowledge.
   d. All of the above.
2. True or false? Technology is an organization that uses industry to produce goods and services.
3. As technological change accelerates, the amount of knowledge in the world __________ (increases, decreases).
4. Which best describes the history of the United States?
   a. Clerk, farmer, laborer.
   b. Laborer, clerk, farmer.
   c. Farmer, laborer, clerk.
   d. Laborer, farmer, clerk.
5. Which of the following is characteristic of the information age?
   a. Steam engines and factories.
   b. Farms and crops.
   c. Computers and telecommunication.
   d. None of the above.
6. List four reasons why you should be technologically literate.
7. Match the following types of technologies with their examples.
   ________ Manufacturing technology.
   ________ Energy, power, and transportation technology.
   ________ Communication technology.
   a. Examples include electricity, electronics, computers and movement of people and products.
   b. Examples include telecommunications, printing, radio, and television.
   c. Examples include producing furniture, cars, televisions, and food.
8. Name the two major categories of communication technology.
9. Which of the following is NOT an example of graphic communication?
   a. Photography.
   b. Telecommunication.
   c. Technical drawing.
   d. Printing.
10. Which of the following academic skills are used in technology courses?
    a. Science and humanities.
    b. Math.
    c. Reading and writing.
    d. All of the above.

APPLYING YOUR KNOWLEDGE

1. Give examples of ways in which technology has improved your way of life.
2. Cite some major changes in the last 50 years that are due to technology. Prepare a time line to share with the class.
3. Visit a local business or industry. Find out how technological changes have affected its operation. Prepare a report and present it to the class.
4. Identify a technology and analyze the technological processes (action steps) for that technology.

3

CHAPTER

THE COMMUNICATION SYSTEM

After studying this chapter, you will be able to:

- *Explain why a system model is useful in understanding technology systems.*
- *Describe the evolution of the systems approach.*
- *Identify the elements of a communication systems model and the purpose of each element.*
- *List the resources needed for communication.*

Communication can be described as a system with the goal of transferring information. Applying a systems approach can help in understanding the communication process.

OUR COMPLEX WORLD

Today, people are surrounded by technology they most likely do not fully understand. A digital clock may wake you in the morning. A telephone call can be made to another city. A television and VCR can be used to play a movie. A compact disc player can provide music for a party. In most cases, people know how to operate the technology, but they have little understanding of how it really works, Fig. 3-1.

The challenge in understanding today s technology lies in its complexity. New technologies utilize older technologies. Therefore, for you to fully understand how something works, you need to understand a variety of technologies.

For example, think about a radio. You know that the radio uses electricity, but what is electricity? You know that the radio operates on alternating current provided by a power company, but do you know why alternating current is used? You know that the radio has an antenna that detects electromagnetic waves from a radio station. However, what are these waves, and how do they reach the radio? You may also be aware that a speaker is used to convert electrical signals into sound waves that you can hear, but how is this done? As you can see, a radio combines several technologies in order to serve its function. This complexity can make technology difficult to understand.

Fig. 3-1. These sailors know how to use the radar systems they are monitoring, but may not understand how the technology actually works. (Official U.S. Navy Photograph)

A SYSTEM APPROACH

A system approach is one way to help us in understanding complex technology. A **system** is simply a number of parts that work together to accomplish a goal. Any technology can be described as a system. By first breaking down a technology into its system components, it becomes easier to understand.

SYSTEMS: A BRIEF HISTORY

Where did the concept of systems originate? First of all, systems have not always been called by that name. Throughout history, people have tried to understand and control the world around them. In order to do this, logic and common sense were often used to explain the world and change it. For example, the Greeks divided the universe into fire, water, air, and earth. This was their system for describing the world. As plants and animals were studied, a system was devised for placing them in categories. Again, this was a system for categorizing and describing the world.

Systems are often used to describe the world, Fig. 3-2. Systems that are used to describe the physical world are known as **natural systems**. The solar system is an example. Descriptions of systems developed by people, such as technology, are called **devised systems.** These are systems that are created by people to serve some purpose. The Roman aqueducts are an historical example. These were pipes that provided clean water for Rome from a mountain river. Aqueducts were built that supplied Rome with over 38,000,000 gallons of water each day. Much planning was involved in developing this system of water supply. Modern examples of devised systems are all around us. These include computer systems, highway systems, government systems, organizational structures in businesses, etc.

Use of the terms *systems* and *system approach* really began in the middle of the Twentieth Century. This was the result of a need to develop complex technological systems as well as a need to describe these systems better. Bell Telephone was one of the first companies to use the approach in the 1940s. There was a need to study the telephone system and determine if microwaves could be used for long distance telephone communication. This study looked at the practicality of this transmission method. A variety of people were utilized to solve this problem, including scientists and engineers. The people involved in this study called it **systems engineering**. This term has continued to be used to describe the planning of complex technological systems.

A system approach has also been used to describe existing systems so they can be studied more carefully. For example, in 1948, Claude Shannon described communication as a system. He developed a model that showed all the elements in the communication process, such as source, signal, and destination. This model was actually developed to determine how to send electrical signals, such as telephone signals, through a wire with the least amount of interference. However, people who worked with other forms of communication began using this systems approach as well to explain the communication process.

A

B

Fig. 3-2. There are natural systems and devised systems. A—The solar system is a natural system consisting of planets. (NASA) B—This communication tower is part of a devised system that is used to transmit messages. (Bud Smith)

Today, the system approach is still being used to describe complex technological systems. It is also being used to help in creating new systems. This approach provides a structure for studying technology so that it can be understood and improved.

THE SYSTEM MODEL

As discussed, a system approach can help us comprehend technology systems. In order to use this approach, you need to understand a system model, Fig. 3-3.

First, it is important to remember that a system has a purpose or goal. This is the desired **output** of the system. The **inputs** are the resources necessary for achieving the output. The **process** is the action part of the system. In this step, the inputs are changed to achieve the output.

In many systems, a feedback loop is added. *Feedback* is a way of determining whether the actual results, or output, is the same as the desired results. A system with feedback is called a **control system** because the feedback loop allows for monitoring the output and making corrections if needed. It is also known as a **closed loop system**.

A home heating system is a good example of a system model. The desired result is to maintain a constant temperature in the home. Inputs necessary for this are a furnace, energy (in the form of electricity and fuel), ductwork, wiring, and a thermostat to monitor the temperature in the home. Processing is the burning of the fuel to generate heat and using a fan to blow it through the ducts. The thermostat provides the control and feedback. When the desired temperature is reached, a signal is sent to turn off the heater.

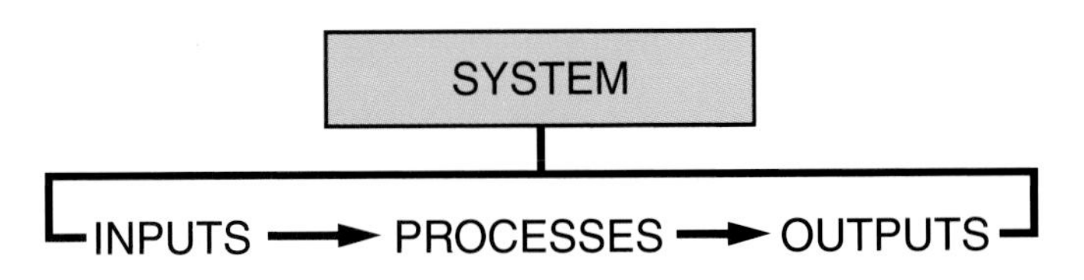

Fig. 3-3. This is a basic system model.

Most systems can be broken down into subsystems. In the previous example, a thermostat is a subsystem. Its purpose, or intended output, is to signal the heater to turn on or off. Inputs, in the form of resources, are electricity and a thermometer. The processing step involves measuring the temperature and signaling the heater.

Any system, whether natural or devised, can be analyzed using the system model. In this way, complex systems can become easier to understand.

COMMUNICATION SYSTEM MODEL

A simple communication model was presented in Chapter 1, consisting of five parts: sender, channel, message, receiver, and feedback. However, we can use the basic systems model and several more elements to explain this system in more detail. Keep in mind that systems can often be described in different ways, so this is just one example.

Beginning with the basic systems model, we need an input, process, and output, as well as a purpose. In the communication system, the purpose of the system is to send a clear message, which we also hope will be the output. Inputs are all the resources needed for sending the message. In the case of a telephone conversation, you will certainly need a phone at each end as a resource, as well as a wire channel between the phones. Processing includes all the actions necessary to communicate. If done correctly, the output will be a clear message as intended.

The **communication system model** is shown in Fig. 3-4. Actions begin with encoding the message so it

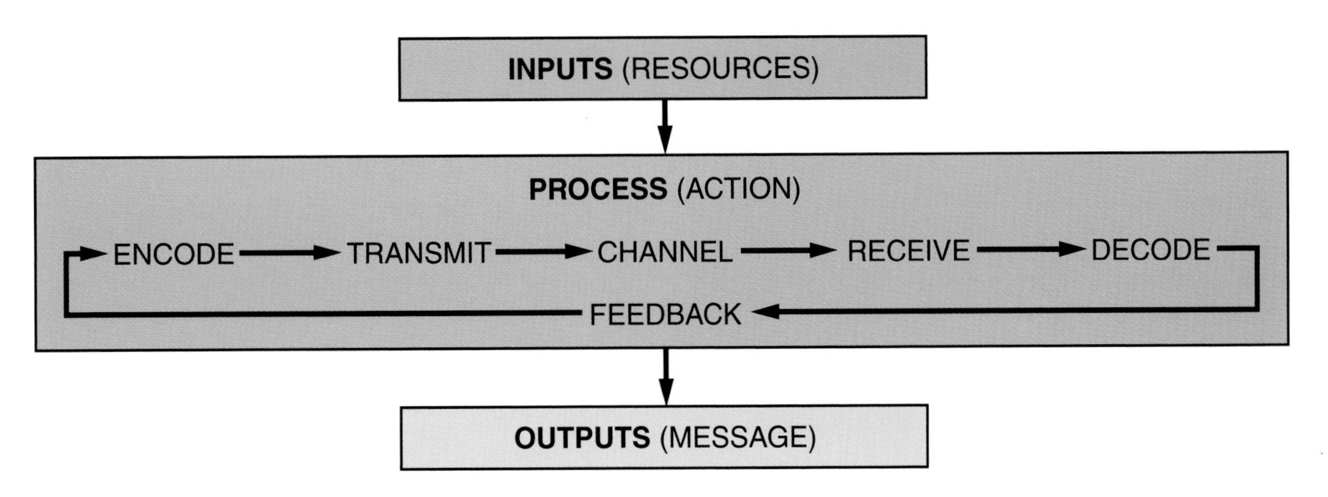

Fig. 3-4. This communication system model shows how messages are sent and received.

can be transmitted through a channel. In the case of a person-to-person conversation, this might simply be using a language you know the receiver will understand. In the cases of a telephone conversation, encoding is also needed to change the voice signal into an electronic signal. Once encoded, the signal is transmitted through a channel to the receiver. The message then needs to be decoded so it can be understood.

Other parts of the system model include interference and feedback. Interference or noise can cause a breakdown in communication. Feedback allows you to determine if the message that was sent was communicated accurately.

In some cases, it is necessary to store the message in the channel. Books and CDs are two good examples. The channel holds the message as it is being sent.

Human-to-human communication can be analyzed using the communication model. See Fig. 3-5. In this case, the message source is a person. The message is designed in the mind and encoded as speech. These words are transmitted through sound waves that are the channel. The receiver then decodes the message. The output of this system is the message that is actually received. Feedback can be achieved by a response from the receiver. In this way, it can be determined whether or not the desired message was received.

In communication technology, machines are often part of the communication process. Use of the telephone is a good example, Fig. 3-6. As you speak into the phone, the microphone changes or encodes your voice into electrical signals. These signals are transmitted through a channel such as phone wires. When the signals reach the other party, the telephone receiver (phone speaker) decodes the electrical signals back into a voice. The receiver (person) can now listen to and understand the message. A thunderstorm can cause electrical interference that could interfere with the communication. By asking *Can you hear me?* you can receive feedback on whether or not communication is occurring.

Fig. 3-6. Machines such as the telephone are often part of the communication process. Sometimes, interference, such as static, can hinder communication.

Fig. 3-5. Both spoken and visual communications are being used by these engineers as they discuss building drawings with a construction superintendent. (Westinghouse Electric Corp.)

In the telephone example, smaller subsystems can also be analyzed. These could include human-to-machine communication, machine-to-machine communication, and machine-to-human communication.

RESOURCES FOR COMMUNICATION

In any communication system, resources are needed in order to send a message, Fig. 3-7. If certain resources are missing, then it may be impossible to send a message, or a different method of communication may be required.

The important resources for communication include the following:
- People.
- Information.
- Materials.
- Capital.
- Finances

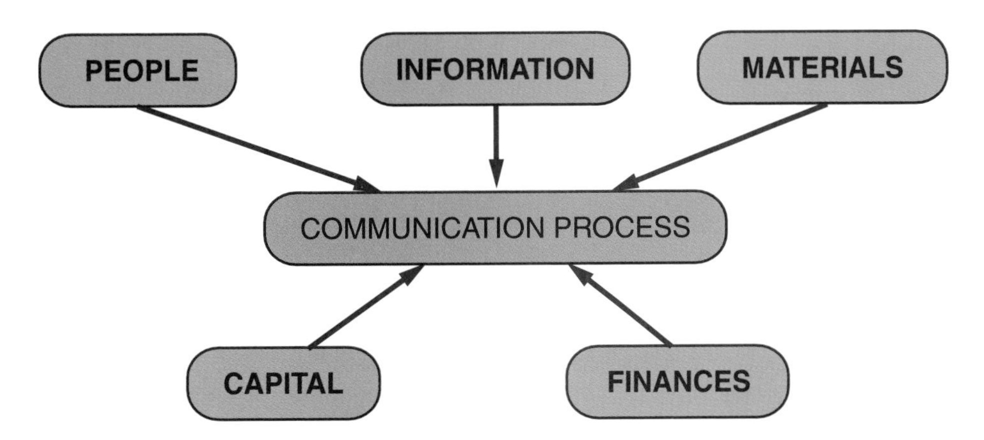

Fig. 3-7. Resources are a necessary part of the communication process.

People

In any communication technology, skilled people are needed to operate the system, Fig. 3-8. People with artistic and technical design skills are needed for sending many types of visual messages. Photographers are needed to produce photos for books, magazines, and newspapers. Graphic arts skills are important in producing large quantities of printed products. Video production requires a number of people with special skills. People skilled in electronics are vital in telecommunications. Of course, leadership, in the form of management, is also needed to coordinate the jobs of many employees. From these examples, you can see that people are a critical element in communication technology.

Information

We often have a great deal of information about how to send a message. However, since new methods are arising constantly, we must keep up with this new information. Schools that provide training in communication skills are just a starting point. We must continually update our skills through individualized study, courses, workshops, apprenticeships, etc.

When using communication technology to send a message, input is needed on the best way to send the message. Considerations in this process are the amount of time and resources available. In order to do this, information is needed about the entire communication system. Knowledge of how best to approach an audience is also part of the needed information. For example, who is the audience, and what is the purpose of the message? What is the best way of sending this message to this audience? How will the communicator know if the right message was received? This is why information is a key resource in the communication process.

Materials

Materials are needed as a resource in communication. Substrates, such as paper and vellum, are needed in graphic communication. Supplies, such as film and chemicals, are needed in photography. Magnetic media, such as videotapes and diskettes, are needed in electronic communication.

A knowledge of the various materials that are available is important in communications. It is also important to keep up with new developments in materials so that communication can be constantly improved.

Capital

The tools and machines used in technological systems are called **capital.** See Fig. 3-9. Examples of communication capital include drawing tools, computers, cables, photography and darkroom equipment, video

Fig. 3-8. People are an important communication resource. Skilled people are needed to work on television transmission systems such as this one. (Zenith Electronics Corp.)

Fig. 3-9. Capital is needed for communication. In this photo, machines, which are capital, are being used to lay a fiber optic cable. Once installed, the fiber optic cable becomes a part of a company's capital. (U.S. Sprint)

and audio devices, television studios, transmitters and receivers for wave transmission, etc. In all these cases, people are needed with the skills to operate and maintain the equipment.

Finances

Money is needed to operate any communication enterprise. Finances are needed for construction of buildings, purchasing equipment and materials, utilities, and paying salaries. Careful selection of people, equipment, and materials is important for keeping quality of work high while keeping expenses at a minimum.

THE PROCESS OF COMMUNICATING WITH TECHNOLOGY

As discussed earlier, the process of communication is the action part of the communication system, and it is often the most visible, Fig. 3-10. A bit more detail may be helpful in better understanding this important part of the communication system. The steps include:

- Encoding.
- Transmitting.
- Receiving and decoding.
- Storing and retrieving.
- Feedback.

Encoding

Encoding involves preparing a message for transmission. As the name implies, this means placing the message in code. Codes are the symbols or signals needed to go through the transmission channel. Words and pictures are examples of codes we use to express our thoughts.

In graphic communication, design is an important part of the encoding process, Fig. 3-11. This involves placing the message in the best format for the intended audience. Important factors in this process are the purpose of the message (inform, instruct, persuade, entertain, control), the kind of audience, and the available resources. The design stage usually begins with the development of alternative ideas, and two methods that might be used for this are brainstorming and thumbnail sketches. The final design is then created. The design chosen should serve the intended purpose and be appropriate for the audience.

Designed messages often go through another encoding stage when machines are utilized in handling the message. In electronic communication, symbols such as words and pictures are encoded by converting them

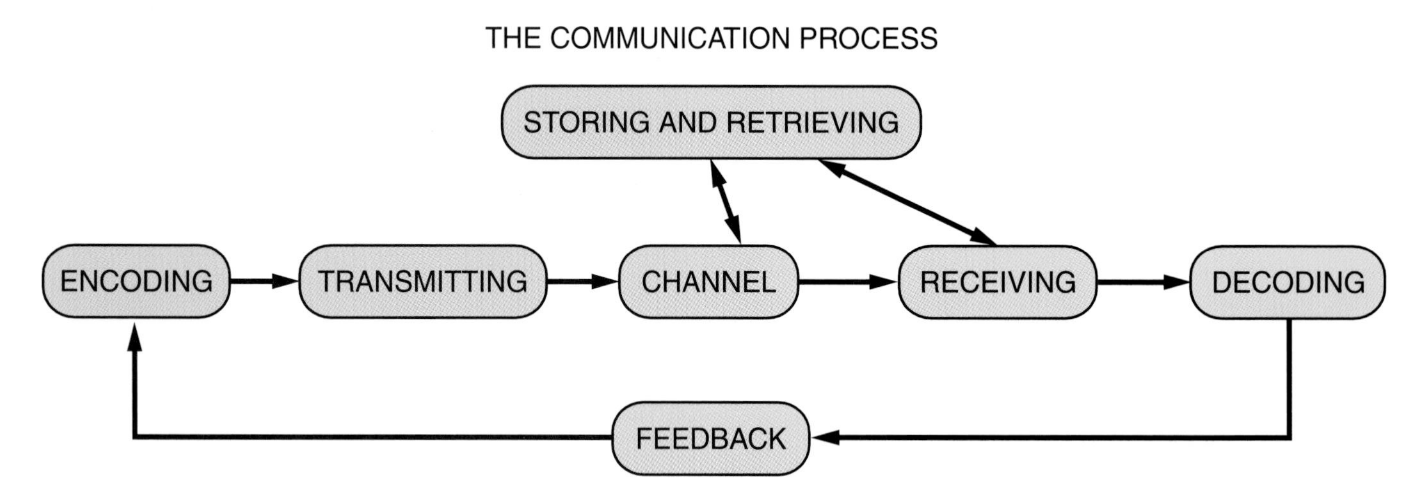

Fig. 3-10. This diagram shows the steps involved in the communication process.

Fig. 3-11. Design is an important part of the encoding process. Here, product managers are designing an advertising campaign. (Colgate-Palmolive Co.)

into electrical signals that can be transmitted. For example, a microphone converts sound into an electrical signal that can be transmitted to a speaker or stored on a magnetic tape or disk. Computer keyboards and video cameras are other encoding devices.

Transmitting

Transmitting is the process of placing a message on a channel so that it can be sent to the receiver. For example, a radio transmitter places the signal on an electromagnetic wave that travels by air to a receiver, such as a car radio.

Typical channels include physical links, such as wire or cable, and air, Fig. 3-12. Sound waves and electromagnetic waves use air as the channel. Some transmitting systems use both types of channels. Telephone communication uses both wire (electric signal) and electromagnetic waves for transmission.

In some instances, the message is transmitted to a physical channel that stores the message. Examples include paper, film, and magnetic disks. For instance, printing involves placing (transmitting) the message onto a substrate such as paper (channel) using a printing press. A camera (transmitter) is used to place an image on film (channel). See Fig. 3-13.

Receiving and decoding

Receiving involves taking a message from the channel. Decoding involves making that message understandable. For example, you may read a newspaper ad. This is reception of the message. Your brain then decodes the message as you read the words and look at the pictures. The advertiser hopes that you receive the message in the same way it was intended.

In electronic communication, reception requires detecting the signal coming through a wire or by air, and then converting these signals into a form that is useful to people. For example, an antenna can detect radio waves, but a speaker is needed to convert these signals back into sound.

Storing and retrieving

Some messages are stored for later use. Books and magazines store messages until someone reads them.

Fig. 3-12. This optical fiber is an example of a physical channel. (U.S. Sprint)

Fig. 3-13. A camera can transmit a message onto a physical channel, such as film, for storage. The message can later be retrieved by making a print from the film.

Video and audio tapes store pictures and sound. Photographs are images that have been stored. Machines are often utilized in the storing process. Examples include tape recorders, cameras, and printing presses.

Special storage methods are available for increasing the life of certain messages. Written documents and pictures can be placed on compact disks. The CD will last longer than paper and requires less storage space.

Retrieval systems facilitate obtaining stored information. This book has a table of contents and an index that can be used for locating certain topics. A counter on a video player can be used to locate a segment on a tape. A word processor can display a list of the files on a disk so the correct one can be opened.

As with storage devices, retrieval systems may involve machines of various kinds. Often, the same equipment may be used for storage and retrieval. A telephone answering machine records and plays back audio messages. A computer disk drive both stores and retrieves information. Another such device is the video cassette recorder that can store and retrieve both picture and sound.

Feedback

Feedback is often a part of the communication process. It helps the sender determine whether or not the message was or is being received as intended. In conversation with another person, this could be done by asking a question. When communicating to a large group (mass communication), a survey might be done to determine the effects of the communication. If the feedback shows that the message is not being received as intended, then a change may be needed in the way it is being presented.

SUMMARY

Some technologies have become very complex. The systems approach, which breaks the system into inputs, processes, and outputs, is one way to make technology easier to understand.

In communication technology, the inputs are the resources needed for communication. Processes are the steps needed for designing the message and sending it to the intended receiver. The output is the message that is received. Feedback provides a way of determining if the message was received as intended.

The systems approach is a tool to help in understanding technology, including communication technology. No matter how complex the technology, this approach can help us break it into its separate elements for better understanding.

WORDS TO KNOW

All of the following words have been used in this chapter. Do you know their meanings?

capital
closed loop system
communication system model
control system
decoding
devised systems
encoding
feedback
input
natural systems
output
process
receiving
retrieval systems
system
systems engineering
transmitting

REVIEWING YOUR KNOWLEDGE

Please do not write in this text. Write your answers on a separate sheet.

1. True or false? In most cases, people know how to operate technology, but they have little understanding of how it really works.
2. Which of the following is NOT a devised system?
   a. Government system.
   b. Highway system.
   c. Solar system.
   d. Organizational structures in businesses.
3. Which of the following is the purpose or goal of a system?
   a. Output
   b. Inputs.
   c. Feedback.
   d Process.
4. Which is NOT a part of the general system model?
   a. Process.
   b. Output.
   c. Product.
   d. Input.
5. The __________ system model is a detailed system model used for explaining information transfer.
6. A book is an example of what part of the communication system model?
   a. Encoding.
   b. Transmitting.
   c. Decoding.
   d. Storage and retrieval.

7. List five important resources for communication.
8. List the five steps involved in the process of communication.
9. Which of the following involves preparing a message for transmission?
   a. Encoding.
   b. Retrieving.
   c. Receiving.
   d. Feedback.
10. True or false? An index can be considered a retrieval system in a book.

APPLYING YOUR KNOWLEDGE

1. Identify examples of systems that you use every day.
2. Investigate the origins of the communication model. Prepare a written report and present it to the class.
3. Divide into groups. Use the communication model to analyze a communication technology such as the radio or telephone. Each group should share its findings with the class.
4. Research how a new technology has been influenced by older technologies. Prepare a written report and present it to the class.

Section ACTIVITIES 1 INTRODUCTION TO COMMUNICATION

BASIC ACTIVITY:
Extending Yourself Through Technology

Introduction:

In many ways, technology extends human senses. You can prove this point. For this activity, your class will form small groups and brainstorm on ways that technology has extended human senses both in the past and today. See Section Fig. 1-1. Following this, your group will choose a technology to help you convey your information to others in the class.

Section Fig. 1-1. How can this technology extend your senses? (Radio Shack)

Guidelines:

- The preferred group size is five to seven students.
- Senses to consider are sight, hearing, touch, taste, and smell.

Materials and equipment:

These will be determined by which technology your group chooses to use.

Procedure:

1. Divide into groups and choose a group leader. The leader is responsible for ensuring that group activities are assigned as equitably as possible.
2. Brainstorm on ways that technology has extended human senses in the past. Choose several examples in each category to share with the other groups. Repeat this process, choosing several examples of how technology is used to extend human senses today.
3. As a group, select a technology that can be used for conveying your findings to the other groups.
4. Develop the presentation. The presentation should include an explanation of how technology is being used to communicate your message and a brief summary of your findings.
5. Give your presentation to the class.

Evaluation:

Considerations for evaluation are:

- Group and individual effort.
- Completion of assigned tasks.
- Effectiveness of presentation.

ADVANCED ACTIVITY:
An Alternative Approach

Introduction:

Communication is so much of your daily life that you may not realize just how important it is. In this activity, you will attempt to communicate in an alternative way. Just think, your new method might become a widely used communication method is the future! See Section Fig. 1-2.

Guidelines:

- Divide into two teams.
- Normal communication methods cannot be used, including speaking and writing in a typical language.
- If a code is developed, it is permissible to give a code sheet to the other team.

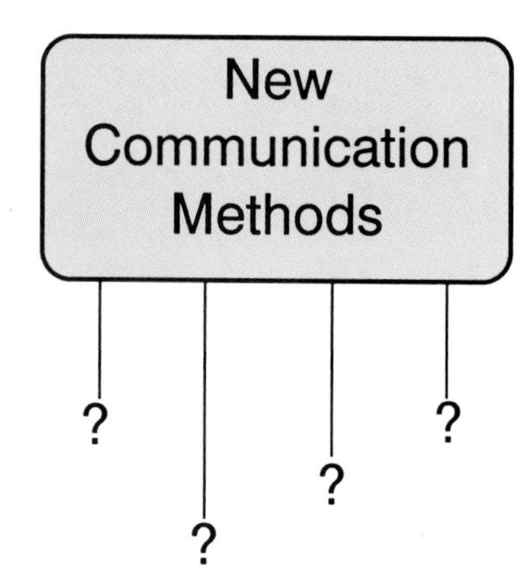

Section Fig. 1-2. What new communication methods can you imagine?

Materials and equipment:

- Available materials in the classroom.

Procedure:

1. As a team, compose a message that your team would like to send. Good examples could include information found in this text or news of a current event.
2. Think of various ways the message could be sent besides the normal methods. Select the most feasible method, taking into consideration the time allowed for the activity and available materials.
3. Assemble all the necessary materials for sending the message.
4. Transmit the message to the rest of the class.
5. Determine how well the other team understood the message by asking for feedback. Ask the other team for their thoughts on ways the system could be improved if you were to communicate another message using this same system.
6. Following the transmission test, develop a simple communication model to show how your new communication system works.

Evaluation:

Considerations for evaluation are:

- Creativity and effort involved in sending the message your group has produced.
- Correctness of the communication model.

ABOVE AND BEYOND ACTIVITY: Communication Technology on the Job

Introduction:

How important is communication technology in the lives of those around you? This activity will help you to decide and give you an opportunity to get to know some people who work at your school. In this activity, you will be interviewing school personnel about communication. See Section Fig. 1-3.

Section Fig. 1-3. How do you think this communication technology teacher would respond to the interview? (Bowling Green State University, Visual Communication Technology Program)

Guidelines:

- Divide into several groups.
- The interview survey should require no more than 10 minutes to complete.

Procedure:

1. As a group, brainstorm on topics to include on a survey about the importance of communication technology for school employees. For example, you may wish to have them discuss past practices as well as current ones. Merge the topics so that there are less than 10.
2. Develop a survey from the topic list. Strive for a logical order. You may want to list some examples of possible answers for each question to help the conversation along.
3. Rough out the interview on a word processor and let your instructor review it. Make any needed changes and print out the final copy. If possible, test the interview survey on someone and make additional changes if needed.
4. Schedule the interviews at least a day in advance.
5. Meet promptly at the time and place designated and complete the interview. Take brief, but legible notes and/or tape the responses. While the information is still fresh in your mind, summarize the interview.
6. Write a short paper summarizing the interview information. Point out similarities among responses and differences.
7. Develop a presentation and present your findings to other class members. Notes are permissible for the presentation, but they should be used as points of reference, not read.

Evaluation:

Considerations for evaluation are:

- The level of effort and completion of all tasks.
- The quality of the interview survey.
- The quality of the paper and presentation.

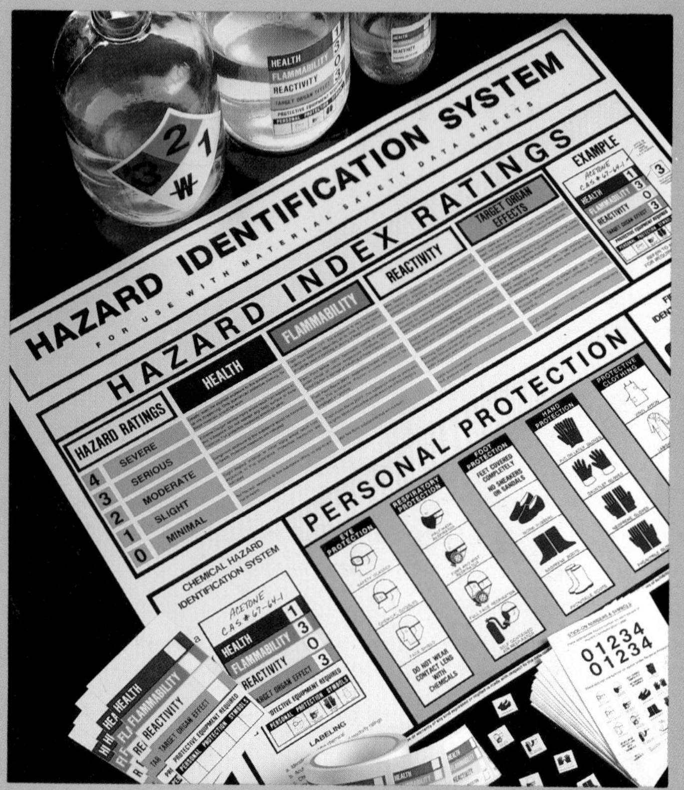

NATIONAL MARKER CO.

BOWLING GREEN STATE UNIVERSITY, VISUAL COMMUNICATION TECHNOLOGY PROGRAM

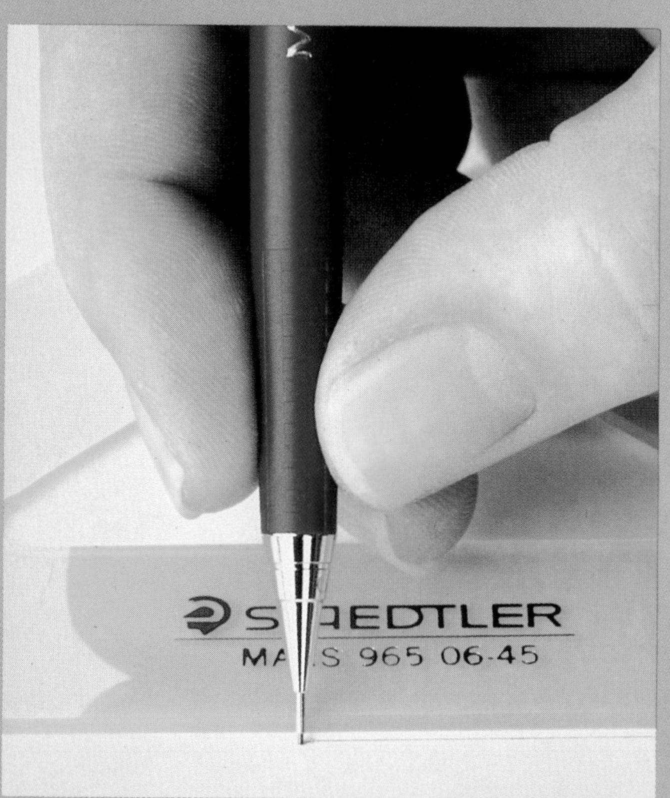
STAEDTLER

section 2

Basic Communication Skills

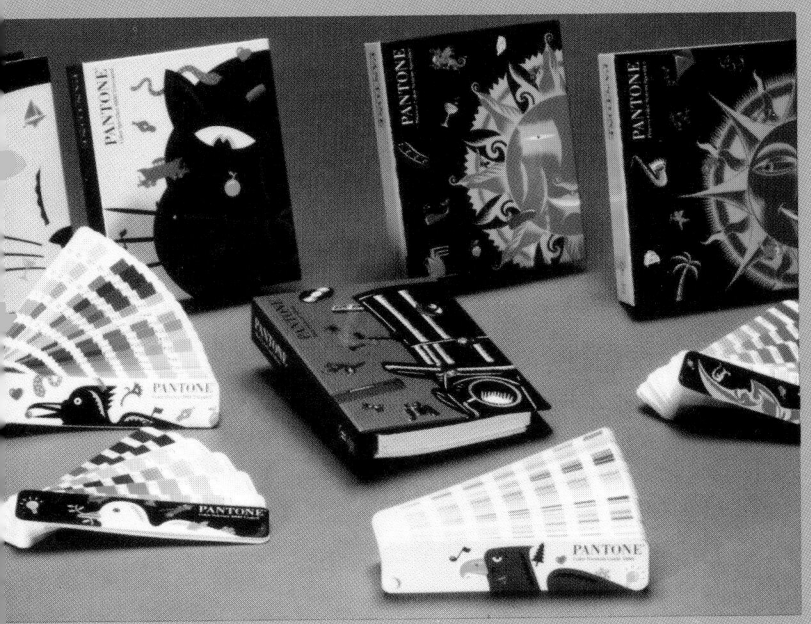

PANTONE

PENTEL

Color is a powerful design element. On these signs, color helps emphasize the importance of the information. (Idesco Corp.)

4

CHAPTER

DESIGN AND PROBLEM SOLVING

After studying this chapter, you will be able to:
- *Apply problem-solving models in finding solutions to problems.*
- *Describe techniques that can be used for generating solutions to problems.*
- *List and describe the elements and principles of design.*
- *Apply the elements and principles of design in creating messages.*

In communication technology, the purpose of design is to develop the best message for the intended audience. A radio broadcast should be designed so the audience gets the message and continues to tune in to that station. An ad should be designed to capture the attention of those looking at it and persuade at least some people to buy the product.

In order to ensure quality results, a system for design is normally used. These are a series of steps you can follow in designing a message. Since **design** is simply finding a solution to a problem, a problem-solving model is a good place to begin when learning about design, Fig. 4-1.

PROBLEM SOLVING

In Chapter 3, the concept of systems was explored. A *system* is a number of parts that work together to accomplish a goal. Systems can be natural, like the human body, or devised, such as a highway system.

A devised system that can be used for solving a variety of problems is the **problem-solving model**. Once learned, this method can be used for solving everyday problems such as deciding what to wear or how to get to work. It can also help solve design problems.

Problem solving plays an important role in the development of technology. In fact, technology can be defined as the technical way in which we solve problems to meet our needs. As the rate of change in the world continues to increase, problem-solving skills take on greater importance. There is a need for people who can find solutions to technological problems. These problems can be big ones, such as pollution, or everyday problems, such as how to use a program on a computer. Problem-solving skills are an important part of technological literacy.

MODELS USED IN PROBLEM SOLVING

There are actually several problem-solving models from which you can choose. However, once they are analyzed, they all have elements in common. One of the

Fig. 4-1. Design and problem-solving techniques are vital in communicating ideas effectively. (Sears, Roebuck, and Co.)

simplest models uses four steps and is easy to remember. This model is shown in Fig. 4-2. The first step, "Hey wait," means to identify that there is a problem to be solved and that the model can help in solving it. The "Think" step involves thinking of different solutions to the problem and deciding on one solution. The "See" step involves actually trying out the solution to see if it works. The "So" step is when you determine whether or not the problem has been solved. If not, then you may need to return to an earlier step and try again. This model is excellent for many everyday problems and contains the essential elements that can be found in more complex models.

For example, suppose your bike needs repair and you need to be at soccer practice in 30 minutes. Certainly, this is an identified problem. In thinking about how to get to practice, alternatives are identified. These might include calling a friend, walking, taking the bus, etc. Those alternatives with the least chance of success are eliminated. For example, the bus may have already passed your stop, and walking may take too long. Based on your analysis, you may decide to call a friend and see if this method will allow you to arrive at practice on time. Your friend arrives late, and you are late for practice. Therefore, next time you may wish to decide on a different method.

A more detailed problem-solving model, which works well for solving technological problems, includes the following six steps:

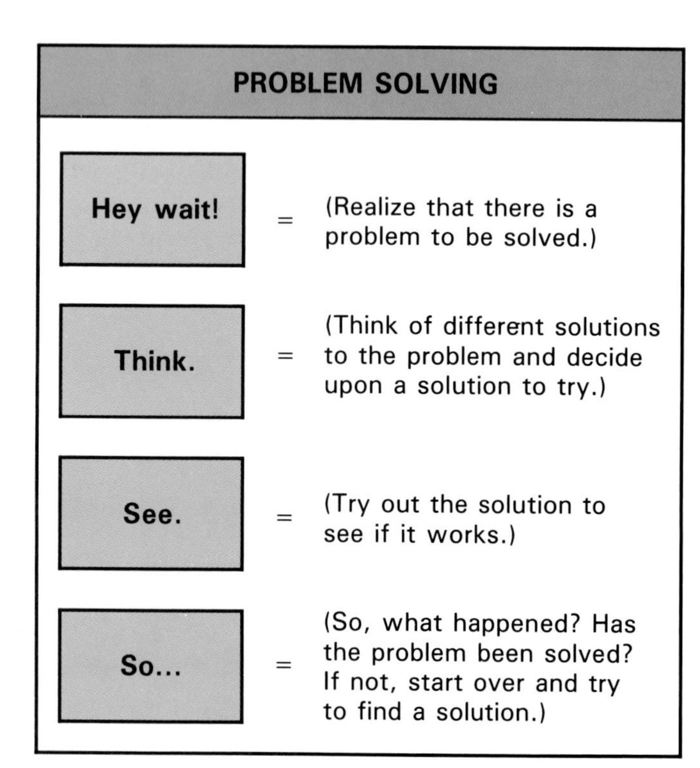

Fig. 4-2. This is a simple problem-solving model. Try to apply it to a design problem that you would like to solve.

1. *Identify the problem and specifications.* In this step, the problem is defined. In addition, any specifications for designing the solution should be identified.
2. *Research the problem.* Investigate how others have solved similar problems.
3. *Generate ideas for solving the problem.* This step involves researching the problem and developing alternative solutions. Sometimes models are made of alternative solutions.
4. *Choose a solution.* Determine the best solution and modify it further, if needed.
5. *Test the solution.* Try out the solution.
6. *Evaluate the solution.* Look at the results to determine if the solution can be further improved or if another solution may be needed.

This model works well for communication technology problems that often involve determining the best design for sending a message. For example, an advertisement needs to be produced for a new product. The problem in this case is the need to develop an ad that will appeal to the intended audience and help sell the product.

After the problem is identified, design requirements or specifications should be considered. For example, what is the purpose of the ad? What is the profile of the intended audience, and how would you best attract that audience? Do resource restrictions, such as finances, limit the size and complexity of this ad? These questions, and any others, should be identified prior to generating ideas.

It is determined that the purpose of the ad is to persuade people in the 18 to 25 age bracket to purchase the product. Because of financial limitations, a small ad (4 in. x 5 in.) will be placed in selected newspapers across the nation. Research is then done on the type of ad that normally appeals to people in this age bracket.

The next step in problem solving is generating ideas. In this step, sketches can be made of alternative designs for the ad. These can be studied, and the best solution is selected. The ad is then prepared so that camera-ready copy can be sent to newspapers.

After the ad appears, the results can be evaluated by keeping track of sales. In addition, subscribers in the appropriate age bracket can be interviewed to obtain further feedback on the ad. This information can be used for further refinement of the ad or possibly a different advertising strategy.

Generating solutions

Generating solutions to problems, an important part of all problem-solving models, can sometimes be difficult. Everyday problems, such as choosing a fruit at the grocery store, may be fairly simple and entail low risk if a mistake is made. However, larger problems may require careful study. For example, a television program that needs to attract many viewers and requires a major financial investment, will need to be carefully analyzed. Techniques for generating ideas can vary depending on the problem. For example, thumbnail sketches are used in graphic arts for quickly sketching different designs. Storyboard cards serve a similar purpose when designing a video production. See Fig. 4-3.

Generating solutions is often a problem of inertia. *Inertia* is defined as the tendency of a body at rest to remain at rest and the tendency of a body in motion to stay in motion unless a force is acted upon it. Although this is a physics principle, the concept works for ideas as well. Often, a technique is needed to overcome idea inertia or the tendency to not have ideas. Thumbnail sketches are an example. Once the ideas begin to be generated, there is a tendency to continue generating ideas.

A special technique for group problem solving is **brainstorming**. This is a method for improving the quantity and quality of ideas generated by a group. A common roadblock when trying to discuss ideas with others is to criticize one another's ideas. Unfortunately, this can stifle the generation of more ideas, and, possibly, the best solution to the problem. Brainstorming enhances group problem solving by setting forth some basic rules to follow:

1. Generate as many ideas as possible.
2. Criticism of ideas is not allowed.
3. Creative solutions are encouraged. (A silly solution may end up being practical, or may help someone else in generating an idea.)
4. Combine ideas as you proceed.
5. Be open-minded about all ideas.

Fig. 4-3. Storyboard cards are used in generating ideas for a video advertising campaign. (Gillette Co.)

A leader normally monitors the brainstorming session and makes sure the rules are being followed. All ideas should be recorded.

Following brainstorming, the ideas can be evaluated. Often, the list of ideas is shortened to several good ideas. Some ideas may be similar, while others may be impractical based on available resources. Gradually, the list is narrowed until the best idea or ideas are chosen by the group.

DESIGNING COMMUNICATION PRODUCTS

Designing communication products is simply another application of the problem-solving model. Since the majority of messages in communication technology are visual or acoustical, special techniques are often needed to help in designing these messages.

Visual design often begins by sketching alternative designs. This makes it possible to quickly develop different designs. Sketching helps one overcome inertia and begin to visualize possible solutions.

Designing written messages also requires you to first develop rough ideas. **Freewriting** is sometimes used for this purpose. After some research on a topic, the writing is done quickly without attention to grammar or spelling. The purpose is to get ideas down on paper without being self-critical. Again, this technique helps to overcome inertia. The freewritten material can then be revised one or more times as the final draft is being developed.

In designing both visual and written material, it is important to delay the evaluation until ideas are developed. Evaluation at an early stage can result in questioning your own ideas, thus resulting in a blank sheet of paper or blank computer screen. Remind yourself that ideas go through stages, and they are usually in rough form in the beginning.

As mentioned earlier in this chapter, design is the process of finding a solution to a problem. The elements and principles of design are tools that can be used in creating an effective design.

ELEMENTS OF DESIGN

What are elements? In chemistry, elements are the basic substances that make up the world, such as iron, oxygen, and hydrogen. Likewise, the **elements of design** are the basic parts that make up the design, Fig. 4-4. The elements of design are color, line, shape, and texture.

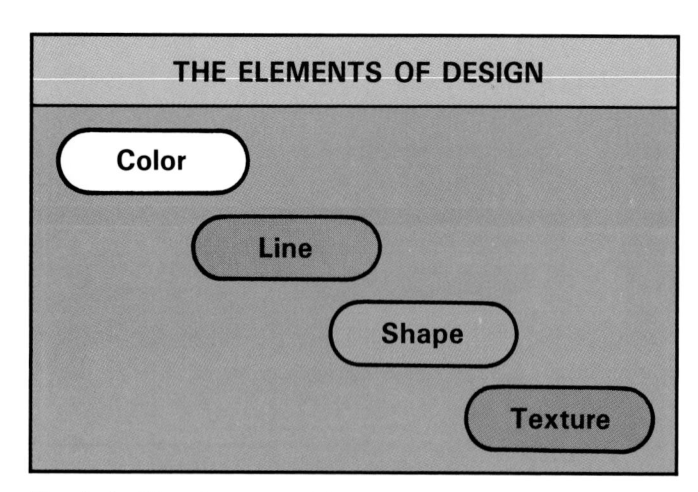

Fig. 4-4. The elements of design are powerful communication tools.

Color

Color is a powerful design element. It has the ability to affect feelings and moods. See Fig. 4-5. Color is often used in conveying messages.

Color can be described as having hue, saturation, and lightness. **Hue** is the name given to a color. Examples include red, green, and yellow. Within a hue, such as green, a variety of colors can be created by changing saturation and lightness. An example would be a mint green and kelly green, which are both of the same hue, but different colors.

Saturation refers to the amount of hue in a color. For example, both pink and scarlet are of a red hue. However, scarlet would be the most saturated since it contains the greatest amount of red. The addition of black, white, or gray to a color lowers the saturation, or purity, of a color.

Colors	Feelings or Moods
Red	Excitement, power, danger, aggression, anger, love.
Orange	Lively, cheerful, friendly, energy, warmth.
Yellow	Cheerful, bright, sympathy, cowardice.
Green	Refreshing, restful, peaceful, luck, envy, hope.
Blue	Calm, serious, reserved, depression, dignified, serenity.
Purple	Dignified, dominating, mysterious, royality.
Black	Sophisticated, somber, despair, death, mourning, wisdom.
White	Innocence, purity, faith, peace.

Fig. 4-5. Although there can be cross-cultural differences, these are feelings or moods that can be suggested by various colors.

Lightness refers to how light or dark a color is. An example would be a light blue and a dark blue. Lightness of a hue can be changed by adding white or black to it.

A change in saturation often changes the lightness of a color and vice versa. For instance, adding white to a hue will make it less saturated and lighter.

Color systems. Since there are so many different colors, **color systems** have been devised that assign a number or code to each color. These codes provide an accurate way to describe colors. For example, if a graphic designer tells a printer that a kelly green is needed, there may be 10 or 20 kelly greens from which to choose. By using a number, the specific color can be described exactly.

Several color systems have been created, so in addition to the color number, the name of the system should be specified. For instance, a Pantone 12-53-30 is a blue color (#53) with a lightness of 12 and a saturation of 30. Other color systems include the Munsell system and the CIE system. A variety of color selection materials and supplies for a color system are shown in Fig. 4-6.

The color wheel. A color wheel is shown in Fig. 4-7. The **color wheel** shows how colors are related to one

Fig. 4-6. This Pantone color system helps to prevent confusion by enabling graphic designers to communicate exactly which colors they want in a design. (Pantone, Inc.)

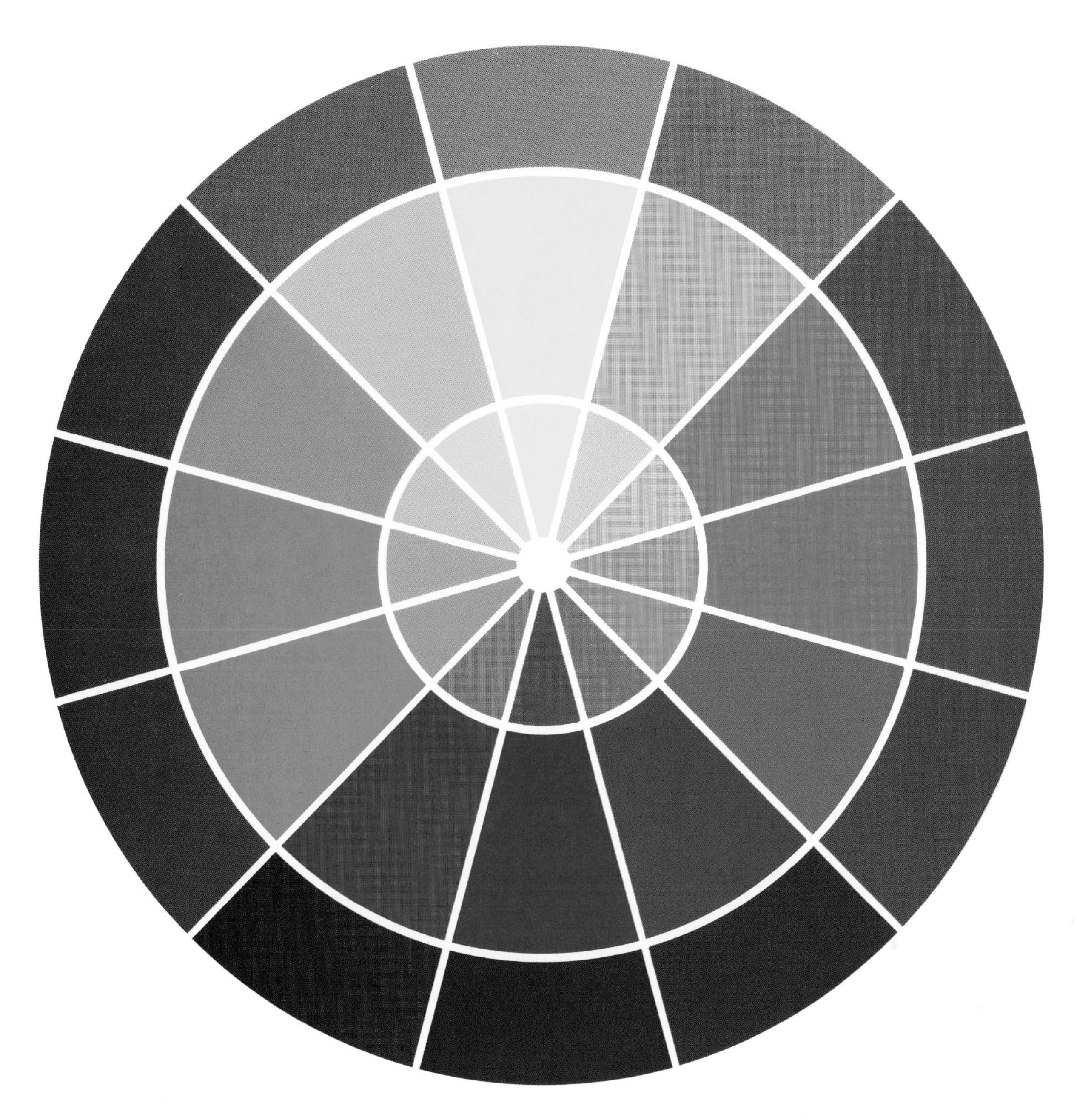

Fig. 4-7. This color wheel shows the relationship among colors.

another. The *primary colors*, used for mixing inks and dyes, are red, blue, and yellow. No other colors can be mixed to produce primary colors. All other colors can be produced from the primary colors. For example, violet (purple) is produced by mixing red and blue; green is produced by mixing yellow and blue; orange is produced by mixing red and yellow. Thus, violet, green, and orange are *secondary colors*. Colors produced by combining a primary color and a secondary color are called *intermediate colors*. Examples of these are blue-green, blue-violet, red-orange, red-violet, yellow-green, and yellow-orange.

Warm and cool colors. Some colors seem to create a feeling of warmth, while others seem to create a feeling of coolness. Colors related to red, orange, and yellow are called **warm colors**. Warm colors are often called advancing colors because they seem to move forward and suggest activity. Colors related to blue, violet, and green are **cool colors**. These colors are restful, relaxing, and calm. They are often called receding colors because they seem to move away or stay in the background.

Neutrals. White, gray, and black are called **neutrals**. They have no color as such. They can be used alone or in combination with other colors. When a neutral is used with a another color in a design, the other color looks brighter. Because white reflects light, it makes objects appear larger. Black absorbs light and makes objects appear smaller.

Line

Line gives direction to a design. Lines can be used to point to an important feature in a design or to convey a feeling. Lines can be vertical (up and down), horizontal (across), diagonal (slanting), or curved (part of a circle). Examples of various types of lines are shown in Fig. 4-8.

Vertical lines carry the eye up and down. They convey a feeling of awe or challenge.

Horizontal lines carry the eye across from side to side. These lines help to convey a calm, peaceful feeling.

Diagonal lines are slanted. They add interest to a design.

Curved lines are lines that are gently bent. They give a soft, relaxed look to a design.

Shape

Shape refers to the form of an object. When lines enclose a space, a shape is formed. The three basic shapes are the circle, square, and triangle. See Fig. 4-9.

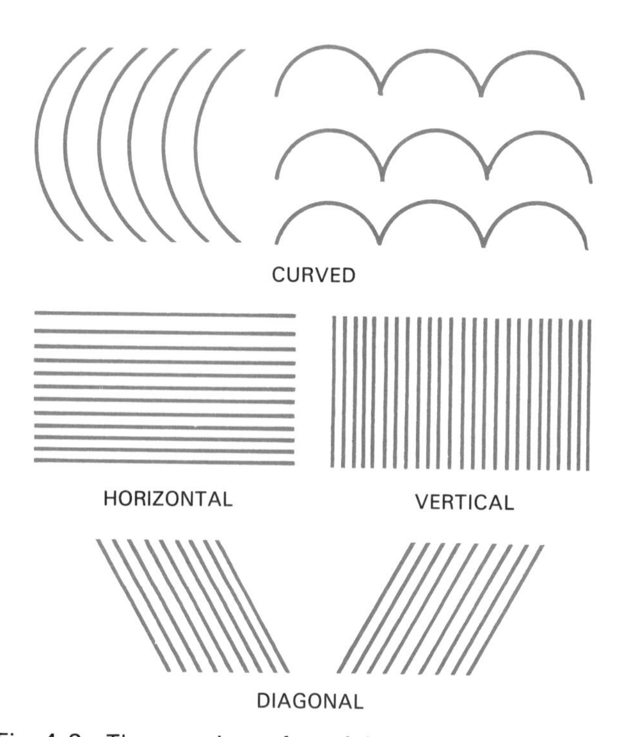

Fig. 4-8. These are just a few of the types of lines used in designs.

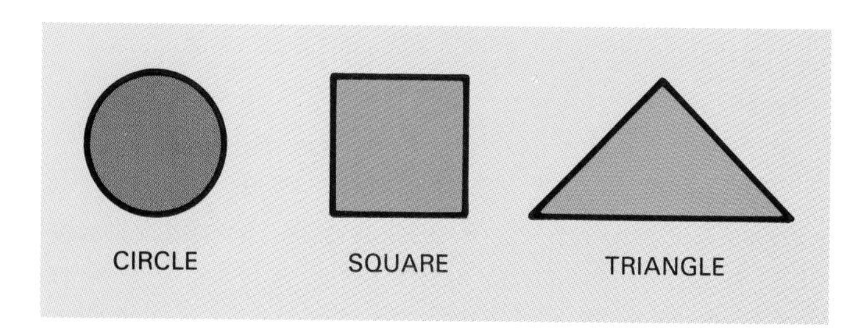

Fig. 4-9. These are the three basic shapes in designs.

Shapes that appear to be at rest are the circle and square. A feeling of movement can be created by extending these into an ellipse and rectangle.

Texture

Texture is how something feels when you touch it and/or how it looks on the surface. Words often used to describe texture include dull, shiny, stiff, fuzzy, smooth, or rough. Texture can be used to create a surface appearance in a design. Dots and lines are often used to create a desired surface texture. Notice how lines and dots convey a look of texture in the shapes shown in Fig. 4-10.

PRINCIPLES OF DESIGN

What is the difference between a principle of design and an element of design? You know that the elements make up the design. However, how do you decide how to fit the elements together into a pleasing design? The **principles of design** provide these guidelines. They tell you how the elements of design should be combined. The principles of design are proportion, balance, rhythm, and emphasis. They are used in combination to create designs, Fig. 4-11.

Proportion

Proportion refers to how the size of one part relates to the size of another part. It also refers to how the size of one part relates to the size of the whole item. For instance, suppose you are making thumbnail sketches for a poster. You will need to have some idea of the proportion of the finished product. In other words, what is the height-to-width ratio of the finished product? For a square, this would be 1:1, and your sketch would also be in a 1:1 space. Pages and photographs are often in certain proportions. See Fig. 4-12. Elements on

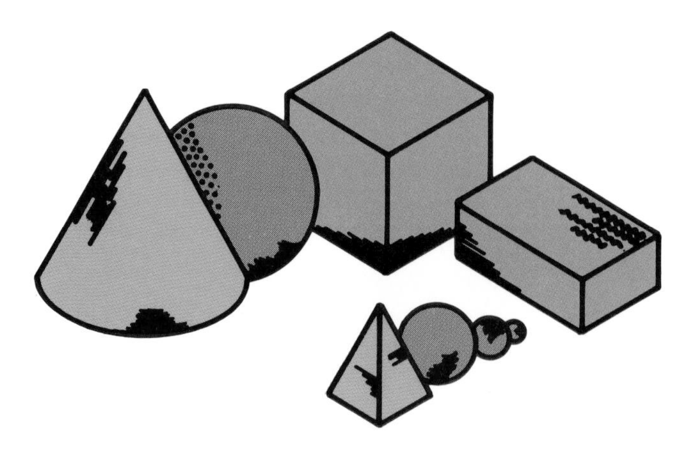

Fig. 4-10. These objects are shaded with lines and dots as a form of surface texture.

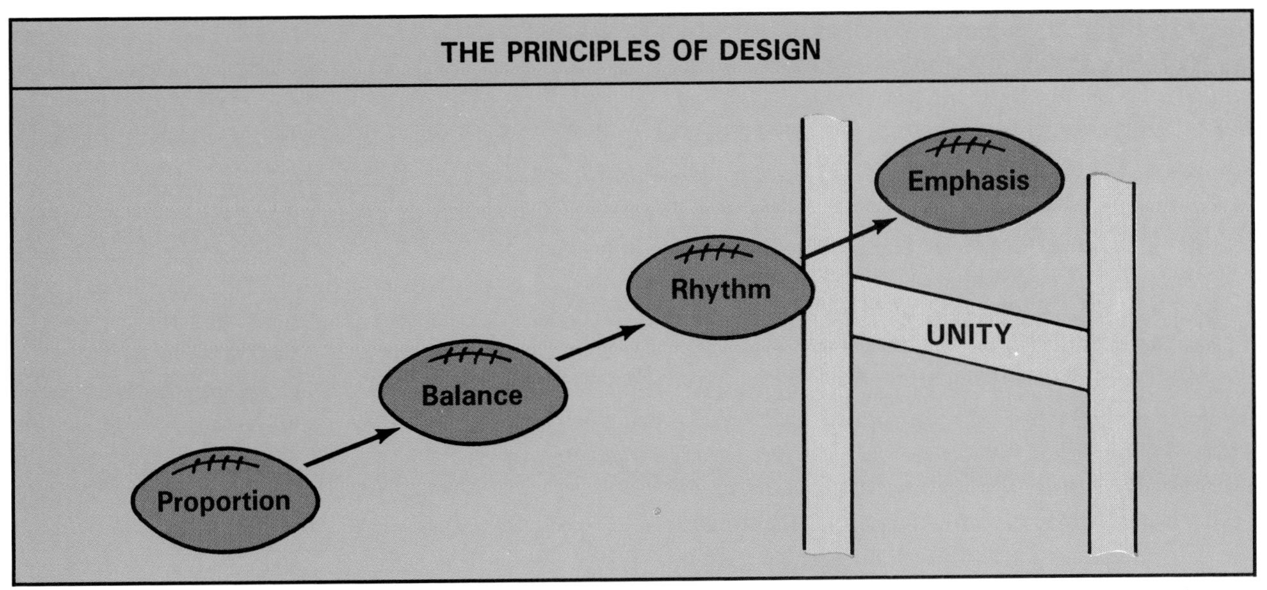

Fig. 4-11. The principles of design serve as guidelines as you use the elements of design. The goal of unity is met when the principles of design are used correctly.

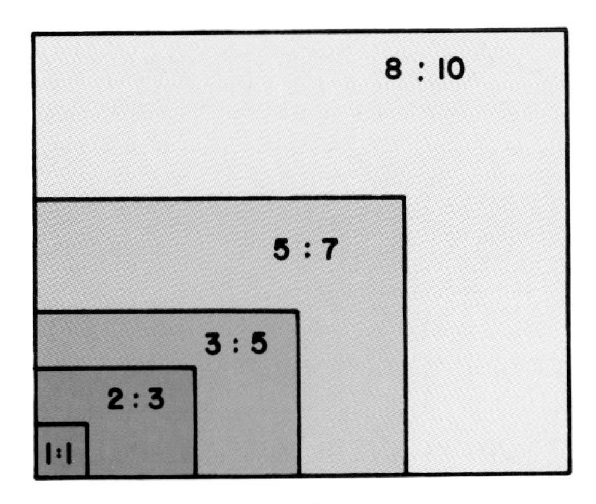

Fig. 4-12. Proportion sometimes refers to page proportions. Some typical proportions are shown here.

the page should also be in proportion. For instance, if you were placing illustrations of a tree and a flower on a page, proportionally, the picture of the tree should be larger than that of the flower.

Balance

Balance refers to how elements are arranged horizontally or vertically in a design. Balance can be achieved with color, line, shape, and/or texture.

Horizontal balance can be formal or informal. See Fig. 4-13. If all elements are equal on both sides of a design, this is known as *formal balance* or symmetrical balance. Balance can also be achieved when the design

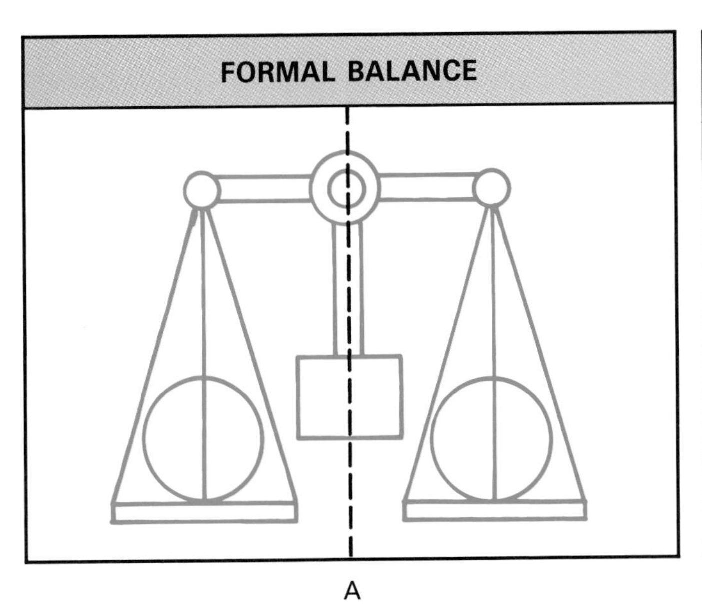

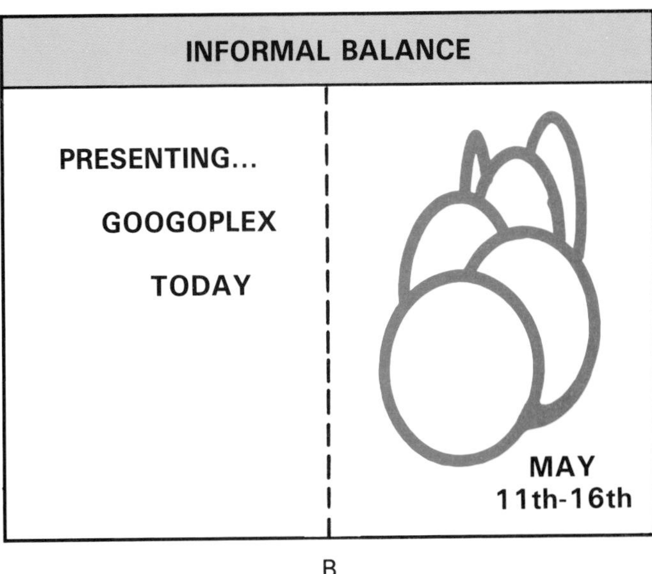

Fig. 4-13. Formal and informal balance refer to the horizontal balance of a design. Formal balance is created when both sides are equal. Informal balance is created when both sides are different, but the design appears to be balanced.

elements are different on both sides. This unequal type of balance is called *informal balance* or asymmetrical balance. Each side is different, but the design still appears to be balanced. Informal balance is often used to make a design appear more exciting.

Formal and informal balance refer to side-to-side or horizontal balance. However, vertical or top-to-bottom balance is also important. Designs are normally more pleasing to the eye if they are slightly above the mathematical center. This visually pleasing center point is known as the **optical center**, Fig 4-14.

Rhythm

In design, **rhythm** is the feeling of movement, Fig. 4-15. Rhythm in music is somewhat like rhythm in design. In music, a sound is repeated, and rhythm is produced. However, in design, you can see rhythm rather than hear it.

You achieve rhythm in a design through *repetition* of lines, colors, shapes, and/or textures. When these elements are repeated, while also changing slightly, rhythm through *gradation* results. *Radiation* is another way of achieving rhythm in a design. Radiation occurs when lines or other elements extend outward from a central point.

Emphasis

Emphasis is the center of interest in a design. It is the first thing you see when you look at a design. For instance, a large, bold headline in a newspaper is a point of emphasis. What is the first thing you notice when you look at the design in Fig. 4-16?

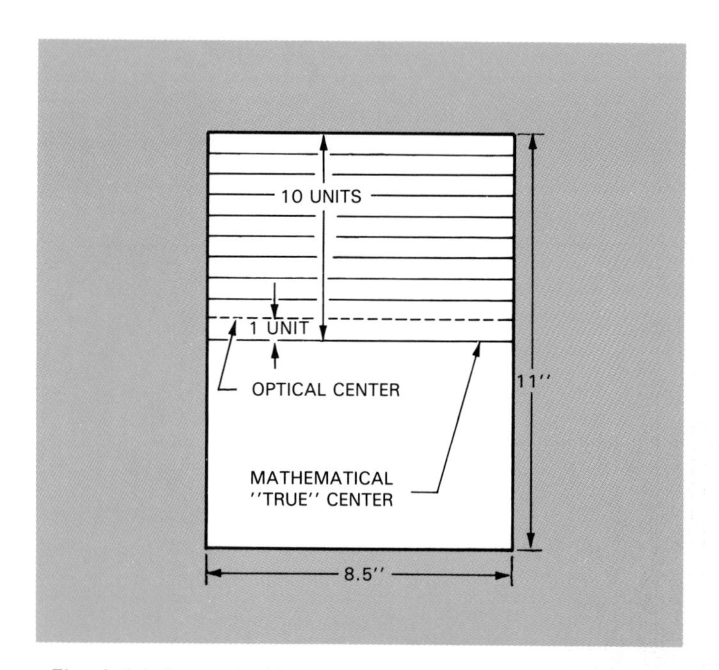

Fig. 4-14. In vertical balance, the optical center is usually more pleasing to the eye than the mathematical center.

ACHIEVING UNITY

Unity, the goal of design, is achieved when all parts of a design look as if they belong together. Refer back to Fig. 4-11. When the elements and principles of design are used effectively, the goal of unity is achieved.

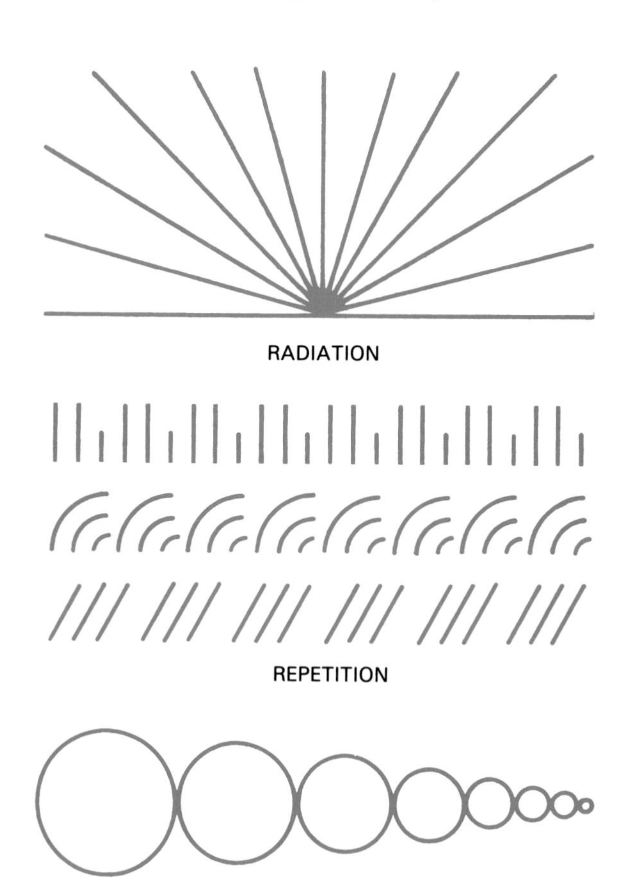

Fig. 4-15. Rhythm creates a feeling of movement. It can be achieved in a variety of ways.

Fig. 4-16. The lines in the background help to emphasize the motorcycle in this design. (Corel Systems Corp.)

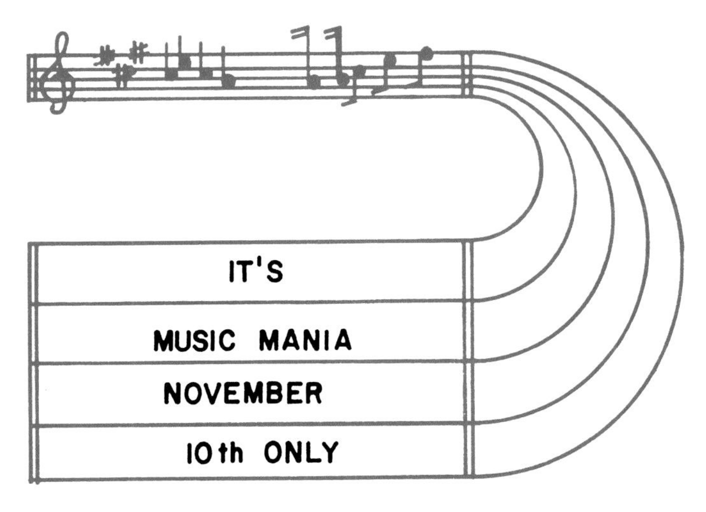

Fig. 4-17. By applying the elements and principles of design, unity was achieved in this design.

In a design that has unity, the design elements of color, line, shape, and texture suit each other. The design is well-balanced, has a sense of rhythm, and is in good proportion. Nothing looks out of place. Everything looks connected. See 4-17.

SUMMARY

Problem-solving techniques provide a structure that can help you in designing messages. Various problem-solving models can be used in generating solutions to problems. Designing messages is another application of problem solving. The elements of design (color, line, shape, and texture) are used as tools in designing messages. The principles of design (proportion, balance, rhythm, and emphasis) provide guidelines for how the elements of design should be designed to create effective messages.

WORDS TO KNOW

All of the following words have been used in this chapter. Do you know their meanings?

balance
brainstorming
color systems
color wheel
cool colors
design
elements of design
emphasis
freewriting
hue
lightness
line
neutrals
optical center
principles of design
problem-solving model
proportion
rhythm
saturation
shape
texture
unity
warm colors

REVIEWING YOUR KNOWLEDGE

Please do not write in this text. Write your answers on a separate sheet.

1. True or false? As the rate of change in the world continues to increase, problem-solving skills will take on greater importance.
2. List six steps often used in solving technological problems.
3. True or false? During a group brainstorming session, all ideas must be criticized, and creative solutions should be discouraged.
4. List the elements of design.
5. If a graphic designer wants a printer to use a specific color:
   a. A color system should be used.
   b. The printer should guess what color the graphic designer wants.
   c. The graphic designer should always trust the printer to choose a color of the correct lightness and saturation.
   d. None of the above.
6. Blue is an example of a:
   a. Warm color.
   b. Cool color.
   c. Neutral.
   d. None of the above.
7. Which of the following lines carry the eye up and down, creating a feeling of awe or challenge?
   a. Curved lines.
   b. Diagonal lines.
   c. Vertical lines.
   d. Horizontal lines.
8. List the principles of design.
9. Designs are normally more pleasing to the eye if they are:
   a. Slightly below the mathematical center.
   b. Slightly above the mathematical center.
   c. In the center.
   d. None of the above.
10. ________, the goal of design, is achieved when all parts of a design look as if they belong together.

APPLYING YOUR KNOWLEDGE

1. Identify some common problems and use the problem-solving model to find solutions.
2. Using the procedures outlined in this chapter, solve a problem by brainstorming.
3. Clip a design from a magazine. Analyze it. Describe how the elements and principles of design are used. Has unity been achieved in this design?
4. Use the elements and principles of design to create a message. Analyze it according to criteria presented in this chapter.

Sometimes problem solving requires a team approach. (Bowling Green State University, Visual Communication Technology Program)

5

CHAPTER

MEASUREMENT SYSTEMS AND QUALITY ASSURANCE

After studying this chapter, you will be able to:
- *Explain the differences between the U.S. Customary System and SI Metric System.*
- *Identify the scale options and their uses.*
- *Demonstrate how to write fractions and decimals correctly.*
- *Describe the American Point System and the use of points and picas.*
- *Explain how measurement can affect the quality of products.*

How important is accurate measurement? Just look around. The clothes you are wearing were measured so that they would fit you. All of the manufactured objects around you required accurate measurement. Hundreds of people had to measure accurately and then communicate these measurements to others. See Fig. 5-1. This chapter will help you learn to measure distances just as accurately. Quality can also be measured.

Measurement is the process of determining a dimension, quantity, or capacity. For example, length, time, temperature, sound, and weight are determined by measurement.

The two most common forms of measurement in the United States are the U.S. Customary (inch-foot) System and the SI Metric System. See Fig. 5-2. Even though the U.S. system has been traditionally used, this is changing because of the fact that the metric system is used throughout most of the world. Many products that are produced in the U.S. are sold in other countries, and if they are not made to metric standards, they may not be acceptable. In addition, many companies have plants in several countries, and a single standard is needed for communication between the plants.

In the United States, the American National Standards Institute (ANSI) establishes standards in the U.S. for measurement systems. This includes both inch-foot and metric measurement. The most recent standard for dimensioning technical drawings is ANSI/AME Y14.5M-1994. The "M" refers to metric. It emphasizes the fact that many of the new standards are related to metric.

At the international level, the International Standards Organization (ISO) establishes standards for measurement systems. When an "SI" is placed on a technical drawing, it means that this standard is being used. This stands for Le Systeme International d'Unites, which is French for "International System of Units."

Since the measure of distance is of primary importance in graphic communication, this chapter will be

Fig. 5-1. People often use measurements to communicate distances to others.

SI METRIC/U.S. CUSTOMARY UNIT EQUIVALENTS AND USES

Multiply:	by:	to get:	Multiply:	by:	to get:	EXAMPLES OF USES
LINEAR						Measurement of paper and other substrates Cutting paper and plastic Scale drawings Page composition
inches	X 25.4	= millimeters(mm)		X 0.0397	= inches	
feet	X 0.3048	= meters(m)		X 3.281	= feet	
yards	X 0.9144	= meters(m)		X 1.0936	= yards	
miles	X 1.6093	= kilometers(km)		X 0.6214	= miles	
inches	X 2.54	= centimeters(cm)		X 0.3937	= inches	
microinches	X 0.0254	= micrometers(μm)		X 39.37	= microinches	
AREA						Paper measurement and cutting Processing photographic materials
inches^2	X 645.16	= millimeters^2(mm^2)		X 0.00155	= inches^2	
inches^2	X 6.452	= centimeters^2(cm^2)		X 0.155	= inches^2	
feet^2	X 0.0929	= meters^2(m^2)		X 10.764	= feet^2	
yards^2	X 0.8361	= meters^2(m^2)		X 1.196	= yards^2	
acres	X 0.4047	= hectares(10^{4}m^2) (ha)		X 2.471	= acres	
miles^2	X 2.590	= kilometers^2(km^2)		X 0.3861	= miles^2	
VOLUME						Measuring chemicals for photography and printing
inches^3	X 16387	= millimeters^3(mm^3)		X 0.000061	= inches^3	
inches^3	X 16.387	= centimeters^3(cm^3)		X 0.06102	= inches^3	
inches^3	X 0.01639	= liters(L)		X 61.024	= inches^3	
quarts	X 0.94635	= liters(L)		X 1.0567	= quarts	
gallons	X 3.7854	= liters(L)		X 0.2642	= gallons	
feet^3	X 28.317	= liters(L)		X 0.03531	= feet^3	
feet^3	X 0.02832	= meters^3(m^3)		X 35.315	= feet^3	
fluid oz	X 29.57	= milliliters(mL)		X 0.03381	= fluid oz	
yards^3	X 0.7646	= meters^3(m^3)		X 1.3080	= yards^3	
teaspoons	X 4.929	= milliliters(mL)		X 0.2029	= teaspoons	
cups	X 0.2366	= liters(L)		X 4.227	= cups	
MASS						Ink measurement and mixing Shipping and postage
ounces(av)	X 28.35	= grams(g)		X 0.03527	= ounces(av)	
pounds(av)	X 0.4536	= kilograms(kg)		X 2.2046	= pounds(av)	
tons(2000 lb)	X 907.18	= kilograms(kg)		X 0.001102	= tons(2000 lb)	
tons(2000 lb)	X 0.90718	= metric tons(t)		X 1.1023	= tons(2000 lb)	
FORCE						Foil stamping Die-cutting
ounces—f(av)	X 0.278	= newtons(N)		X 3.597	= ounces—f(av)	
pounds—f(av)	X 4.448	= newtons(N)		X 0.2248	= pounds—f(av)	
kilograms—f	X 9.807	= newtons(N)		X 0.10197	= kilograms—f	
ENERGY OR WORK (watt-second = joule = newton-meter)						Long distance communication Electric power consumption
foot-pounds	X 1.3558	= joules(J)		X 0.7376	= foot-pounds	
calories	X 4.187	= joules(J)		X 0.2388	= calories	
Btu	X 1055	= joules(J)		X 0.000948	= Btu	
watt-hours	X 3600	= joules(J)		X 0.0002778	= watt-hours	
kilowatt-hrs	X 3.600	= megajoules(MJ)		X 0.2778	= kilowatt-hrs	
LIGHT						Photography Electro-static copying Exposure for light-sensitive materials
footcandles	X 10.76	= lumens/meter^2 (lm/m^2)		X 0.0929	= footcandles	
PRESSURE OR STRESS (newton/sq meter = pascal)						Printing press adjustments Die-cutting Air pressure
inches Hg(60 °F)	X 3.377	= kilopascals (kPa)		X 0.2961	= inches Hg	
pounds/sq in	X 6.895	= kilopascals (kPa)		X 0.145	= pounds/sq in	
inches H_2O(60 °F)	X 0.2488	= kilopascals (kPa)		X 4.0193	= inches H_2O	
bars	X 100	= kilopascals (kPa)		X 0.01	= bars	
pounds/sq ft	X 47.88	= pascals (Pa)		X 0.02088	= pounds/sq ft	
POWER						Electric moters Equipment driven by moters
horsepower	X 0.746	= kilowatts (kW)		X 1.34	= horsepower	
ft-lbf/min	X 0.0226	= watts(W)		X 44.25	= ft-lbf/min	
TORQUE						Maching adjustment Bolt tightening
pound-inches	X 0.11298	= newton-meters(N-m)		X 8.851	= pound-inches	
pound-feet	X 1.3558	= newton-meters(N/m)		X 0.7376	= pound-feet	
VELOCITY OR SPEED						Automatic processing equipment Print press speed
miles/hour	X 1.6093	= kilometers/hour(km/h)		X 0.6214	= miles/hour	
feet/sec	X 0.3048	= meters/sec(m/s)		X 3.281	= feet/sec	
kilometers/hr	X 0.27778	= meters/sec(m/s)		X 3.600	= kilometers/hr	
miles/hour	X 0.4470	= meters/sec(m/s)		X 2.237	= miles/hour	
TEMPERATURE						Comfort level of room Mixing chemicals Processing film
(°Fahrenheit −32) X 1.8 = °Celsius						
(°Celsius X 1.8) + 32 = °Fahrenheit						

Fig. 5-2. In the field of communications, you may encounter both SI Metric and U.S. Customary measurements.

devoted to that form of measurement. In addition to the U.S. Customary and SI Metric Systems, the American Point System will be introduced. This system is primarily used in printing and publishing. Quality assurance and its relationship to accurate measurement will also be discussed.

U.S. CUSTOMARY MEASUREMENT

The **U.S. Customary Measurement System** is one form of linear measurement. Distance is measured in inches, feet, yards, and miles. The symbol for feet is ' and the symbol for inches is ″. Common or decimal fractions are used for measurements smaller than an inch.

When you use common fractions, the denominator (at the bottom) shows how many times the number is divided. Sometimes, a fraction can be reduced further. For example, 8/16 = 4/8 = 2/4 = 1/2.

There is a simple way to use fractions that does not require reductions. When measuring, find the largest fraction that "fits." In the example in Fig. 5-3, the fraction is not 1/2, 1/4, or 1/8. Therefore, it must be sixteenths. Counting, the line is 1 15/16" long.

Decimal fractions of an inch are also used for measurement. Using this method, the inch is divided into 10 parts (.1), 100 parts (.01) or 1,000 parts (.001). For example, 1 1/2″ = 1.5″.

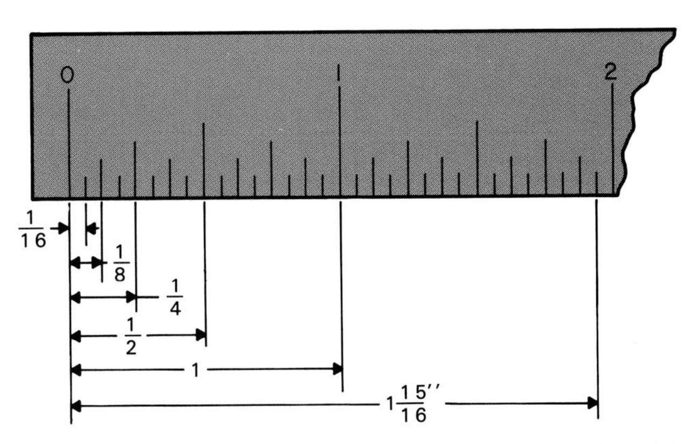

Fig. 5-3. Common fractions of an inch are marked on this scale.

SI METRIC MEASUREMENT

The **SI Metric System** originated in France, but has now been adopted by most countries in the world. As discussed earlier, the SI stands for Le Systeme International d'Unites, which is French for "International System of Units." The organization that has developed rules for using the metric system is the International Standards Organization (ISO).

In the SI Metric System, the meter is the basic unit for length, Fig. 5-4. The meter is subdivided into decimeters (1/10 meter), centimeters (1/100 meter), and millimeters (1/1000 meter). For long distances, the kilometer is used, which is equal to 1 000 meters. Some common prefixes are shown in Fig. 5-5.

In metric measure, decimals are always used for measurements less than a whole number, Fig. 5-6. For example, 12 1/2 mm is written as 12.5 mm. If the

COMMON METRIC PREFIXES

Prefix	Multiplication factor	Symbol
mega	1 000 000 = 10^6	M
kilo	1 000 = 10^3	k
hecto	100 = 10^2	h
deka	10 = 10	da
deci	0.1 = 10^{-1}	d
centi	0.01 = 10^{-2}	c
milli	0.001 = 10^{-3}	m
micro	0.0001 = 10^{-6}	μ

Fig. 5-5. Study these common metric prefixes.

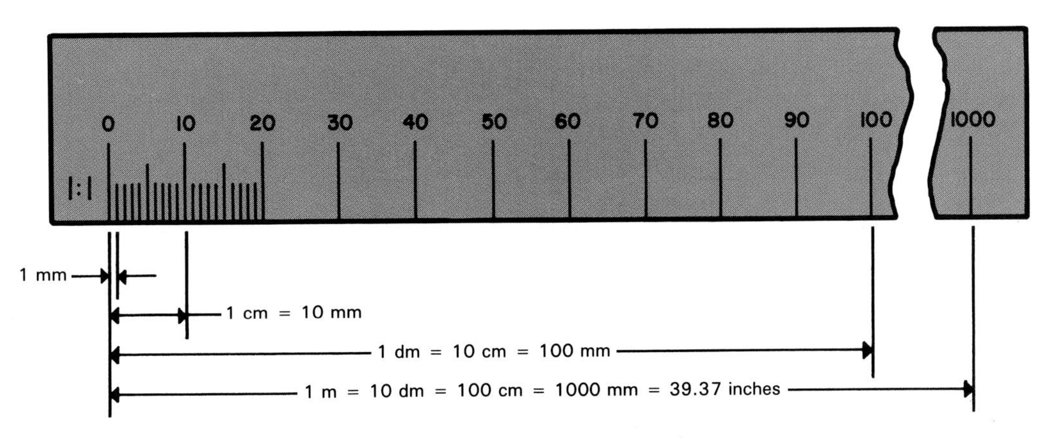

Fig. 5-4. These are subdivisions of a meter.

$12\frac{1}{2}$mm NO	.45mm NO	25.0mm NO
12.5mm YES	0.45mm YES	25 mm YES

Fig. 5-6. These are correct and incorrect forms for metric fractions.

measurement is smaller than one, then a 0 precedes the decimal point, such as 0.45 m.

SCALES

A **scale** is a device used for measuring distances on drawings. It comes in a variety of shapes and is usually around 12″ long. The 12″ or 300 mm triangular scale is very popular in schools.

Regardless of shape, scales are made so the marks lie directly against the paper. When a measurement is found, it is marked with a short, light dash at the scale marking, Fig. 5-7.

Scales are used to draw an object full-size, reduce it, or enlarge it. See Fig. 5-8. This is often called *"drawing to scale."* The correct scale to use depends on the type of drawing needed and the size of the finished drawing. For example, a skyscraper and a watch could both be drawn on the same size paper. They would just require different scales.

The scale that is used is almost always shown on the drawing. The first number is the drawing size and the second is the object size. For example, 1/4″ = 1″ means that 1/4″ on the drawing equals 1″ on the actual object.

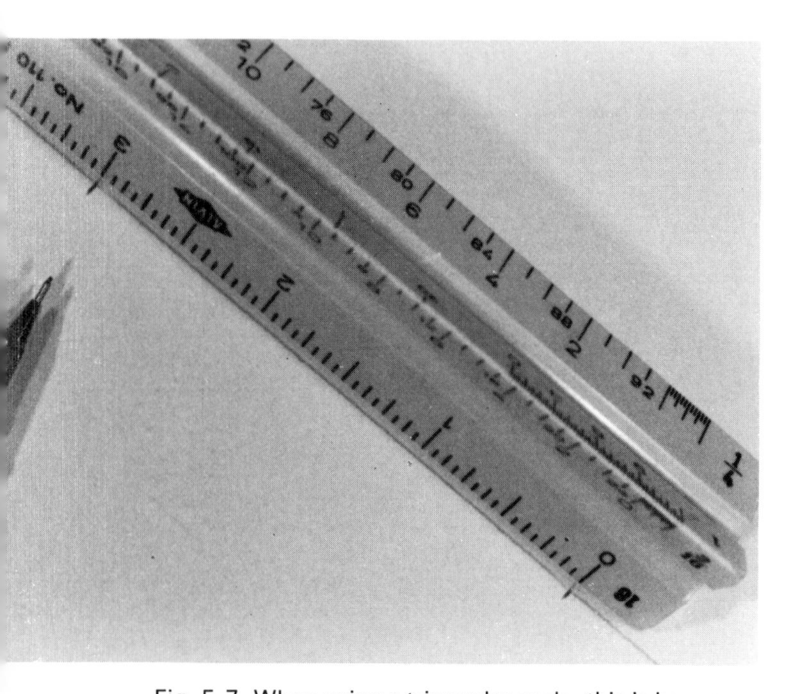

Fig. 5-7. When using a triangular scale, this is how a measurement is marked.

MECHANICAL ENGINEER'S SCALE

The **mechanical engineer's scale** is often used for drawing machines and parts of machines. This scale uses inch measurements to make drawings 1/16 size, 1/8 size, 1/4 size, 1/2 size, or full size. For example, using the 1/4″ = 1″ (1/4 size) scale, 1/4″ on your drawing would actually represent 1″ on the object, Fig. 5-9.

To measure, place the correct scale over the line to the nearest whole inch along the scale. Any remainder is found at the end of the scale in fractions of an inch. A scale with small divisions at the end is known as an open-divided scale.

ARCHITECT'S SCALE

The **architect's scale** is most often used for drawing buildings. Since very large objects have to be drawn very small, large reductions in size are necessary. Therefore, marks on the scale are used to represent feet. For example, the scale marked 1/4 means that 1/4″ = 1′ 0″.

The architect's scale is an open-divided scale. It looks similar to the mechanical engineer's scale. However, the divisions at the end of this scale represent inches, not fractions of an inch. To measure, find the nearest whole number of feet first, Fig. 5-10. Find the remaining number of inches at the end of the scale. Measurements are usually shown in feet and inches, even if zero. For example, measurements might be 7′-5″, 0′-3″, or 5′-0″.

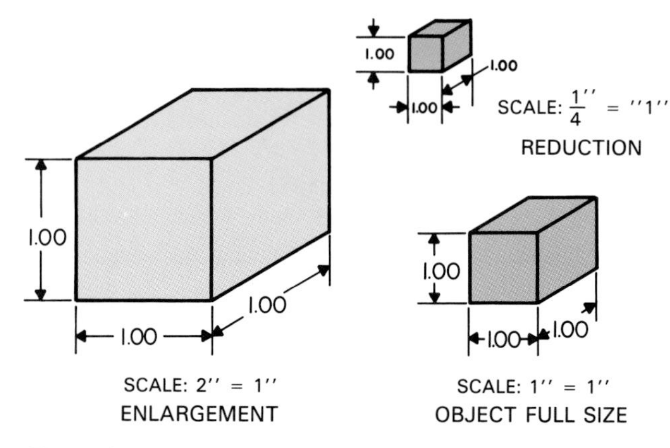

Fig. 5-8. This is an example of an object drawn to scale. The object is drawn full-size, reduced, and enlarged.

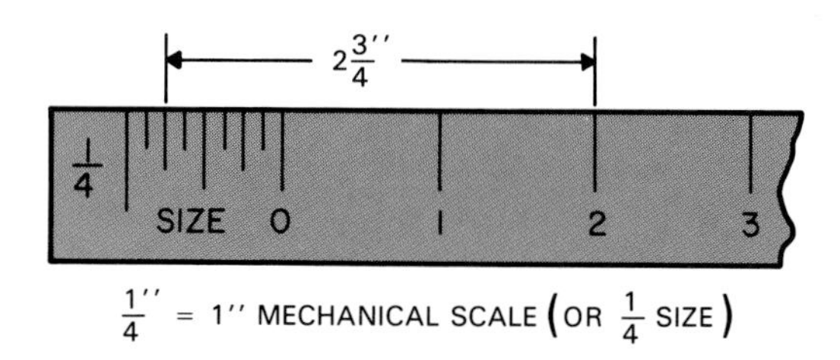

Fig. 5-9. A mechanical engineer's scale can be used for drawing machines and parts of machines.

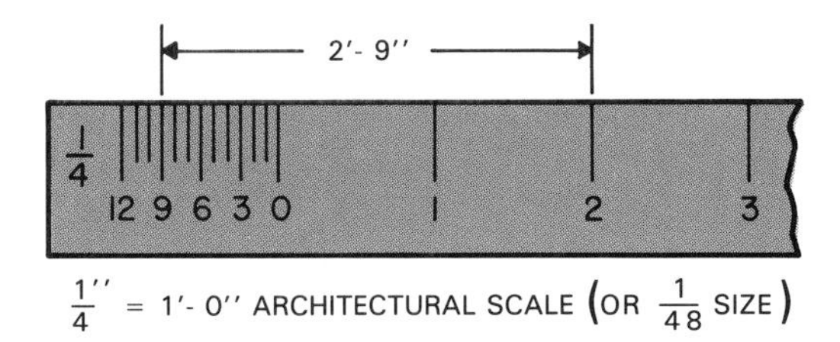

Fig. 5-10. An architect's scale can be used for drawing buildings and floor plans.

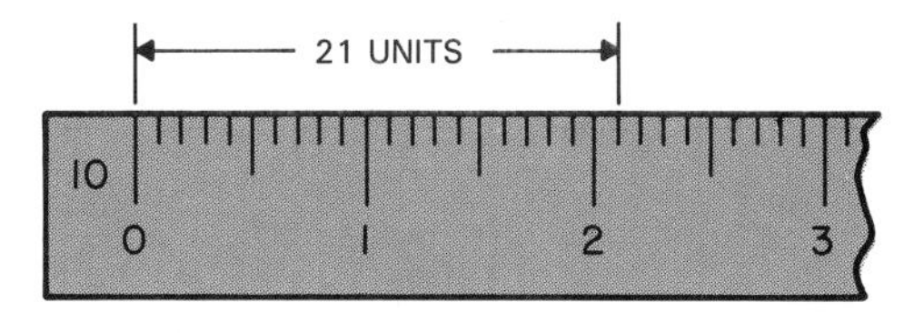

Fig. 5-11. A civil engineer's scale can be used for drawings of highways and maps.

CIVIL ENGINEER'S SCALE

The **civil engineer's scale** is a fully divided scale. It is often used for drawings of highways and maps. It is a decimal scale that includes scales that divide an inch into 10 parts, Fig. 5-11, or multiples of 10 such as 30, 50, or 60. These divisions can be used to represent common distances, such as feet, yards, or miles. For example, the 40 scale could be used to stand for 1″ = 40 miles. In this case, the smallest division, (1/40″), would equal 1 mile.

The 50 scale is useful for measuring in decimal inches, Fig. 5-12. The smallest division on this scale would equal .02 of an inch. When decimal inches are used, inch marks are omitted as well as the zero for dimensions of less than 1″. For example, .25 and 1.30 would be the correct format.

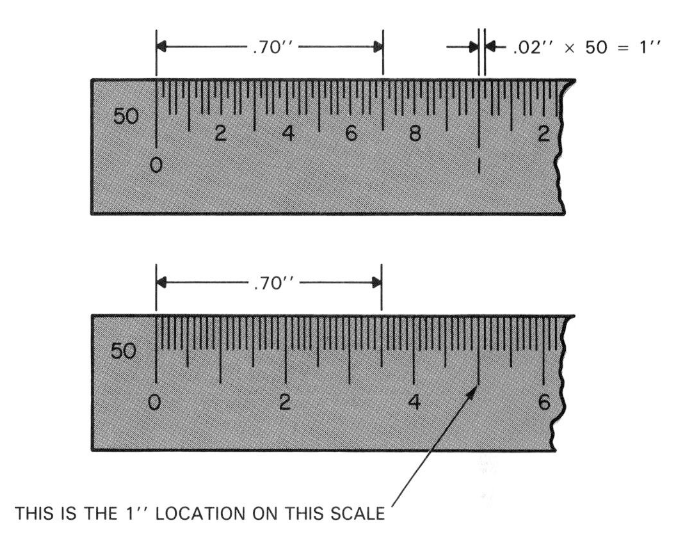

Fig. 5-12. These are two common types of scales for measuring in decimal inches. The bottom example is a civil 50 scale

METRIC SCALE

The **metric scale** is used to make drawings with dimensions in metric units. Metric scales, in general, are shown as ratios, such as 1:1 for full-size. See Fig. 5-13. A 1:2 reduction scale means that 1 mm on the drawing equals 2 mm on the object being drawn. In other words, with this scale ratio, the drawing is half the size of the object. A 2:1 scale means that 2 mm on the drawing equals 1 mm on the object; the drawing is twice the size of the object at this scale or ratio. One triangular scale will usually have at least six different ratios. Typical ratios include 1:1, 1:2, 1:20, 1:50, and 1:100.

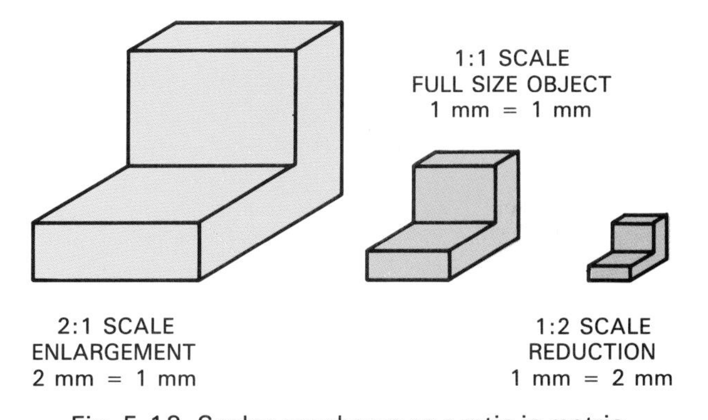

Fig. 5-13. Scales are shown as a ratio in metric.

One metric scale can be used for various reductions and enlargements, Fig. 5-14. For example, a 1:2 scale can be used for ratios of 1:2, 1:20, 1:200, etc. Multiply the measurement on the 1:2 triangular scale by 10 for each zero to be added to the scale ratio.

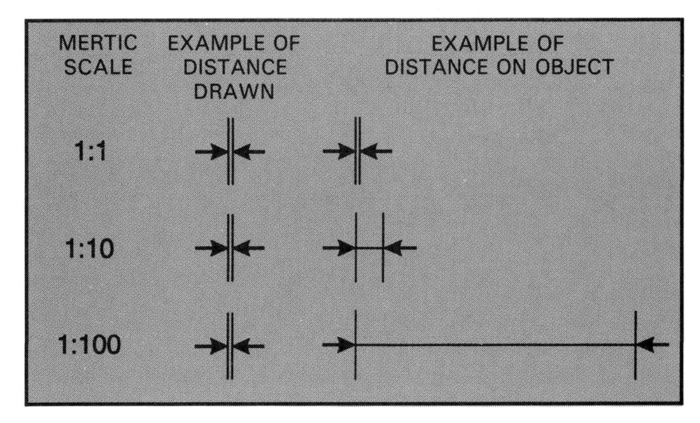

Fig. 5-14. A single metric scale can be used for these ratios.

Suppose, for example, a 200 mm line is to be drawn using a 1:2 metric scale. Using the numbers directly from the 1:2 scale, a line representing 200 mm (one-half actual size) may be drawn. If, instead, a 1:20 scale ratio is desired, the same 1:2 scale is used, but the values of the divisions are multiplied by 10. Therefore, what appears on the scale as 20 mm, now has a value of 200 mm

(20 x 10 = 200). This would be the length representing the 200 mm line (now, one-twentieth actual size).

In order to change from a smaller to larger ratio, (i.e. 1:20 to 1:2), divide the measurement on the scale by 10 for each zero to be removed. For example, 1000 mm (1 m) represented on a 1:20 scale would represent 100 mm (1 dm) if used for a 1:2 scale; 1000 mm (1 m) represented on a 1:100 scale would represent 10 mm (1 cm) if used for a 1:1 scale, etc.

Before using a metric scale, always be sure you know what the numbers represent. On smaller ratio scales, such as 1:3 or 1:5, the numbers are usually in units of centimeters. For larger ratio scales, such as 1:500 or 1:2500, the numbers are usually in units of meters.

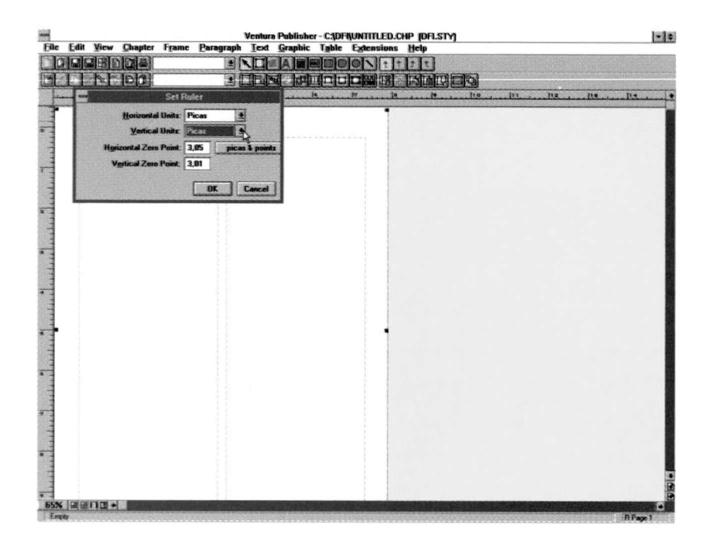

Fig. 5-17. When using a desktop publishing program, picas can be used for page measurement.

AMERICAN POINT SYSTEM

The **American Point System** is a measurement that is used in the printing and publishing fields. Measurements are made in **points** and **picas**, and a **line gauge** is often used to make the measurements, Fig. 5-15. There are 72 points in one inch. Twelve points are equal to one pica. Since points are very small, the numbers on the side of the line gauge represent picas.

Type size is measured in points, Fig. 5-16. Type is measured from the top of a letter with an ascender (d,b) to the bottom of a letter with a descender (g,p). This is greater than the height of any single letter, including capitals.

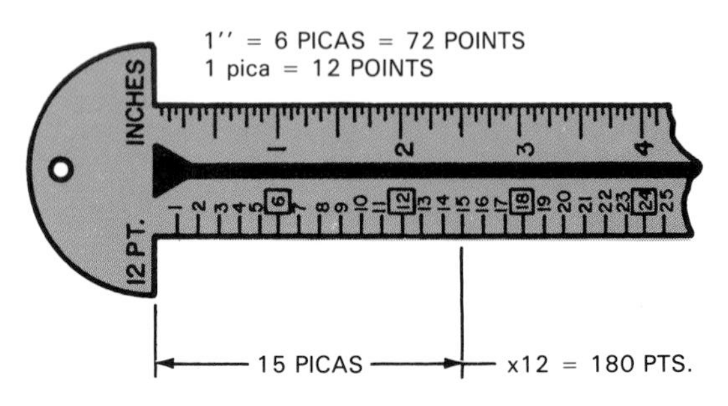

Fig. 5-15. A line gauge is used in the printing and publishing field.

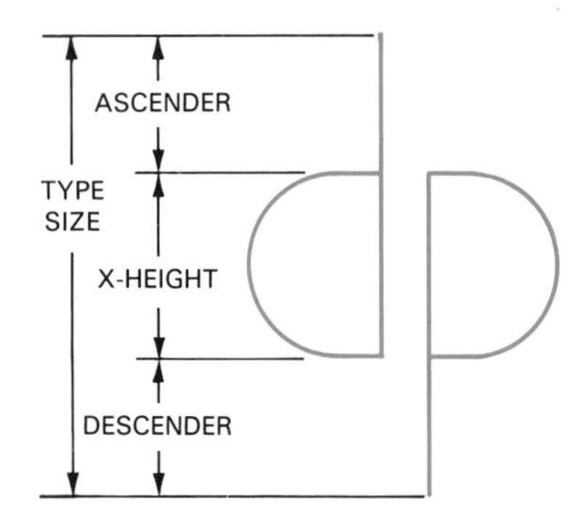

Fig. 5-16. This is how type size is measured correctly.

Picas are often used for line lengths, Fig 5-17. For example, a newspaper might use 20 pica columns and 14 point type.

One side of the line gauge is numbered in picas. Refer again to Fig. 5-15. At the top, the points in the smallest scale marking and points in a pica are shown. The other side of the gauge allows measurement in inches or millimeters. Some gauges are divided in agates that are approximately 5 1/2 points each.

Column height is sometimes designated by the **agate line**. There are 14 agates in one inch.

QUALITY ASSURANCE

Measurement directly affects the quality of a product or service. **Quality assurance** or quality control, is a method for ensuring that products or services meet a certain standard or level of quality. Careful measurements are taken each step of the way in the production of a product, from the design stage to the customer acceptance stage.

A long distance phone service is an example of the importance of quality control. In order to keep customers, the quality of the service must be consistently high, including courteous operators, a clear signal, and fast and easy connections. Film processing is another example. Consistently good prints and customer service are necessary for satisfied customers. In any business, product and service quality is a key element for success. Quality control provides a way to continuously measure that level of quality.

The system model can be applied to quality assurance. Quality can be controlled by monitoring inputs, processes, and outputs. For example, raw materials must be free of defects to prevent problems at the input stage. Equipment used in the production of a product

Fig. 5-18. This product is being inspected for correct thickness.

must be adjusted correctly to prevent problems in the processing stage. The output, which is the finished product or service, must be of high quality in order to keep customers satisfied.

Quality control of inputs and outputs in a technology system is normally done by inspection. Inspection consists of checking to ensure that a service or product meets a certain level of quality, Fig. 5-18. Since it can be very costly and time-consuming to inspect every item produced (100 percent inspection), a technique known as **acceptance sampling** is used. This involves inspecting a certain percentage of the goods or services. On the basis of inspecting this percentage, a decision can be made about whether or not to accept or reject the entire lot.

Acceptance sampling, a form of inspection, can be used to measure inputs and outputs. Raw materials, such as paper, plastics, and inks, can be inspected for quality. In addition, outputs, such as printed products and telecommunication services, can be sampled in order to make decisions about the overall level of quality. Usually, the smaller the number of items, the greater the percentage that should be sampled. For example, it is very easy to inspect 50 percent of 40 items produced in a day. However, inspecting 50 percent of 150,000 items can be very costly and time-consuming, so a smaller sample is used. Charts are available that provide suggestions for sample percentages based on the number of items.

Checking to be sure that processes are performed to a certain standard is known as **process control**. Since a form of math, known as *statistics*, is often used to help in carrying out process control, the term **statistical process control** is also used.

As with acceptance sampling, statistical process control involves sampling. However, in this case, sample readings of a process are taken. For example, a machine is adjusted to make the perfect widget. Every 30 minutes of operation, the adjustment of the machine and the widget quality are checked. If the widget is not correct when inspected, then a change must be made. This might involve readjusting the machine, sharpening blades, or repairing a broken part.

Although there may be an ideal quality level for each process, most processes result in some variation in quality. So, with most processes, some range in measurement of quality is allowed. For example, the perfect widget might be 40 mm long. However, even when set perfectly, the widget machine might produce widgets in a range of sizes from 38 mm to 42 mm long. If these sizes are allowable, then these would be the upper and lower limits for acceptable widgets. As the widgets are being processed and inspected, if their size is within the set limits, then the process is proceeding correctly. If not, then the process must be changed in some way.

Charts are often used that graphically display the measurements that are taken for a process. These are known as **control charts**. See Fig. 5-19. These devices provide a way to analyze a process and continually improve the output.

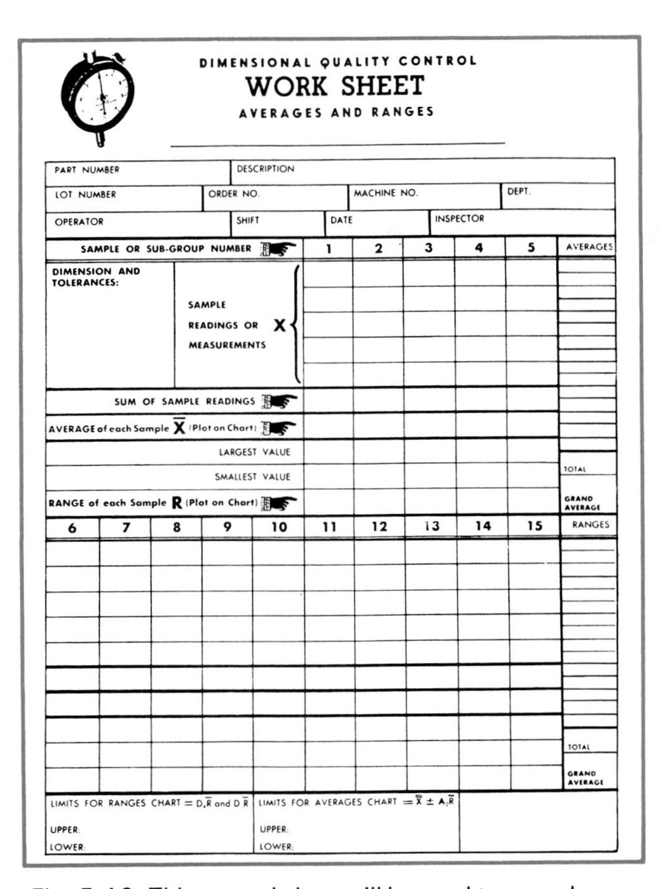

DIMENSIONAL QUALITY CONTROL
WORK SHEET
AVERAGES AND RANGES

PART NUMBER | DESCRIPTION
LOT NUMBER | ORDER NO. | MACHINE NO. | DEPT.
OPERATOR | SHIFT | DATE | INSPECTOR

SAMPLE OR SUB-GROUP NUMBER	1	2	3	4	5	AVERAGES
DIMENSION AND TOLERANCES: SAMPLE READINGS OR X MEASUREMENTS						
SUM OF SAMPLE READINGS						
AVERAGE of each Sample $\bar{X}$ (Plot on Chart)						
LARGEST VALUE						
SMALLEST VALUE						TOTAL
RANGE of each Sample R (Plot on Chart)						GRAND AVERAGE

6	7	8	9	10	11	12	13	14	15	RANGES
										TOTAL
										GRAND AVERAGE

LIMITS FOR RANGES CHART = $D_4\bar{R}$ and $D_3\bar{R}$
UPPER:
LOWER:

LIMITS FOR AVERAGES CHART = $\bar{\bar{X}} \pm A_2\bar{R}$
UPPER:
LOWER:

Fig. 5-19. This control chart will be used to record measurements each time an inspection is made.

SUMMARY

In this chapter, you have learned about the U.S. Customary, SI Metric, and American Point systems of measurement. Becoming familiar with these systems of measurement and scales will be useful in communicating your own designs as well as understanding the designs of others.

Quality assurance, a method for ensuring that a product or service meets a certain standard is also an important measurement consideration. Acceptance sampling and statistical process control are often used in measuring quality.

WORDS TO KNOW

All of the following words have been used in this chapter. Do you know their meanings?

acceptance sampling
agate line
American Point System
architect's scale
civil engineer's scale
control charts
inspection
line gauge
measurement
mechanical engineer's scale
metric scale
picas
points
process control
quality assurance
scale
SI Metric System
statistical process control
U.S. Customary Measurement System

REVIEWING YOUR KNOWLEDGE

Please do not write in this text. Write your answers on a separate sheet.

1. True or false? Many products that are produced in the United States are sold in other countries, and if they are not made to metric standards, they may not be acceptable.
2. True or false? In the United States, the International Standards Organization establishes standards for measurement systems.
3. In the U.S. Customary Measurement System, distance is measured in:
   a. Millimeters, centimeters, meters, and kilometers.
   b. Inches, feet, yards, and miles.
   c. Picas and points.
   d. None of the above.
4. True or false? In metric measure, decimals are always used for measurements less than a whole number.
5. Scales are used to draw an object:
   a. Full-size.
   b. Reduced size.
   c. Enlarged size.
   d. All of the above.
6. Which of the following is used for drawings of highways and maps?
   a. Mechanical engineer's scale.
   b. Architect's scale.
   c. Civil engineer's scale.
   d. None of the above.
7. A 1:2 reduction scale means that 1 mm on the drawing equals:
   a. 1/2 mm on the object being drawn.
   b. 2 mm on the object being drawn.
   c. 3 mm on the object being drawn.
   d. None of the above.
8. Explain how type size is measured in points.
9. _________ consists of checking to ensure that a service or product meets a certain level of quality.
10. True or false? Statistical process control involves inspecting a certain percentage of goods or services, and on the basis of this inspection, a decision is made about whether or not to accept or reject the entire lot.

APPLYING YOUR KNOWLEDGE

1. Research the conversion to metric. Outline the advantages and disadvantages.
2. Visit a drafting firm and find out what scales are used, and for what purposes. Share this information with the class.
3. Using a page from a newspaper or magazine, determine the point size of type, and the line length in picas.
4. Visit a local industry or business. Ask how they use quality control methods and how accurate measurements and other product design aspects affect the quality of products.

COMMUNICATING IDEAS BY SKETCHING

After studying this chapter, you will be able to:

- *Explain why sketching is important in the design process.*
- *Describe tools needed for sketching.*
- *Demonstrate techniques for sketching lines and shapes.*
- *Differentiate between sketching pictorial and two-dimensional views of an object.*
- *Describe the role of the computer in sketching.*

Sketching is a quick method of communicating ideas visually. You can communicate with yourself by placing your mental image on paper. You can also share these sketches with others for further ideas. When you know how to sketch well, your ideas for products can be developed and communicated more accurately and quickly. See Fig. 6-1.

In industry, sketches are used in the product planning stage. By making sketches first, a variety of ideas can be created and shared quickly. In addition, changes can be made before large amounts of time and money are spent on instrument drawings. For example, a floor plan for a home is sketched and shared with a customer. The customer discusses changes, and revised sketches are made. Eventually, a complete set of house plans is produced.

Computer graphics is another area where sketching is important. A sketch is developed and used as input for the computer. The computer then assists in producing the finished drawing.

MATERIALS NEEDED FOR SKETCHING

Very little equipment is needed for sketching. In fact, many sketches are made with any available pencil and paper. (Drawing tools are discussed in Chapter 8 of this text.)

PENCILS

Use a drafting pencil with a soft lead or an ordinary writing pencil. For thin lines, use a pencil with a sharpened point. For thick lines, use a pencil with a conical

Fig. 6-1. Sketches are needed to plan a home and to communicate your ideas to others.

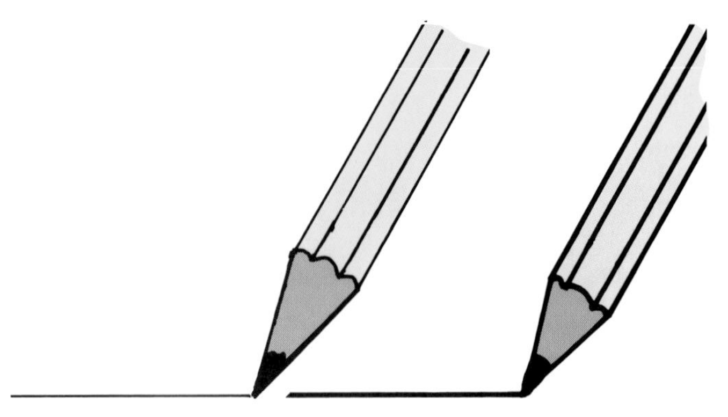

Fig. 6-2. A sharpened pencil point produces a thin line. A rounded pencil point produces a thicker line.

point, Fig. 6-2. The pencil eraser or a drafting eraser is adequate for erasures.

PAPER

Plain or graph paper can be used. Sketching is usually easier on graph paper. **Graph paper** has a grid that aids you in keeping lines straight, parallel, or of equal length. Graph paper is available in square and isometric grids, Fig. 6-3. Various grid sizes are available. If you use plain, translucent paper, graph paper can be placed behind the translucent paper to be used as a guide and removed for use again. Often, the paper is left loose on the surface so that it can be moved to the most comfortable sketching position. For transferring a distance from one part of a sketch to another, the edge of a sheet of paper can be used. Simply mark the distance needed and transfer it.

SKETCHING PROCEDURE

Sketches can be produced using a variety of methods. You can sketch a picture that is in your mind. This is known as designing. Another method is to sketch an object you see or a picture of the object. In either case, the following procedure is recommended:

1. Decide how many sketches are needed and how much time will be spent on each one. For example, sketching ideas at random will take less time than producing a final sketch.
2. Hold the pencil in a writing position.
3. If the object or drawing is nearby, study it. Notice the size relationship of the parts. Begin with height, then width, and finally, depth. What is the approximate angle of lines to be drawn in relation to an imaginary horizontal or vertical line?
4. Relax and have confidence. Sketch the outline of the object using light construction lines. Draw a new line for corrections without erasing the first line.
5. Darken the correct lines. See Fig. 6-4.
6. Repeat for details, if necessary.

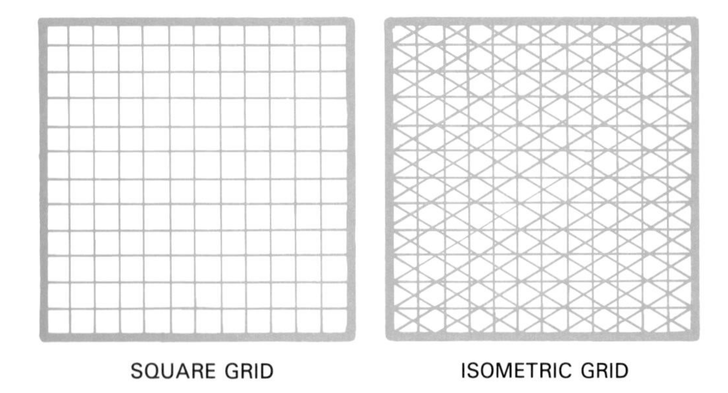

Fig. 6-3. Graph paper is available in square and isometric grids.

ACHIEVING PROPORTION

Proportion can be accomplished by drawing each part of an object so that it is the correct size in relation to the whole object. For instance, if you were sketching a person and a building, the person should be smaller in proportion to the building.

In order to draw an object in proportion, study it first and determine the size relationship of the parts to one another. For example, a rectangle might be twice as long as it is high. Start by looking at the height, width, and depth of the object. Sketch so that proportion appears visually correct, Fig. 6-5. Always think about proportion before and during sketching.

SKETCHING LINES AND SHAPES

Lines and shapes are the basic elements in technical drawing. Shapes have two dimensions. These are width and height. Sketching lines and shapes properly will

Fig. 6-4. When sketching, sketch the outline and then darken the lines.

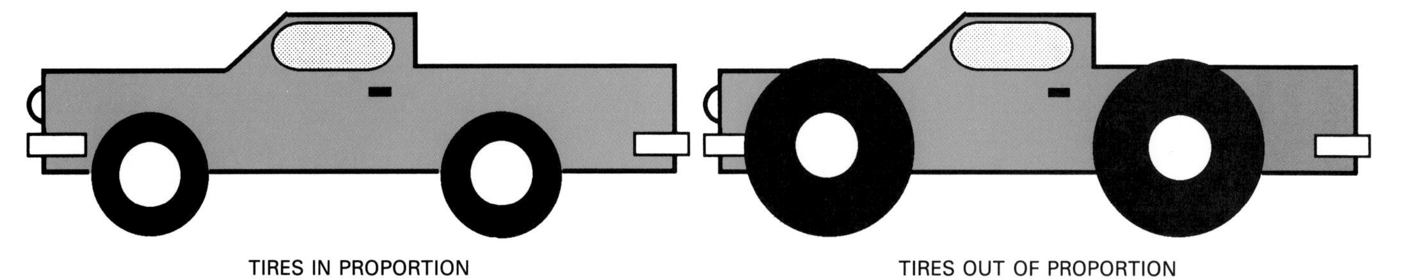

Fig. 6-5. These sketches, done with the help of a computer, illustrate the principle of proportion.

enhance the appearance of the drawing as well as speed up the sketching process. In fact, sketches are simply just a group of lines and shapes.

Sketching straight lines

When producing a straight line, locate the beginning and ending points with a dot. Draw short construction lines in between to connect the dots as shown in Fig. 6-6. Now draw in the visible line. Short lines can be drawn in one stroke. Draw horizontal, vertical, and slanted lines in the directions as shown in Fig. 6-7.

Sketching irregular curves

An **irregular curve** is a curved line with no single center. A string tossed on the ground would form an irregular curve. First make light dots on the curve. Connect the dots smoothly with short construction lines. Then darken the curve. See Fig. 6-8.

Sketching angles

An **angle** consists of two lines intersecting at a point. A **right angle** is 90°. **Acute angles** are less than 90°, and **obtuse angles** are greater than 90°. See Fig. 6-9. For an acute angle, draw a 90° angle first, if necessary. Now draw a line inside the 90° angle to form the acute angle. For example, half of 90° is 45°. For obtuse angles, draw two right angles and then draw the correct angle.

Sketching squares and rectangles

Draw the axes with construction lines and mark the width and height. If plain paper is being used, use the edge of a sheet of paper to obtain an approximate measurement. Lightly sketch the sides, and then darken the lines. See Fig. 6-10.

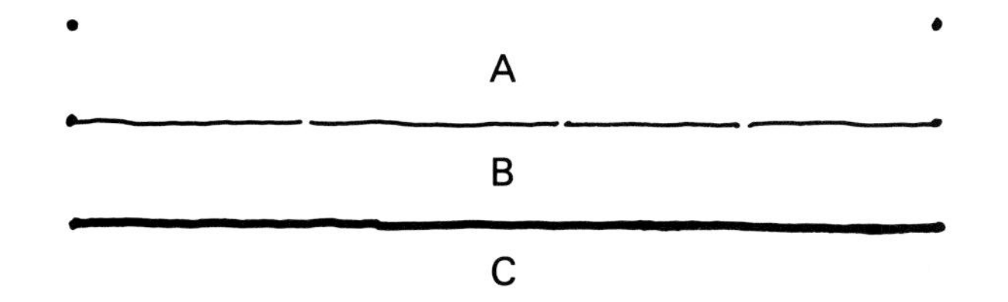

Fig. 6-6. Sketching a straight line is a three-step process: A–Locate the beginning and ending points with dots. B–Draw short construction lines in between to connect the dots. C–Draw in the visible line.

Sketching circles

A **circle** is a 360° line with one center point. Draw a horizontal and vertical axis for the circle. Now draw two axes at 45° to the ones already drawn. Mark the

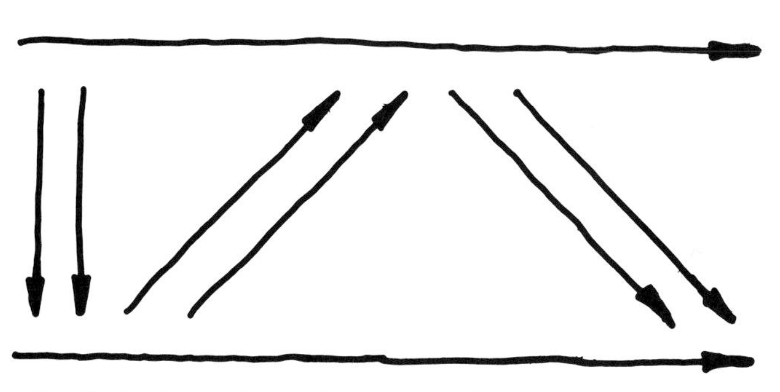
Fig. 6-7. Generally, horizontal, vertical, and slanted lines are drawn from your non-drawing hand towards your drawing hand. These lines are examples of how a right-handed person might draw different types of lines.

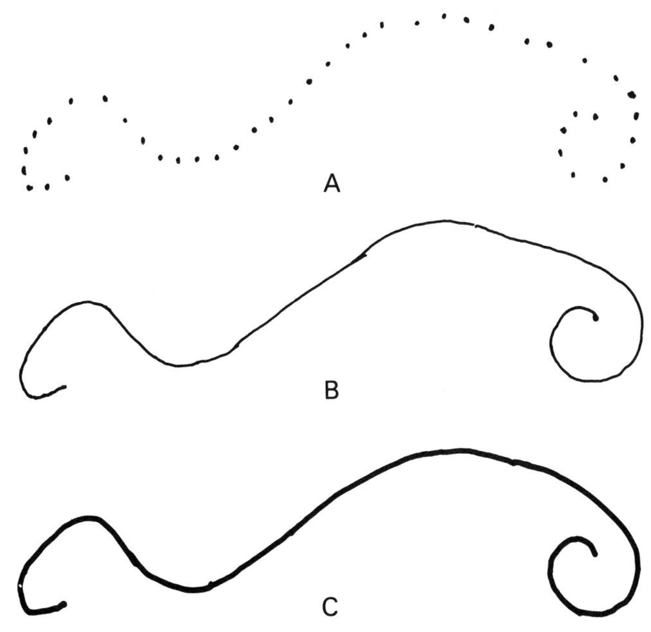

Fig. 6-8. Sketching an irregular curve is a three-step process: A–Make light dots on the curve. B–Connect the dots with short construction lines. C–Darken the curve.

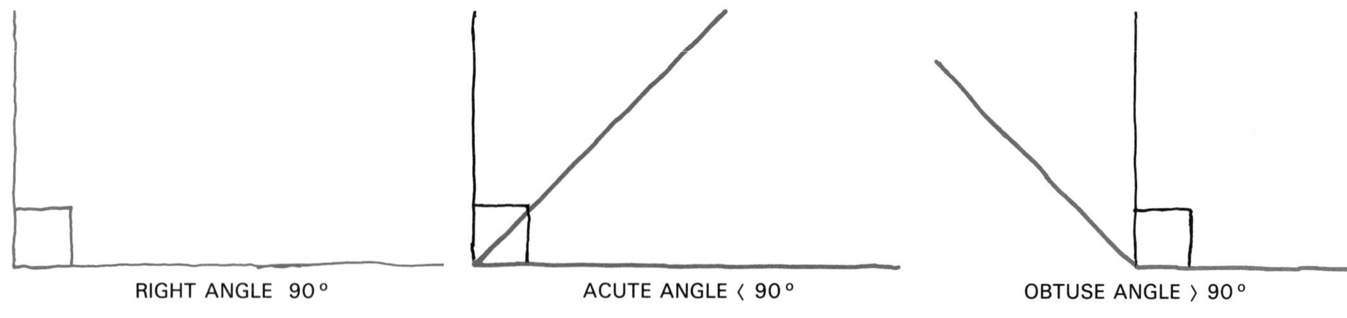

Fig. 6-9. A right angle equals 90°. An acute angle is less than 90°. An obtuse angle is greater than 90°.

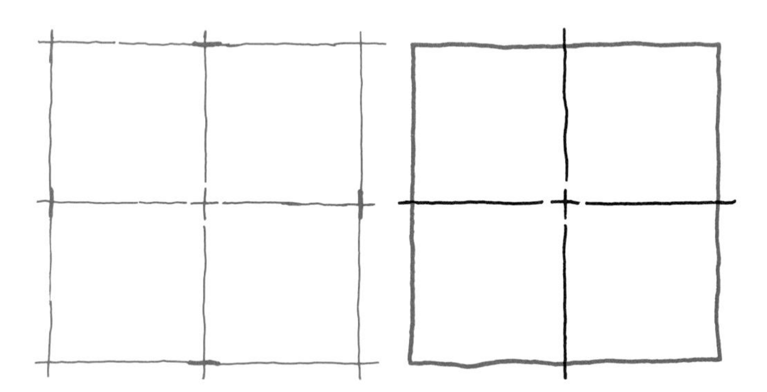

Fig. 6-10. When sketching a square, lightly sketch the lines and then darken them.

radius from the center point out on each axis. Sketch the circle lightly, turning the paper as necessary. See Fig. 6-11. Smaller circles require only two axes.

Sketching arcs

Arcs are curved lines, less than 360°, with a single center point. Small arcs are usually sketched lightly and then darkened. Arcs are often used to round corners, or connect two straight lines, Fig. 6-12. Draw a parallel line to one of the two lines, a distance apart equal to the radius of the arc. Repeat on the other line, and make sure the two lines intersect. Connect the line intersections with a light line. Mark the radius on this line, and sketch in the arc.

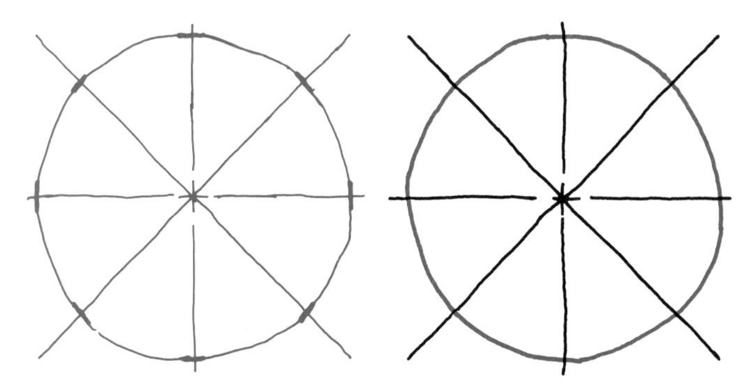

Fig. 6-11. When sketching a circle, first, draw a horizontal and vertical axis. Then draw axes at 45° to the angles already drawn. Mark the radius from the center point out on each axis. Finally, sketch the circle and darken the lines.

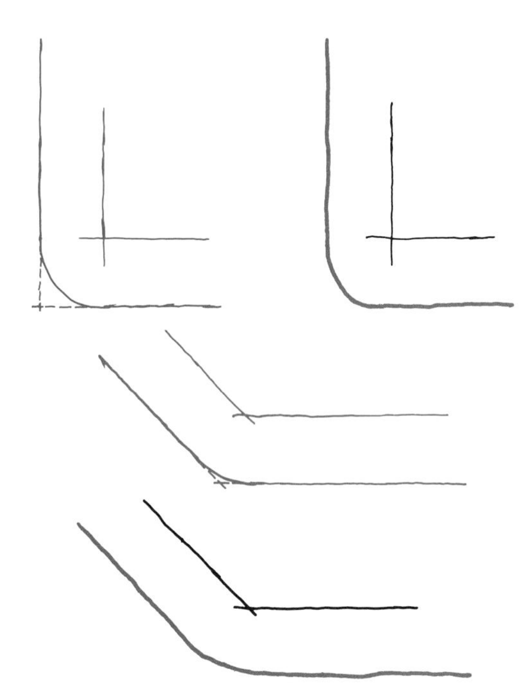

Fig. 6-12. Arcs are often used to round corners.

Sketching ellipses

An **ellipse** is an elongated circle. Unlike a circle, the arcs that make an ellipse have two different radii. These are on a major and minor axis. Start by drawing the axes, Fig. 6-13. Now, sketch a rectangle with the length equal to the minor axis. Sketch diagonals. Mark two-thirds of the distance out on each diagonal from the center. The marks on the diagonal are the target points of the different arcs. Sketch the ellipse, making sure the arcs meet smoothly.

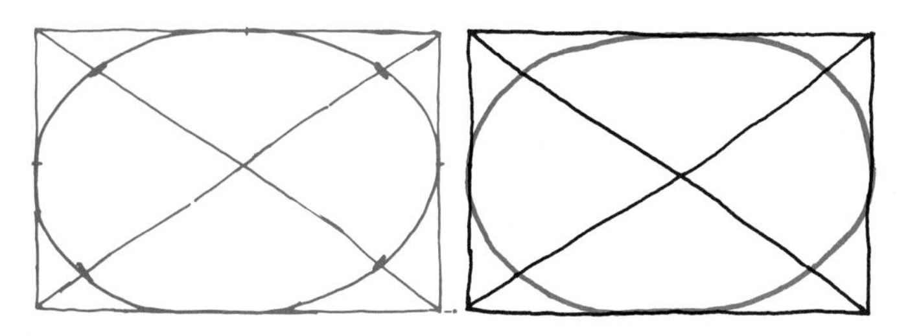

Fig. 6-13. Sketching a rectangle first will aid you in sketching an ellipse.

PICTORIAL SKETCHES

Pictorial sketches show height, width, and depth, Fig. 6-14. The three major types (all of which will be discussed in more detail in Chapter 13) are:

- **Axonometric sketches:** Sketches in which the lines of sight are perpendicular to the plane of projection.
- **Oblique sketches:** Sketches in which only one plane of projection is used and, while the lines of sight are parallel to each other, they meet the plane of projection at an oblique angle.
- **Perspective sketches:** Sketches in which parallel lines tend to converge as they recede from a person's view.

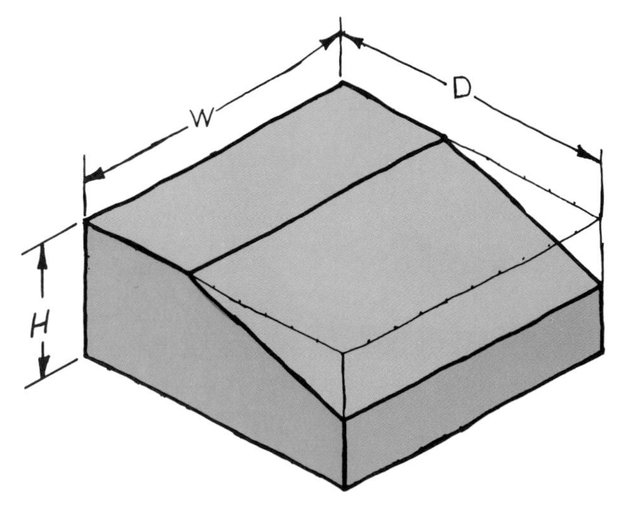

Fig. 6-14. This pictorial sketch shows height, width, and depth.

Isometric sketch

The most common pictorial sketch is the **isometric sketch**, a type of axonometric sketch. Isometric grid paper can be used for this type of sketch. Start by drawing an isometric box using the maximum dimensions of the object. See Fig. 6-15. Horizontal lines on the object are drawn at 30°, and vertical lines remain vertical. These are called **isometric lines**. Draw other isometric lines as necessary for details. Non-isometric lines are found by first locating points on isometric lines. Darken the sketch when completed as was done in Fig. 6-14.

A circle appears as an ellipse in an isometric sketch, Fig. 6-15. Start by drawing an "isometric square" with the sides equal to the circle diameter. Now draw isometric center lines that cross at the midpoint of each side. These midpoints are the tangents for the two arcs. Sketch the ellipse lightly and darken the lines when they appear to be correct. See Fig. 6-16.

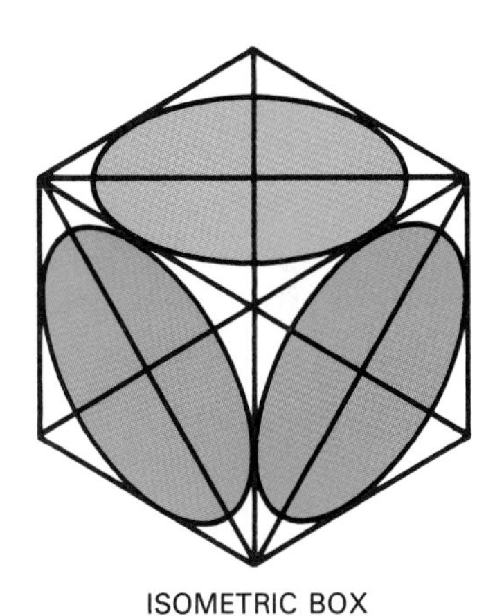

Fig. 6-15. This shows the orientation of ellipses on an isometric box.

TWO-DIMENSIONAL SKETCHES

Solid objects have three dimensions. These are width, height, and depth. Sketches of solids can be made that show two dimensions only. This is known as **multiview sketching** or **orthographic projection** because more than one view is often needed to describe the object.

In multiview sketches or orthographic projection, various sides of the object are drawn to accurately communicate what the object looks like and how it is made. These views are in a specific place in relation to one another as shown in Fig. 6-17.

A variety of lines, known as the *Alphabet of Lines*, are used in multiview or orthographic sketching. As with instrument drawings, these can all be sketched as thin or thick lines. The Alphabet of Lines is discussed in Chapter 9.

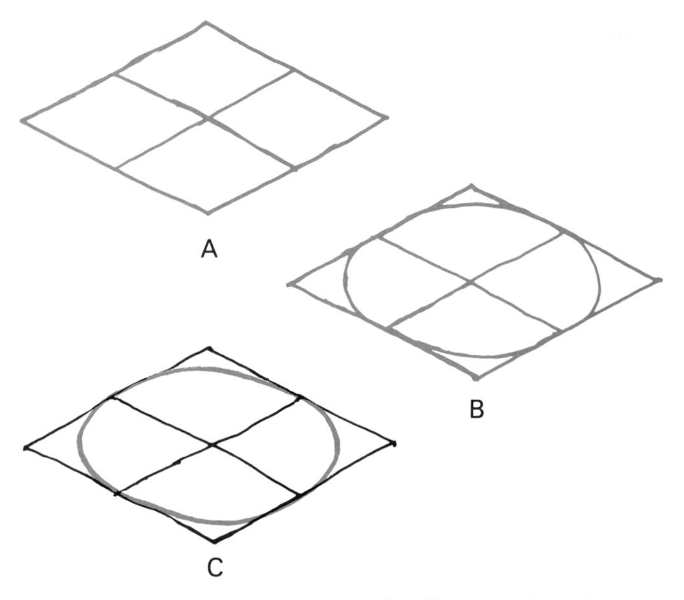

Fig. 6-16. Sketching an isometric ellipse involves three steps: A–Drawing an "isometric square." B–Drawing isometric center lines that cross at the midpoint of each side. C–Sketching the ellipse lightly and darkening the lines when they appear to be correct.

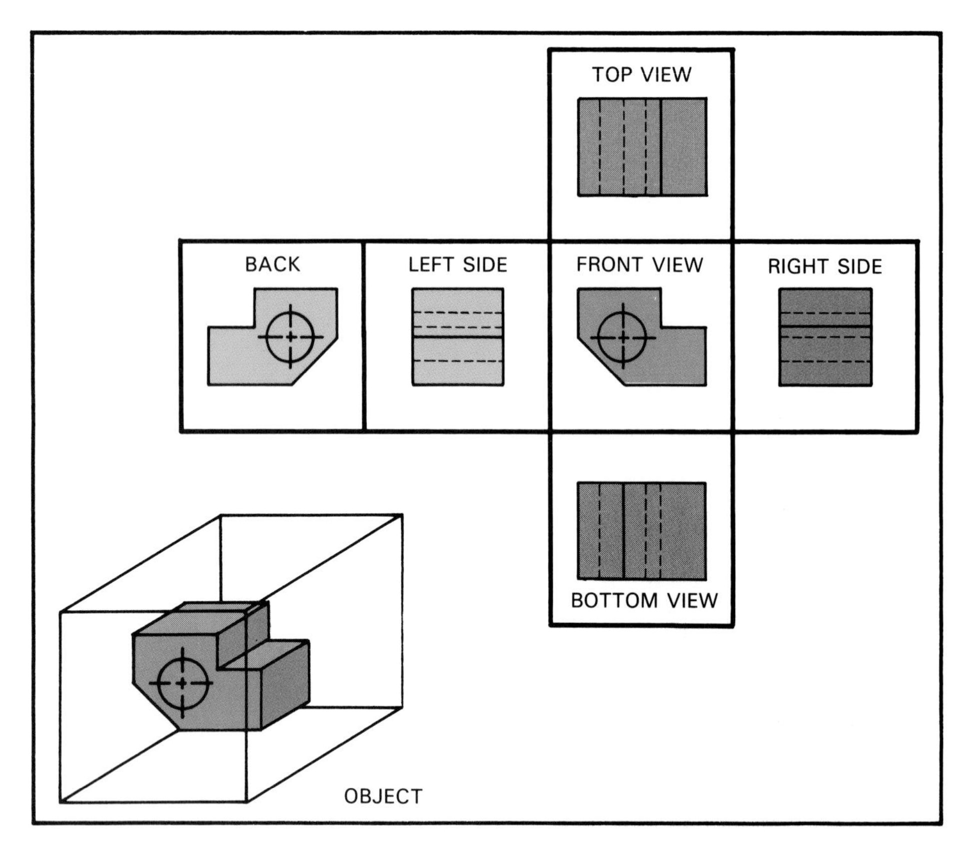

Fig. 6-17. This is an example of a multiview sketch or an orthographic projection.

Multiview drawings are usually dimensioned. Dimensions can be easily sketched, Fig. 6-18. Extension lines are drawn that extend the dimensions off the drawing. A dimension line with an arrowhead at each end is used to show the measurement. The dimension is placed inside a break in the line. Dimensions will be further discussed in Chapter 12.

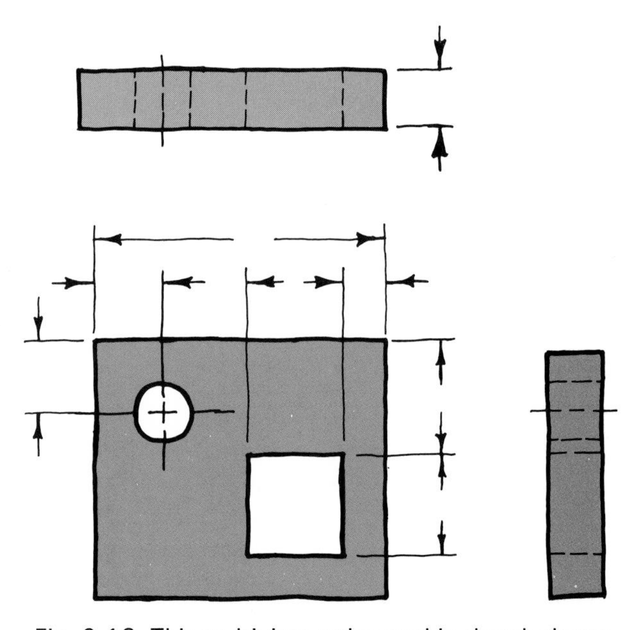
Fig. 6-18. This multiview orthographic sketch shows dimension lines.

SKETCHING BY COMPUTER

In some cases, the computer is used for drawing designs quickly. Sketching by hand is then eliminated entirely. Graphics software is used for this task and is explained in Chapter 32 of this text.

Software programs that most closely simulate freehand sketching are those meant to be used by artists and others adept at freehand drawing. An example of a drawing done by such a program is shown in Fig. 6-19. Some tasks, such as designing a notepad, might be done completely with this type of program, from initial designs to the final design. In other cases, a different software program may be needed for the finished drawing.

A computer-aided drafting program provides another option for quickly generating alternative designs. This technique more closely simulates using drawing tools such as a T-square, triangle, protractor, and compass. The drawings are fairly precise, even at the design stage. When drawings are being generated that require precision, such as designs for industrial products, then this type of program is a good choice for the design as well. This eliminates having to use one program for design and another for the final product.

Desktop publishing software, which is used for page layout, works well for designing pages. This eliminates the need to sketch alternative page designs.

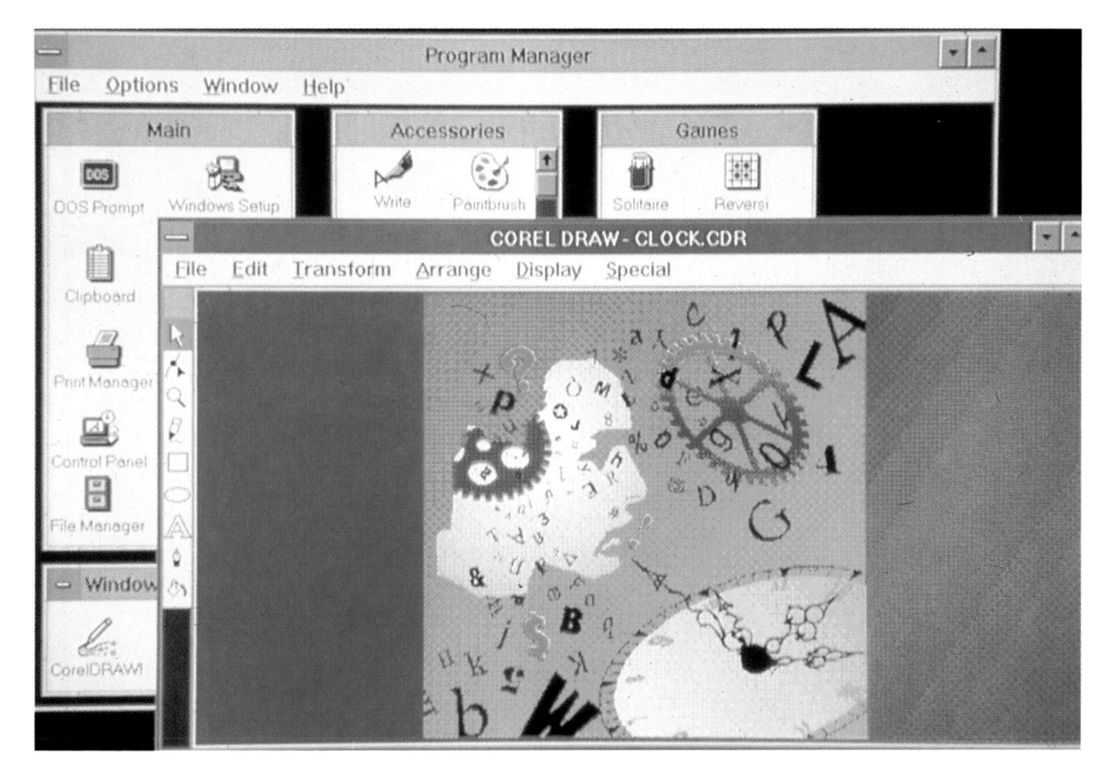

Fig. 6-19. Some software programs can be used to produce freehand sketches such as this one. (Corel Systems Corp.)

As can be seen from these examples, computers can provide a method of sketching alternative designs. The decision on which system to use (freehand or computer) is a personal choice.

SUMMARY

The purpose of sketching is to:

- Create alternative designs for a product.
- Refine a design before expensive instrument drawings are made.
- Quickly communicate technical ideas.

Sketching can help you just as it helps people in industry. By sketching first, your ideas take visual form and can be shared with others for further ideas. This can be done freehand or by using the computer. Sketching allows you to refine your design quickly before continuing with a final drawing.

WORDS TO KNOW

All of the following words have been used in this chapter. Do you know their meanings?

acute angles
angle
arcs
axonometric sketches
circle
ellipse
graph paper
irregular curve
isometric lines
isometric sketch
multiview sketching
oblique sketches
obtuse angles
orthographic projection
perspective sketches
pictorial sketches
right angle
sketching

REVIEWING YOUR KNOWLEDGE

Please do not write in this text. Write your answers on a separate sheet.

1. In industry, sketches are used in the:
   a. Product production stage.
   b. Product evaluation stage.
   c. Product planning stage.
   d. None of the above.
2. Name two items of equipment needed for sketching.
3. For thin lines, use a pencil with a:
   a. Dull point.
   b. Sharpened point.
   c. Conical point.
   d. None of the above.

4. When studying an object prior to sketching it:
   a. Begin with height, then width, and, finally, depth.
   b. Begin with depth, then height, and, finally, width.
   c. Begin with width, then depth, and, finally, height.
   d. None of the above.
5. __________ can be accomplished by drawing each part of an object so that it is the correct size in relation to the whole object.
6. True or false? An obtuse angle is less than 90°.
7. Which of the following types of sketches are sketches in which the lines of sight are perpendicular to the plane of projection?
   a. Axonometric sketches.
   b. Oblique sketches.
   c. Perspective sketches.
   d. None of the above.
8. True or false? A circle appears as an ellipse in an isometric sketch.
9. What type of sketches would you use to show various sides of an object?
10. List three purposes of sketching.

APPLYING YOUR KNOWLEDGE

1. Draw two objects that are near you in the room. Draw them from the perspective of where you are sitting. Briefly describe the drawings in terms of proportion.
2. Using graph paper, practice drawing lines and shapes as shown in this text. If you prefer, combine these to draw an object.
3. Use isometric graph paper to produce a pictorial drawing. Draw an isometric circle on two sides of the object.
4. Produce a multiview sketch of an object of your choice. Use hidden and center lines as needed.

7
CHAPTER

SAFETY AND HEALTH CONCERNS

After studying this chapter, you will be able to:

- *Describe the purpose of safety and health rules and OSHA.*
- *Explain how to protect your eyes and ears in the laboratory.*
- *Demonstrate the proper way to lift and carry objects.*
- *Describe appropriate conduct and personal attire for the laboratory.*
- *Demonstrate correct equipment and tool use.*
- *Describe considerations for working at computer workstations.*
- *Discuss how to handle chemicals properly.*
- *Identify various types of fires and appropriate extinguishers to use.*

Safety and health rules and regulations should be understood and practiced by everyone, whether in the classroom laboratory, in the workplace, or at home. **Safety** is defined as action taken to reduce the likelihood of personal injury. **Health** is defined as freedom from physical disease or pain. The roles of safety and health in the classroom laboratory are the same as in the workplace, to assure safe and healthy working conditions. Safety and health rules help protect you and those around you from injury. Safety and health are everyone's responsibilities. See Fig. 7-1.

This chapter will explore federal and state safety regulations, eye safety, hearing protection, and lifting and carrying. Working at computer workstations, general health and safety practices, laboratory maintenance, fire safety, and what to do in case of an accident will also be discussed.

FEDERAL AND STATE SAFETY REGULATIONS

Safety is such a serious matter that there are both federal and state laws to enforce safety standards and laws. The Occupational Safety and Health Act of 1970 is a law that governs federal safety regulations. This act gave the **Occupational Safety and Health Administration (OSHA)** the mission"to assure as far as possible every working man and woman in the nation safe and healthful working conditions and to preserve our human resources."

Employers have a legal obligation to inform employees of OSHA safety and health standards for their companies. This includes displaying an OSHA poster and

Fig. 7-1. When you work in a laboratory, you are responsible for following certain safety and health rules and regulations. This person is following safety practices regarding eye protection, ventilation, proper attire, and use of chemicals.

having an employee training program for safety and health. Employers should also have a copy of OSHA standards available for employees to review.

As a student, you should be aware of OSHA regulations because they are the laws that will apply when you move from the school setting to the workplace.

State safety regulations vary from state to state. It is your responsibility to know the laws in the state in which you will be working.

OSHA FUNCTIONS

Setting standards and inspecting workplaces are some of the functions of OSHA. OSHA also is responsible for leveling citations and penalties for violations, making consultations, and conducting training and educational programs.

A **Compliance Safety and Health Officer (CSHO)** is an OSHA employee who performs company inspections. This person can determine if a company is meeting OSHA standards and what changes may be needed.

HAZARD COMMUNICATION STANDARD

An important OSHA standard is the **Hazard Communication Standard**. This is a set of rules for businesses that use chemicals and other substances that can be a safety or health risk.

This standard requires that a company keep on file a **Material Safety Data Sheet (MSDS)** for each hazardous substance. See Fig. 7-2. This form should contain the name of the chemical (common and scientific), safe exposure levels, explosion and health hazard information, and precautions for safe use. Also, containers should be properly labeled as well.

Training programs covering proper use of chemicals should be conducted by companies. These training programs should be documented in the written hazard communication program. The company should keep this documentation as a record of what has been done regarding hazard communication.

Fig. 7-2. An information center, such as this one, can be used by companies to provide employees access to the Material Safety Data sheets that are required by OSHA. (National Marker Co.)

PROTECTING YOUR EYES

Much of what you learn is by sight. Since your eyes are so important, eye protection should be used in the laboratory whenever necessary, Fig. 7-3.

Standards for safety eyewear have been developed by the **American National Standards Institute (ANSI)**. The standard for safety eyewear is ANSI Z87. Eyewear that can be used to meet this standard is normally stamped Z87.

One of the most common times when eye protection is necessary in a visual communication laboratory is when you are handling and mixing chemicals. Chemical goggles are used for this purpose. They should be splash-proof.

When there is a danger of small, flying objects, goggles or spectacles with side shields should be worn. Your

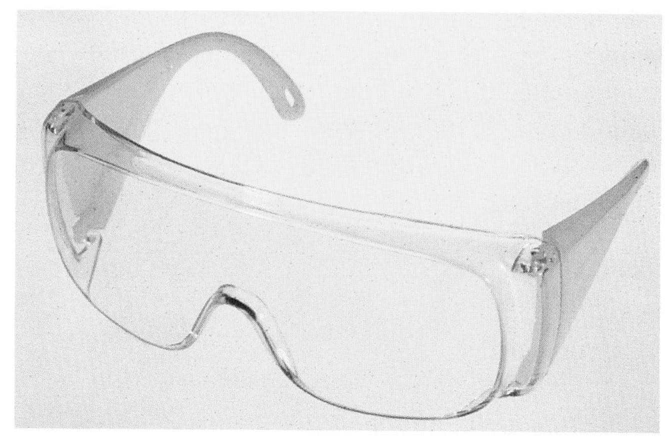

Fig. 7-3. Proper eye protection should be worn when working with various types of chemicals and equipment. (Willson Safety Products)

choice should be based on the task you will be performing. For example, you should wear goggles if you are working on an operation that may result in small, flying objects, such as stitching (the process of binding books with staples).

When lasers are used, it should be determined what type of eyewear is necessary. An ANSI standard for eye protection when using lasers is Z136.1-1980. You should never look directly into a laser beam or project the beam off a reflective surface without first being sure it can be done safely.

Several types of equipment can be sources of intense light. These are normally exposure units of various types, such as process cameras and plate exposure equipment in printing operations and whiteprint machines in technical drawing. You should never look at the lights when exposure is taking place. If this is necessary, then protective eyewear must be used.

It is important to be aware of first aid for eye emergencies. See Fig. 7-4. This information should be posted in the laboratory.

Avoid eyestrain. Make certain that light is adequate for the task you are doing. Avoid shadows and bright sunlight.

FIRST AID FOR EYE EMERGENCIES

Chemical Burns

Eye damage from chemical burns may be extremely serious, as from alkalis or caustic acids; or less severe, as from chemical "irritants."

In all cases of eye contact with chemicals:

DO flood the eye with water immediately, continuously and gently, for at least 15 minutes. Hold head under faucet or pour water into the eye using any clean container. Keep eye open as widely as possible during flooding.

DO NOT use an eye cup.

DO NOT bandage the eye.

SPRAY CANS are an increasing source of chemical eye injury, compounded by the force of contact. Whether containing caustics or "irritants," they must be carefully used and kept away from children.

Specks in the Eye

DO lift upper eyelid outward and down over the lower lid.

DO let tears wash out speck or particle.

DO —if it doesn't wash out—keep eye closed, bandage lightly and see a doctor.

DO NOT rub the eye.

Blows to the Eye

DO apply cold compresses immediately, for 15 minutes; again each hour as needed to reduce pain and swelling.

DO —in case of discoloration or "black eye," which could mean internal damage to the eye—see a doctor.

Cuts and Punctures of Eye or Eyelid

DO bandage lightly and see a doctor at once.

DO NOT wash out eye with water.

DO NOT try to remove an object stuck in the eye.

Fig. 7-4. Follow these first aid procedures if an eye injury occurs. (National Society to Prevent Blindness)

PROTECTING YOUR HEARING

OSHA has established standards for noise exposure. The two considerations are loudness (intensity) and duration (time).

The loudness of a sound is measured in **decibels**. See Fig. 7-5. The louder a noise is, the less time a person should be exposed to it. However, sounds over 115 decibels are unsafe regardless of the duration and can cause hearing damage.

When the noise level is being measured in decibels, it is given an "A" weighting (adjustment) that simulates

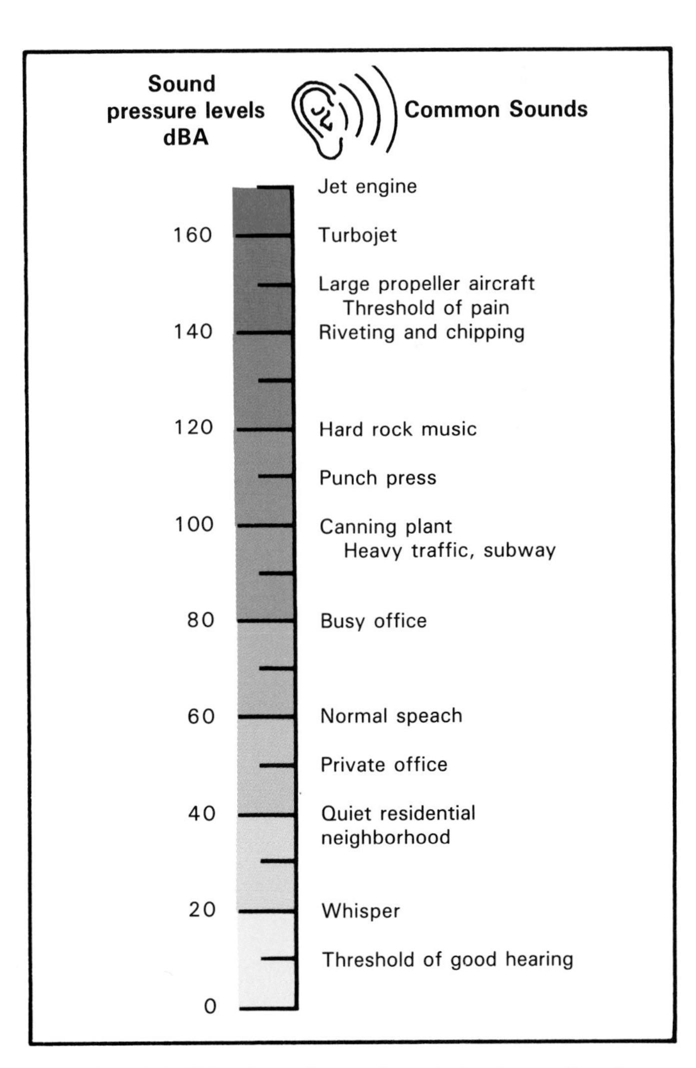

Fig. 7-5. This chart shows the relative intensity of common sounds.

the human ear. This weighting takes into account the fact that humans lack the sensitivity to certain high and low frequencies. This "A" weighting of the decibel rating is written as dBA. According to OSHA regulations, a noise level of 85 dBA is allowed for 16 hours, but a noise level of 115 dBA is allowed for only 15 minutes, Fig. 7-6.

Since there is often no indication of hearing damage, noise protection can be overlooked. Unfortunately, many common devices, such as audio tape and disk players, can pose a risk of hearing loss if played too loud. A general rule of thumb for loud noise is that if you must shout in order to be heard from three feet away, the noise is too loud, and hearing protection should be used.

Some indications of too much noise exposure include ringing in the ears, muffled hearing, and talking too loud. However, it is possible to have hearing damage and not experience any of these symptoms.

Common forms of ear protection include earplugs and earmuffs. Examples are shown in Fig. 7-7.

OSHA STANDARDS FOR NOISE EXPOSURE

Noise Level	Time Allowed
85 dBA	16 hours
90 dBA	8 hours
95 dBA	4 hours
100 dBA	2 hours
105 dBA	1 hours
110 dBA	.5 hours
115 dBA	.25 hours

Fig. 7-6. OSHA has established standards regarding time limits for various amounts of noise exposure. These are per day limits.

LIFTING AND CARRYING

Back injuries are one of the most common injuries on the job. Lifting and carrying heavy objects correctly can help to prevent many back injuries. See Fig. 7-8.

Fig. 7-8. Following proper lifting techniques can prevent back injuries. (Pennsylvania Department of Education)

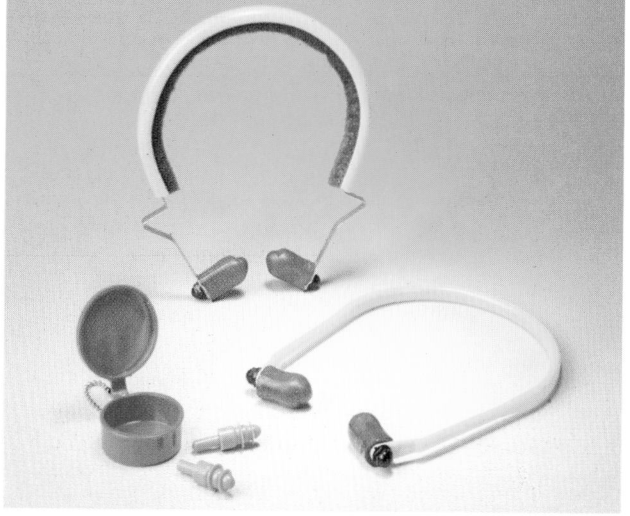

Fig. 7-7. A variety of hearing protection devices is available. (Willson Safety Products)

The first step in lifting is sizing up the load. What is the best way to carry it? Sometimes a hand truck or dolly may be necessary. How heavy is the object? If it is heavy or bulky, another person may be needed to help.

Before lifting an object, use any protection that is necessary. If a crate is made of rough boards, gloves should be worn. Safety shoes are also sometimes necessary to prevent foot injuries.

When lifting a load, lift using the muscles in your legs, not your back. Make sure your footing is solid and that you have a good grasp. Keep your back as straight as possible. When lowering the load, again use the muscles in your legs instead those in your back.

The key to safe lifting is to think before you lift and obtain help, if necessary. Many injuries could be avoided if people would simply ask for help.

COMPUTER WORKSTATIONS

Since many jobs now require the use of a computer, it is important to know how to set up a workstation. A well-designed workstation will increase your comfort and reduce fatigue. See Fig.7-9. This will increase your productivity.

The computer keyboard should be at a height so that your upper and lower arms form about a 90° angle. This should be about 24 to 28 inches. At this height, the upper arms are at rest and do not strain your back.

The computer monitor should be set so that the top is no higher than eye level. It should also be about 18 inches from your eyes. Tilting the screen back 10°-20° will make the screen easier to see, as long as glare is not increased. Closing blinds or using an anti-glare filter can help prevent glare problems.

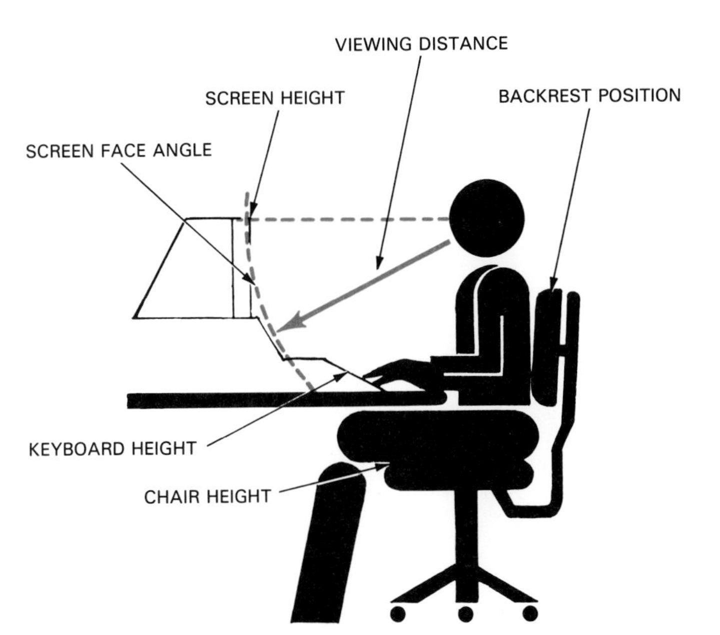

Fig. 7-9. You can work better and feel less fatigue if your computer workstation is well-designed and adjusted for your size and the way you move. (National Safety Council)

Chairs should also be appropriate for the workstation. The height should be adjusted so feet are flat on the floor. In addition, the backrest should fit comfortably at the small of the back for good back support.

GENERAL SAFETY PRACTICES

General safety practices involve using common sense. However, it is a good idea to review and become acquainted with some basic safety practices. Use caution when working in the laboratory. The safety guidelines presented in this chapter can be applied in most laboratories. Your instructor will provide additional rules that are to be followed in your school.

Be sure to let your instructor know if you have any handicap that might cause you difficulty in the laboratory. Your instructor can then make any necessary adaptations to enable you to work safely.

CONDUCT

Conduct refers to correct behavior in the appropriate place and at the appropriate time. For instance, correct behavior on the soccer field is not the same as correct behavior in the laboratory. The way you conduct yourself in various situations involves using common sense and good judgment. Unsafe conduct can endanger everyone in the class.

Several safety rules related to conduct are listed here.

- Know the safety rules to be followed in the laboratory in which you will be working.
- The laboratory should be used only when authorized personnel are present.
- Disruptive behavior should never be tolerated in any laboratory.
- Report unsafe conditions immediately.
- Unsafe actions such as running or throwing will not be permitted.
- Chairs or stools should never be leaned or tilted back while sitting on them. Falls can result in serious head or back injuries.
- Keep drawers in drafting tables closed.
- Request approval from your teacher for special machine set-ups.
- Avoid dangerous operations when ill or overtired.

PERSONAL ATTIRE

Clothing and protective devices should be appropriate for the job you are doing. Accidents can result when clothing becomes caught in machinery or when chemicals splash. Keep the task you will be doing and safety in mind when determining what you will wear or will not wear in the laboratory.

Safety rules related to personal attire are listed below.

- Use approved safety glasses or goggles when needed and required.
- Wear ear protection when around excessive noise.
- Wear a lab coat or apron, if needed, to protect ypur clothing.
- Wear shoes that fully enclose the foot. Pointed or sharp tools can be accidentally dropped.
- Remove jewelry when working with equipment. This includes rings, watches, and necklaces.
- Secure loose clothing and hair. For example, rollup sleeves, tuck in shirttails, and take off ties. Tie long hair back or wear a hat.
- Use rubber gloves when working with materials that can irritate your skin.

EQUIPMENT AND TOOL USE

Equipment and tools can make a job easier. However, you must understand how to use them properly. These are some safety tips on proper equipment and tool use.

- Use tools and equipment only after a demonstration and/or safety check.
- If you forget how to use a piece of equipment, ask your instructor for help before starting.
- Never use equipment that is out of order. Also, if a machine does not seem to be working properly, turn it off and tell the instructor.
- Lighting should be adequate for using tools and equipment.
- Only one operator at a time should use a machine. Before starting a machine, make sure others are a safe distance back.
- When using power equipment, concentrate on your work. Keep talking to a minimum.
- Use guards at all times, Fig. 7-10.
- Make adjustments with the machine turned off.
- Be careful when using sharp tools (knives and razor blades) or pointed tools (dividers, compass, pencil). Keep points turned away from you and others.
- Be careful when using an air hose. Make sure others are out of the way before you start.
- Keep handles of all tools clean and free of oil and grease.

Fig. 7-10. Guards on machines are there for your protection. Make sure they are in place before you operate a machine.

- Avoid placing fingers into openings of machines where paper is normally fed.

USING ELECTRICITY

Since many machines use electricity, a knowledge of electrical safety is important. An electric shock can be fatal. That is why it is important to follow these safety rules carefully.

- Do not use any equipment with damaged or exposed electrical wire. Report it.
- Equipment should be grounded, Fig. 7-11.
- Use only heavy duty, U.L. approved extension cords.
- Never use electrical equipment on a wet surface.
- Do not touch electrical switches or equipment with wet hands.
- Keep electrical equipment dry.
- Disconnect cords when portable tools are not in use.

Fig. 7-11. Electrical grounding can prevent you from being shocked or electrocuted.

HANDLING CHEMICALS AND OTHER LIQUIDS

Chemicals are necessary in most work environments. Understanding some basic safety rules can help you use them safely. These are some common guidelines to follow when working with chemicals and other liquids.

- Spilled liquids should be wiped up immediately to avoid slipping.
- Rags that have been used with flammable solvents should be placed in safety containers, Fig. 7-12.
- Flammable liquids should be stored in appropriate containers in approved cabinets.
- Chemicals should be stored in glass bottles on low shelves in case they fall.
- Some chemicals should only be used with a ventilating fan or in a well-ventilated area. If a filter mask is used, make sure the filter type is correct.
- Use chemical goggles when mixing and handling chemicals.
- Keep your hands away from your eyes when using chemicals. Wash your hands after using chemicals.
- Know the procedures for washing the eyes in case they are exposed to a chemical.
- Keep flammable liquids away from heat or flame.
- Never shake a bottle containing a strong acid or alkali liquid.
- Identify the contents of bottles and containers by label; never trust taste or smell. See Fig. 7-13.
- Pour acid into water; never reverse the procedure.
- Know the formula and mixing procedure before handling chemicals.

Fig. 7-12. To prevent fires, place rags soiled with flammable solvents in a covered container.

MAINTENANCE

Maintenance improves the appearance of your work area, but more importantly, it makes it safer. Accidents often occur in laboratories that are dirty, cluttered, and unsanitary. Some basic maintenance guidelines are given below.

- Thoroughly clean your work area at the end of each lab period.
- Tables, counters, and sinks should be clean.
- Store projects, equipment, and materials in approved locations.
- Floors must be kept free of objects or materials that might cause falls.
- Aisles should be kept clear at all times.
- The laboratory should be well-ventilated.

COLOR CODE

Colors have a special meaning in the laboratory. A typical color code is described below.

- *Red.* This is often used for danger signs, fire equipment, emergency stop buttons, and safety cans.
- *Orange.* This is often used for equipment safety guards or to indicate dangerous parts on equipment.
- *Yellow.* This often means to use caution, for instance, to avoid falling, or tripping or to avoid injury. See Fig. 7-14.
- *Green.* This is often used for marking safety or first aid equipment.
- *Blue.* This often means equipment should not be used, usually because of needed repairs.
- *Black and white.* These often designate traffic zones and housekeeping markings.
- *Purple.* This is used for radiation hazards.

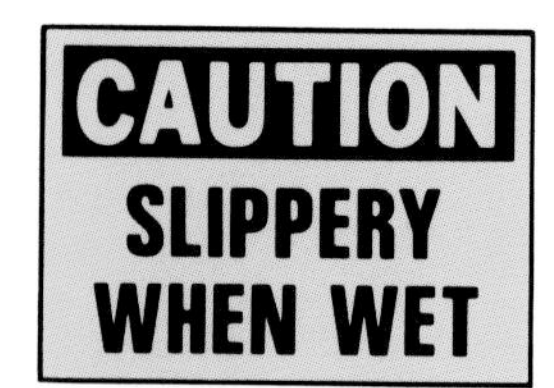

Fig. 7-14. In addition to the words on them the yellow color of these signs indicates caution. (National Marker Co.)

FIRE SAFETY

You should be familiar with the procedures to follow in case of a fire. Your instructor or other person in authority should be notified immediately. Steps for exiting the building should be followed.

The three most common types of fires are classified as A, B, or C. The letter is always shown with a symbol to help avoid confusion. See Fig. 7-15.

- **Type A fires** involve combustible materials such as paper, cloth, or wood.
- **Type B fires** involve flammable liquids such as solvents and oil.
- **Type C fires** involve electrical equipment or electric wires.

There is a fourth type of fire. **Type D fires** are flammable metals that have been ignited. These are rare types of fires in the school and require a special fire extinguisher.

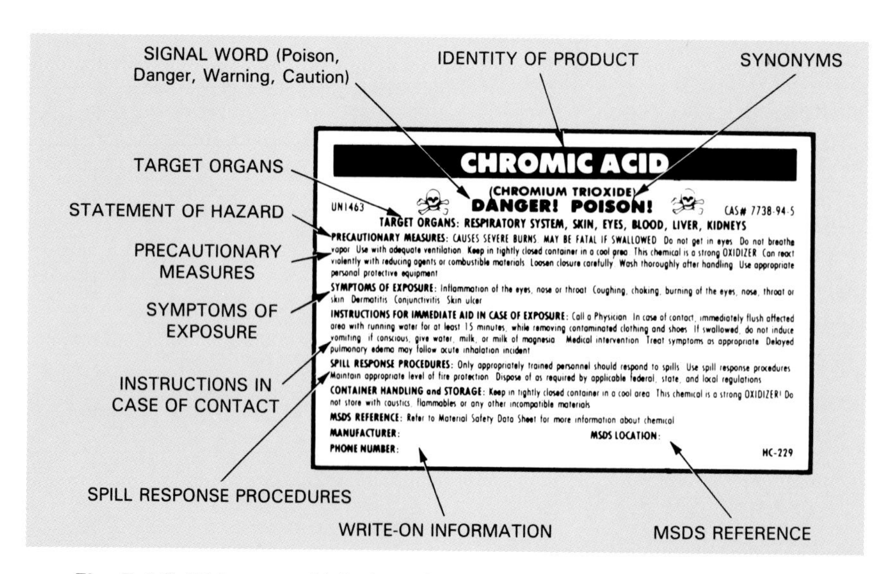

Fig. 7-13. This type of label can be purchased to place on hazardous chemical containers. (National Marker Co.)

Your extinguisher must fit the fire.

Type A: Ordinary combustibles, such as wood, cloth, paper, rubber, any plastics, and other common materials.

Type B: Flammable liquids, such as gasoline, oil, grease, tar, oil-based paint, lacquer, and flammable gas.

Type C: Energized electrical wiring, fuse boxes, circuit breakers, machinery, and appliances

Fig. 7-15. These are the three most common types of fires. (National Fire Protection Association)

Fire extinguishers are labeled to show the type of fire that can be extinguished, Fig. 7-16. Using the wrong type of fire extinguisher on the wrong type of fire can be dangerous. For example, a liquid should not be used to extinguish an electrical fire. People standing in the wet area may be electrocuted. If the symbol for a type of fire is in black with a red diagonal line, the extinguisher should not be used for that type of fire.

Fig. 7-16. Use the correct type of fire extinguisher for the type of fire you are fighting. (Ansul Fire Protection)

Before deciding to fight a fire, be sure that people are leaving the building and that the fire department has been called. Also, you need to be sure you have the correct extinguisher and that you know how to operate it properly. Never fight a fire that is spreading and that could block your escape route.

If you must fight a fire using a portable extinguisher, use the **PASS Method**. See Fig. 7-17. This stands for:

Pull the pin.
Aim low.
Squeeze the handle.
Sweep from side to side.

IN CASE OF AN ACCIDENT

Make sure you are familiar with steps that must be followed in case of an accident. This includes knowing the location of emergency telephone numbers and first aid equipment.

If you do fight the fire, remember the word **PASS**

PULL the pin: Some extinguishers require releasing a lock latch, pressing a puncture lever, or other first step.

AIM low: Point the extinguisher nozzle(or its horn or hose) at the base of the fire.

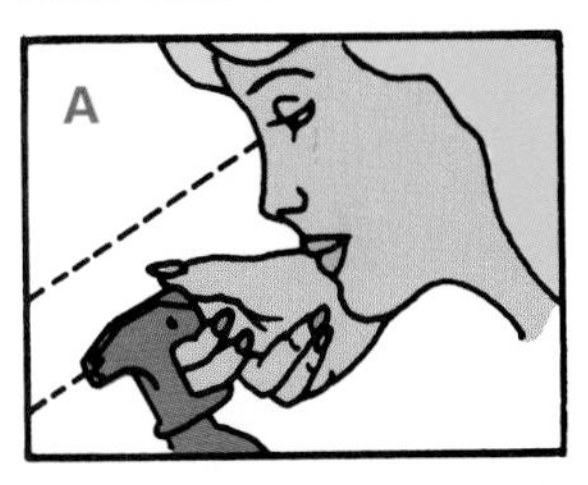

SQUEEZE the handle: This releases the extinguishing agent.

SWEEP from side to side: Keep the extinguisher aimed at the base of the fire and sweep back and forth until it appears to be out. Watch the fire area. If the fire breaks out again, repeat the process.

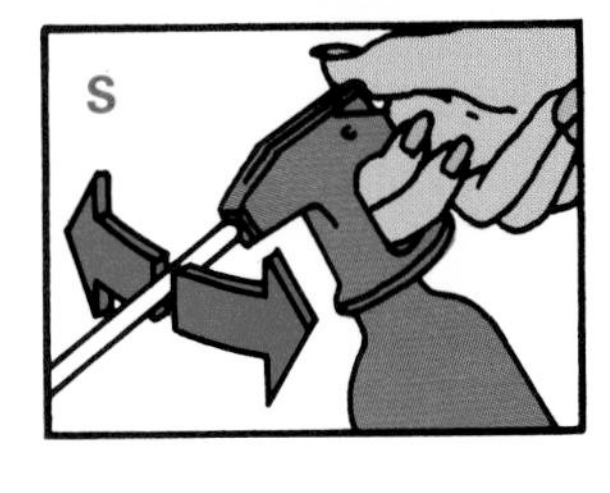

Most portable extinguishers work according to the directions, but some do not. Follow the directions on the model you're using.

Fig. 7-17. The PASS Method should be used when fighting a fire using a portable fire extinguisher. (National Fire Protection Assoc.)

Any accident, no matter how slight, should be reported to the instructor immediately. Any injuries should be treated as soon as possible by the school nurse or another qualified person.

Following an accident, find out how it happened. As a class, discuss ways to prevent future accidents.

SUMMARY

This chapter has covered some of the basic safety information you should know. Since laboratories are different, your instructor will provide other safety rules as needed.

The safety rules that you learn at school can be applied both at home and on the job. Remember that good safety and health habits can prevent accidents that can result in injury, death, or loss of property.

WORDS TO KNOW

All of the following words have been used in this chapter. Do you know their meanings?

American National Standards Institute (ANSI)
Compliance Safety and Health Officer (CSHO)
conduct
decibels
Hazard Communication Standard
health
Material Safety Data Sheet (MSDS)
Occupational Safety and Health Administration (OSHA)
PASS Method
safety
Type A fire
Type B fire
Type C fire
Type D fire

REVIEWING YOUR KNOWLEDGE

Please do not write in this text. Write your answers on a separate sheet.

1. Safety and health are:
   a. Your responsibilities.
   b. Your teacher's responsibilities.
   c. Everyone's responsibilities.
   d. Your employer's responsibilities.
2. Describe the purpose of OSHA.
3. In which of the following situations should eye protection be worn?
   a. When handling and mixing chemicals.
   b. When lasers are used.
   c. When there is a danger of small, flying objects.
   d. All of the above.
4. Name the two considerations for which OSHA has established standards for noise exposure.
5. True or false? When lifting a load, lift using the muscles in your back, not your legs.
6. Which of the following is a safe practice in the laboratory?
   a. Leaning back on chairs and stools.
   b. Wearing jewelry and/or loose clothing.
   c. Avoiding dangerous operations when you are ill or overtired.
   d. Both a and b.
7. True or false? When using equipment, it is best to make adjustments with the machine turned off.
8. True or false? If you are in doubt about what a chemical is, it is a good idea to taste it or smell it.
9. If there was a fire involving combustible materials such as paper, cloth, or wood, which type of fire extinguisher would be the best choice to use to fight the fire?
   a. Type A.
   b. Type B.
   c. Type C.
   d. Type D.
10. Describe the procedures to be followed in your classroom laboratory in case of an accident.

APPLYING YOUR KNOWLEDGE

1. Inspect the laboratory in which you will be working and correct any safety hazards or make sure they are corrected.
2. Discuss the steps to follow in case of an accident. Role-play various types of situations and discuss how to handle them properly.
3. Ask a fire fighter to discuss fire prevention and demonstrate how to extinguish various types of fires.
4. Using safety equipment catalogs, find examples of color coding.

Understanding and following safety and health rules is an important part of any job. Notice that this worker is wearing proper ear protection. (McCain Manufacturing Corp.)

Section ACTIVITIES 2 BASIC COMMUNICATION SKILLS

BASIC ACTIVITY:
Reverse Engineering

Introduction:
A good way to figure out how to do something is to study how others have done it. In this activity, you will study and determine how an advertisement was created. The process of deciding how something was made is sometimes known as "reverse engineering."

Guidelines:
- The advertisement should contain both graphics and words. See Section Fig. 2-1.
- Do not remove ads from books or magazines without permission.

Section Fig. 2-1. This is an example of an ad that might be used for this activity. (Flying Color Graphics)

Materials and equipment:
- One advertisement.
- Scissors.
- Grid paper.
- Pencil.

Procedure:
1. Find a large ad that deals with a topic of interest to you.
2. Write a short summary of how the graphic designer applied the elements and principles of design in the ad. For example, if balance was used in creating the ad, is it formal or informal?
3. Reverse engineer the advertisement. This will include stating the purpose of the ad, and developing thumbnail sketches that you think the designer might have created. Use grid paper for the thumbnail sketches and review the technique for making them before proceeding.
4. If time allows, develop thumbnails and a rough layout of a new and improved ad.

Evaluation:
Considerations for evaluation are:
- Effort in completing the assigned task.
- Correct procedure used for making the thumbnail sketches.
- Quality of the written explanation.
- Creativity used in the process.

ADVANCED ACTIVITY:
Think Safety, Work Safely

Introduction:
One of the most important things to remember in technology education is to work safely at all times. Sometimes, people need to be reminded just how important safety is. This activity involves designing safety reminders using the basic communication skills you have studied in the last four chapters. Once your instructor has approved them, the safety reminders may be posted in lab and workstation areas.

Guidelines:
- The reminders should be developed with the audience (the students) in mind.

Materials and equipment:
- This will vary depending on the product.

Procedure:
1. As a class or in small groups, brainstorm on the best way to remind students of safe practices in the laboratory. Make a list of the most important rules to remember. Research how others have handled this problem. In groups or individually begin working on the project.
2. Develop thumbnail sketches and a rough layout. Have your instructor review and sign the rough layout before continuing.

3. Gather needed materials and complete the product.
4. Make a short presentation to the rest of the class on your product and describe the steps taken to complete it. Discuss why you think this is a good way to communicate safety information. In what ways did problem-solving techniques help you in completing the task?

Evaluation:

Considerations for evaluation are:

- Group and individual effort.
- Completion of all parts of the assignment.
- Correct thumbnail sketches and rough layout.
- Quality of the final product.
- Presentation quality.

ABOVE AND BEYOND ACTIVITY: Taking Control

Introduction:

Have you ever noticed that two products that are supposed to be the same actually are not? There are always slight differences in any two objects. They may look the same, but using precise measuring tools you can usually find a difference. This difference is known as variation. Control charts are used in quality control to monitor variation and determine if a process is "in control." These charts are set up so that a certain amount of variation is allowed, but when the variation becomes too great, the chart shows that there is a problem. In this activity, you will produce a simplified control chart, keeping track of some aspect in your life to see if it stays in control.

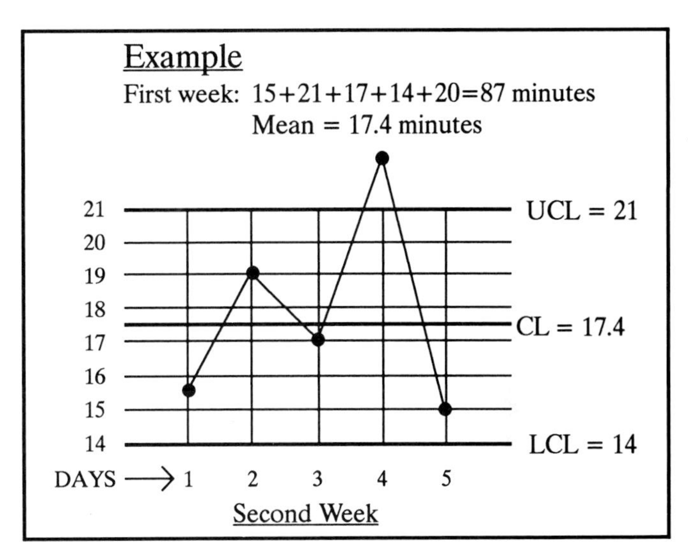

Section Fig. 2-2. This control chart example has points on it that designate the times recorded in Week 2. Note that one point is outside the control limits.

Guidelines:

- The process to be tracked should be a daily occurrence, such as the time it takes to get to school.

Materials and equipment:

- Layout tools for the chart.

Procedure:

1. Select some aspect of your life you wish to monitor over a two-week period. For example, you might time how long it takes you to get to school or the amount of studying you do each night.
2. Keep a daily record for one week. Use these numbers to set up the control chart. For example, let's say that you have been keeping track of the time it takes to get to school. First, discard any times that did not allow you to get to school on time. Now take the rest of the numbers, find the sum, and divide by the total of numbers added. This is the mean for the numbers that will be the center line on your control chart. Now find the smallest and largest numbers. The smallest number is the lower control limit, and the larger number is the upper control limit.
3. Set up the control chart with an upper control limit (UCL), center line (CL), and lower control limit (LCL). Make lines on the control chart that signify minutes. See Section Fig. 2-2.
4. Keep a record for a second week and place marks on the chart at about 1-inch intervals. Connect the dots with straight lines. If the graph stays within the control limits, the process is in control. That means that the variation observed is normal. If you have marks that go outside the control limits, then there is usually some assignable cause. For example, if a wreck occurred and slowed down traffic, you might be late for school. This would be an assignable cause.
5. Discuss how industry would use control charts to maintain product quality. If possible, have an industry representative talk to the class about this concept.

Evaluation:

Considerations for evaluation are:

- Completion of assigned tasks.
- Quality of control chart.

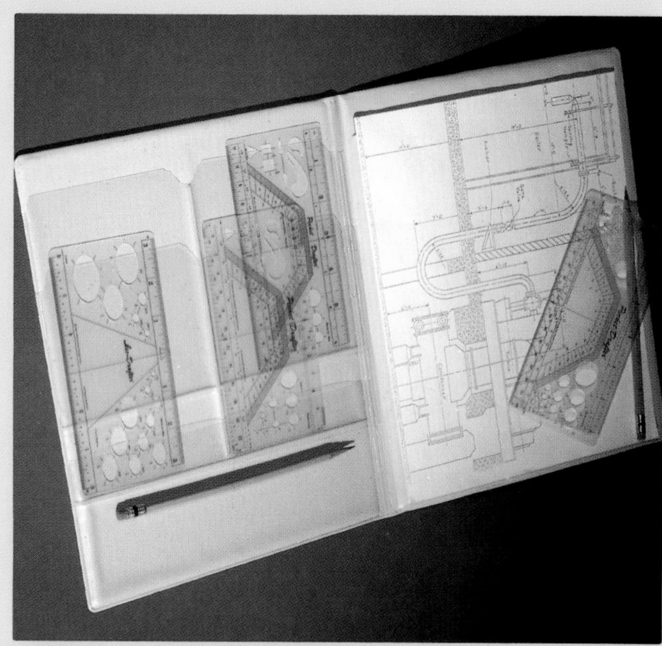

INTERNATIONAL DESIGN CORP.

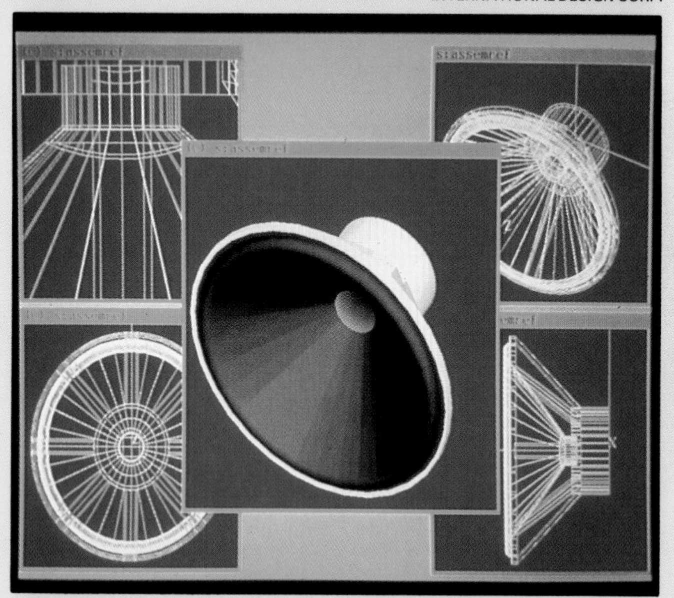

AUTODESK, INC.

PETER KANE

COMPUTERVISION

ZENITH ELECTRONICS CORP.

section 3

Technical Drawing and Illustration

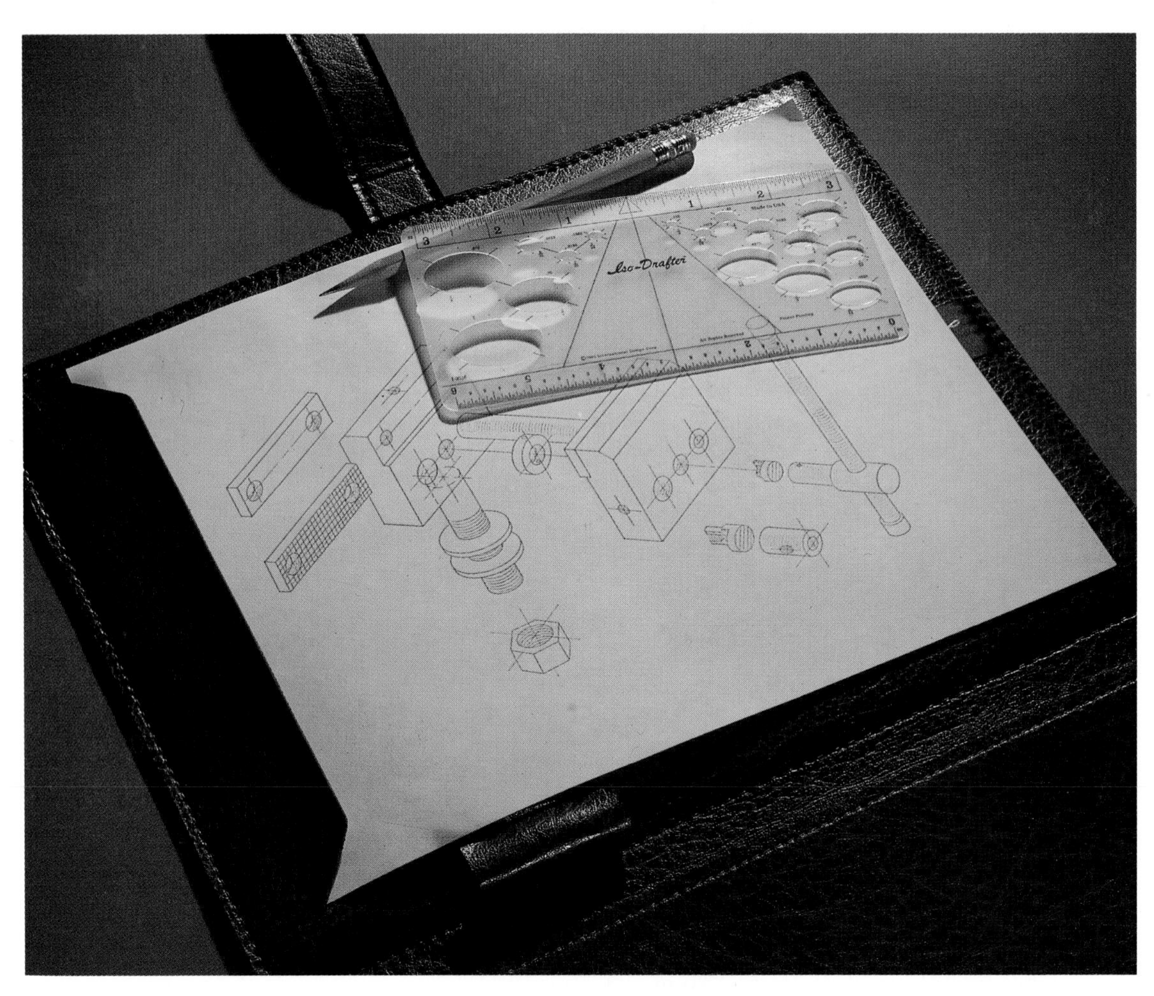

An important aspect of making a quality drawing is knowing how to use your drawing tools. (International Design Corp.)

8
CHAPTER

TECHNICAL DRAWING TOOLS

After studying this chapter, you will be able to:
- *Identify common drawing tools.*
- *Describe uses of common drawing tools.*
- *Demonstrate how to use some common drawing tools.*
- *Distinguish between various types of drawing papers and films.*
- *Describe the proper care of drawing tools.*
- *Explain important safety rules to keep in mind as you use drawing tools.*

Designing a message or product usually begins with sketching. However, in the final stages, instrument drawings are often needed. An **instrument drawing** provides the detailed size and shape information necessary for production. Since manual tools are used, this is a good, inexpensive way of communicating technical information about a product in visual form.

In this chapter, you will learn how to use basic technical drawing or drafting tools, Fig. 8-1. This knowledge and skill will allow you to communicate technical information in the same graphic form used in industry.

DRAFTING PENCILS

Drafting pencils contain special leads. For drawing on paper, these leads are graphite or graphite and plastic. Drafting pencil leads come in 18 degrees of hardness, from very hard to very soft, Fig. 8-2. In general, the medium leads are used most often in drafting.

Since lead grades vary by manufacturer, it is sometimes necessary to draw with the lead before purchasing. A general rule is to choose the hardest lead that makes an acceptable line with moderate (not heavy) pressure. If the lead is too soft, the lead will wear rapidly and your drawing will be smudged. If the lead is too hard, you will have to use heavy pressure when drawing with it.

Most lines can be made with an F or H pencil. Construction lines, which are very light, can be made with a harder lead such as a 5H or 6H pencil.

Fig. 8-1. Drawing with instruments helps you to communicate ideas accurately.

When drawing on polyester film, a special plastic lead is used. Grading systems may vary by manufacturer. In general, there are five or six degrees of hardness available for plastic leads.

There are three types of drafting pencils with which you should be familiar. These are:

- Wooden drafting pencils.
- Mechanical pencils.
- Fineline pencils.

WOODEN DRAFTING PENCILS

Wooden drafting pencils, Fig. 8-3, are often used because of their low cost. However, they do require extra work in sharpening.

- *Sharpening*: For this pencil, remove the wood first and then sharpen the lead. Use a knife or drafting pencil sharpener to remove the wood from the end opposite the pencil grade. After exposing 3/8" of lead, point the lead by rotating the pencil on a sandpaper pad or 220 grit sandpaper. A lead pointer may also be used. Finish the desired point by drawing on scrap paper until the correct line thickness is obtained. Clean graphite off the point with a cloth.
- *Use*: When using this pencil with drawing tools, always draw away from yourself. Tilt the top of the pencil slightly in the direction of travel. Rotate the pencil slowly while drawing to wear the point evenly, Fig. 8-4.

MECHANICAL PENCILS

Mechanical pencils (commonly called leadholders) hold 2 mm diameter leads that are bought separately. See Fig. 8-5. Any lead grade can be used in this pencil. For inserting and extending leads, a release button is pressed on the end of the pencil. This opens a chuck inside the pencil.

- *Sharpening*: Sharpen the lead only. A sandpaper pad, Fig. 8-6, or lead pointer can be used. Expose 3/8" of lead and sharpen like the wooden pencil.
- *Use*: This pencil is used exactly like the wooden drafting pencil.

FINELINE PENCILS

Fineline pencils, Fig. 8-7, are purchased according to the line diameters needed. The pencil and lead are both

Fig. 8-3. A wooden drafting pencil looks like a normal writing pencil, except it usually has no eraser. (J.S. Staedtler, Inc.)

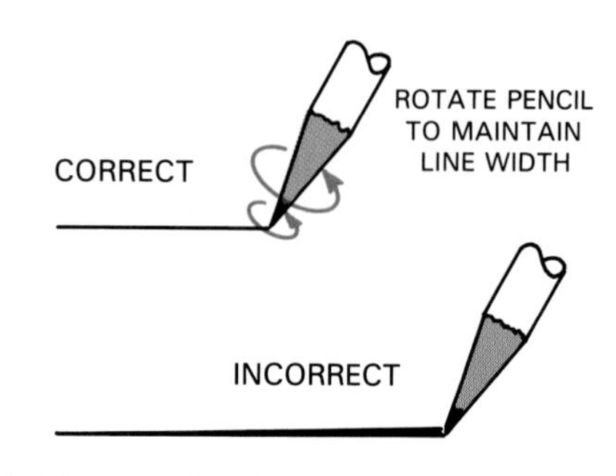

Fig. 8-4. The pencil point will wear more evenly if you slowly rotate the pencil as you draw.

Fig. 8-5. A mechanical pencil holds various grades of leads. Unlike a wooden drafting pencil that becomes shorter each time it is sharpened, the mechanical pencil does not change in length. (Staedtler Inc.)

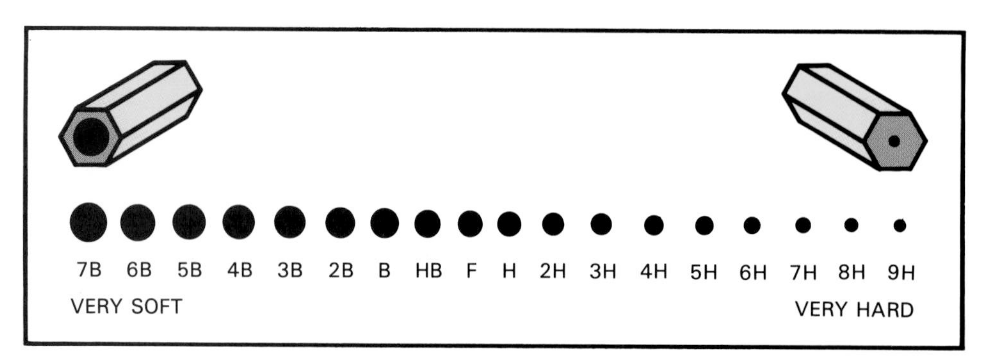

Fig. 8-2. Drafting pencil leads range from very soft to very hard. These leads are used for various purposes. For example, you might choose an HB or F pencil lead for lettering or sketching. For general use, you may choose an H or 2H pencil lead. For drawing construction lines, you may choose pencil leads ranging from 3H to 7H.

classified in a millimeter size. For example, a 0.3 mm would be used for thin lines and a 0.7 mm would be used for thicker lines. The lead is inserted and extended in the same way as the mechanical pencil. A wire comes with the pencil for cleaning clogged points.

You have a choice of tips when purchasing fineline pencils. See Fig. 8-8. The tip is the part of the pencil that presses against the straightedge as you draw.

The fixed tip, which does not move, works well for most applications. It has an advantage over sliding tips in that it is easier to use with templates, especially those used for lettering. A sliding sleeve is sometimes more difficult to place in the template since it can slide up above the template.

The advantage of the semi-sliding sleeve is that the lead is exposed as you draw. This protects smaller leads, such as a 0.3 mm, from breakage. Another advantage is that more lead is available for use. Thus, the lead is advanced less often.

The full-sliding sleeve retracts completely into the pencil as the lead is used. This can be a disadvantage when using a straightedge since it can retract and cause the lead to be against the straightedge. This type of pencil is most often used for writing rather than drafting purposes.

- *Use*: Hold the pencil vertically for the correct line diameter, Fig. 8-9. Pull the pencil toward you with medium pressure. Rotation is not required. If you are using the sliding tip, be sure it does not "ride up" on the straightedge being used.

Fig. 8-6. A sandpaper pad can be used to point or sharpen lead in either a mechanical or wooden drafting pencil.

Fig. 8-7. A fineline pencil uses very thin leads. This type of pencil does not require sharpening. (Staedtler, Inc.)

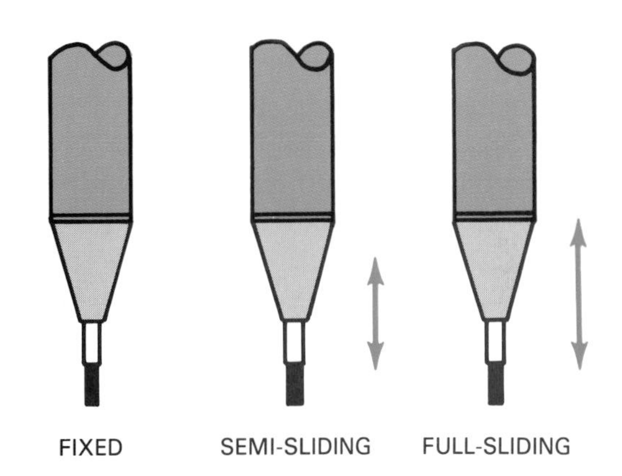

Fig. 8-8. You have several tip options from which to choose when selecting fineline pencils.

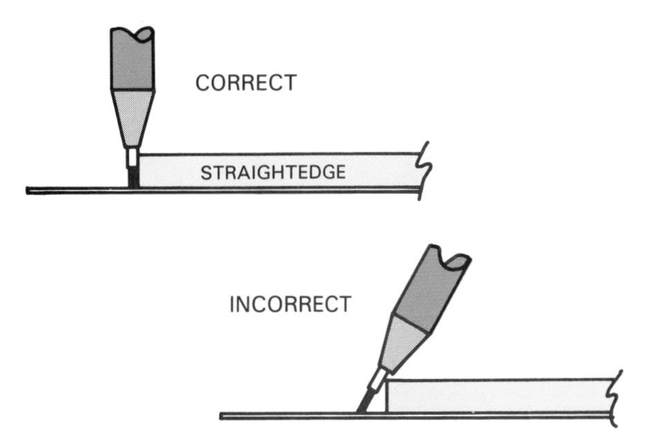

Fig. 8-9. A fineline pencil is held nearly vertical.

ERASING PENCIL MARKS

When erasing errors on paper, a rubber or vinyl eraser is used. The more abrasive rubber eraser is good for removing dark lines, and the less abrasive vinyl or gum eraser is good for removing light lines and smudges. A nonabrasive eraser is used on polyester film. This type of eraser is used to avoid smudging or scratching the surface of the film. If this happens,it will not accept pencil lines.

The procedure for correcting errors is as follows:

1. Hold the paper around the error to avoid tearing the paper while erasing. If possible, use an erasing shield and choose the hole that best fits the error. See Fig. 8-10. Make sure that correct lines are covered completely with the eraser shield.
2. Choose the correct eraser and make sure it is clean. Clean it on scrap paper if necessary.
3. Erase with the least amount of pressure possible. If too much pressure is used, you could rub through the paper or leave a "ghost" (graphite embedded in

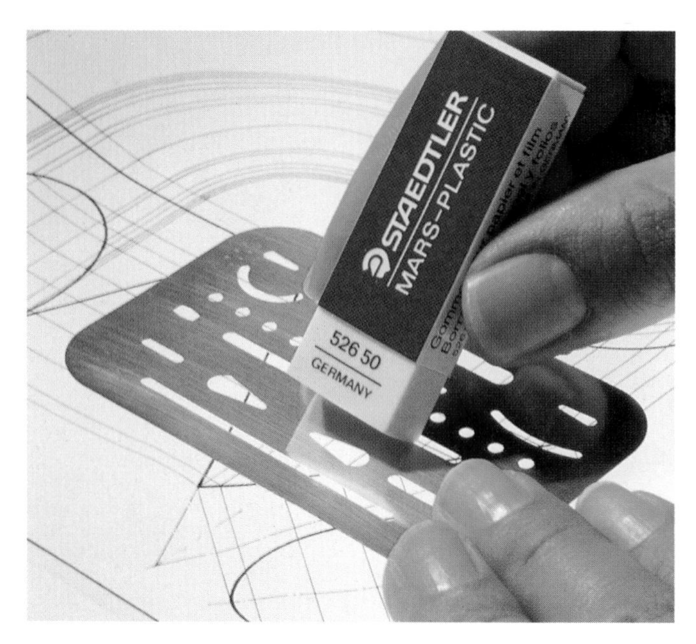

Fig. 8-10. To avoid erasing correct lines, use an erasing shield. (Staedtler, Inc.)

the paper). An electric eraser can be used for large mistakes. See Fig. 8-11.

4. Brush away particles with the drafting brush.

INKING WITH DRAFTING PENS

Inking involves writing or drawing with ink. Inking produces a better line density than pencil. This is important for technical illustrations that will be copied by photographic methods. Improved pens and erasable drafting film have resulted in an increased use of ink drafting. If inking on paper, always use a treated paper, such as vellum, so the ink will not be absorbed into the paper fibers. The two types of pens that are available for inking are the adjustable ruling pen and the technical fountain pen.

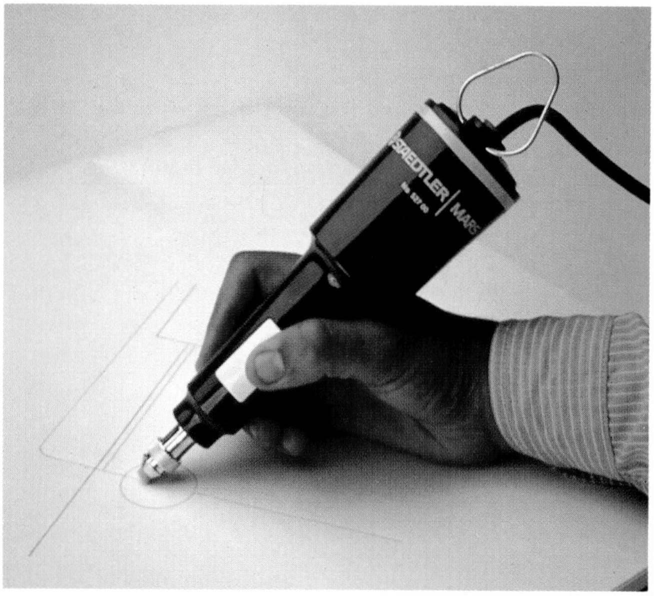

Fig. 8-11. An electric eraser can be used for large mistakes. (Staedtler, Inc.)

ADJUSTABLE RULING PEN

The **adjustable ruling pen** consists of two nibs that can be adjusted for a variety of line widths. This tool is used with a straightedge or as a compass attachment when drawing circles and arcs. It is not used freehand.

Fill the pen over scrap paper up to 1/4″ between the nibs. On scrap paper, draw lines and adjust the nibs with the thumbwheel until the line width is correct. Place the stiff nib (opposite the thumbwheel) lightly against the straightedge and draw the line, Fig. 8-12. If you hesitate at the beginning or end of the line, the line may thicken. Try to complete all the inking of one line width before changing the width setting. Refill the pen as needed.

Clean the pen by passing paper through the nibs. Some ink left on the pen will help the ink flow. Use an ink solvent for thorough cleaning.

TECHNICAL FOUNTAIN PEN

The **technical fountain pen**, like the fineline pencil, comes in a variety of point sizes according to the line width needed. The pen can be used with a straightedge, as a compass attachment, with templates, or freehand. Because the pen has an ink cartridge, frequent refilling is not necessary.

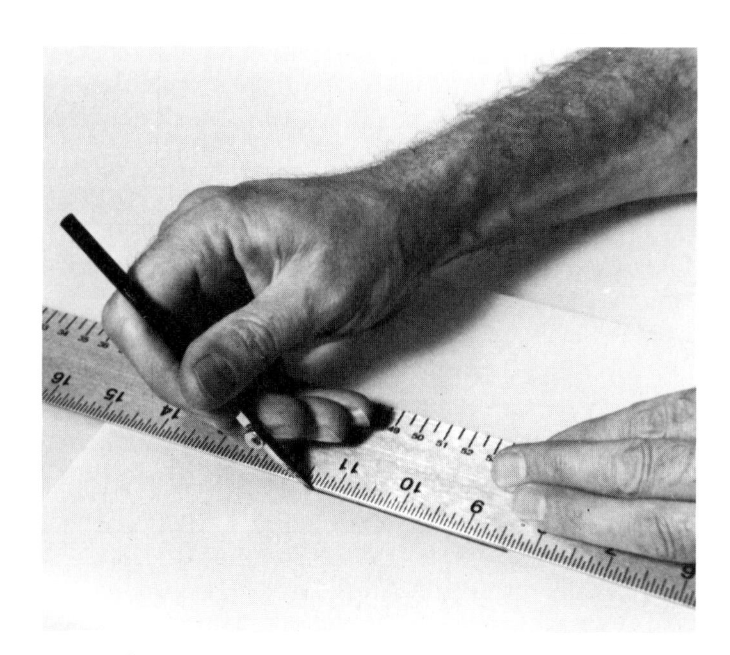

Fig. 8-12. This is the proper way to use an adjustable ruling pen.

Fig. 8-13. Each technical fountain pen provides a different line width. (Staedtler, Inc.)

Most technical fountain pens work in a similar manner, Fig. 8-13. Ink flows down from the cartridge to the point or nib. The nib contains a wire connected to a valve that allows ink to flow to the sheet when the wire is pushed up by the sheet. As ink is used, air enters the cartridge from around the nib to replace the ink. If air cannot enter the cartridge from around the nib to replace the ink, a vacuum is created and the ink flow will stop.

Pens come in American Standard line widths, Metric widths, or a combination of the two. A standard set might begin with 0000 (0.13 mm) for thin lines and progress to 7 (2 mm) for thick lines. See Fig. 8-14.

To use the pen, hold it horizontally and shake it gently to ensure the valve is free and clicking. Alternate shaking and drawing until the ink flow starts. Hold the pen vertically when drawing. Keep the pen moving when drawing and do not apply pressure on the point as you do with a pencil. Recap the pen when finished to keep the ink from drying in the nib.

To fill the pen or clean it, do the following:

1. Remove and fill the cartridge 3/4 full, if needed. Slant the cartridge when filling so bubbles will not form. See Fig. 8-15.
2. Take the point section apart and wash and dry each part.
3. Attach the cartridge to the pen body before you reassemble it.
4. Screw the nib on and tighten gently.
5. Alternate shaking the pen horizontally and drawing until a constant flow of ink begins.
6. If the point is clogged even after washing, disassemble and soak the point section in ink solvent or an ultrasonic cleaner.
7. After cleaning the pen, store it nib up, if possible, until it is to be used. If the pen will not be used for a while after cleaning, you can refill the cartridge with distilled water. This will help keep the parts in good working condition. Before storing the pen, shake the pen so that the point section fills with water.

Fig. 8-14. This pen set has four pens, so four line widths can be drawn. (Staedtler, Inc.)

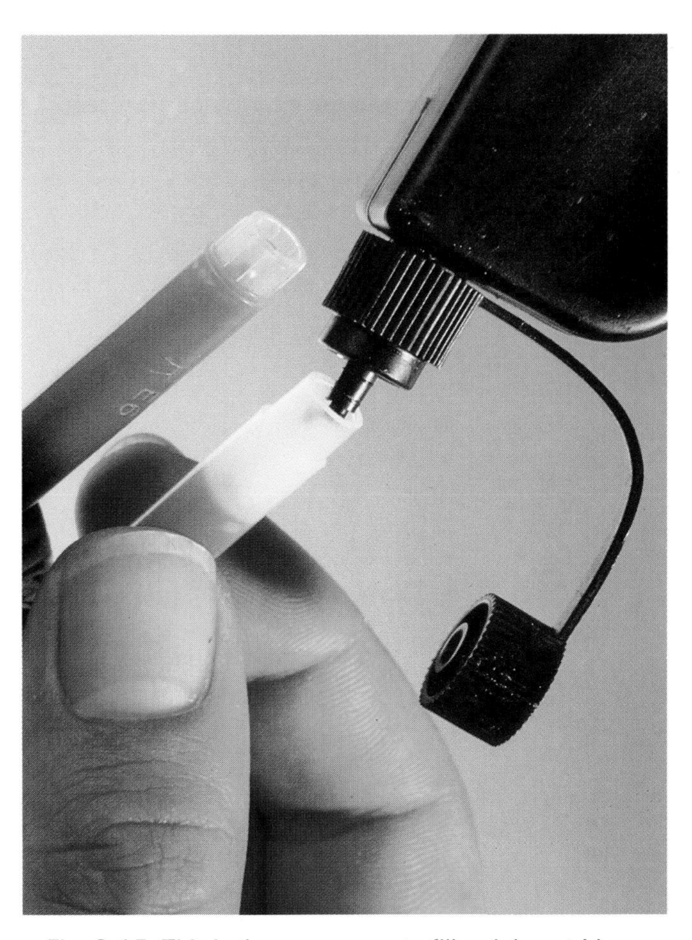

Fig. 8-15. This is the proper way to fill an ink cartridge. (Staedtler, Inc.)

CORRECTING INKING ERRORS

Ink deletion on paper can be made by scraping, erasing, or opaquing. For erasing, use an eraser shield with an eraser and rub gently to remove the error. Use an artist knife when scraping and tilt the blade. Scrape gently. White opaque fluid can be used if a translucent sheet is not needed for copying.

Errors are corrected on drafting film with a vinyl eraser and deletion fluid. Special erasers with ink solvent inside are also available, Fig. 8-16. Do not use a highly abrasive eraser on film. This will smooth the rough surface and make it less receptive to ink.

DRAWING SURFACES

The two most common drafting surfaces are the drawing board and the drawing table. The board and table top are made of softwood or synthetic materials.

The **drawing board** is portable and normally made to be used with a T-square. Two edges are very straight and have a metal or plastic edge. These are the "working edges." While drawing, it is usually more comfortable to raise the far edge of the board slightly.

A **drawing table** has a drawing board surface, Fig. 8-17. This top can be tilted as needed. Many tables can also be raised and lowered for greater comfort.

Drawing surfaces are usually covered to protect the surface. For a low-cost cover, poster board can be taped over the surface. A special board cover paper is also available that has a waterproof finish.

Special vinyl covers are made that "self-heal" when tacks and compass points are used. One vinyl cover has ferrite particles in it that attract magnetic strips. These strips are used in place of tape for attaching the sheet.

Fig. 8-16. Inking errors can be corrected with deletion fluid or ink erasers.

Fig. 8-17. An adjustable drawing table can be raised, lowered, or tilted as needed.

DRAWING SHEETS

There are a variety of drawing papers and films available for drafting. With a basic knowledge of these, you should be able to decide which one is best for your needs.

Drafting sheets come in roll and sheet form. There are two systems for designating sheet size: American Standard and Metric. In the American System, the letters A through E show size, Fig. 8-18. Each sheet size up the alphabet is twice the area of the previous letter. For example, size A = 8 1/2 x 11 or 9 x 12. Size B = 11 x 17 or 12 x 18, which is twice the area of the A size. The metric system is also shown in the chart.

Drafting sheets usually have a smooth side and a rough side. The rough side is usually up for most drafting. This information is usually provided on the package label.

OPAQUE PAPER

Opaque papers are those that you cannot see through. Many, such as bond, are inexpensive and can be used for general drafting. A color should be chosen that is easy on the eyes, such as pale green, blue, or beige. Points to remember when using this type of paper are:

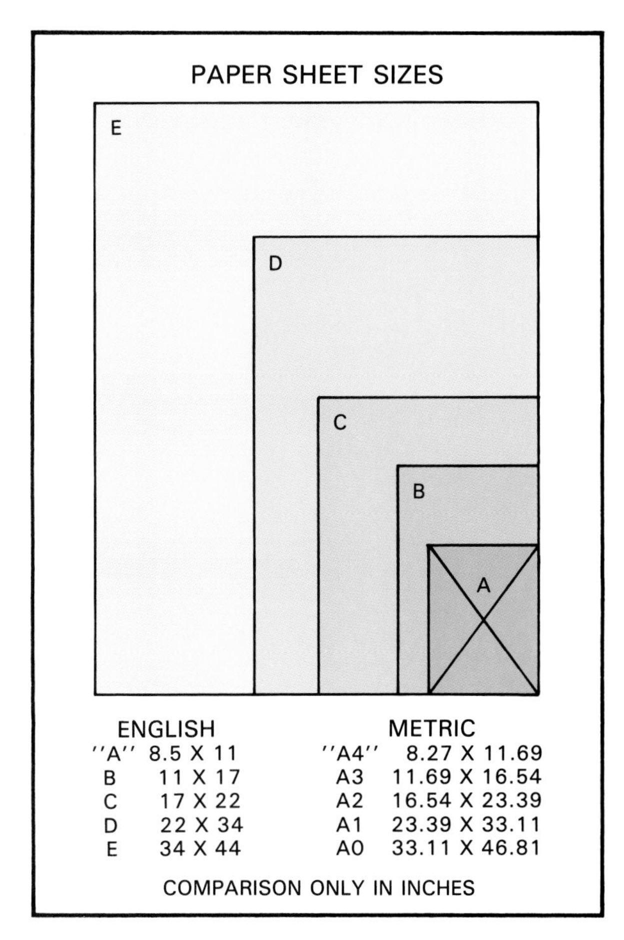

Fig. 8-18. Drafting sheets come in a variety of sizes.

- Errors should be erased carefully to avoid tearing the paper.
- Any copying methods that require translucent paper cannot be used.
- The size of the paper may change slightly with moisture conditions.
- Inking may be difficult because this paper can absorb and spread the ink.
- These papers may color and deteriorate with age.

TRANSLUCENT PAPER

Translucent paper transmits light and can be used for most copying methods. The two major categories in this group are tracing paper and vellum. **Tracing paper** is thin, untreated paper that can be used for pencil and ink drafting. Precautions for opaque paper would also apply to this paper. Treated tracing paper, or **vellum**, contains oils and waxes that improve the quality of the paper. Advantages of vellum are:

- It has a hard surface that is not grooved as deeply with pencils as untreated tracing paper.
- It is good for inking because it resists moisture and will not absorb the ink.
- It resists discoloring and deterioration.

DRAFTING FILM

Drafting film is a plastic sheet with one side roughened (dull or matte side) to accept ink and pencil lines. It has all the advantages of vellum with the following additions:

- It is moisture-proof so drawings are safer and do not change size with moisture conditions.
- Inked errors can be deleted with special erasers and ink solvent without damaging the sheet.

Since ink can be erased on drafting film, it is very popular for inking in industry. Erasing should be done with a special eraser that will not smooth the film surface.

Drafting film can also be purchased with two matte (dull) sides. This is used in computer plotters because it is less likely to slip.

TRACING CLOTH

Tracing cloth is cotton cloth that is treated with plastic so that it is transparent. Like drafting film, the dull surface is used for drawing. Tracing cloth works well for ink and pencil lines. It will resist deterioration and moisture. Tracing cloth is rarely used today.

ATTACHING PAPER TO A DRAWING SURFACE

There are a variety of ways to attach drawing paper to the drawing surface. However, the method used should align the paper with the T-square and provide a wrinkle-free surface. See Fig. 8-19.

Special **drafting tape** is best for taping the sheet to the drawing surface. This tape has low adhesion and is not

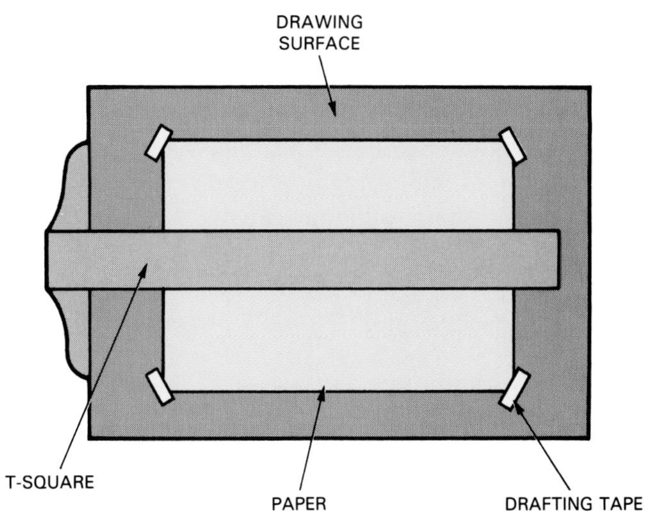

Fig. 8-19. When attaching paper to a drawing surface, the paper should be aligned with the T-square, taped with drafting tape, and provide a wrinkle-free surface.

likely to tear the paper when removed. Use a small piece of tape and attach it diagonally across the corners.

SHEET POSITION

If right-handed, place the sheet in the bottom left corner of the board. Leave enough room for your hands to rest on the board below the drawing. If left-handed, place the sheet in the bottom right corner.

ATTACHMENT

Slightly bend the bottom two corners of your sheet up without actually creasing the paper. Slide the paper down against the top of the blade of the T-square, parallel straightedge, or drafting machine. Holding the T-square and paper with the palm and fingers of one hand, tape the top two corners. Slide the straightedge out of the way, smooth the sheet, and tape the bottom corners.

ALIGNING THE DRAWING

Sometimes the sheet and drawing on it are not aligned or square with each other. In this case, align to a horizontal line on the drawing. Then tape the paper down as done previously.

TOOLS FOR DRAWING STRAIGHT LINES AND ANGLES

A combination of tools is normally used when drawing straight lines and angles. The most basic combination is the T-square, triangle, and protractor. Other options include the parallel straightedge and drafting machine. Each tool is described below.

T-SQUARE

The **T-square** consists of a head attached at a 90° angle to a blade. See Fig. 8-20. The blade is used for horizontal lines and as a guide for other instruments. To use the T-square, pull the head against the working edge of the board with the fingers on the blade. Hold it in this position when drawing horizontal lines. For vertical or slanted lines, hold a triangle against the top edge of the blade. See Fig. 8-21. To slide the T-square, first lift the blade by pushing down on the head. This will keep the blade from smudging lines.

TRIANGLE

A **triangle** is a tool for drawing vertical and inclined (slanted) lines. Triangles are described by height and angle. The two most useful triangles are the 45° triangle and the 30°-60° triangle. See Fig. 8-22. Both triangles can be purchased in various heights, such as 8″ or 10″.

To use the triangle for vertical lines, hold it against the top of the T-square with your fingers as explained

Fig. 8-20. A T-square and triangle are being used to make a drawing. (Staedtler, Inc.)

Fig. 8-21. A T-square and triangle are being used here to draw a vertical line.

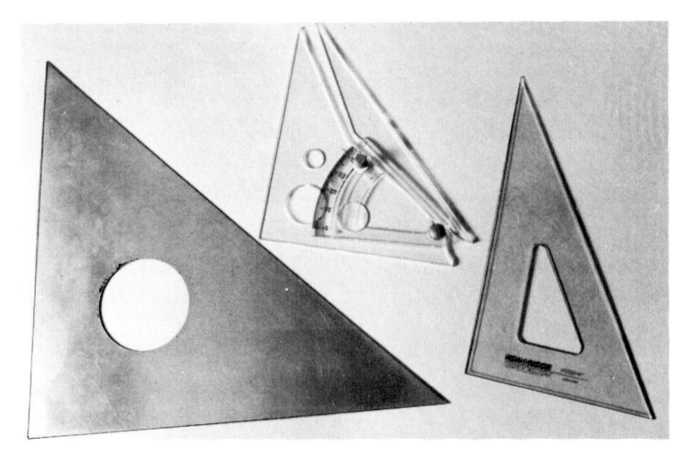

Fig. 8-22. A 45° triangle, adjustable triangle, and 30°-60° triangle are shown here.

earlier. If you are right-handed, have the inclined side to the right. Your hand then rests on the triangle when drawing and cannot smudge the paper. Also, when possible, use this same procedure for inclined lines. This procedure is reversed if you are left-handed. The general rule is to always position the triangle so your hand rests on it instead of your paper as you draw.

Any inclined line at 30°, 45°, or 60° can be drawn with the 45° and 30°-60° triangle. It is also possible to draw other angles by stacking these two triangles in different configurations on the T-square. Adjustable triangles are also available for drawing any angle. See Fig. 8-23.

PROTRACTOR

A **protractor** is used for drawing inclined lines and angles. It is usually made of clear plastic and has a semicircular shape. When using the protractor, align the base line of the protractor with a reference line already drawn. Set the center point at the vertex of the angle. Now, mark the correct degree shown on the edge of the protractor starting at zero on the left or right side, Fig. 8-24. Use a triangle as a straightedge to connect the marks.

PARALLEL STRAIGHTEDGE

The **parallel straightedge** takes the place of a T-square. It is attached to the drawing surface by wires and pulleys and remains horizontal when raised and lowered, Fig. 8-25. This is an advantage over the T-square, which must be held tight against the edge of the drawing surface. Triangles and the protractor are used with the parallel straightedge in the same manner as the T-square. This tool is especially useful for long drawings since it does not flex like a T-square.

Fig. 8-24. You can use a protractor to draw inclined lines and angles.

DRAFTING MACHINE

The **drafting machine** takes the place of the T-square, triangle, protractor, and scale. It is attached to the drawing table. A horizontal and vertical

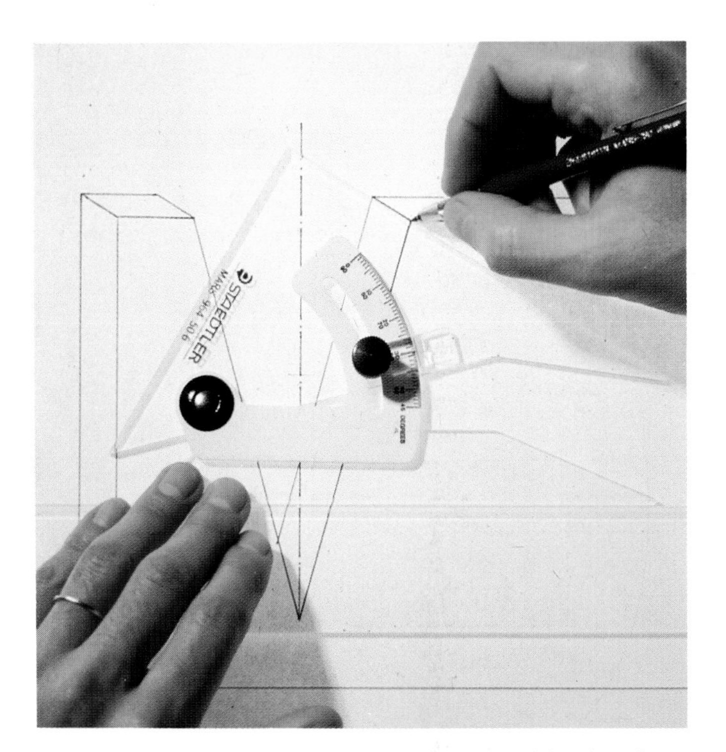

Fig. 8-23. An adjustable triangle can be used for drawing any angle. (Staedtler, Inc.)

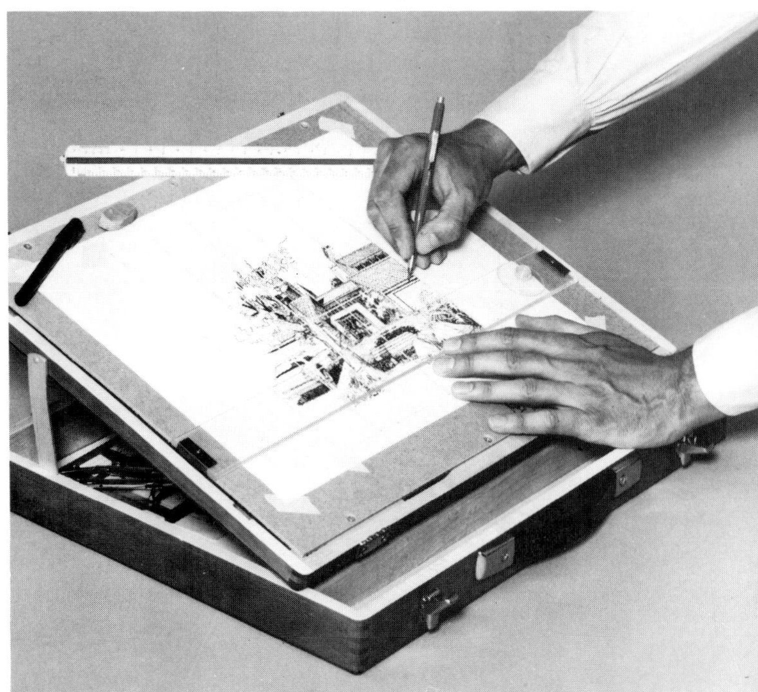

Fig. 8-25. A parallel straightedge is used on this portable drawing board. (Olson Manufacturing and Distribution, Inc.)

scale/straightedge are attached to a protractor head that can be set for any angle. When drafting, the drafting machine can be moved freely to any place on the board.

The two types of drafting machines are the elbow type and track type, Fig. 8-26. The track type is preferred for larger drawings.

Some drafting machines have electronic components. Some of the common functions are:

- Digital readout in degrees.
- Digital readout of linear measurements.
- Converting graphic information into digital values for computer-aided drafting.

TOOLS FOR DRAWING CIRCLES AND CURVES

A compass or circle template is used for drawing circles and curves. For those curves not having a common center, a French curve is used.

COMPASS

A **compass** is a tool for drawing circles and arcs. The most common type is the bow compass. Often, a small and large compass are needed to draw a variety of circles up to six inches in diameter. See Fig. 8-27.

One leg of the compass holds a pin that is placed in the center of the circle to be drawn. The pin should have a shoulder to reduce the hole size in the paper. The other leg often holds a wedge-shaped lead one to two grades softer than the pencil normally used. With the legs together, the pin point should be slightly longer than the pencil point so the length will be even when the pin is inserted in the paper. See Fig. 8-28.

Some compasses have a holder for a fineline pencil lead or fineline pencil. These can also be purchased as

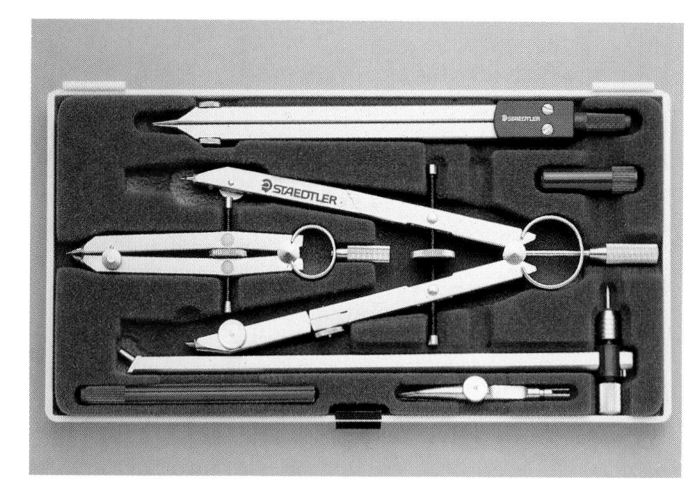

Fig. 8-27. This drafting set comes with two bow compasses. (Staedtler, Inc.)

A

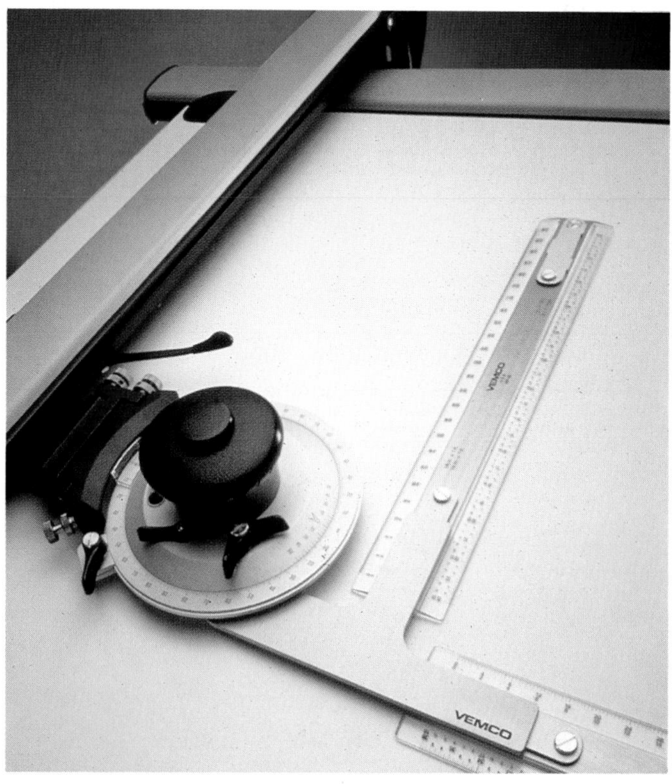

B

Fig. 8-26. Two types of drafting machines are shown here: A–The elbow type. B–The track type. (Vemco Corp.)

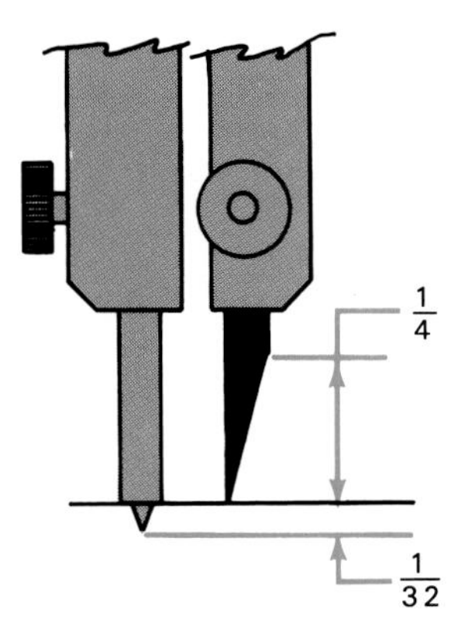

Fig. 8-28. When using a compass, the pin point should be slightly longer than the pencil point so that the length will be even when the pin is inserted in the paper.

attachments. This lead does not require sharpening. In addition, attachments are also available for using adjustable ruling pens or technical fountain pens in the compass.

An attachment is available for compasses that allows the use of drawing pencils. See Fig. 8-29. In addition, the attachments are available for using adjustable ruling pens or technical fountain pens in a compass. See Fig. 8-30.

To use the compass, first mark the radius on the drawing. Never measure by placing the compass on the scale. Insert the pin in the center point of the circle and move the other leg to the desired radius. Tilt the compass in the direction of travel and rotate it between the thumb and index finger to draw the circle, Fig. 8-31. Rotate more than one revolution, if necessary, to get the correct line density. It may be helpful to practice on scrap paper first.

For large circles, compass extensions can be purchased. A beam compass can also be used.

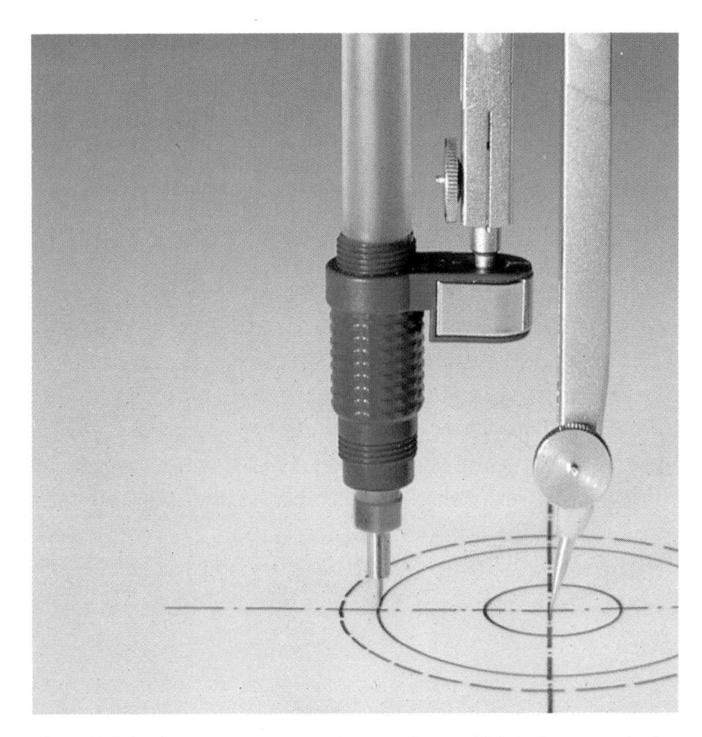

Fig. 8-30. A compass attachment is available for a technical fountain pen. (Staedtler, Inc.)

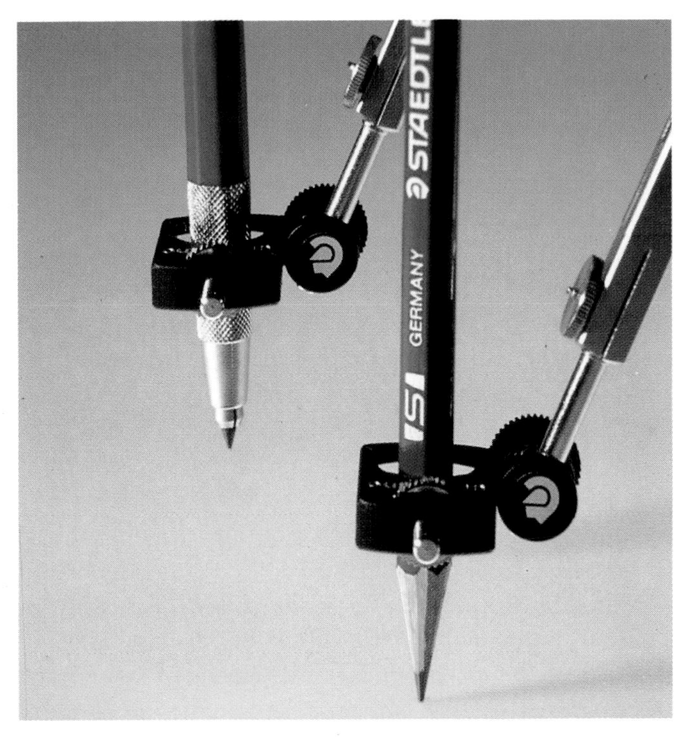

Fig. 8-29. A compass attachment is available for drawing pencils. (Staedtler, Inc.)

Fig. 8-31. A compass is used to draw accurate circles.

If required, sharpen the compass lead on a sandpaper pad to a wedge shape as shown in Fig. 8-32. The beveled or slanted surface should be on the outside when in the compass.

IRREGULAR OR FRENCH CURVE

An **irregular** or **French curve** consists of a variety of curves that can be used when arcs are not satisfactory. To use the instrument, first mark points along the curve to be drawn. Find a part of the irregular curve that aligns with three or more points and draw the line lightly. Continue this procedure with other points on the curve being sure to overlap each curve with the last one drawn. See Fig. 8-33. When this task is completed, use the same location on the irregular curve and darken the curve. Be sure the finished curve has a "smooth" appearance.

DIVIDER

A **divider** looks like a compass, but both legs have a steel point at the ends. See Fig. 8-34. This tool is often used for measurement purposes. For example, if a line needs to be divided into 8 mm intervals, the divider can simplify the task. Set the divider at 8 mm and mark off the distances on the line.

When setting a divider, it is recommended that measurements be taken off the drawing and not directly off the scale. Otherwise, the scale can be scarred by the divider points.

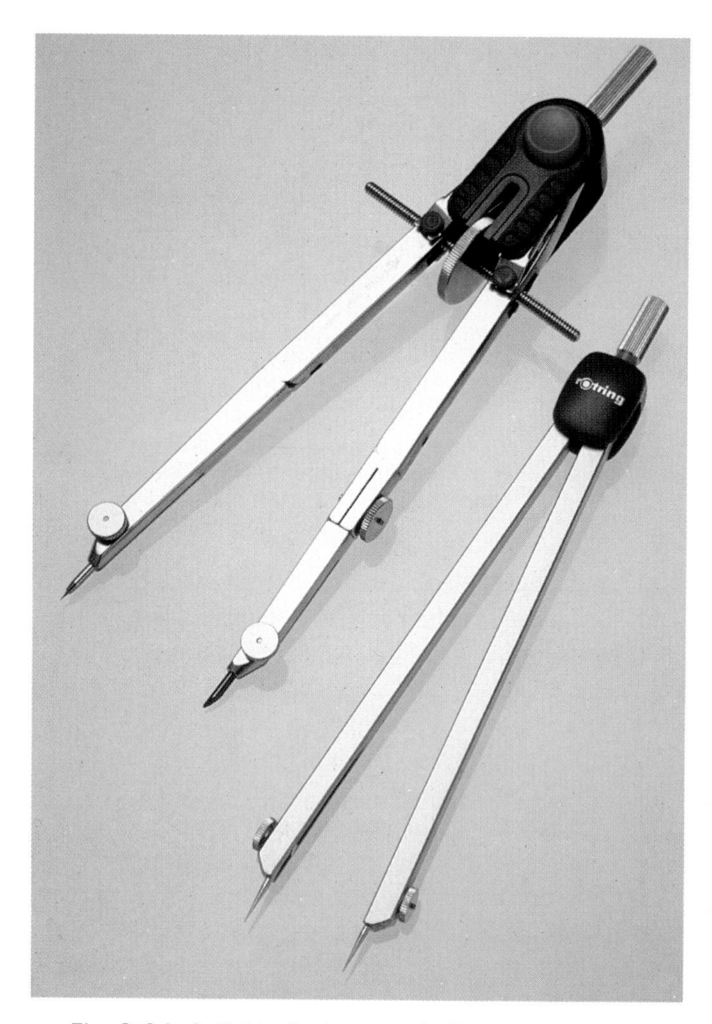

Fig. 8-34. A divider looks very similar to a compass. (Koh-I-Noor, Inc.)

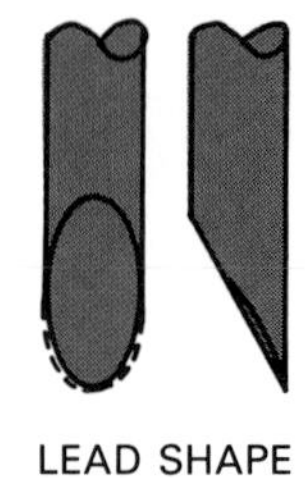

Fig. 8-32. This is the correct compass lead shape.

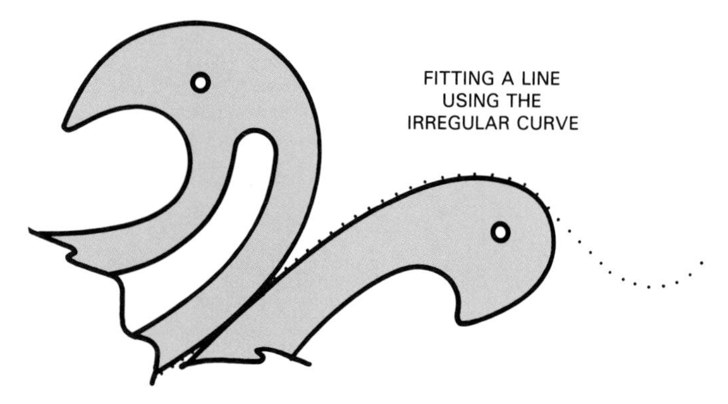

Fig. 8-33. An irregular or French curve is useful in drawing curves. Points along the curve are marked. Then they are connected by aligning the curve with three or more points.

TEMPLATES

A **template** is used to help in drawing shapes and symbols. There are templates for producing squares, ellipses, triangles, etc. Templates can also be used for drawing circles and arcs. Unlike the compass, you are limited to specific circle diameters with the **circle template**. Choose the diameter needed and trace the circle, Fig. 8-35. There are also special templates for each industry, such as electronic symbols and architectural symbols. See Fig. 8-36.

To use a template, place it on your drawing in the desired location and trace the object.

When the template is used for inking, the bottom side can be raised slightly with drafting tape. Otherwise, the ink may flow under the template. This can also be done by using special inking templates or by purchasing ink risers (plastic disks) that attach beneath the template.

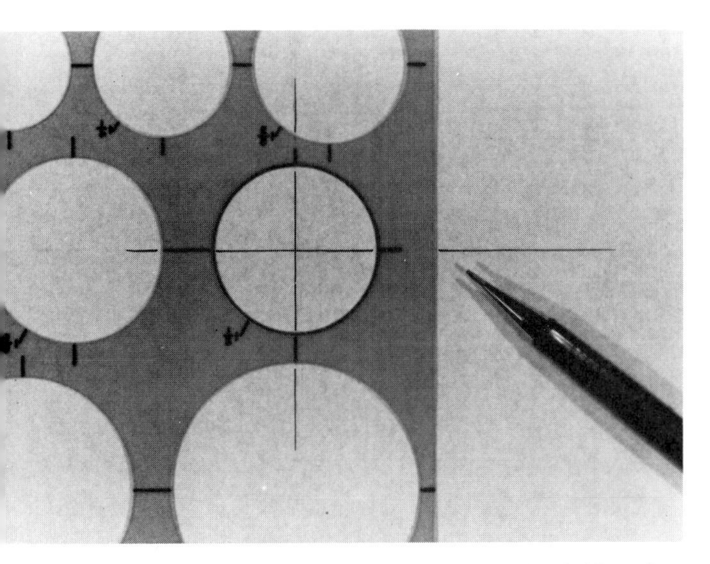

Fig. 8-35. Circles can be produced easily and quickly using a circle template.

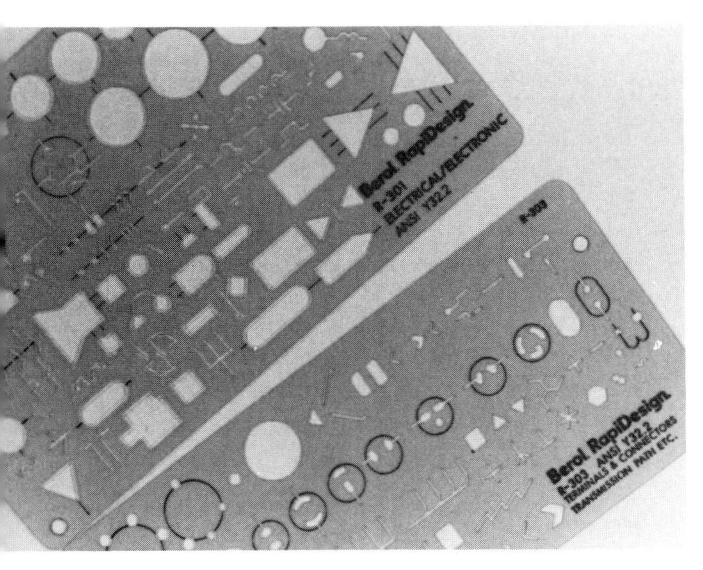

Fig. 8-36. These templates for electrical/electronic symbols are often used in industry drawings.

CARE OF TOOLS

If properly cared for, technical drawing tools should last many years. Some important points to remember are as follows:

- Store tools in their cases when not in use.
- Use a rag to wipe tools off after use.
- Clean smudges with a vinyl eraser or erasing solvent.
- Do not cut against the plastic edges of tools.
- Do not use an abrasive cleanser on clear plastic tools.
- Do not measure directly off the scale with the compass or dividers.
- Keep wooden tools and boards away from high moisture conditions.
- Occasionally check T-squares for rigidity and blade alignment.

DRAWING INSTRUMENT SAFETY

Some important safety reminders with drawing instruments are as follows:

- Pass, do not throw, tools.
- Use tools with points, such as the compass and dividers, only as directed.
- If strong cleaning solvents are used, protect your skin and eyes.
- Use artist knives as directed and store them in their proper containers.

APPLICATION IN INDUSTRY AND BUSINESS

All of the tools you studied in this chapter are also used in industry. In industry, however, there is a greater need to use instruments that will reduce drawing time. Drafting machines and computer-aided drafting equipment are two examples of equipment used to meet this need. See Fig. 8-37. However, it should be remembered that the quality of the drawing depends upon the expertise of the drafter, not the complexity of the equipment being used. Knowing how to use basic technical drawing tools properly will allow you to make quality technical drawings.

Fig. 8-37. Computer-aided drafting (CAD) equipment provides another option for making technical drawings. (Computervision)

SUMMARY

Instrument drawing provides the detailed size and shape information necessary for production. Instrument drawings are used to communicate technical information about a product in visual form. The skills you develop as you use technical drawing tools can be applied as you use more complex drafting equipment.

Drafting pencils, pens, erasers, drawing surfaces, and drawing sheets are the basic tools required for technical drawing. A T-square, triangle, protractor, parallel straightedge, and drafting machine are used for drawing straight lines and angles. For drawing circles or curved lines, a compass or an irregular curve can be used. Templates are often used to simplify the drawing of shapes and symbols.

Technical drawing instruments should be cared for properly. Safety should also be kept in mind when using technical drawing instruments.

WORDS TO KNOW

All of the following words have been used in this chapter. Do you know their meanings?

adjustable ruling pen
circle template
compass
divider
drafting film
drafting machine
drafting tape
drawing board
drawing table
fineline pencils
French curve
inking
instrument drawing
irregular curve
mechanical pencils
opaque paper
parallel straightedge
protractor
technical fountain pen
template
tracing cloth
tracing paper
translucent paper
triangle
T-square
vellum
wooden drafting pencils

REVIEWING YOUR KNOWLEDGE

Please do not write in this text. Write your answers on a separate sheet.

1. Explain why it is important to know how to use basic technical drawing tools properly.
2. List three types of drafting pencils.
3. A _________ (less, more) abrasive eraser is best for removing light pencil lines.
4. True or false? The adjustable ruling pen is used freehand.
5. True or false? Opaque paper transmits light and can be used for most copying methods.
6. Which of the following is good for inking?
   a. Opaque paper.
   b. Tracing paper.
   c. Vellum.
   d. None of the above.
7. True or false? Drafting film is moisture-proof.
8. Explain why it is best to use drafting tape for taping drawing paper to a drawing surface.
9. Which of the following would you use to draw a circle or arc?
   a. Protractor.
   b. Compass.
   c. Parallel straightedge.
   d. Triangle.
10. The drafting machine takes the place of a:
    a. T-square.
    b. Protractor.
    c. Triangle.
    d. All of the above.
11. True or false? It is a good idea to use an abrasive cleaner on clear, plastic tools.
12. Which of the following is a safe practice?
    a. Passing, not throwing, tools.
    b. Using tools with points, such as a compass, only as directed.
    c. Protecting your skin and eyes when using strong cleaning solvents.
    d. All of the above.

APPLYING YOUR KNOWLEDGE

1. Practice making light and dark lines with various types of drafting pencils.
2. Compare using bond paper and vellum. Try drawing, erasing, and inking on the paper. Prepare a report outlining your results.
3. Demonstrate how to use a T-square, triangle, and protractor.
4. Compare drawing circles with a compass to drawing them using a template.
5. Trace a drawing with an adjustable ruling pen and a technical fountain pen. Write a report comparing the results.

9

CHAPTER

GETTING STARTED IN DRAFTING

After studying this chapter, you will be able to:
- *Describe steps involved in producing attractive freehand lettering.*
- *Identify lettering options.*
- *Recognize typical lines used in drafting.*
- *Describe a sheet layout in drafting.*
- *Discuss options available for copying and preserving drawings.*

Drafting or technical drawing is a technique used for drawing items that are to be produced. Since instruments are used, the term **mechanical drawing** can also be used.

If you lived in a foreign country, you would need to know the language in order to communicate effectively. In industry, a technical drawing is the language that explains how a product is to be made. Since the drawing is understood by people within the company, designs can be discussed and improved. See Fig. 9-1.

Fig. 9-1. In industry, the technical drawing is the ''language'' a drafter uses to communicate ideas.

This chapter will cover some important aspects of the language of drafting. Topics will include:
- Lettering.
- Alphabet of Lines.
- Sheet layout.
- Copying drawings.

LETTERING

The accurate drawing of letters and numbers on a drawing is known as **lettering**. Lettering requires practice because it is often done freehand. A 0.5 mm pencil with a soft lead, such as an F, can be used.

Single-stroke Gothic is a style of lettering commonly used because it is easiest to draw and read, Fig. 9-2. In

ABCDEFGHIJKLMNOPQRSTUV
abcdefghijklmnopqrstuvwxyz
WXYZ 0123456789

ABCDEFGHIJKLMNOPQRSTUV
abcdefghijklmnopqrstuvwxyz
WXYZ 0123456789

Fig. 9-2. Sans serif, single-stroke, Gothic letters are normally used in drafting.

graphic arts, this is known as a **sans serif letter**. ("Sans" is French for without. Serifs are the thickened cross strokes at the tips of letters. Thus, a sans serif letter is a letter without serifs.) Sans serif letters and numbers are made with single strokes that connect, Fig. 9-3. Letters vary in width, Fig. 9-4. Often, all capital letters are used on drawings.

Gothic lettering can be vertical or inclined, Fig. 9-2. When vertical, the upright parts of letters are straight up or vertical. Inclined letters are tilted at 68°. Either style can be used, but styles should not be mixed.

Lettering is usually 3 mm (1/8"). However, larger letters can be used on larger drawings, Fig. 9-5. It may be necessary to increase the line width of the letters as the drawing size increases.

Spacing is very important when lettering, Fig. 9-6. **Letterspacing**, the space between letters in a word, should be done by eye to appear equal. **Word spacing**, or space between words, should be equal to the height of the letters. **Line spacing**, or space between lines, should be half or full letter height. For example, 3 mm letters might have a 3 mm space between words and a 3 mm space between lines.

Guidelines are always used for precise lettering, Fig. 9-7. They are very light construction lines. Horizontal guidelines are used for letter height and space between lines. They are also used for the height of lowercase letters when needed. Vertical guidelines are used to keep letters straight and are drawn randomly. Inclined lines are used to keep inclined letters at 68°. Guidelines can be made with a scale and normal drafting tools. For

Fig. 9-3. When making sans serif letters, follow the directions indicated by the arrows.

Letters and numerals have
PROPORTION as follows

BCDEFGHJLNPRUZ 234567890

5 UNITS WIDE X 6 UNITS HIGH

AKMOQSTVXY

6 UNITS WIDE X 6 UNITS HIGH

W

8 UNITS WIDE X 6 UNITS HIGH

Fig. 9-4. Letters vary in proportion. Note the various widths.

Drawing size can change
lettering size

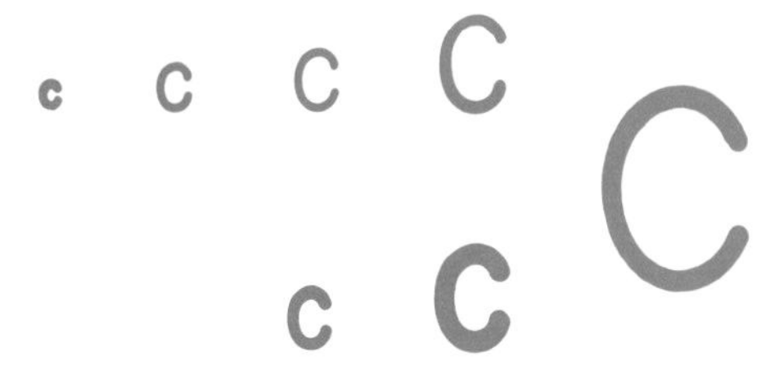

Fig. 9-5. As a drawing size increases, it may be necessary to increase the line width of the letters.

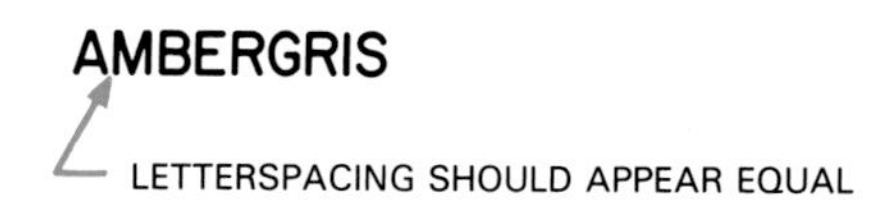

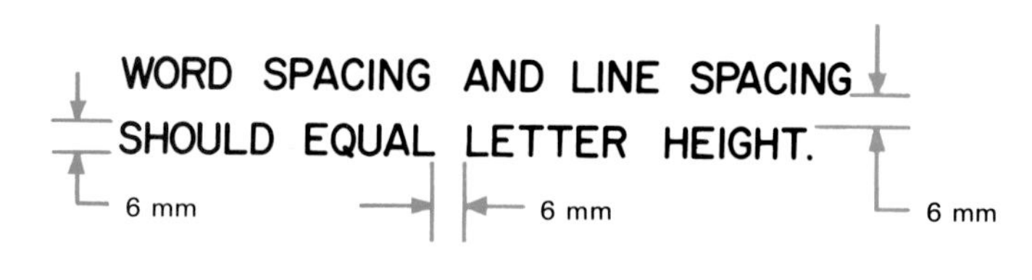

Spacing between sentences equals
2 x the letter height. OK ?
12 mm

Fig. 9-6. Letterspacing, word spacing, and line spacing are important concepts in lettering.

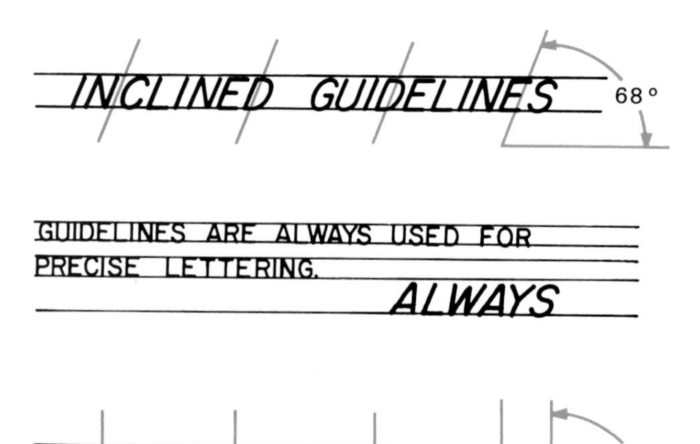

Fig. 9-7. Using guidelines, which are very light construction lines, can aid you in doing precise lettering.

convenience, special guideline instruments can be purchased, Fig. 9-8. They are especially helpful for inclined letters and when using capital and lowercase letters.

Although freehand lettering is often used on drawings, other techniques are available. Lettering templates allow you to trace letters through a template, Fig. 9-9. These are useful for large letters or when inking. Another style of template is used with a tracer for inking. While you trace the letter with a steel point, the letter is drawn in ink, Fig. 9-10.

Other lettering techniques require no drawing by hand. Transfer letters are rubbed off onto the drawing, Fig. 9-11. These are useful for bold lettering and un-

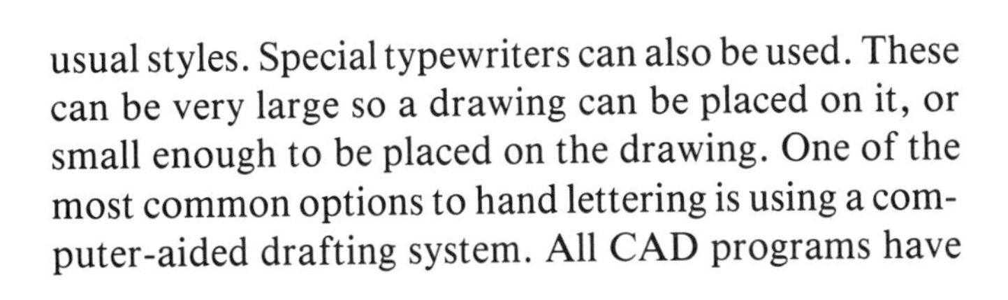

usual styles. Special typewriters can also be used. These can be very large so a drawing can be placed on it, or small enough to be placed on the drawing. One of the most common options to hand lettering is using a computer-aided drafting system. All CAD programs have

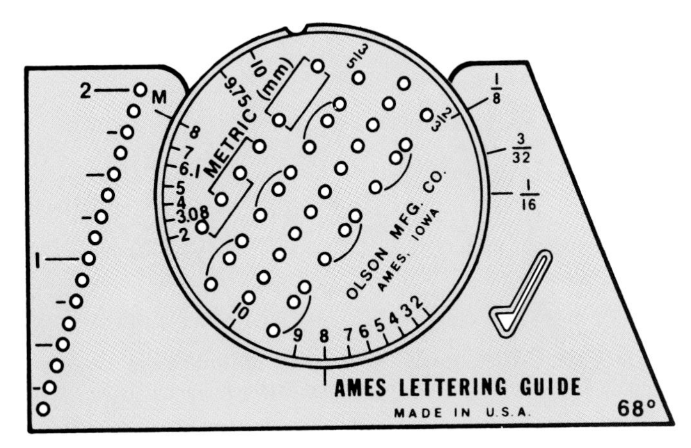

Fig. 9-8. This lettering guide is an example of a guideline instrument. (Olson Manufacturing and Distribution, Inc.)

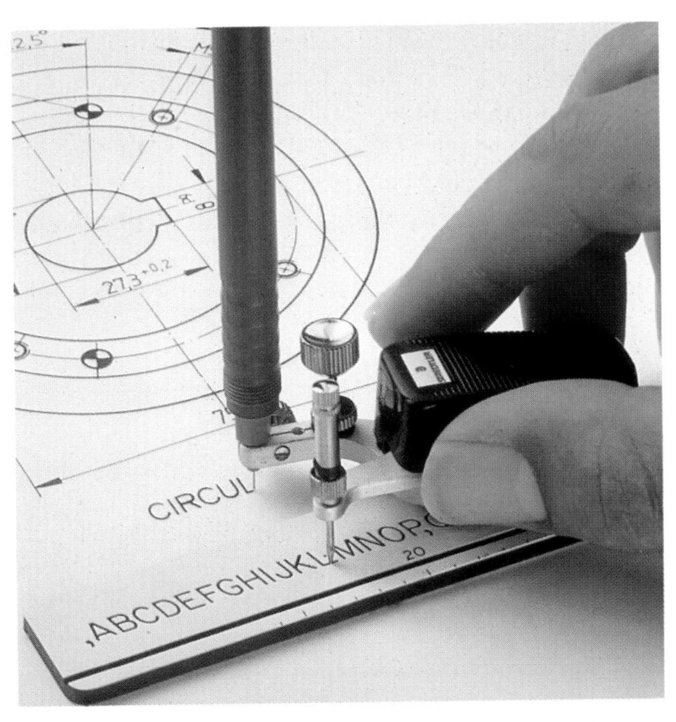

Fig. 9-10. A template and tracer can be used for inking. (Staedtler, Inc.)

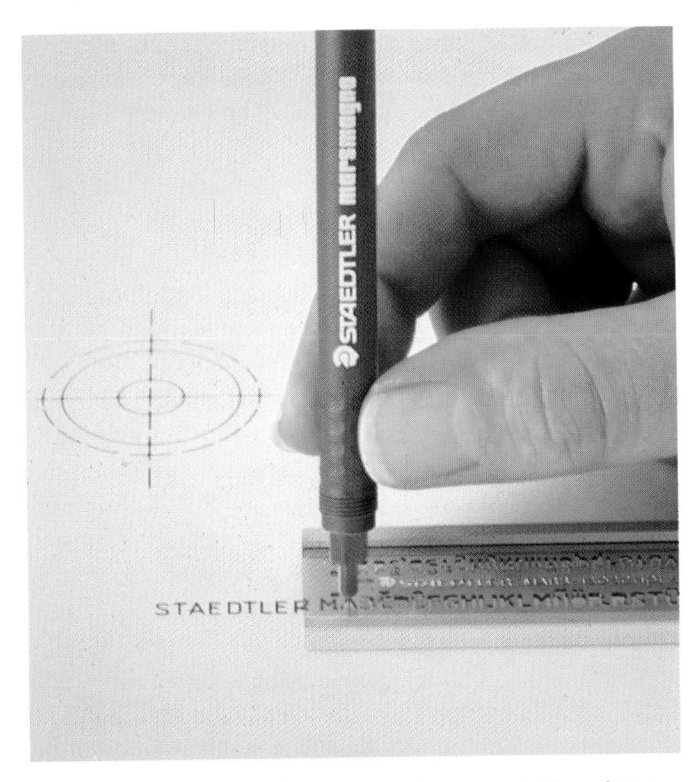

Fig. 9-9. Letters can be made easily and quickly using a template. (Staedtler, Inc.)

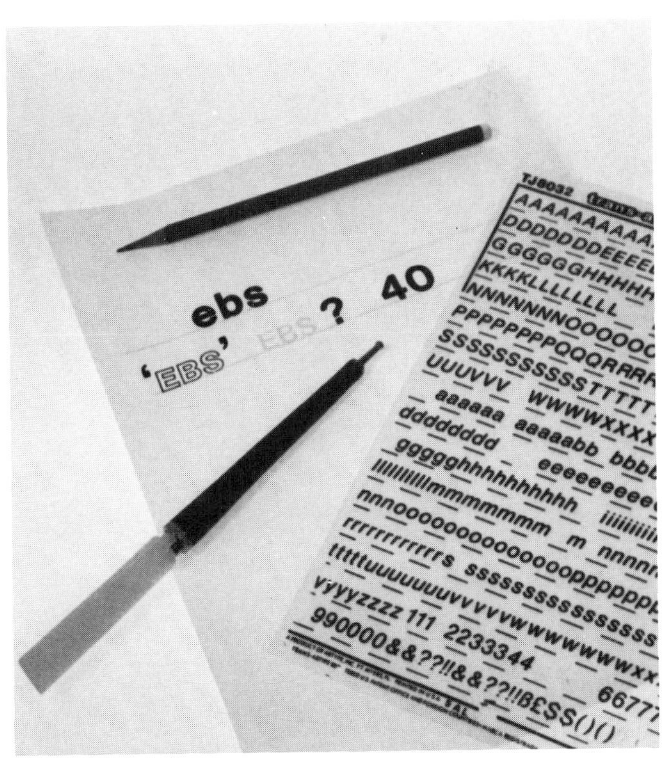

Fig. 9-11. Using transfer letters that are rubbed off is a lettering technique that does not require drawing.

lettering capabilities. A plotter is typically used for producing output.

ALPHABET OF LINES

The **Alphabet of Lines** is a standard for lines that was developed by the American Society of Mechanical Engineers. The lines, used in drafting, can be grouped into three categories, Fig. 9-12. These are:

- Very thin, light lines.
- Dark, thin lines.
- Dark, thick lines.

Very thin, light lines are drawn so that they are barely visible. They are made with a 0.3 mm pencil or a sharpened drafting pencil with a hard lead. These lines can be used as construction lines to lay out a drawing. These same lines can be used as guidelines for lettering. In either case, the lines should be light enough not to show on copies.

Dark, thin lines are drawn with a soft lead such as an F. The 0.5 mm pencil can be used for most drawings. This line is used for all other drafting lines, such as hidden lines, center lines, extension lines, and dimension lines. These are shown in Fig. 9-13.

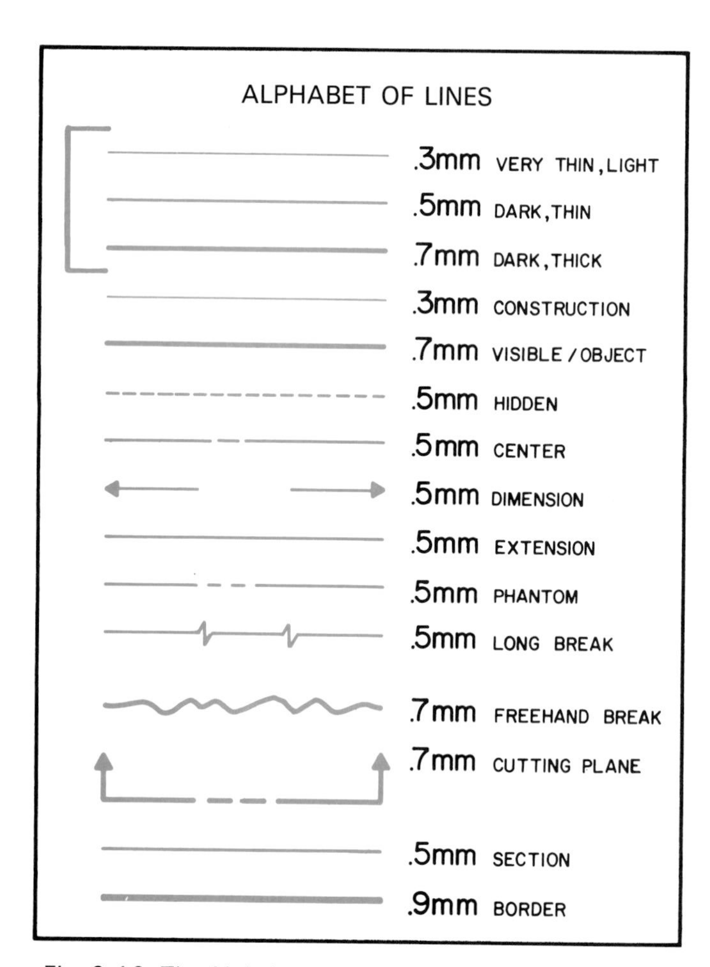

Fig. 9-12. The Alphabet of Lines is made up of three types of lines, as indicated above.

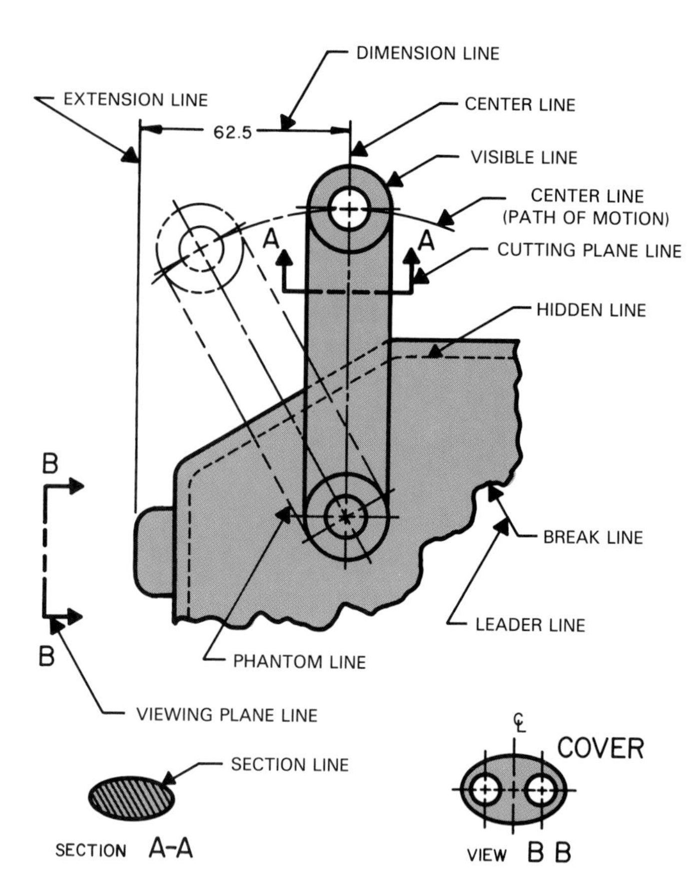

Fig. 9-13. Notice how the Alphabet of Lines was used in producing this diagram.

Dark, thick lines are dense and drawn with a soft lead such as an F. If the lead is too soft, the lines will smudge, and if the lead is too hard, lines will be gray and fuzzy. The 0.7 mm or 0.9 mm pencil can be used for these lines, depending on the drawing size. These lines are used as visible lines, cutting plane lines, and break lines as shown in Fig. 9-13.

SHEET LAYOUT

Technical drawings have a border around the edge and a title block. In industry, the **sheet layout** is printed on the sheet to save time. See Fig. 9-14.

The sheet layout for a drawing varies by company or school. A typical layout has a 6 mm (1/4″) border. The **title block** is usually a rectangle in the lower right corner or a 13 mm (1/2″) strip along the bottom of the sheet. The title block should always include the drawing title, drafter's name, scale, and date. A **revision block** may be placed above this which is a record of changes to the original drawing. A typical layout is shown in Fig. 9-15.

Begin a drawing by laying out the border and title block with light construction lines. Also, place guidelines within the title block. After the drawing is completed, darken the border and title block using dark, thick lines. Also, letter the title block.

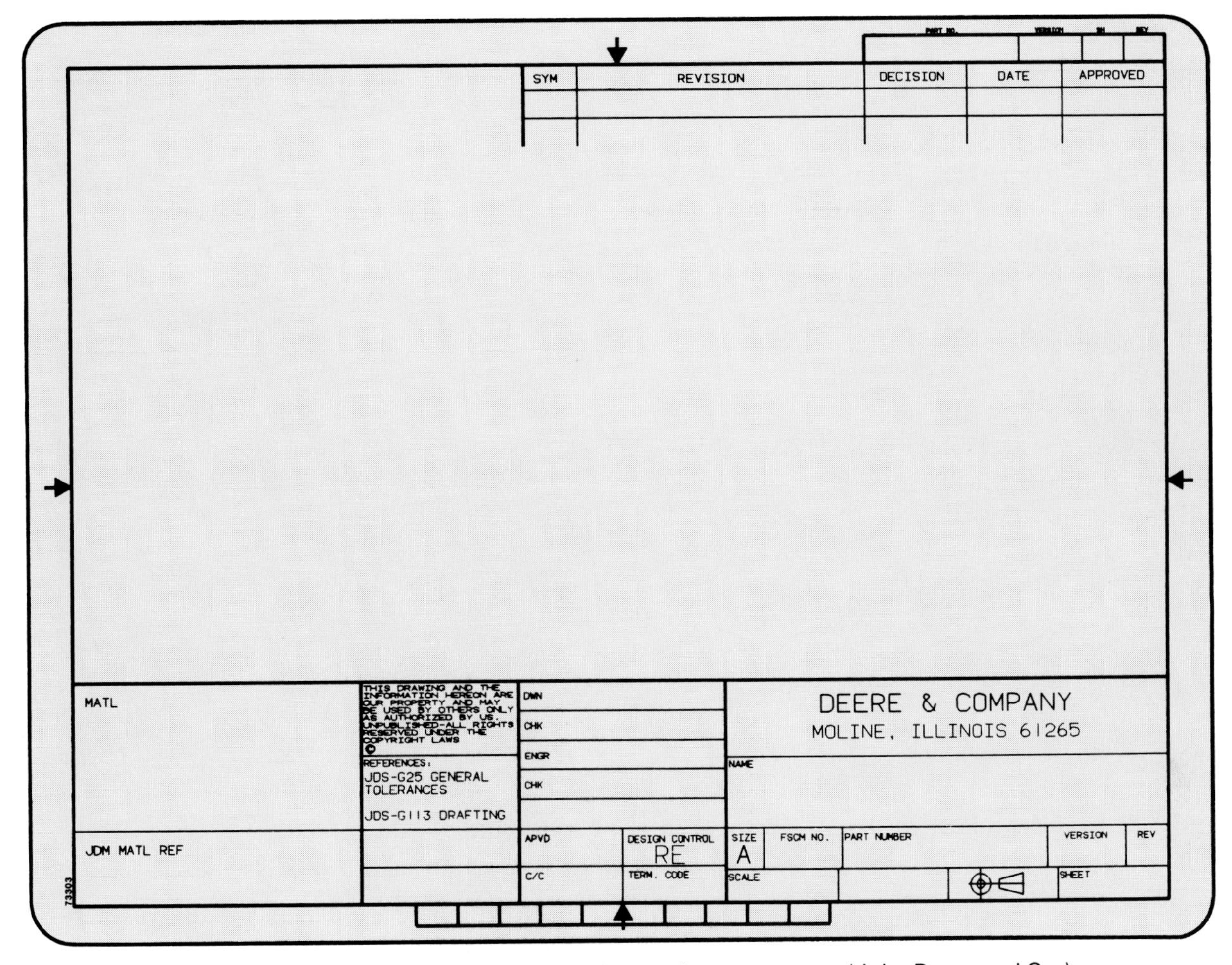

Fig. 9-14. This is an example of a sheet layout for a company. (John Deere and Co.)

COPYING AND PRESERVING DRAWINGS

Most technical drawings are made on translucent materials so they can be copied. In industry, drawing copies are used for marking corrections. By reviewing a copy, you can determine whether or not line weight is correct. For example, if construction lines are visible on the copy because they were drawn too dark, the original can then be revised.

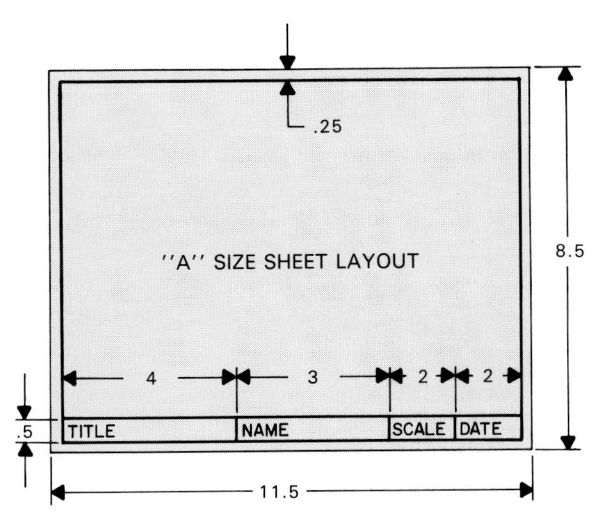

Fig. 9-15. A typical sheet layout has a title block that includes the drawing title, name, scale, and date.

Diazo prints are most often used for reproducing technical drawings, Fig. 9-16. They are also called **whiteprints** because they have a white background with blue image lines. To make a whiteprint, the original is placed over the diazo paper. When the paper is exposed to ultraviolet light, the diazo dye is destroyed in the nonimage area. The paper is then developed with an ammonia vapor that causes the dye left in the image area to turn blue.

Blueprints have a blue background and white image lines. Like whiteprints, a blueprint is made with an exposure and development step. Water is used for development. This washes away the unexposed dye leaving the white paper. A fixer is used to make the print last longer.

Electrostatic copies can be made from the drawing, Fig. 9-17. A translucent original is not needed with this process. A plain paper copier is an example. A powder is fused to the paper by heat. This forms the image. Some copiers can enlarge and reduce the drawing.

Fig. 9-16. Diazo prints have a white background with blue image lines. This machine is a whiteprinter.

Photographic methods can be used for very high quality copies. Reduction and enlargement is possible using this technique. When a drawing is to be printed in a book or magazine, a photographic copy is most often used. Special copy cameras are used in this process.

Fig. 9-17. A plain paper copier is used to make electrostatic copies of drawings.

Another photographic technique is **microfilming**. This is a method for reducing a drawing on film for storage. Microfilm requires much less space than original drawings, and the film will last a long time. Drawings are stored on **aperture cards, roll microfilm**, or **microfiche**, Fig. 9-18. Computerized methods can be used for automatically filing and retrieving drawings on microfilm.

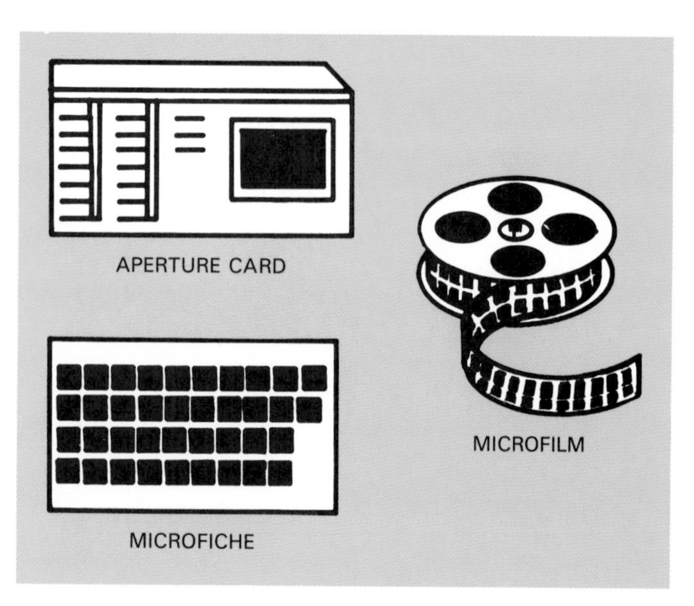

Fig. 9-18. Drawings can be stored on microfilm on aperture cards, microfilm rolls, or on microfiche.

Microfilm is viewed with a reader that enlarges the drawing. A printer can be used to print an enlarged drawing from microfilm. This print is called a **blowback**. Some machines are **reader/printers**, Fig. 9-19.

Fig. 9-19. This reader/printer allows you to view microfilm and print an enlarged drawing from the microfilm. (Minolta Corp.)

SUMMARY

This chapter has introduced you to some basics of drafting. Good freehand lettering is important for a nice drawing appearance. The Alphabet of Lines is used as the language of drawing. Each type of line must be drawn carefully since it has a special meaning. Sheet layout, which includes the border and title block, should be consistent. Also, a knowledge of copying methods is helpful as you begin drafting. By reviewing a copy, you can determine if your line weight is correct. Also, corrections can be made on the copy. In future chapters, you will have an opportunity to incorporate these basics into a finished drawing.

WORDS TO KNOW

All of the following words have been used in this chapter. Do you know their meanings?

Alphabet of Lines
aperture cards
blowback
blueprints
diazo prints
drafting
electrostatic copies
guidelines
lettering
letterspacing
line spacing
mechanical drawing
microfiche
microfilming
reader/printers
revision block
roll microfilm
sans serif letter
sheet layout
title block
whiteprints
word spacing

REVIEWING YOUR KNOWLEDGE

Please do not write in this text. Write your answers on a separate sheet.

1. True or false? In industry, a technical drawing is the language that explains how a product is to be made.
2. True or false? Sans serif letters have thickened cross strokes at the tips of letters.
3. True or false? When producing lettering, styles should be mixed.
4. Letterspacing should be:
   a. Equal to the height of the letters.
   b. Done by eye to appear equal.
   c. Be equal between each letter.
   d. None of the above.
5. Word spacing should be:
   a. Twice the height of the letters.
   b. Half the height of the letters.
   c. Equal to the height of the letters.
   d. None of the above.
6. True or false? Guidelines are very dark construction lines.
7. The ________ is a standard for lines developed by the American Society of Mechanical Engineers.
8. What information should be included in the title block or on a sheet layout?
9. Diazo prints are also called:
   a. Blueprints.
   b. Whiteprints.
   c. Electrostatic copies.
   d. None of the above.
10. Name two advantages of copying a drawing onto microfilm.

APPLYING YOUR KNOWLEDGE

1. Practice lettering sentences with correct letterspacing, word spacing, and line spacing.
2. Using any scale, copy an orthographic drawing. Make sure lines are of the correct width and density.
3. Request some drafting sheet layouts from local industries. Find out what title block information is required and why it is needed.
4. Visit a local industry and find out how the drawings are stored and reproduced. Write a report and present it to the class.

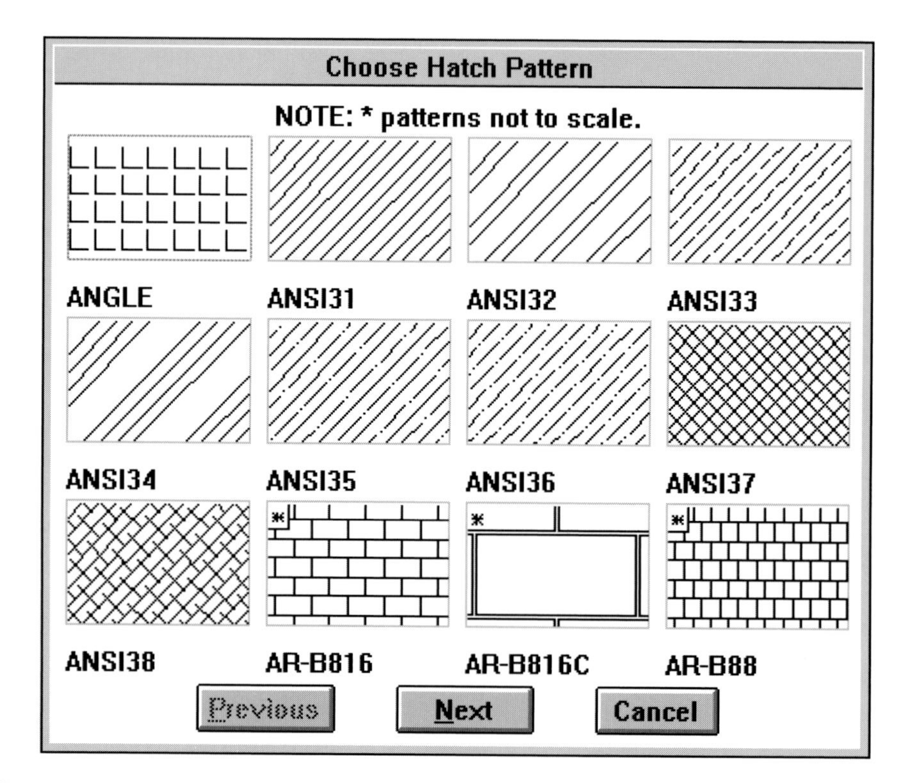

Basic drafting concepts, such as the alphabet of lines, are important whether you are drawing on a drawing table or by computer. (Autodesk, Inc.)

10

CHAPTER

GEOMETRY FOR DRAWING

After studying this chapter, you will be able to:
- *Explain the purpose of geometry.*
- *Identify common shapes.*
- *Demonstrate various geometric construction techniques.*
- *Identify common solids.*
- *Describe how to draw various geometric constructions involving solids.*

Geometry is used in the design and construction of almost every object that is produced. A building looks the way it does because of geometry. Furniture is created using geometry. Even the clothes you are wearing were designed using geometry. So what is geometry?

Thousands of years ago, geometry was a technique for laying out buildings and fields so that they had square, 90° corners. In fact, the term geometry was formed from the Greek "geo" (earth) and "metry" (measure). This meant to "measure the earth." From this beginning, **geometry** has evolved into a branch of mathematics dealing with lines, shapes, and solids, Fig. 10-1. In this form of math, however, the correct answer can be shown graphically as a drawing instead of a number.

All technical drawings use geometry for their construction. In fact, the special drawing tools described in Chapter 8 were invented to help make geometric construction easier.

As you read this chapter, remember that you will be using light, construction lines to do geometric construction. When the shape is developed, you can then darken it with visible lines.

LINES

In a technical drawing, lines are drawn to a specific width and length. They are drawn straight using tools such as a T-square and triangle or a drafting machine.

Fig. 10-1. By using geometry, you can communicate ideas through lines, shapes, and solids.

BISECTING A LINE

A line can be bisected or divided in half using a compass. **Bisecting** a line means dividing it in half. A method for bisecting a line is shown in Fig. 10-2.

1. Swing an arc from each end of the line that intersects at (I).
2. Repeat on the other side.
3. Draw a line through the intersections (I) to bisect the line.

EQUALLY DIVIDING A LINE

When a line must be divided into equal parts it can usually be done with a scale or dividers. However, another method may be useful for lines not easily divided into equal parts.

Follow the steps shown below in Fig. 10-3 to divide line (AB).

1. Draw a line (AC) at any angle to (AB).
2. Mark the needed number of divisions on (AC) using a convenient measurement. Start measuring from (A).
3. Draw (DB) and parallel lines to find the divisions on (AB).

Another method of dividing a line requires a 90° angle. This can be used to make graphs that have 90° corners. This is illustrated in Fig. 10-4.

1. Draw a line (AC) at 90° to (AB).
2. Using a scale, place the zero on (B) and a measurement equal to the divisions needed along (AC).

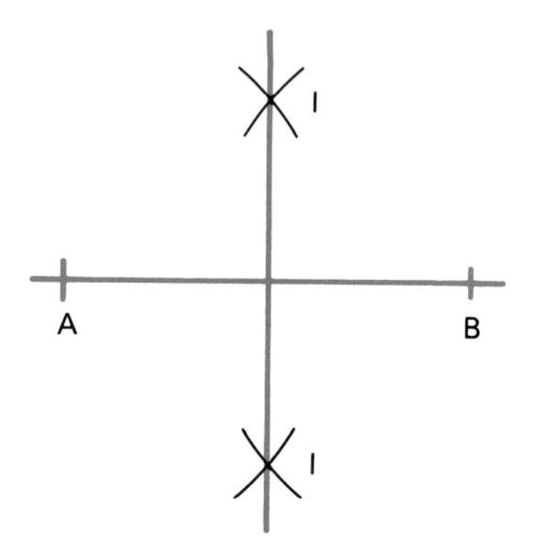

Fig. 10-2. A compass can be used to bisect the line (AB) as shown.

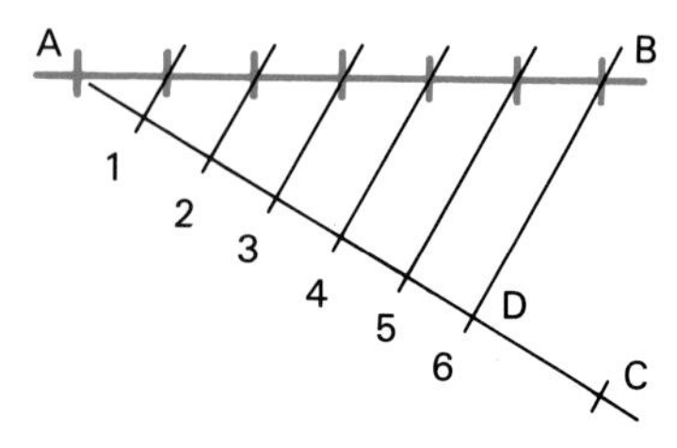

Fig. 10-3. The line (AB) is being divided into six equal parts.

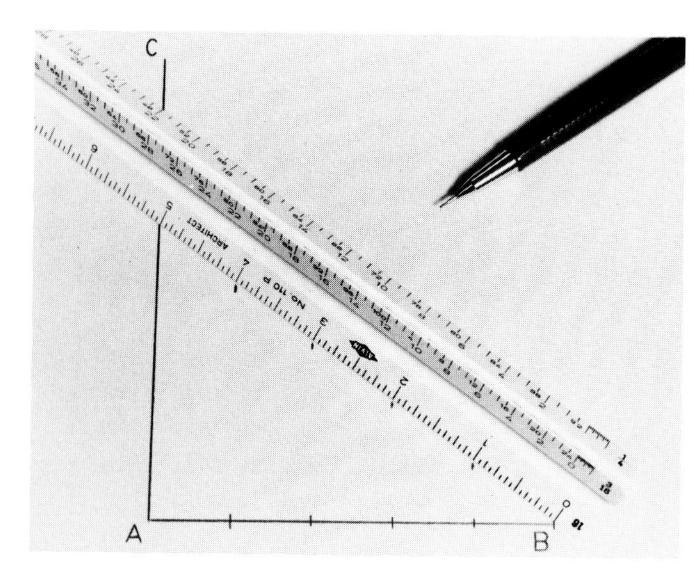

Fig. 10-4. The line (AB) is being divided into five equal parts by first drawing a line at 90° to it and then using a scale.

3. Mark the divisions against the scale and draw 90° lines to (AB) in order to equally divide the line.

ANGLES

When two lines meet at a point, an **angle** is formed. The intersection of these lines is called the **vertex**. Angles are measured in degrees. For greater precision, a degree can be divided into 60 minutes ('), and each minute can be divided into 60 seconds ("). An example would be 35°10'30". For most purposes, degrees will be the smallest division used.

As shown in Fig. 10-5, a 90° angle formed by two perpendicular lines is called a **right angle**. Angles less than 90° are called **acute angles**, and angles greater than 90° are called **obtuse angles**. Both types of angles are shown in Fig. 10-6. An angle can be divided in half (bisected) as shown in Fig. 10-7.

1. Mark equal distances (C) on (VA) and (VB) using a compass.
2. Using (C) as the center, swing two arcs with a compass that intersect at (I).
3. Draw the bisector from (V) to (I).

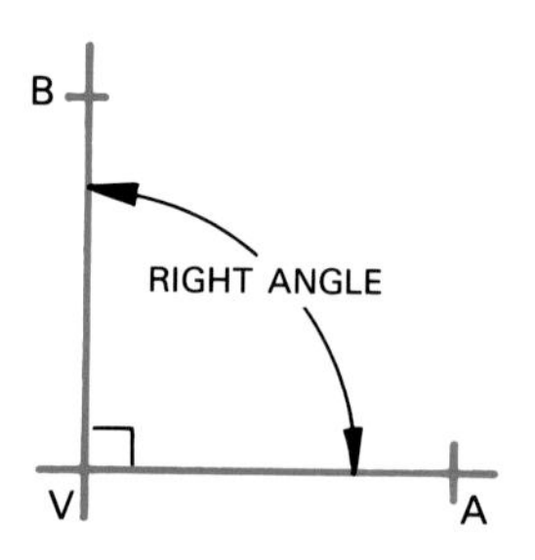

Fig. 10-5. A right angle is formed when two lines meet at 90°.

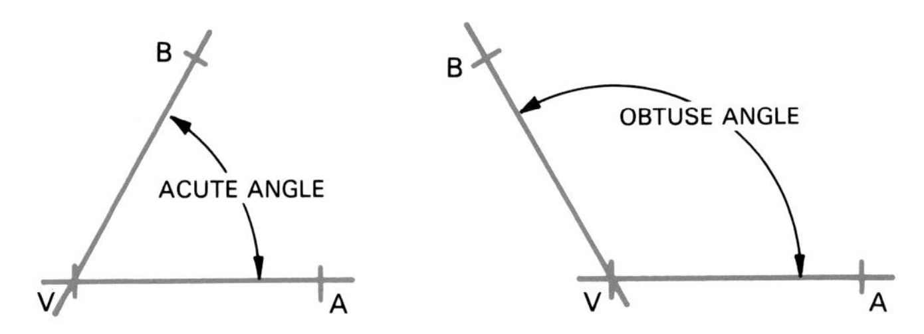

Fig. 10-6. Acute angles are less than 90°, and obtuse angles are more than 90°.

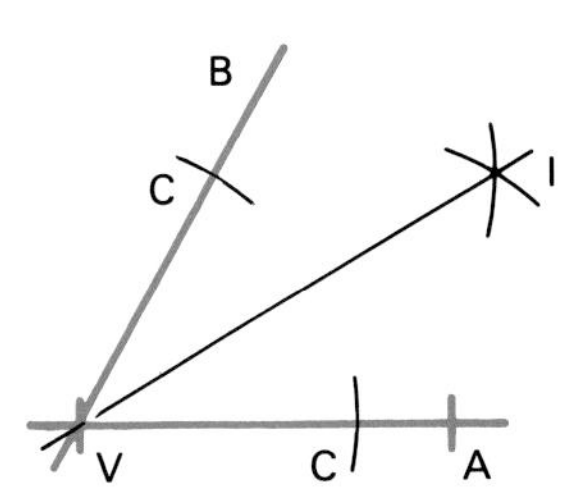

Fig. 10-7. A compass can be used for bisecting an angle.

SHAPES

A **shape** is an area enclosed by straight or curved lines. Other names for shapes are "surfaces" and "planes." Shapes have two dimensions, height and width. They have no depth.

Common types of shapes include:

- Circles and arcs.
- Ellipses.
- Polygons.

CIRCLES AND ARCS

A **circle** is a 360° curve equally distant from a center. As shown in Fig. 10-8, a circle has a *diameter* that passes through the center of the circle. The *radius* is half the diameter. The *circumference* is the distance around a circle. A *quadrant* is one fourth of a circle. An **arc** is a part of the circle circumference.

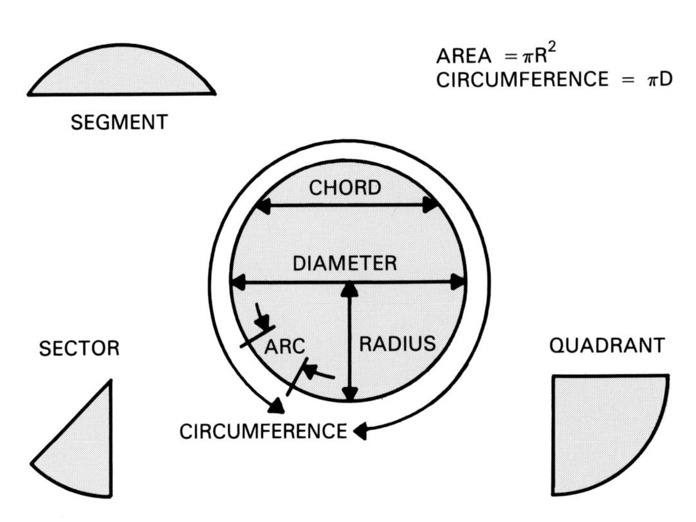

Fig. 10-8. These are important terms and formulas often used when working with circles.

ELLIPSES

An **ellipse** is an elongated circle with opposite arcs that are equal. Since it is elongated, an ellipse has a major diameter (largest) and a minor diameter (smallest), Fig. 10-9. When a circle is tilted to the line of sight, an ellipse is formed, as shown in Fig. 10-10.

An ellipse can be drawn using the trammel method as shown in Fig. 10-11.

1. Draw the major axis and minor axis.
2. On the edge of a piece of paper (trammel), mark half the length of the major axis of the ellipse on the edge as (BE).

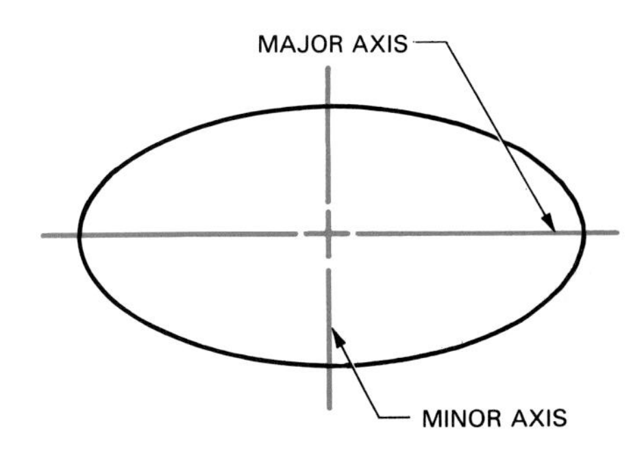

Fig. 10-9. Every ellipse has a major axis (diameter) and minor axis.

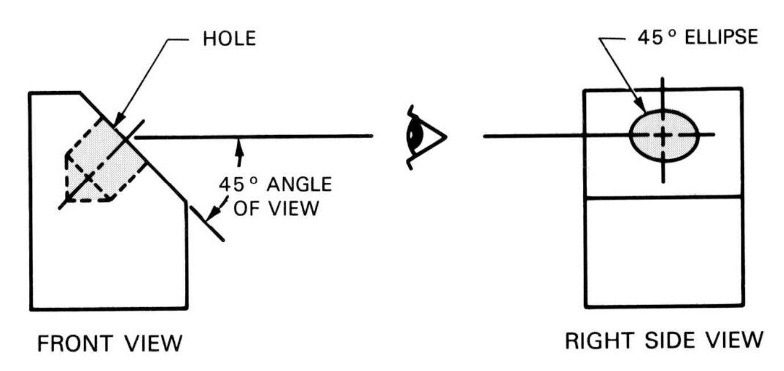

Fig. 10-10. A circle is seen as an ellipse if it is tilted to the line of sight. Try it.

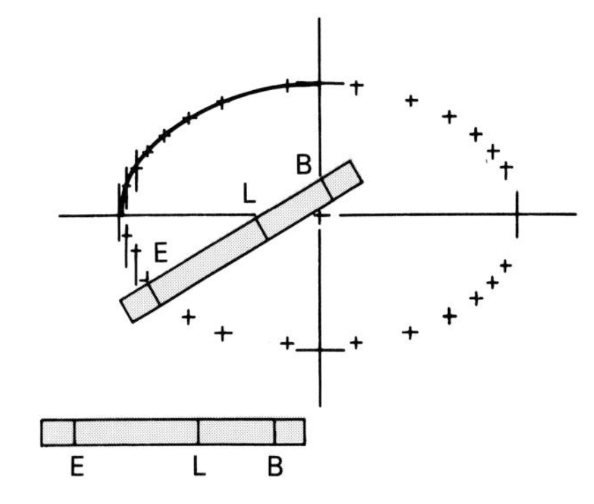

Fig. 10-11. An ellipse can be drawn using the trammel method.

3. Measuring from (E), mark (L) at half the minor axis.
4. With (B) on the minor axis and (L) on the major axis, mark points at (E) along the ellipse.
5. Darken in each quadrant with the same part of an irregular curve.

Isometric ellipse

In isometric drawing, circles are also drawn as ellipses. If they are on a vertical or horizontal surface, they are 30° ellipses. See Fig. 10-12. The major axis is tilted differently for each plane.

An approximate isometric ellipse can be drawn as illustrated in Fig. 10-13.

1. Draw an isometric square at 30° as shown, with each side being equal to the circle diameter.
2. From each obtuse angle, draw two lines that bisect the opposite sides. These are tangent points (T).
3. Using the obtuse angle vertex (0) as the center, set the compass lead at the tangent point and swing a large arc to the other tangent point. Repeat on the opposite side.
4. Using the intersecting bisectors (I) as the center, swing the small arcs at each end to complete the ellipse.

Ellipse templates

A template simplifies the construction of ellipses. To use the template, you must first know the circle diameter and ellipse angle. For example, in multiview drawing, a hole on a sloping surface will appear as an ellipse. As shown in Fig. 10-14, the angle of view is the ellipse angle.

To use the template, first draw the major and minor axis. Find the correct ellipse and align it on the axes.

An isometric ellipse template is used for isometric ellipses. Draw isometric center lines for aligning the template as shown in Fig. 10-15.

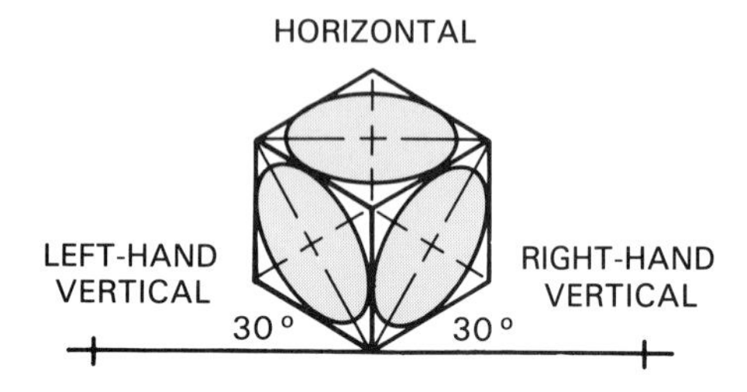

Fig. 10-12. Circles on an isometric cube are drawn as ellipses. The major axis of the ellipse is tilted differently for each plane as shown.

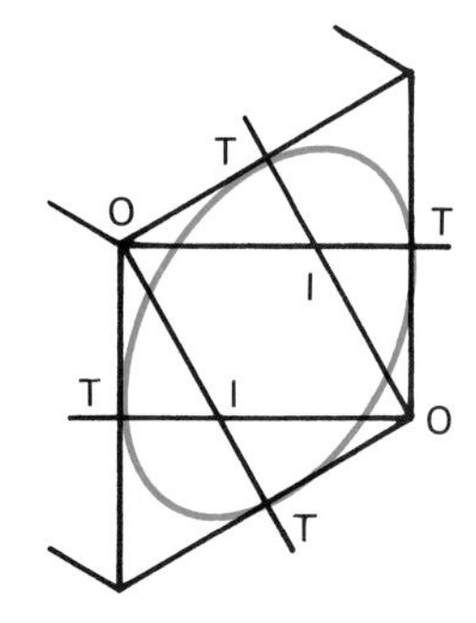

Fig. 10-13. An approximate ellipse can be drawn by using a compass. The ellipse is constructed by drawing two large arcs and two small arcs.

POLYGONS

A **polygon** is a shape made from straight lines. At least three lines are required to make a polygon. Common polygons include:

- Triangles.
- Quadrilaterals.
- Regular polygons.
- Hexagons.
- Octagons.

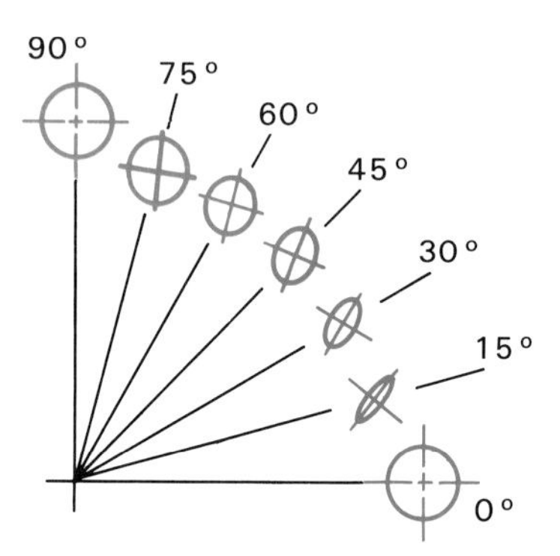

Fig. 10-14. The ellipse angle is determined by the angle at which a circle is tilted to the viewer.

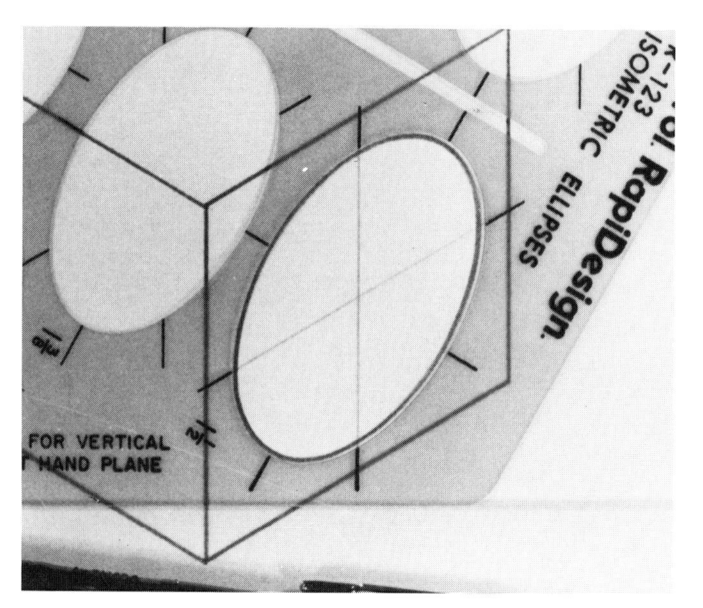

Fig. 10-15. Isometric center lines on a drawing are used for aligning the isometric ellipse template.

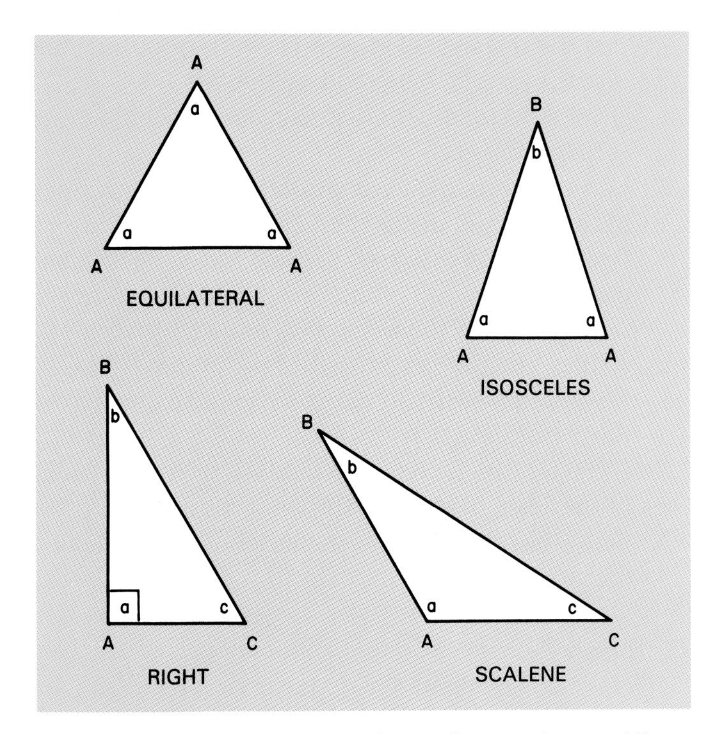

Fig. 10-16. Various kinds of triangles are shown. All triangles have inside angles that total 180°.

Triangles

Triangles are three-sided polygons. Various kinds of triangles are shown in Fig. 10-16. The inside angles of a triangle equal 180°.

Quadrilaterals

Quadrilaterals are four-sided polygons. Several types of quadrilaterals are shown in Fig. 10-17. If all angles equal 90°, the quadrilateral is a *square* or *rectangle*. A *rhombus* has equal sides like a square, but the sides are not 90° to each other. A *rhomboid* has opposite sides that are equal like a rectangle, but the sides do not form a right angle (90°). A *trapezoid* is formed when two sides are parallel, but the other two sides are not parallel.

Regular polygons

Regular polygons are shapes with equal sides and angles. The equilateral triangle and square are examples of regular polygons. Other regular polygons are shown in Fig. 10-18. Regular polygons with more than

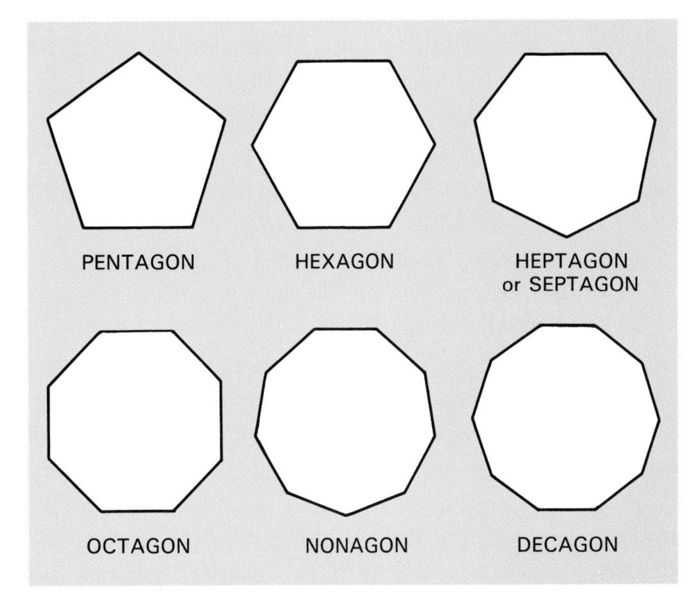

Fig. 10-18. Regular polygons are shapes with equal sides and angles.

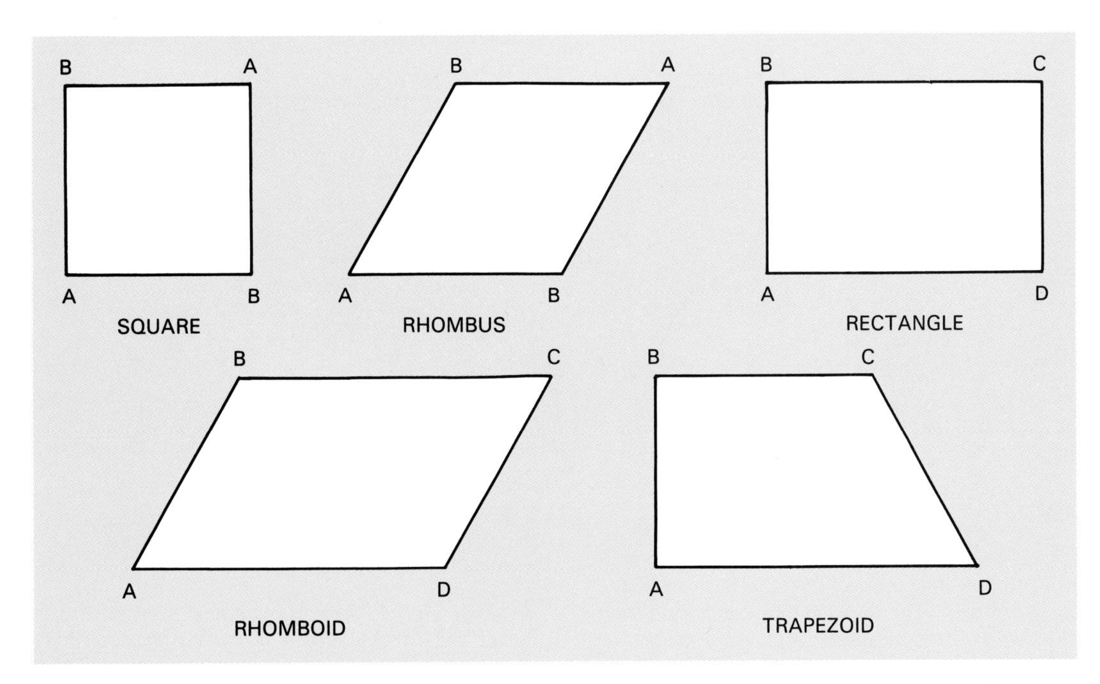

Fig. 10-17. Various types of quadrilaterals are shown. A quadrilateral is a four-sided polygon.

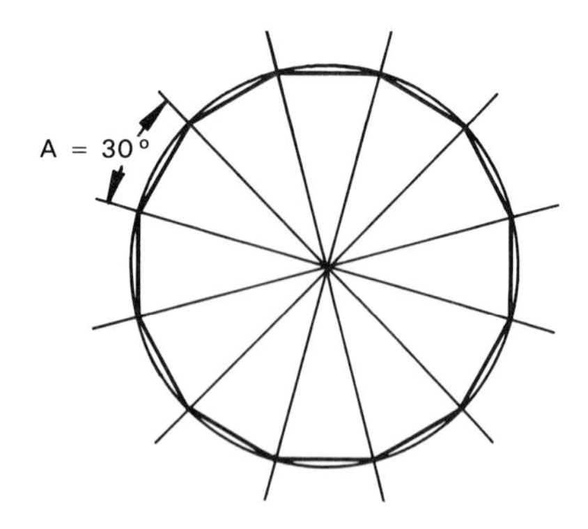

Fig. 10-19. A circle can be used to draw a regular polygon. Begin by dividing the circle into the same number of angles as the number of sides on the polygon.

four sides are often drawn inside a circle (*inscribed*) or outside a circle (*circumscribed*).

As shown in Fig. 10-19, a regular polygon with any number of sides can be drawn inside a circle.

1. Divide 360° by the number of sides of the polygon to find angle (A).
2. Draw a circle the desired size.
3. Using a protractor, divide the circumference into equal arcs using angle (A).
4. Connect the ends of each arc with a straight line to form the polygon.

When a regular polygon has an even number of sides (4,6,8,10, etc.), it can be drawn using an *"across corners"* or *"across flats"* distance, Fig. 10-20. This refers to whether the size of the polygon is known from opposite sides or corners. When "across corners" is used, the polygon is inscribed inside the circle. When "across flats" is used, the polygon is circumscribed (outside the circle).

Hexagons

Hexagons are six-sided polygons that can be easily drawn with a 30°-60° triangle. This is illustrated below in Fig. 10-20.

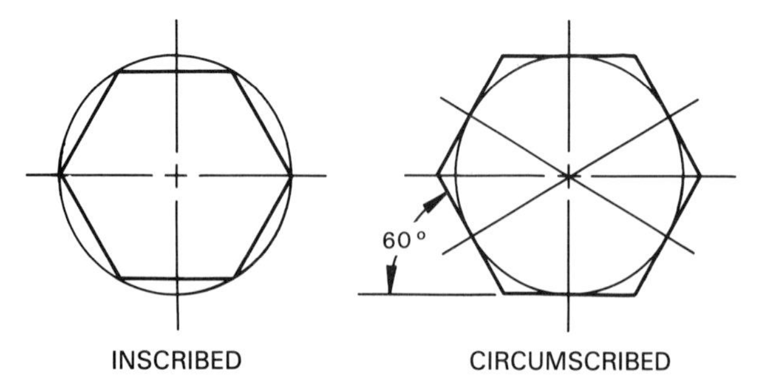

Fig. 10-20. Regular polygons with an even number of sides can be drawn inside a circle (inscribed) or outside a circle (circumscribed). The method used depends on whether the size of the polygon is known between corners or sides. Hexagons are shown being drawn by each method.

When the distance is known between opposite corners, use the inscribed method as follows:

1. Draw a circle with the diameter equal to the "across corners" distance.
2. Use a 60° angle (left and right) to draw four sides of the hexagon inside the circle.
3. Draw the other two sides inside the circle as horizontal lines.

When the size of the hexagon is known between opposite sides, use the circumscribed method as follows:

1. Draw a circle with the diameter equal to the "across flats" distance.
2. Use a 60° angle (left and right) to draw four sides of the hexagon outside the circle.
3. Draw the other two sides as horizontal lines outside the circle.

Octagons

Octagons are eight-sided polygons. They can be drawn similarly to the hexagon as shown in Fig. 10-21. A 45° angle can be used to produce octagons.

TANGENT POINTS

A **tangent point** is the exact point where a line or arc joins another arc. Tangents are often used in technical drawing. It is important to learn how to make accurate tangents.

Whenever an arc needs to be drawn tangent to other lines and arcs, a circle template can be used. Since the correct size is not always available on the template, it is also helpful to be able to use the compass method.

LINE TANGENT TO A CIRCLE

The simplest tangent is a line from a point tangent to a circle or arc. This tangent can be drawn by eye as shown in Fig. 10-22.

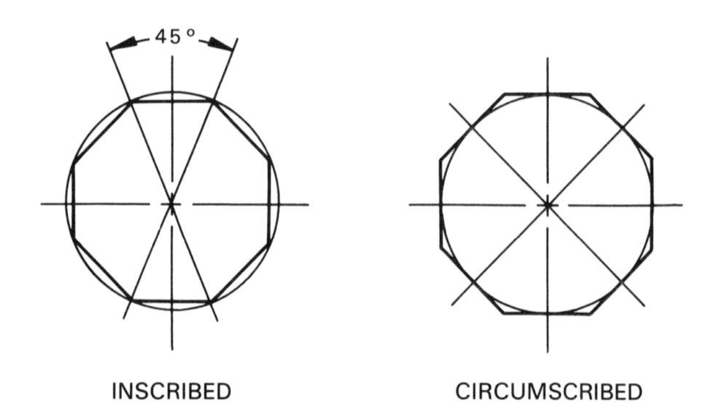

Fig. 10-21. An octagon is drawn with the inscribed method if the distance is known across flats (sides). It is drawn with the circumscribed method if the distance is known across corners.

ARC TANGENT TO TWO LINES AT 90°

Producing an arc tangent to two lines at 90° is used for rounding the corners of an object with sides at right angles. A circle template can be used as shown in Fig. 10-23. The steps for using a compass are illustrated in Fig. 10-24.

1. Measure the arc radius (T) on both lines from the point where they intersect. A compass can also be used for this.
2. Swing two intersecting arcs from (T) to locate the tangent arc center (C).
3. Draw the arc to each tangent point (T). Use a compass for this step.

ARC TANGENT TO TWO LINES NOT AT 90°

Producing an arc tangent to two lines not at 90° is shown in Fig. 10-25 and Fig. 10-26 for two lines meeting at an obtuse and acute angle.

1. Swing an arc (R) inside each line equal to the radius of the arc that will connect the two lines.
2. Draw parallel lines (P) tangent to these arcs which intersect at (I).
3. Find the tangent points (T) on the lines by drawing right angle lines from (I).
4. Using (I) as the center, swing the connecting arc to each tangent point (T).

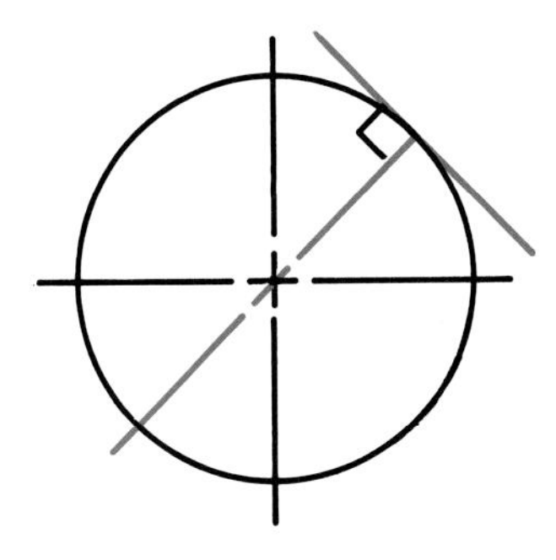

Fig. 10-22. A line can be drawn tangent to a circle by eye. A line tangent to a circle will be at 90° to a line drawn from the center of the circle as shown.

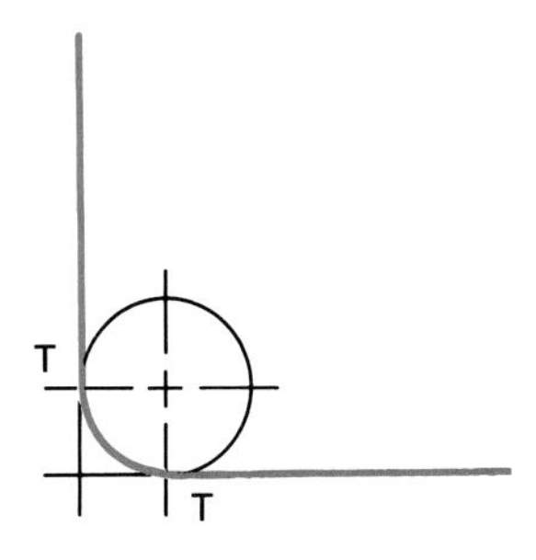

Fig. 10-23. An arc can be drawn tangent to two lines at 90° to each other using a circle template.

ARC TANGENT TO A LINE AND ARC OR CIRCLE

Producing an arc tangent to a line and arc or circle is illustrated in Fig. 10-27.

1. Draw a parallel line to (AB) using, as a measurement, the radius (R) of the arc that will connect the line and arc or circle.
2. Swing an arc from the center of the circle or arc equal to the sum of the circle radius (C) and the connecting arc radius (R).

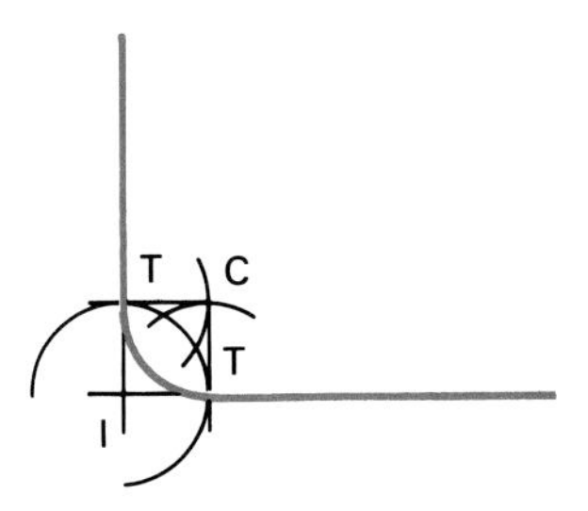

Fig. 10-24. A compass can be used to draw an arc tangent to two lines at 90° to each other.

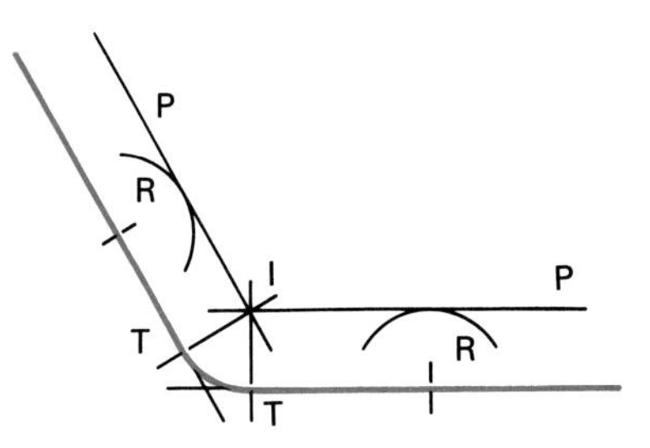

Fig. 10-25. An arc that connects two lines forming an obtuse angle can be drawn as shown.

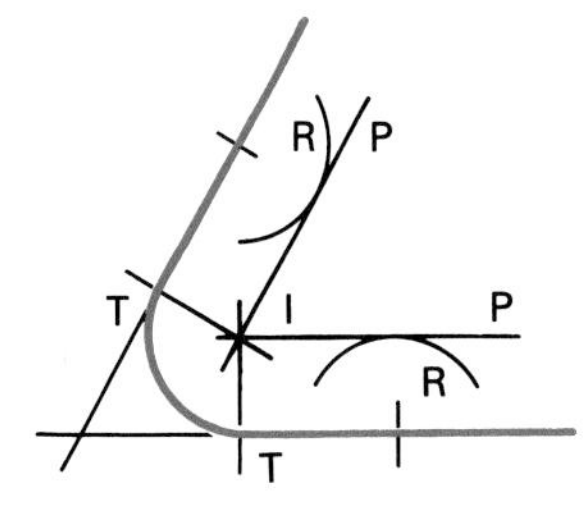

Fig. 10-26. This method is used to draw an arc that connects two lines meeting at an acute angle.

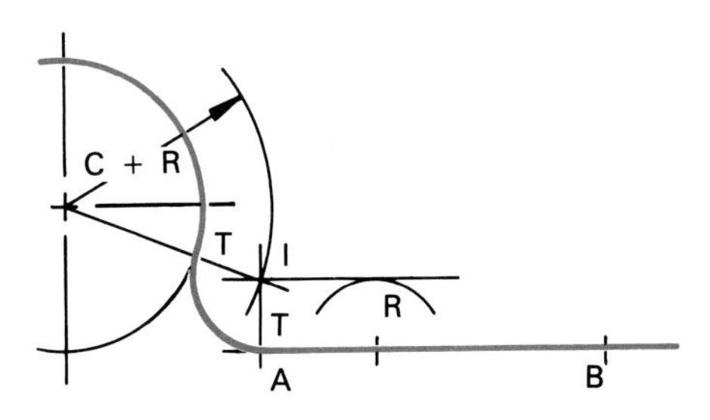

Fig. 10-27. An arc is drawn that connects the larger arc and line.

3. From the intersection (I), find one tangent point (T) by drawing a line to the center of the arc. Find the other tangent (T) by drawing a line from (I) at a right angle to (AB).
4. Use the intersection (I) as a center to swing an arc tangent to the line and arc.

ARC TANGENT TO TWO ARCS OR CIRCLES

The compass technique for producing an arc tangent to two arcs or circles is shown in Fig. 10-28.

1. Swing an arc from the center of one circle with a radius equal to the sum of the circle radius (A) and connecting arc (R).
2. Follow the same procedure with the second circle using a radius equal to the circle radius (B) and connecting arc (R). The two arcs should intersect at (I).
3. From (I), draw a line to the center of each circle to find tangent points (T) on the circumference.
4. Using (I) as a center, swing an arc with radius (R) tangent to the two circles or arcs.

OGEE CURVE TANGENT TO PARALLEL LINES

An **ogee curve** is a double, or reverse curve. The ogee curve can be used to connect two parallel lines as shown in Fig. 10-29.

1. Draw a line (AB) that intersects the ends of the parallel lines to be joined.
2. Select a point (P) on (AB) where the connecting arcs will meet. An equal ogee curve will be formed if (P) divides (AB) in half.
3. Using a compass, bisect (AP) and (BP).
4. Draw a 90° line at (A) that intersects the nearest bisector at (I). Repeat at (B).
5. Using (IA) as the radius, swing the first arc to (P). Repeat for (IB) to finish the ogee curve.

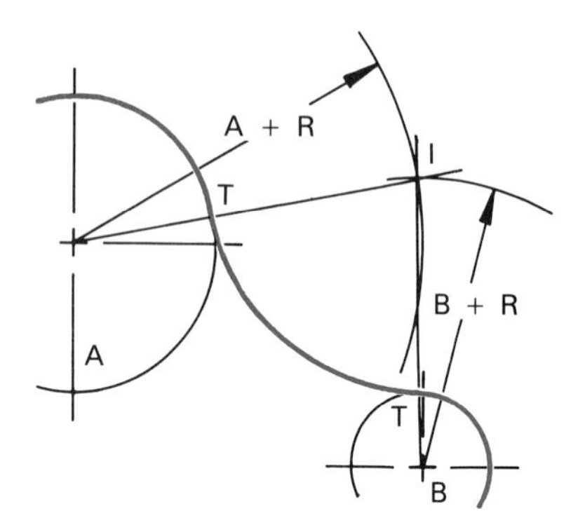

Fig. 10-28. Using a compass, an arc is drawn tangent to two other arcs or circles.

SOLIDS

Solids are three-dimensional objects. They have height, width, and depth. Some of the common solids are shown in Fig. 10-30. A *prism* has two parallel sides

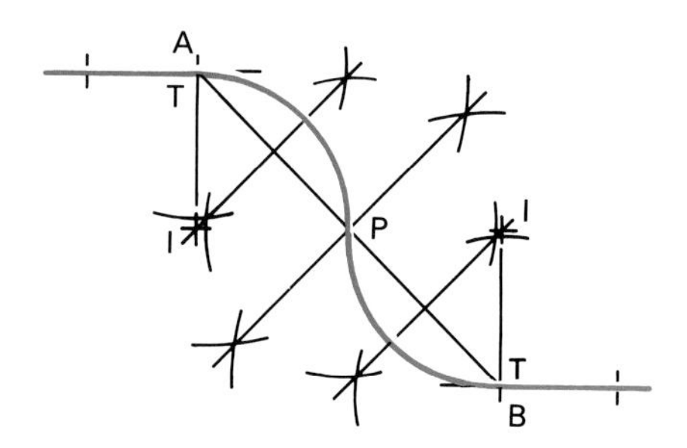

Fig. 10-29. An ogee curve is drawn that connects two parallel lines.

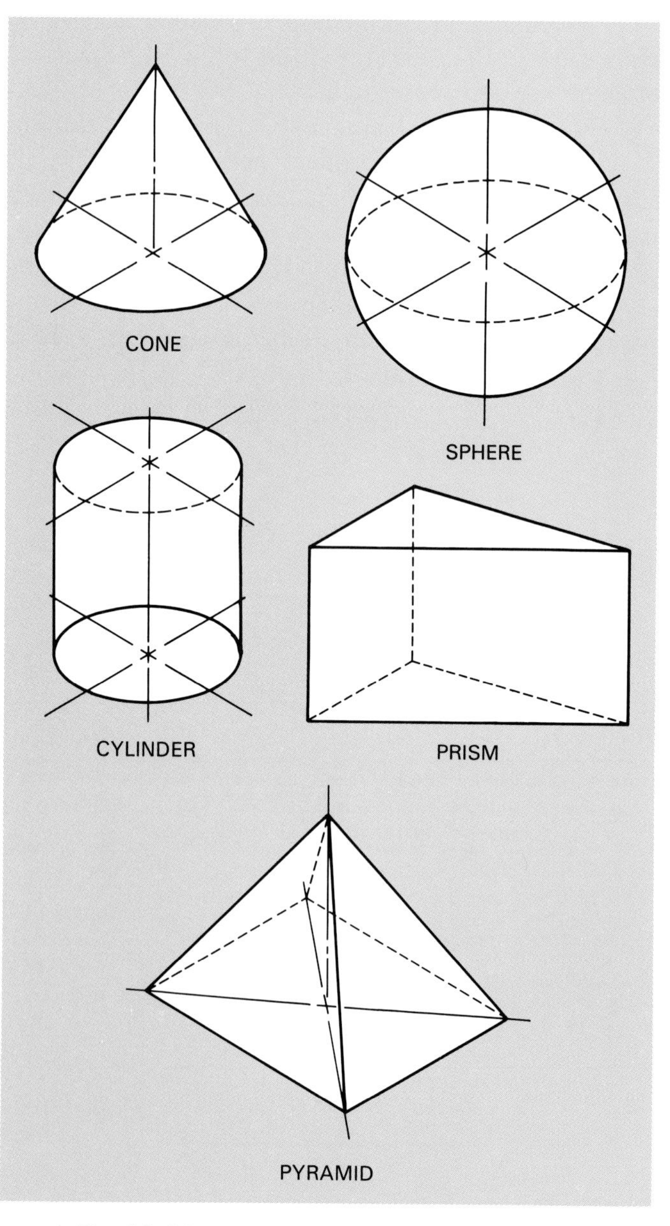

Fig. 10-30. Some common solids are shown.

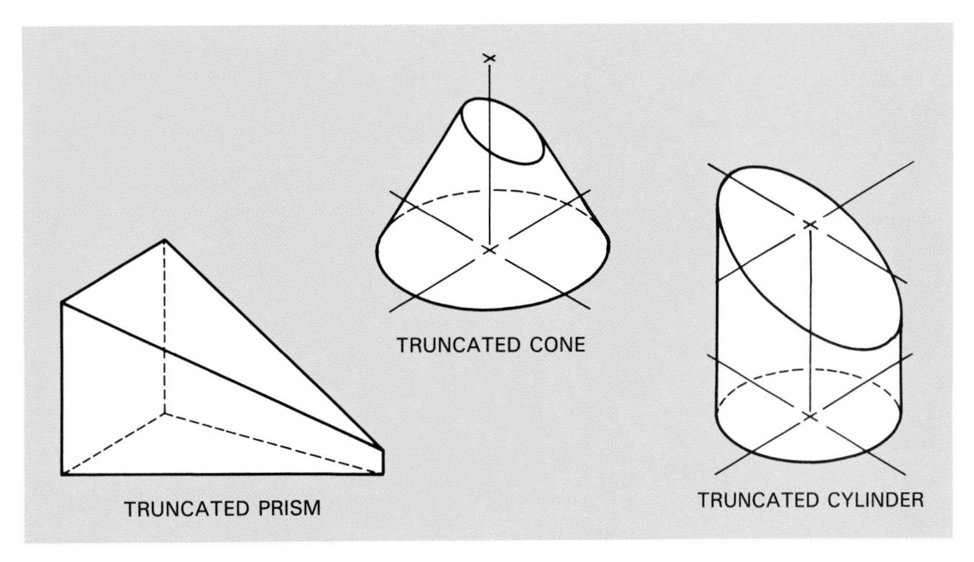

Fig. 10-31. A truncated solid has the end removed so it is not parallel to the base. Several examples are shown.

and three or more faces joining the sides. A *pyramid* has a polygon at one end, a point (vertex) at the other end, and faces connecting them.

A *sphere* is a round solid, like a ball. A *cylinder* is formed by a line revolving around a central axis. A *cone* is like a pyramid, except one end is a circle instead of a polygon.

When the end of a solid is removed so the resulting face is not parallel to the base, the solid is called **truncated**, Fig. 10-31. When solids, such as pyramids and cones, have the end removed parallel to the base, the remaining solid is called a **frustum**, Fig. 10-32.

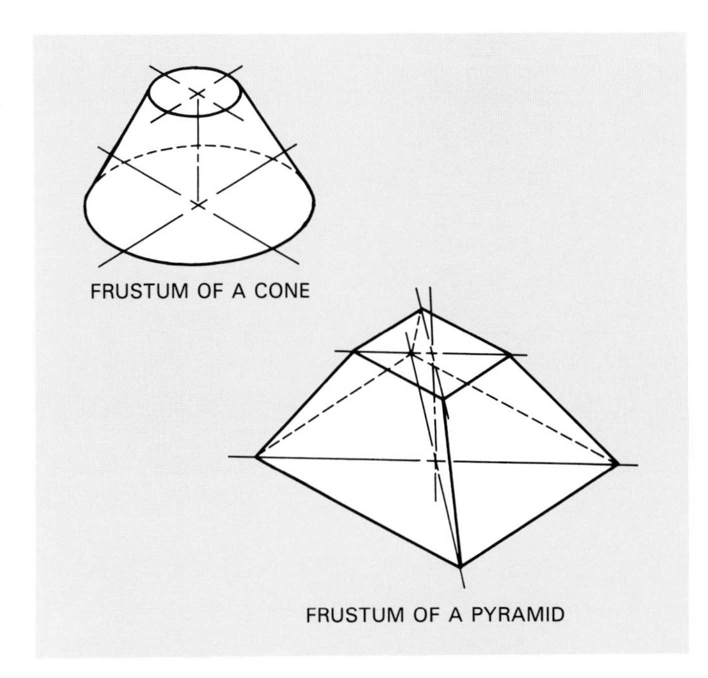

Fig. 10-32. A solid that is created by removing the end parallel to the base is called a frustum. Several examples are shown.

SUMMARY

In this chapter, you learned techniques for using geometry. Although it is difficult to memorize all the geometric construction methods, you should be able to find the construction method you need for your drawing. You should then be able to follow the directions for completing the drawing and solving drawing problems.

WORDS TO KNOW

All of the following words have been used in this chapter. Do you know their meanings?

acute angles
angle
arc
bisecting
circle
ellipse
frustum
geometry
hexagons
obtuse angles
octagons
ogee curve
polygon
quadrilaterals
regular polygons
right angle
shape
solids
tangent point
triangles
truncated
vertex

REVIEWING YOUR KNOWLEDGE

Please do not write in this text. Write your answers on a separate sheet.

1. Geometry is a branch of mathematics dealing with:
   a. Lines.
   b. Shapes.
   c. Solids.
   d. All of the above.
2. True or false? A line can be bisected using a compass method.
3. True or false? Acute angles are angles greater than 90°.
4. Name three common types of shapes.
5. Which of the following is the distance around a circle?
   a. Diameter.
   b. Circumference.
   c. Radius.
   d. Quadrant.
6. When a circle is tilted to the line of sight, an __________ is formed.
7. Name five polygons.
8. A __________ point is the exact point where a line or arc joins another arc.
9. True or false? When pyramids and cones have the vertex end removed parallel to the base, the solid is called a frustum.
10. Explain how solids are different than shapes.

APPLYING YOUR KNOWLEDGE

1. List various objects you see. Then identify the shapes and solids needed to produce them.
2. Draw an isometric cube with a hole on each visible plane. Use the geometric construction method for making the ellipse. Draw another cube and use the template method for drawing the ellipses. Compare the two methods.
3. Create a display containing shapes and solids. Label each of them.

11

CHAPTER

MULTIVIEW DRAWINGS

After studying this chapter, you will be able to:

- *Identify various types of orthographic drawings.*
- *Describe multiview drawings and explain how they are produced.*
- *Determine the best combination of views to use to communicate information about an object.*
- *Explain how to draw hidden and center lines.*
- *Demonstrate how to produce a multiview drawing correctly.*

When you shop for clothes, you may stand in front of a three-way mirror. You want to see multiple views of an outfit to see how it looks on you, Fig. 11-1. If you stood in front of a single mirror, the view would be limited. Like a three-way mirror, a **multiview drawing** gives you another view or a combination of views of an object.

Fig. 11-1. Like a three-way mirror, a multiview drawing shows different views of a person or an object.

Hold an object in your hand. Notice that it has three dimensions. These are width, height, and depth. Pictorial drawings, like a photograph, show these three dimensions. However, it isn't easy to make an object from a pictorial drawing. This is because pictorials do not show the true size and shape of an object. For instance, compare the pictorial of an object and the actual shape of one side in Fig. 11-2. Multiview drawings show true shape and size and so they are often used for objects that are produced in industry.

In a multiview drawing, an object is drawn as a combination of views. Each view of an object shows one side of the object or two dimensions. Since objects have three dimensions (height, width, and depth), more than one view is usually needed to fully describe an object. See Fig. 11-3.

ORTHOGRAPHIC PROJECTION

Multiview drawings are produced through **orthographic projection**. The orthographic projection technique involves showing each view of an object in a

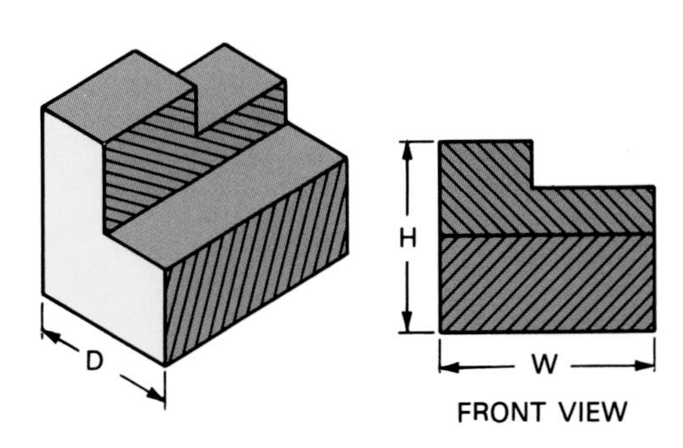

Fig. 11-2. All sides of the pictorial drawing (left) are drawn distorted. The orthographic view (right) shows the true shape and size of one side of the object.

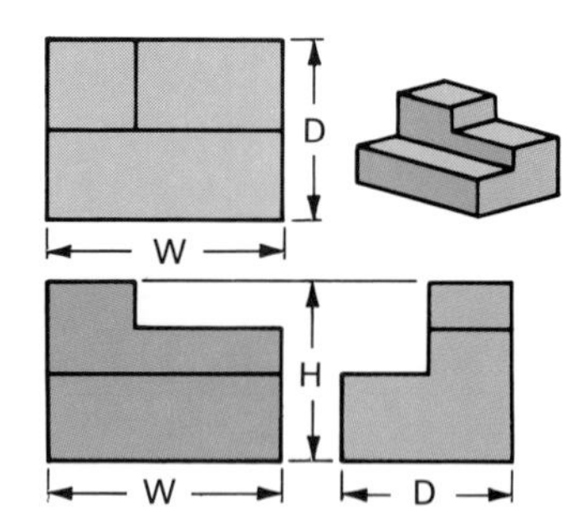

Fig. 11-3. Since only two dimensions are shown in each view of an object, several views are often needed to fully describe the object.

specific location on a drawing in relation to the other views. (Orthographic means right angle drawing.)

To better understand orthographic projection, pretend an imaginary glass box is placed around the object to be drawn, Fig. 11-4. A *projector*, like a laser beam, transfers each point on the object to the side of the glass box. These projectors are at right angles to the glass box. Each side of the box is known as a *projection plane*, much like a movie screen. In this way, each side of the object is projected to the side of the glass box and flattened into two dimensions.

When all the views are projected onto the sides of the glass box, it can be unfolded. See Fig. 11-5. Notice that the views are aligned with each other. Just as there are six sides of the glass box, there are six *normal* or *principal* views of an object.

FRONT VIEW

How do you decide which view is the front view? Often, the front view provides the best understanding of an object. The front view should be the most descriptive view. Once you decide on the front view, you can decide on the number of views needed. The front view is always shown on multiview drawings.

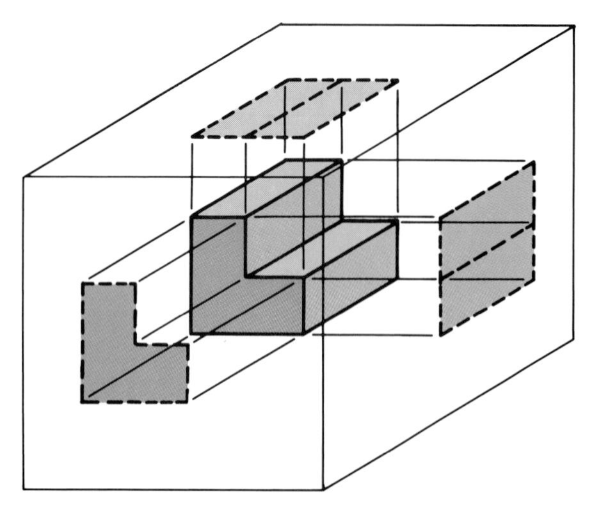

Fig. 11-4. Three views of this object have been projected onto projection planes. Each view is shown as a dashed line on the projection plane.

OTHER VIEWS

How many views are necessary? In multiview drawing, the fewest number of views that will fully describe the object should be drawn. The views you should choose are the ones that best describe the object and have the least number of details hidden from the eye, Fig. 11-6.

The views of an object that are most often used are the front, right side, and top views, Fig. 11-7. Because of the position of the projector planes, the top is also known as a **horizontal view**, and the right side is a **profile view**.

For some objects, one or two views are satisfactory, as shown in Fig. 11-8 and Fig. 11-9. On the other hand, more than three views can also be used to describe an object.

HIDDEN LINES

Since each view of a multiview drawing shows only two dimensions, **hidden lines** are used to help others

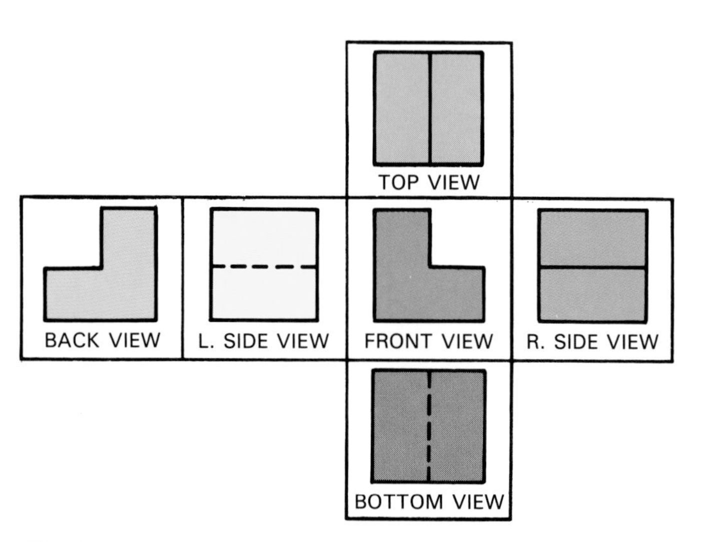

Fig. 11-5. These are the six principal or normal views of an object. Notice how the views are aligned with each other.

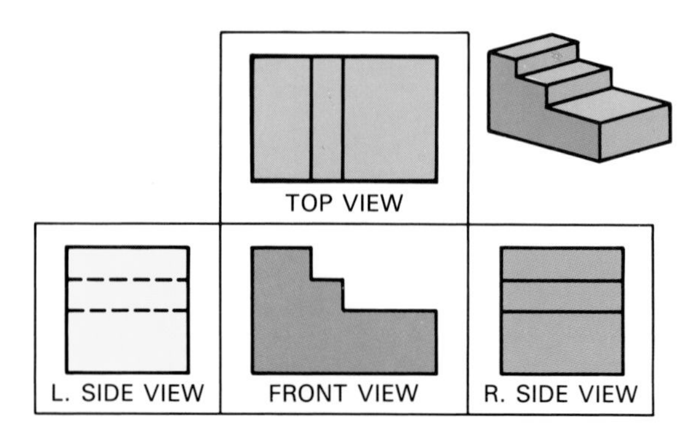

Fig. 11-6. Notice how the front view best describes the object and has the least number of hidden details.

better understand the views of an object. These lines show edges that are hidden from view inside an object.

Hidden lines are drawn as thin (0.5 mm) lines. The lines are made by drawing 3 mm (1/8 in.) dashes about 1.5 mm (1/16 in.) apart, as shown in Fig. 11-10. The length of the dash can vary depending upon the size of the drawing.

Hidden line rules are described here and illustrated in Fig. 11-11 and Fig. 11-12.

- Hidden lines touch a visible line (A) unless they appear to extend the line (B) or they are passing under a visible line (C).
- Hidden lines that form corners touch (D). However, hidden lines that cross (two different features) do not touch (E).
- Hidden arcs begin and end with a dash at the tangent points (F) unless they are tangent to a visible line (G).
- Parallel hidden lines are staggered like bricks (H).
- The order of importance for lines is visible, hidden, and center lines (I).

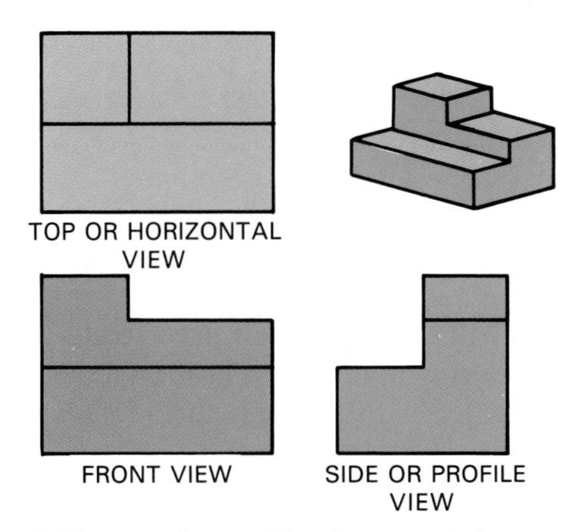

Fig. 11-7. Three orthographic views are being used to describe this object. The name for each view is shown.

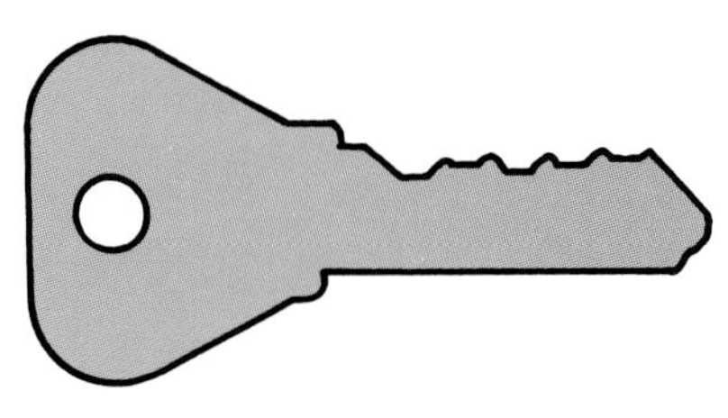

1/8'' THICK; BRASS

Fig. 11-8. A single view is satisfactory for describing this object. A note may be used to indicate thickness.

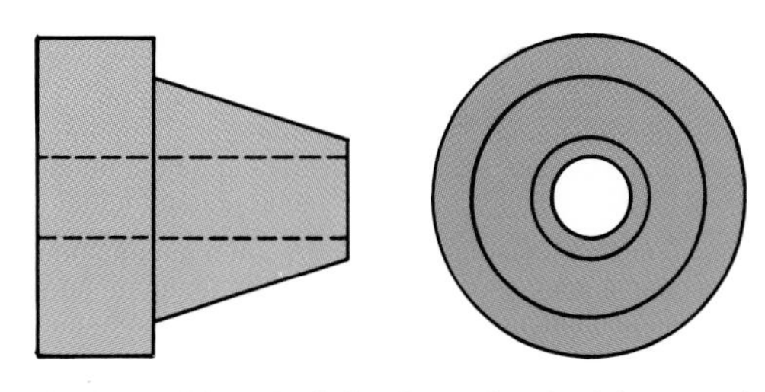

Fig. 11-9. This object is fully described with two views. If the top view were shown, it would be the same as the front view.

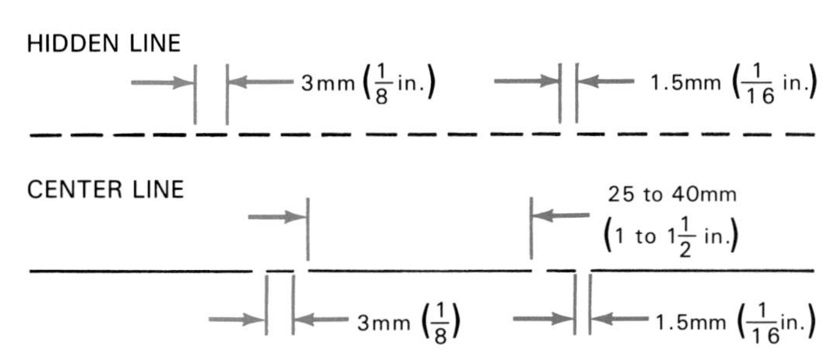

Fig. 11-10. The correct dimensions for hidden and center lines are shown.

CENTER LINES

Center lines are used to show the center of round features on a multiview drawing. Center lines are thin (0.5 mm) lines. They are produced by drawing a long dash and a short dash by eye. The short dash is about 3 mm (1/8 in.) and the long dash is about 25 to 40 mm (1 to 1 1/2 in.), as shown in Fig. 10-10. Dash length can vary depending on the size of the drawing.

Rules for using center lines are listed below and illustrated in Fig. 11-13.

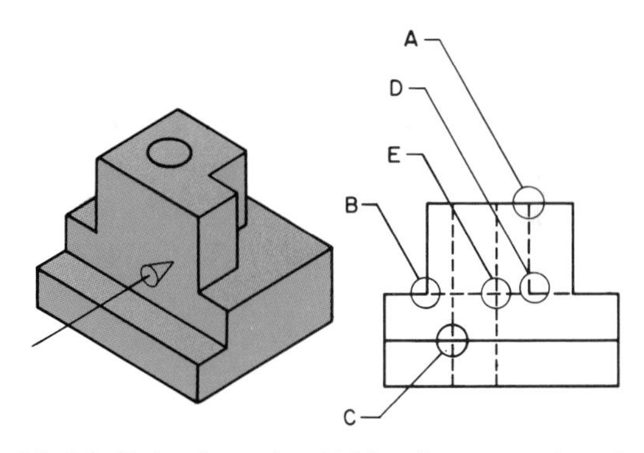

Fig. 11-11. Rules for using hidden lines on a drawing are shown. The arrow indicates how the object is being viewed.

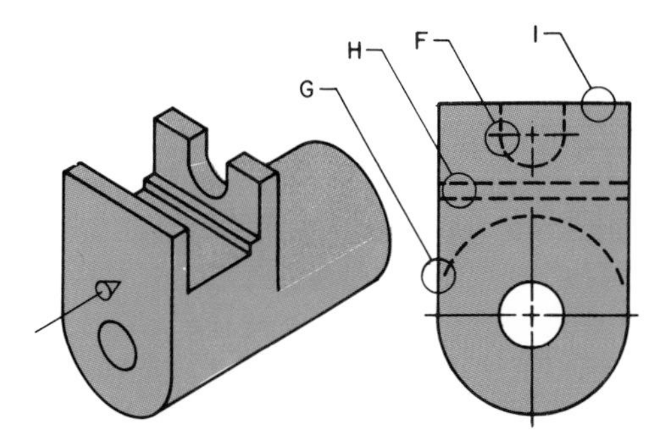

Fig. 11-12. Additional rules for using hidden lines on a drawing are shown. The arrow indicates the line of sight.

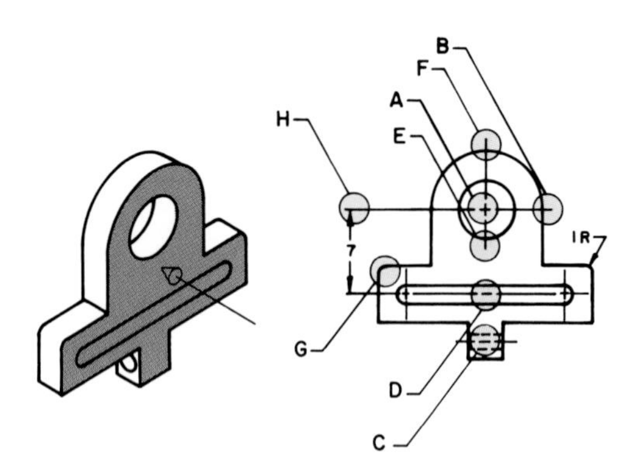

Fig. 11-13. The correct use of center lines is shown in the orthographic view of the object.

- When a view shows circular features in round form, two center lines are placed at right angles (A). The short dashes cross in the center.
- Center lines extend about 3 mm (1/8 in.) beyond the last round feature (B).
- When a side (longitudinal) view of a round feature such as a hole is shown, a single center line passes through the center (C). The short dash is centered on the line.
- For long center lines, a short dash is inserted every 25-40 mm (1 - 1 1/2 in.) on the line (D).
- Center lines extend about 3 mm (1/8 in.) beyond the last round feature, and so they may stop inside the object (E).
- A center line can be used for more than one round feature if the features are concentric (F).
- Center lines are not used for small arcs (G).
- Center lines can be used as extension lines (H).

PRODUCING THE DRAWING

When producing a multiview drawing, follow these guidelines:

1. Lightly sketch the views. Make the sketch on grid paper if it is available.
2. Center the drawing in the drawing space provided. For example, in a three-view drawing, two views have to be centered horizontally and two views have to be centered vertically. To center vertically, add the vertical dimensions of the views (height and depth) and subtract this amount from the available space. Divide by three for the top, middle, and bottom spacing. Follow the same procedure for centering horizontally. In this case, add the horizontal dimensions of the two views (width and depth).If dimensions are to be placed on the drawing, then spacing may be slightly different. Make sure that space is adequate between the views for dimension lines.
3. Block in the views with construction lines using the overall width, height, and depth. Make sure the views are aligned and in correct position. Some dimensions can be projected from view to view by extending construction lines.
4. Lay out the rest of the drawing detail. To save time, transfer distances from one view to another by using dividers.
5. Finish the drawing by darkening the lines. Start with circles and arcs first.

UNDERSTANDING OTHER ORTHOGRAPHIC DRAWINGS

A **section view** is sometimes used to show internal features. A **cutting plane line** is used to show where the section is located. A full section and a revolved section are shown in Fig. 11-14. Section lines can be drawn as thin, diagonal lines about 2 mm apart. An **auxiliary view** is used when a normal projection plane does not show a surface true size and shape. An example is shown in Fig. 11-15.

An **assembly drawing** shows how parts of an object fit together. See Fig. 11-16. Exploded assemblies are

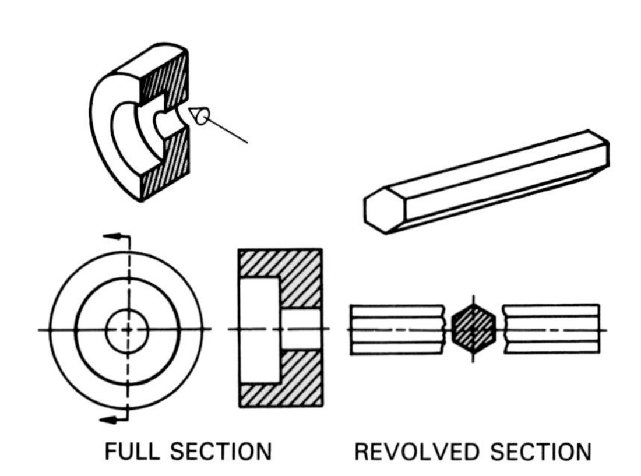

Fig. 11-14. Two types of section views are shown. A cutting plane line is used on the full section view to show where the section is located.

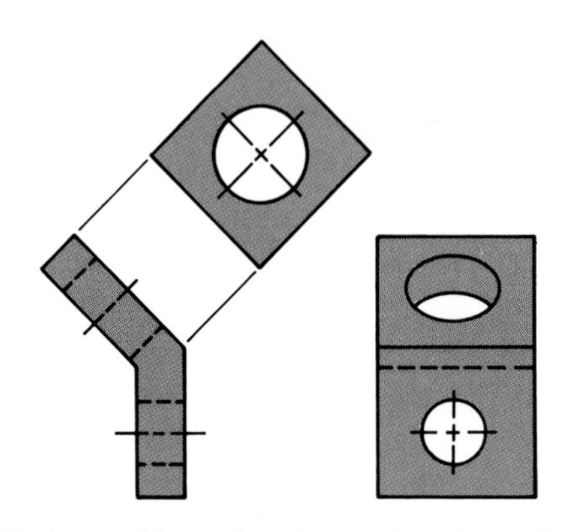

Fig. 11-15. An auxiliary view is used to show the top hole as true size and shape. Notice that the hole is shown as an ellipse in the side view.

Fig. 11-16. This assembly drawing shows how the parts fit together. A sectioned assembly is being used so internal features can be shown.

sometimes used that show the parts separated along a center line.

SUMMARY

Multiview drawings show a combination of views of an object. Various types of lines also have different meanings. By learning how to produce multiview drawings, you can create your own designs for projects and understand designs you find in books and magazines. In the following chapter, you will learn techniques for dimensioning a multiview drawing.

WORDS TO KNOW

All of the following words have been used in this chapter. Do you know their meanings?

assembly drawing
auxiliary view
center lines
cutting plane line
hidden lines
horizontal view
multiview drawing
orthographic projection
profile view
section view

REVIEWING YOUR KNOWLEDGE

Please do not write in this text. Write your answers on a separate sheet.

1. Since they show the true shape and size, _________ drawings are often used for objects that are produced in industry.
2. In a multiview drawing, how many dimensions are shown in each view?
   a. One.
   b. Two.
   c. Three.
   d. Four.
3. True or false? Multiview drawings are produced by using orthographic projection.
4. The front view of a multiview drawing should be the:
   a. Most descriptive view.
   b. First view you see.
   c. Top view.
   d. None of the above.
5. True or false? Because of the position of projection planes, the top is also known as the profile view, and the right side is a horizontal view.
6. The order of importance for lines is:
   a. Center, visible, hidden.
   b. Hidden, center, visible.
   c. Center, hidden, visible.
   d. Visible, hidden, center.
7. What type of lines would you use to show the center of round features on a multiview drawing?
8. Which of the following shows how parts of an object fit together?
   a. Section view.
   b. Auxiliary view.
   c. Assembly view
   d. None of the above.

APPLYING YOUR KNOWLEDGE

1. Make a model of an object from wood or plastic. Paint each orthographic view of the model a different color. Use the model to help in understanding multiview drawing.
2. Ask a manufacturing firm for some outdated multiview drawings and corresponding parts. Study how the drawings were used to produce the parts.
3. Research the history of orthographic drawing. Prepare a written report.

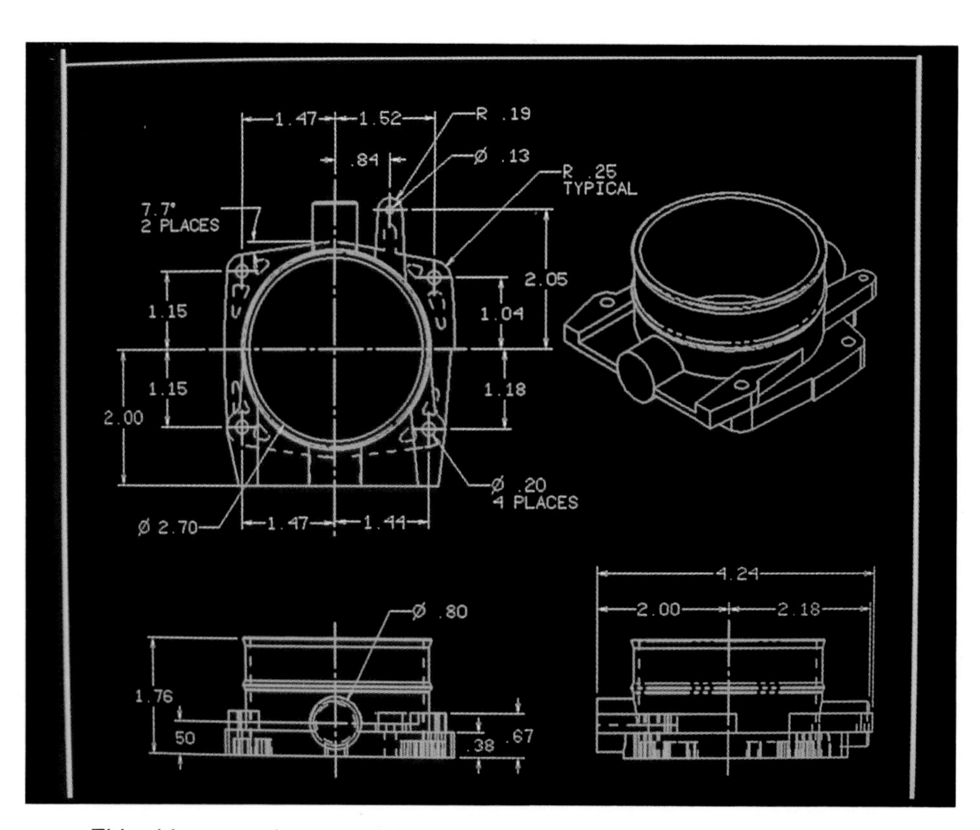

This object was drawn and dimensioned with a CAD program. (IBM Corp.)

12

CHAPTER

DIMENSIONING

After studying this chapter you will be able to:

- *Describe how dimensioning communicates information about a drawing.*
- *Identify and demonstrate how to use dimension lines, extension lines, leaders, numbers, and notes.*
- *Identify aligned and unidirectional dimension systems.*
- *Explain how to dimension round features.*
- *Describe how finished surfaces and tolerances are indicated on a drawing.*

In Chapter 11, you learned how to produce a multiview drawing that describes the shape of an object. However, to produce an object, you need information about the size of the object, Fig. 12-1. You also need to know the location of features. For instance, if there is a hole in the object, you need to know its location, Fig. 12-2. **Dimensioning** provides the size and location information on a drawing.

Dimensioned drawings are also known as *working* or *detail* drawings. This is because they provide the "details" from which to "work" as a product is being made.

DIMENSIONING STANDARDS

Drawing standards help to ensure consistency in the way a drawing is drafted and interpreted. Suppose a company has plants throughout the world. By using a common graphic language, drawings created in company plants in the United States would be the same as those produced in New Zealand, Singapore, or Norway. A common graphic language is important to ensure that there are no misinterpretations.

Fig. 12-1. Dimensions are needed to determine the correct size of an object.

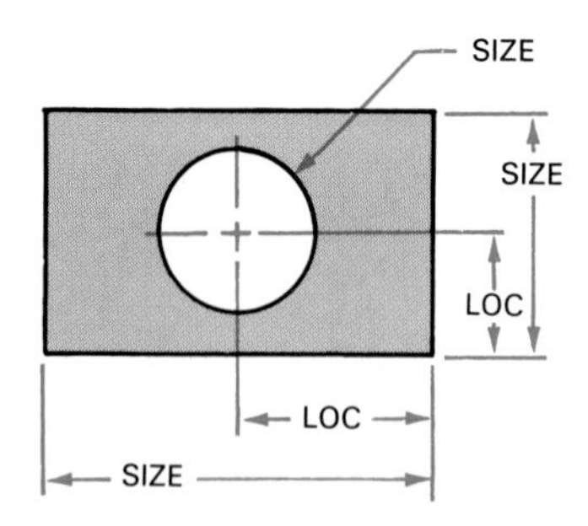

Fig. 12-2. Dimensioning provides size and location information on a drawing. Notice how both are needed for the hole.

One of the most important areas for drawing standards is dimensioning. In the United States, the American National Standards Institute (ANSI) establishes the standards for dimensioning technical drawings. This standard is known as ANSI/AME Y14.5M-1994.

The "M" refers to metric. This is to emphasize the fact that many of the new standards are related to metric. At the international level, the International Standards Organization (ISO) establishes standards for measurement systems. When an "SI" is placed on a technical drawing, it means that this standard is being used. This standard is always in metric. Some of the symbols used with these standards are shown in Fig. 12-3.

DIMENSIONING LINES

Three types of lines are used in dimensioning. They are all drawn as thin (0.5 mm), dark lines. These lines include:

- Extension lines.
- Dimension lines.
- Leaders.

BASIC DIMENSIONING	ANSI Y14.5	METRIC ISO
1. DIAMETER	Ø	Ø
2. RADIUS	R	R
3. NUMBER OF TIMES/PLACES	2X	2X
4. COUNTERBORE/SPOTFACE	⌴	NONE
5. COUNTERSINK	⌵	NONE
6. DEEP/DEPTH	↧	NONE
7. SQUARE	□2	□2
8. DIMENSION NOT TO SCALE	2 (underlined)	2 (underlined)
9. CONICAL TAPER	▷	▷
10. SLOPE	◺	◺
11. ARC LENGTH	⌒2	NONE
12. SPHERICAL RADIUS	SR	NONE
13. SPHERICAL DIAMETER	SØ	NONE
14. BASIC DIMENSION	[2]	[2]
15. REFERENCE DIMENSION	(2)	(2)

h = letter height

Fig. 12-3. ANSI and ISO drawing symbols are compared in this chart. For example, when using either standard, a radius on a drawing is shown with an "R," such as R16. (Hearlihy and Co.)

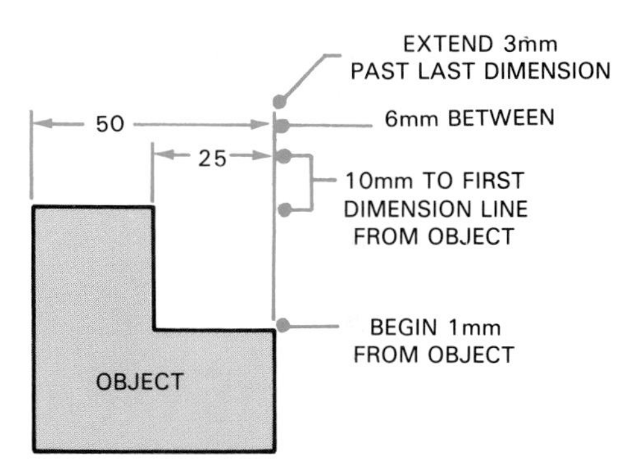

Fig. 12-4. This is the correct method for drawing extension and dimension lines.

EXTENSION LINES

Extension lines are used to move dimensions off a drawing so it is not cluttered, Fig. 12-4. These lines begin about 1.5 mm (1/16 in.) from the object. They extend 3 mm (1/8 in.) past the last dimension line. Some extension line rules are described below and are shown in Fig. 12-5.

- Extension lines can cross (A).
- Extension lines can cross a visible line, if necessary (B).
- Extension lines never connect views (C).
- Centerlines can be used as extension lines (D).

DIMENSION LINES

Dimension lines are placed between extension lines to show measurement. Refer again to Fig. 12-4. They are parallel to the side dimensioned. An arrow on each end touches the extension line. The arrow length is the same as the number height and one-third as wide, Fig. 12-6. A break along the line is used for inserting the dimension. The line does not touch the number. Some dimension line rules are described below and shown in Fig. 12-7.

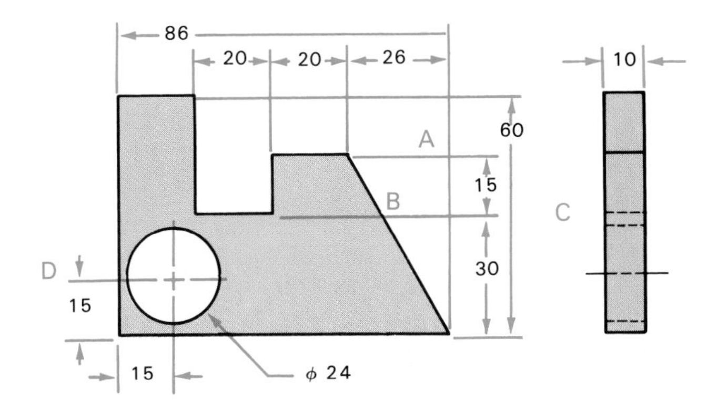

Fig. 12-5. Rules for using extension lines on a drawing are shown here.

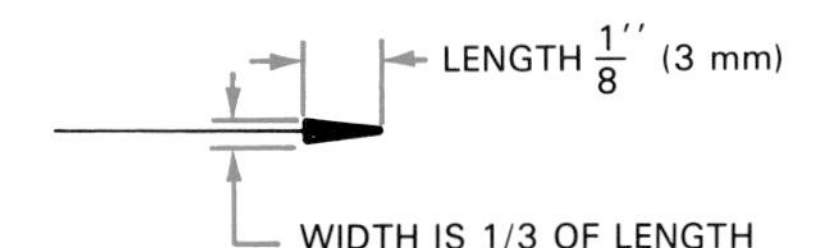

Fig. 12-6. Arrows on dimension lines are one-third as wide as they are long. If the number height on the drawing is 3 mm, then the arrow length is also 3 mm.

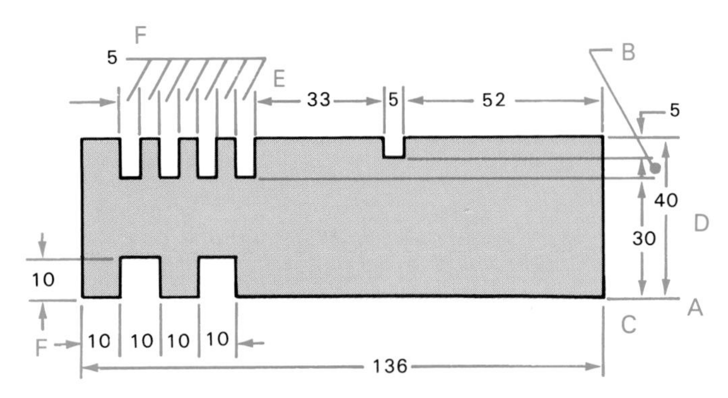

Fig. 12-7. Rules for using dimension lines are shown.

- Detail dimensions are placed next to an object and overall dimensions outside (A).
- Leave one detail dimension out when the overall dimension is provided (B).
- The first dimension line should be at least 10 mm (3/8 in.) from the drawing. Other lines should be at least 6 mm (1/4 in.) from the first line (C).
- Breaks in dimension lines should be staggered to avoid cluttered dimensions (D).
- Dimension lines do not cross dimension, extension, or visible lines (E).
- For small spaces, dimensioning techniques are slightly different (F).

LEADERS

Leaders are angled lines with an arrow on one end and a shoulder (short, horizontal line) on the other. The arrow touches a feature on the drawing. A shoulder on the other end points to a dimension or note off the drawing. The shoulder is about 6 mm (1/4 in.) long. Notes that are used with dimensions are often called **callouts**. Callouts supplement the dimension. Rules for leaders are described below and shown in Fig. 12-8.

- Draw leaders at a convenient angle, such as 30°, 45°, or 60°, but never vertically or horizontally (A).
- Place leaders parallel, if possible (B).
- Leaders do not cross (C).
- When leaders are used to show the size of arcs and circles, they must point to the center of that arc or circle (D).
- The shoulder points to the center of the first or last word of a note (E).

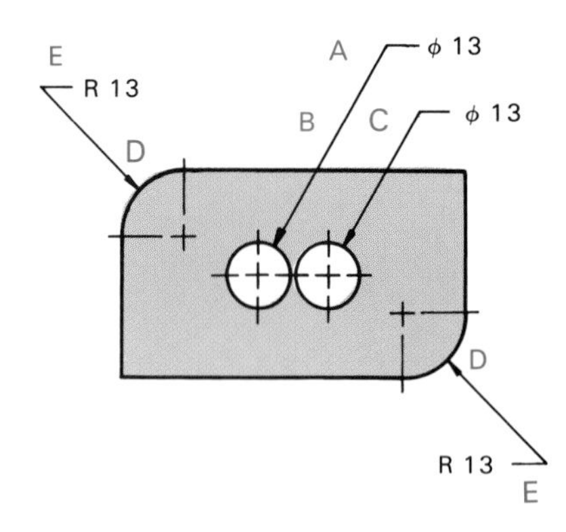

Fig. 12-8. The correct method for drawing leaders is shown in this illustration.

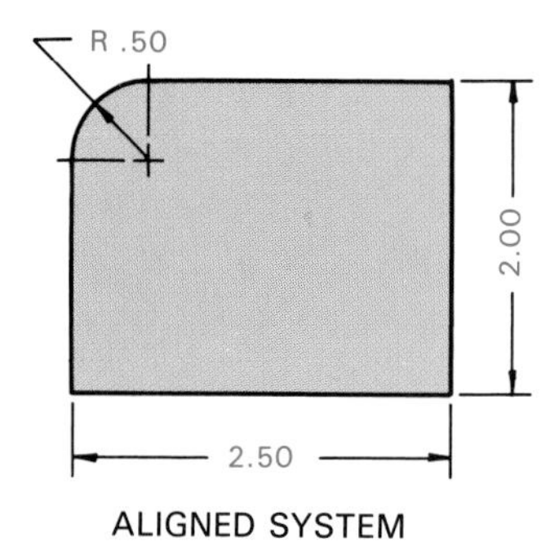

Fig. 12-10. The aligned system of dimensioning is shown here. Notice that the height dimension is read from the right.

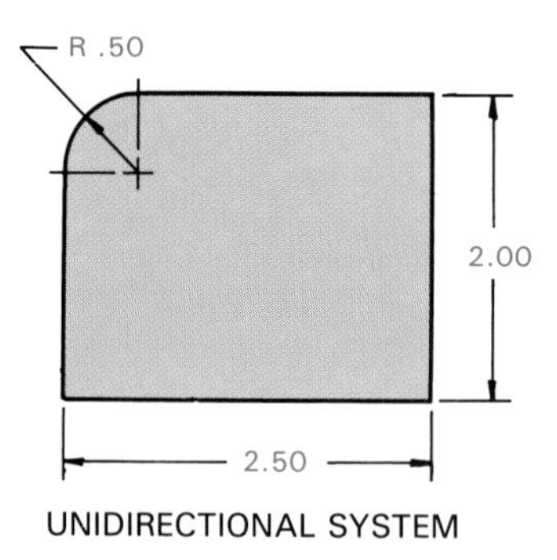

Fig. 12-11. In the unidirectional system of dimensioning, all dimensions are read from the bottom of the drawing.

DIMENSIONING NUMBERS AND NOTES

Dimension numbers and notes are usually of a uniform size, such as 3 mm (1/8 in.). Since the system of measurement is given in a note for the drawing, mm and in. abbreviations are left off individual dimensions. Other rules are described below and shown in Fig. 12-9.

- Fractions are twice the height of numbers (A).
- In metric, dimensions are usually rounded to the nearest whole millimeter (B).
- For decimal inches, a minimum of two decimal places are often used, even if it is zero (C).
- A zero precedes a dimension of less than 1 mm, but not for less than one decimal or fractional inch (D).

DIMENSIONING SYSTEMS

There are two dimensioning systems for placing dimensions on a drawing. These are the aligned system and unidirectional system. In the **aligned system**, dimensions are placed to read from the bottom and right side. See Fig. 12-10. In the **unidirectional system**, all dimensions are placed to read from the bottom of the drawing. See Fig. 12-11.

The unidirectional system is most often used in industry because of certain advantages. First, it is easy to read. Second, it is more easily applied to the drawing when using lettering machines and CAD systems.

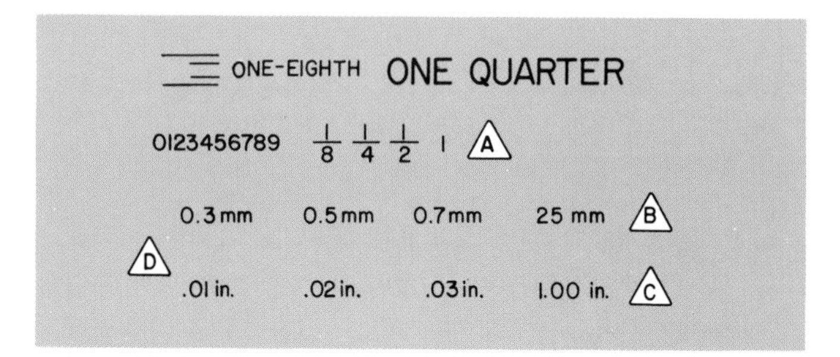

Fig. 12-9. Rules for dimension lines and numbers are shown in this illustration and described in the text.

In both systems, dimensions with leaders are read from the bottom of the drawing. Also, notes are always placed to read from the bottom of the drawing.

GENERAL DIMENSIONING RULES

No two people will dimension in exactly the same way. However, general rules should be followed. These are described below and shown in Fig. 12-12.

- Place most dimensions between the views where they apply to both views (A). Remember to have an overall width, height, and depth.
- Dimension the front view first since it is the most descriptive (B).
- Show a dimension one time only.
- Dimension where true size and shape are shown with visible lines (C). Do not dimension to hidden lines.
- Whenever possible, align detail dimensions (C).
- Angles are dimensioned by size in degrees or location (D). Degrees can be placed horizontally within the dimension line.

DIMENSIONING ROUND FEATURES

Round features include arcs, holes, and cylinders. There are differences in dimensioning these that will be explained in this section.

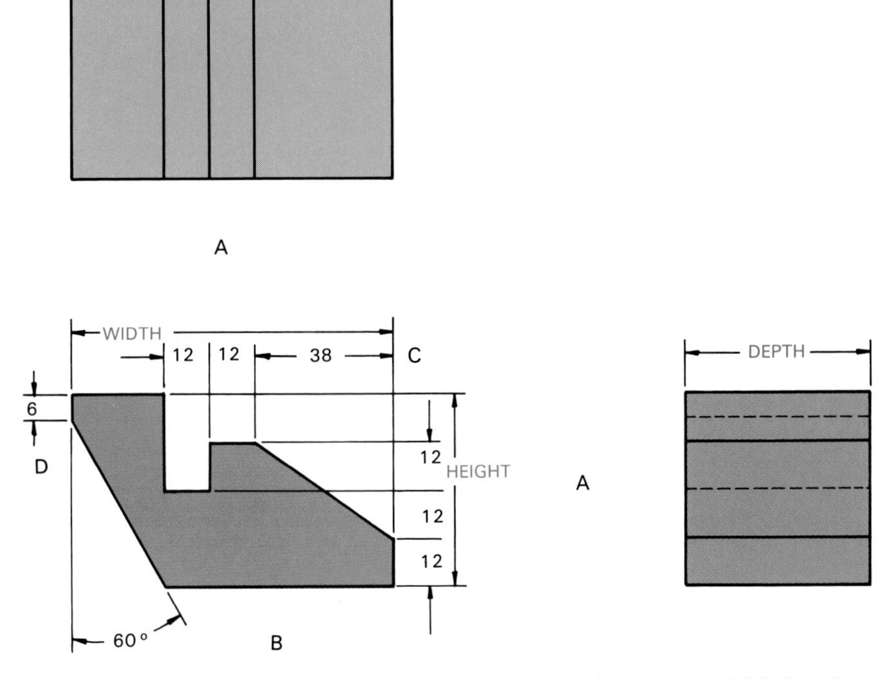

Fig. 12-12. General dimensioning rules are illustrated here. The width, height, and depth would normally be described by numbers.

ARCS

Arcs are dimensioned with a radius, shown by R, before or after the size. The new ANSI standards specify that the R appear before the size. As shown in Fig. 12-13, for large arcs, the radius is placed within the arc. The number can be placed to read from the bottom. For smaller arcs, a leader is used.

Fillets and rounds are often found on cast parts, Fig. 12-14. A **fillet** is a small arc found on inside corners. A **round** is found on outside corners. If fillets and rounds are the same size, a note can be used such as: ALL FILLETS AND ROUNDS R4. Another option is to add information after the radius, as shown in Fig. 12-14. TYP stands for typical.

HOLES

Holes are dimensioned with a diameter. The symbol for diameter is DIA after or before the number. See Fig. 12-15. When you are following ANSI standards (ANSI Y14.5M-1982), the diameter symbol should be used. For holes, the size and location are both needed. Holes are located by their center lines in the circular view. Size is also shown in the circular view. For large holes, the diameter is placed inside. A leader is used for small holes. See Fig. 12-16.

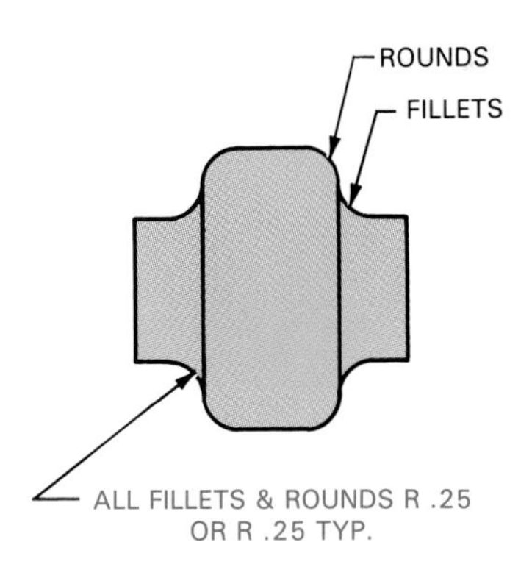

Fig. 12-14. Fillets and rounds are typically found on cast parts. If all fillets and rounds are the same size, a note can be used as shown at the bottom.

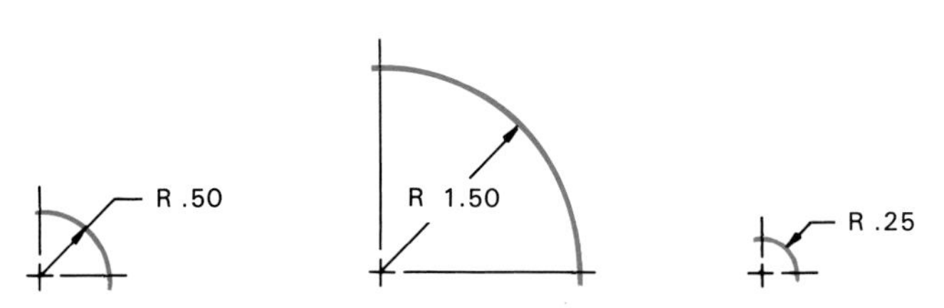

Fig. 12-13. Techniques for dimensioning large and small arcs are shown here. As the arc becomes smaller, less information is placed inside the arc.

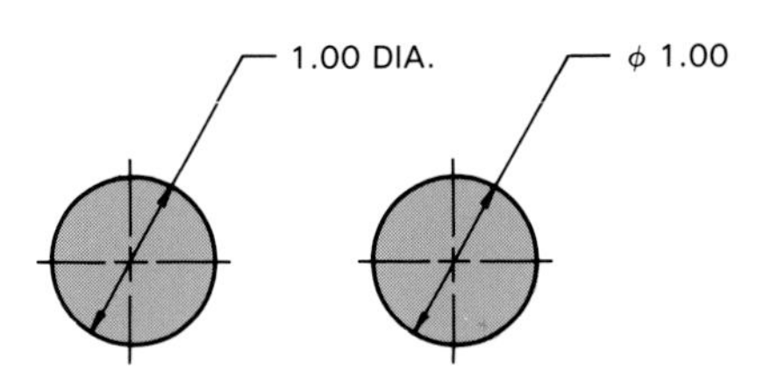

Fig. 12-15. Two options for describing a diameter are shown here. The diameter symbol is used in the example on the right.

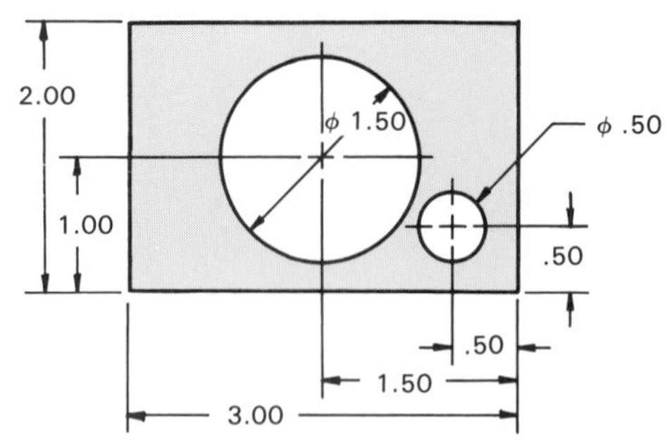

Fig. 12-16. This is the technique for dimensioning large and small arcs.

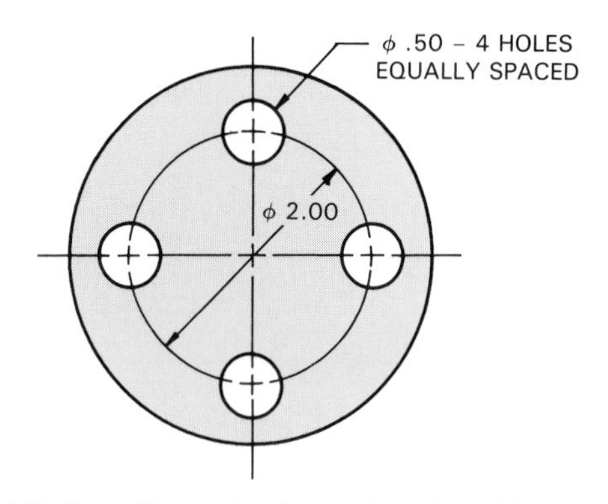

Fig. 12-19. One dimension is used to describe the holes on the bolt circle.

If a hole does not pass through an object, the depth is needed. This is called a **blind hole**, Fig. 12-17. It is measured to the shoulder. Using ANSI standards, the symbol for depth precedes the dimension.

When more than one hole is the same size, one dimension can be used for all of them, Fig. 12-18. If holes are equally spaced in a row or circle, this can also be shown with a single dimension. When holes are equally spaced in a circle, a circular center line called a **bolt circle** is used. The diameter of the bolt circle must be shown as in Fig. 12-19.

Spotfacing levels the area around a hole, Fig. 12-20. This provides a better seat for fasteners such as bolts. A spotface can be shown with a leader to the area to be spotfaced, or as a separate note on the drawing.

Different types of holes require different dimensioning techniques. These kinds of holes include:

- Countersunk holes.
- Counterdrilled holes.
- Counterbored holes.

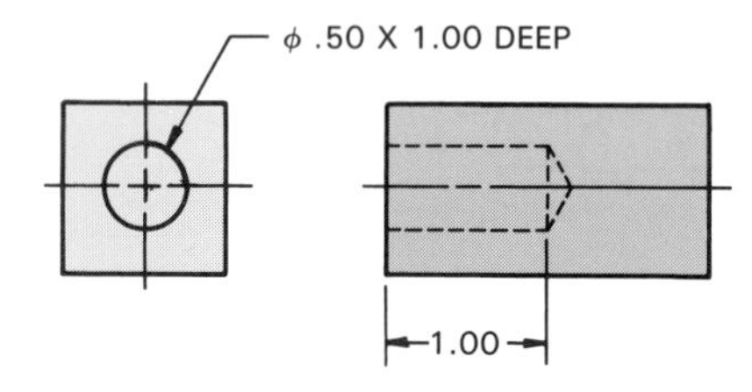

Fig. 12-17. Depth is shown for a blind hole.

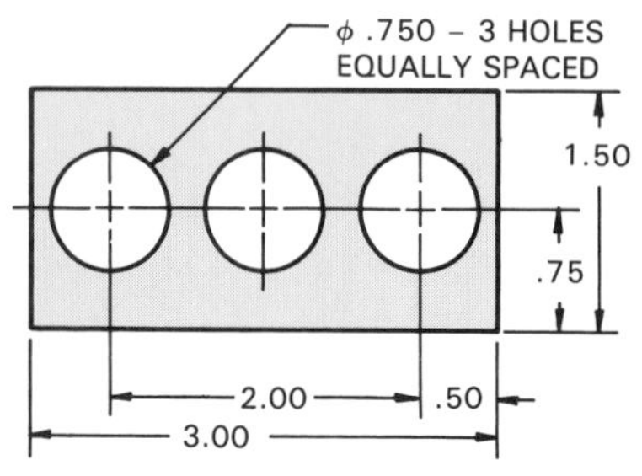

Fig. 12-18. When holes on an object are the same size, one dimension can be used as illustrated. Notice that spacing information can also be provided.

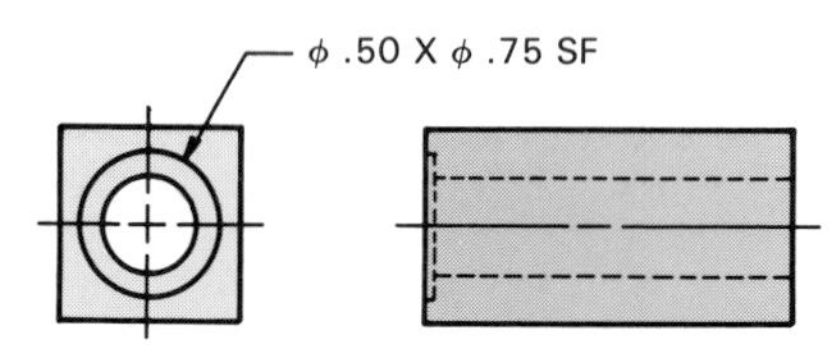

Fig. 12-20. A spotfaced hole can be designated directly on a drawing as shown, or as a separate note.

Countersunk holes

Countersunk holes are used when flat head fasteners must be even with the surface. As shown in Fig. 12-21, the leader touches the outside circle. The dimension begins with the hole diameter. When using ANSI standards, the countersink symbol precedes the diameter symbol.

Counterdrilled holes

Counterdrilled holes are formed when a large hole is drilled part way into a smaller one, Fig. 12-22. The leader touches the outside circle and the small hole diameter is listed first. Depth of the counterdrill can be

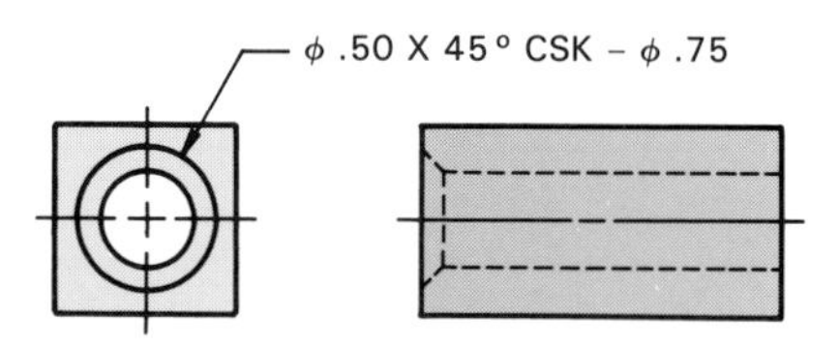

Fig. 12-21. The technique for dimensioning a countersunk hole is shown here. Notice that the leader touches the outside circle.

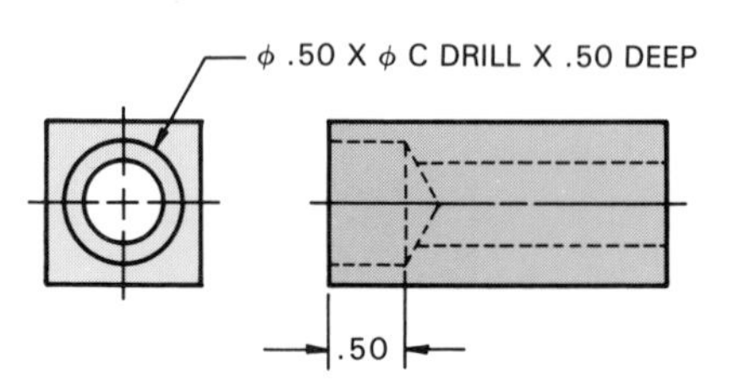

Fig. 12-22. Like most holes, a counterdrilled hole is typically dimensioned in the round view.

shown by the depth symbol, and it is placed before the dimension.

Counterbored holes

Counterbored holes are similar to counterdrilled holes, but the counterbored hole has a flat bottom, Fig. 12-23. Again, the leader touches the large hole, and the small hole diameter is listed first. When using ANSI standards, the counterbore symbol precedes the diameter symbol.

CYLINDERS

Like holes, cylinders are located in the circular view with centerlines. However, the size is shown in the edge view, Fig. 12-24. The size includes the diameter and height of the cylinder.

FINISHED SURFACES

Finished surfaces are surfaces that have been machined flat. This is needed when parts must fit together. When the surface is shown as a visible or hidden edge, it is marked with a V, Fig. 12-25. The V is at a 60° angle. Holes are located from finished surfaces when available. When an object is finished all over, a general note, FAO, can replace the finish marks.

TOLERANCING

A **tolerance** shows the amount a size can vary. When a product is mass-produced in industry, there are usually slight differences. You can see this by inspecting two supposedly identical products in the store. Slight differences always exist. These variations are present because of differences in machines, differences in operators, and even differences in the materials used to make the products. Therefore, a tolerance is often needed to show how much a size can vary and still be acceptable.

For example, many products are made up of smaller parts that must be assembled. If one part is too big or too small, it will not fit correctly with the other parts. In this case, the part is out of tolerance.

Tolerances are often given in a note for a whole drawing, as shown in Fig. 12-26. This can include angle tolerances also. Exceptions are placed on the drawing where needed. A typical exception is parts that must fit together, called mating surfaces. A shaft that fits into a hole is an example. Mating surfaces usually require different tolerances than other parts of the product.

Tolerances on a drawing can be shown in bilateral or limit form. These are shown in Fig. 12-27 and Fig. 12-28.

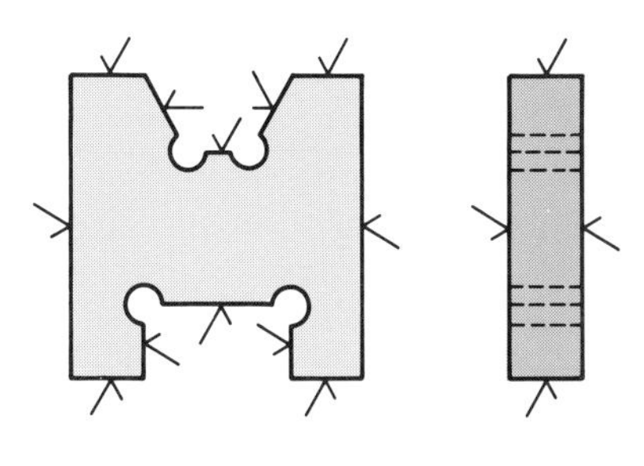

✓ OR NOTE: FINISH ALL OVER (FAO)

Fig. 12-25. Finished surfaces can be designated with a symbol as shown here. A note can be used (bottom) in place of the symbols that might be found on a drawing.

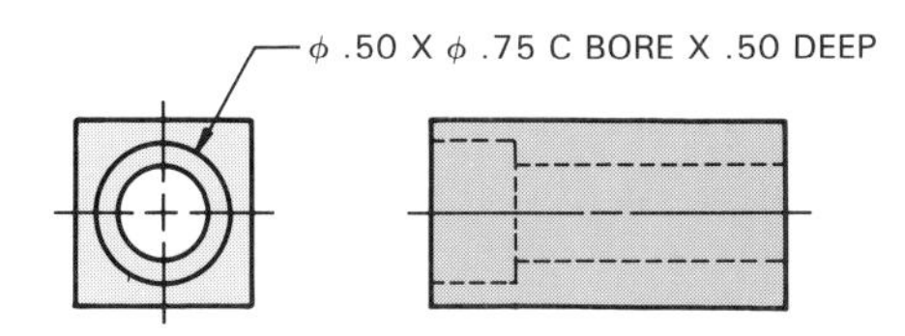

Fig. 12-23. A counterbored hole is dimensioned in this illustration.

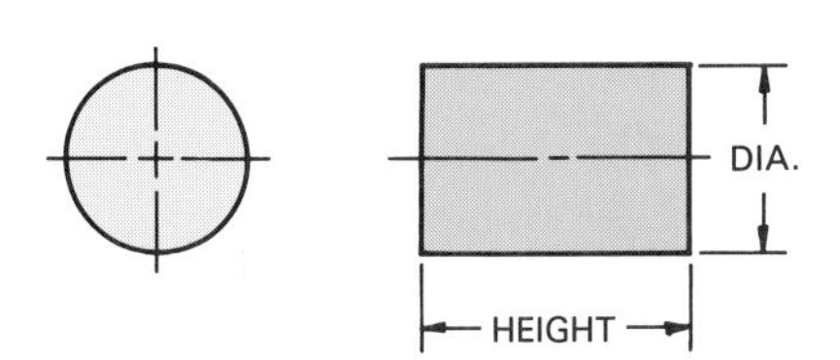

Fig. 12-24. Unlike a hole, the size of a cylinder is shown in the edge view. This is helpful since cylinders and holes can look similar on drawings with a lot of detail.

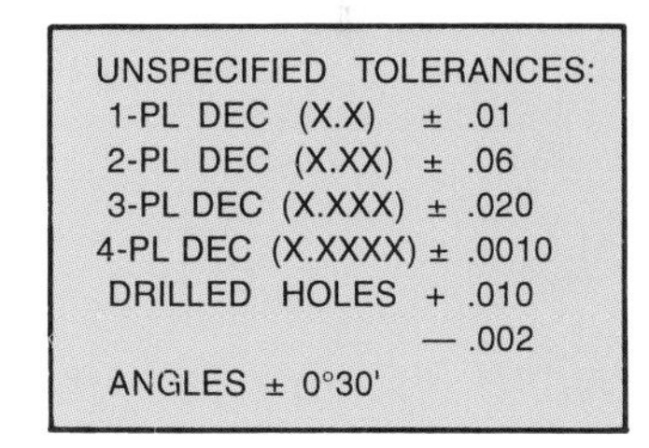

UNSPECIFIED TOLERANCES:

1-PL DEC (X.X)	± .01
2-PL DEC (X.XX)	± .06
3-PL DEC (X.XXX)	± .020
4-PL DEC (X.XXXX)	± .0010
DRILLED HOLES	+ .010 — .002
ANGLES	± 0°30'

Fig. 12-26. This is an example of a note for tolerances that might be found on a drawing.

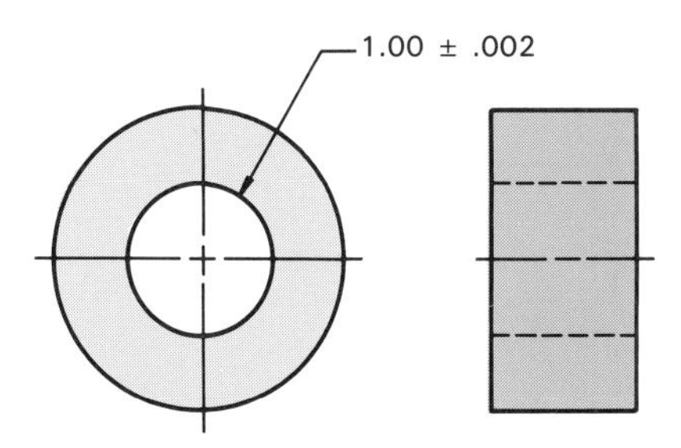

Fig. 12-27. A bilateral tolerance is shown here. In this case, the size can vary equally in either direction from the specified dimension.

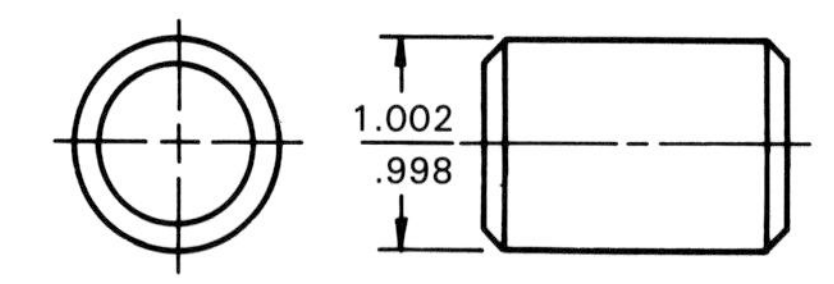

Fig. 12-28. A tolerance in limit form is shown. The high limit (largest acceptable size) is placed above the low limit (smallest acceptable size).

SUMMARY

Dimensioning enables you to place size and location information on a drawing. This provides the information necessary to communicate to others how an object is made. Dimensioning standards help to ensure consistency in the way a drawing is drafted and interpreted.

WORDS TO KNOW

All of the following words have been used in this chapter. Do you know their meanings?

aligned system
blind hole
bolt circle
callouts
counterbored holes
counterdrilled holes
countersunk holes
dimension lines
dimensioning
extension lines
fillet
finished surfaces
leaders
round
spotfacing
tolerance
unidirectional system

REVIEWING YOUR KNOWLEDGE

Please do not write in this text. Write your answers on a separate sheet.

1. True or false? Dimensioning provides the size and location information for a drawing.
2. In the United States, what organization establishes the standards regarding dimensioning technical drawings?
3. Name and describe three types of dimensioning lines.
4. True or false? In the aligned dimensioning system, all dimensions are placed to read from the bottom.
5. Which of the following would be dimensioned with a radius?
   a. Arcs.
   b. Squares.
   c. Triangles.
   d. None of the above.
6. When dimensioning holes, if more than one hole is the same size:
   a. Each one must be dimensioned separately.
   b. One dimension can be used for all of them.
   c. Every other one must be dimensioned.
   d. None of the above.
7. When a finished surface is shown as a visible or hidden edge, it is marked with a(n):
   a. O.
   b. V.
   c. X.
   d. None of the above.
8. A _________ shows the amount a size can vary.

APPLYING YOUR KNOWLEDGE

1. Visit a local industry to find out how they use dimensioning standards. Prepare a written report.
2. Discuss the importance of accurate dimensioning. If possible, ask someone from a local industry to recall some instances of incorrect dimensioning and the consequences.
3. Make a multiview drawing, with dimensions, of an existing product.
4. Review some of the automatic dimensioning features on a CAD system. Report your findings to the class.

13

CHAPTER

PICTORIAL DRAWINGS

After studying this chapter, you will be able to:
- *Identify and describe the three major types of pictorial drawings.*
- *Demonstrate how to accurately produce pictorial drawings.*
- *Determine when dimensions are needed on pictorial drawings.*
- *List and describe the shading options used in technical illustration.*

A pictorial drawing is like a picture or photograph of an object. See Fig. 13-1. It shows height, width, and depth in one view, Fig. 13-2.

There are three types of pictorials. These are
- Axonometric.
- Oblique.
- Perspective.

AXONOMETRIC DRAWINGS

Axonometric drawings are identified by the angles between the lines joining the front corner or axis, Fig. 13-3. There are three types of axonometric drawings. These include:
- Trimetric.
- Dimetric.
- Isometric.

In **trimetric drawings**, the angles are unequal. In **dimetric drawings**, two angles are equal. In **isometric drawings**, all angles are equal on the front corner. The isometric drawing is most often used and the easiest to draw.

An isometric drawing is produced from isometric and non-isometric lines. Isometric lines are parallel to the isometric axis, Fig. 13-4. These lines are vertical, 30° to the left, or 30° to the right. They represent the vertical and horizontal lines on an object. Measurements are true length on these lines.

Non-isometric lines are those lines that are not parallel to isometric lines. They are created by connecting

Fig. 13-1. Notice how the picture of the object appears to have three dimensions, just like the actual object.

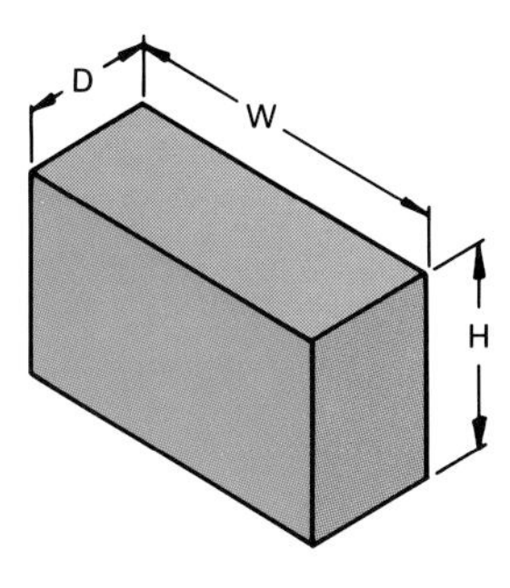

Fig. 13-2. A pictorial drawing shows height, width, and depth dimensions in one view.

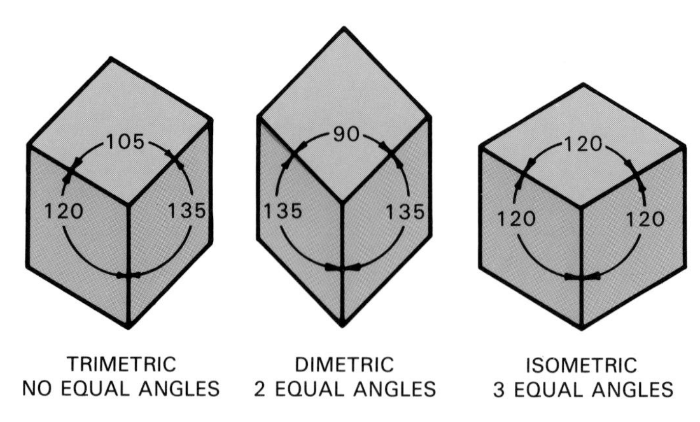

Fig. 13-3. The three types of axonometric drawings are shown here. Each type is identified by the angles created by the three visible faces.

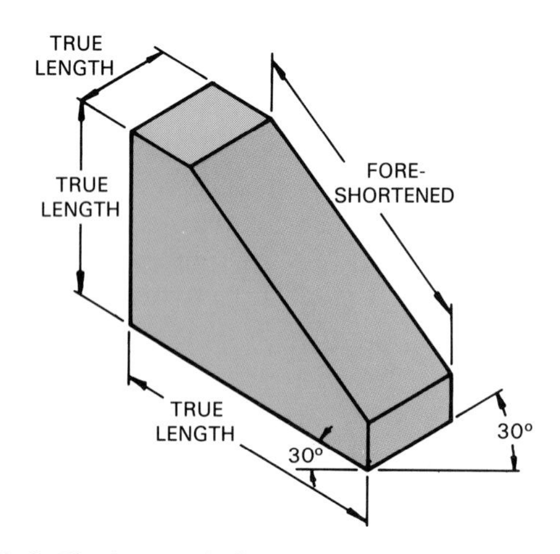

Fig. 13-4. The isometric lines on this object are drawn true length. The non-isometric lines are drawn foreshortened.

points on isometric lines, Fig. 13-4. These lines will be **foreshortened** (made shorter than the actual length).

Circles appear as ellipses on isometric drawings. This is because every face on this type of drawing is distorted. Ellipses can be drawn with a compass or template. This technique was described in Chapter 10. Remember that the ellipse will be oriented differently on the three visible faces, Fig. 13-5.

Irregular curves on isometric drawings are made by first locating points along the line, Fig. 13-6. These are then connected to form the irregular curve.

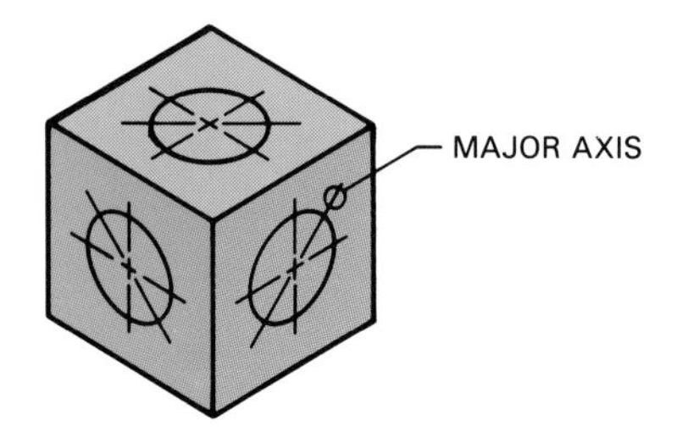

Fig. 13-5. The orientation of isometric circles (drawn as ellipses) is different on the three visible faces of this cube.

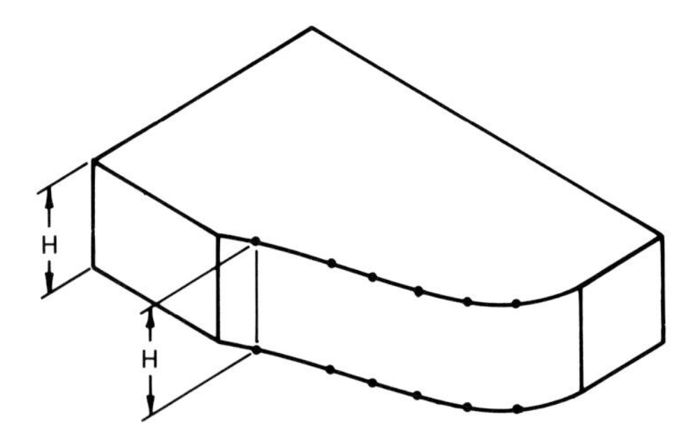

Fig. 13-6. Irregular curves on an isometric drawing are made by first locating points along the curve.

PRODUCING THE ISOMETRIC DRAWING

The box-construction method is often used in producing an isometric drawing. This method is described below.

1. Center the drawing by eye or use the following technique for greater accuracy, Fig. 13-7:
   a. Locate the drawing space center with diagonal lines.
   b. Draw a 30° line to the right, half the total width of the drawing.
   c. Draw a 30° line to the left, half the total depth of the object.
   d. Draw a vertical line half the height of the object to locate the front corner.
2. Draw isometric lines from the front corner. See Fig. 13-8. Mark the total width, height, and depth dimensions on these lines. Width is usually placed on the vertical plane to the left.
3. Finish the isometric box.
4. Layout details within the box using isometric and non-isometric lines. See Fig. 13-8.
5. Add isometric circles (ellipses) and irregular curves, if needed.
6. Darken the visible lines.

OBLIQUE DRAWINGS

An **oblique drawing** has a front face that is shown in its true size and shape like a multiview drawing, Fig. 13-9. The top and side faces recede back from the front,

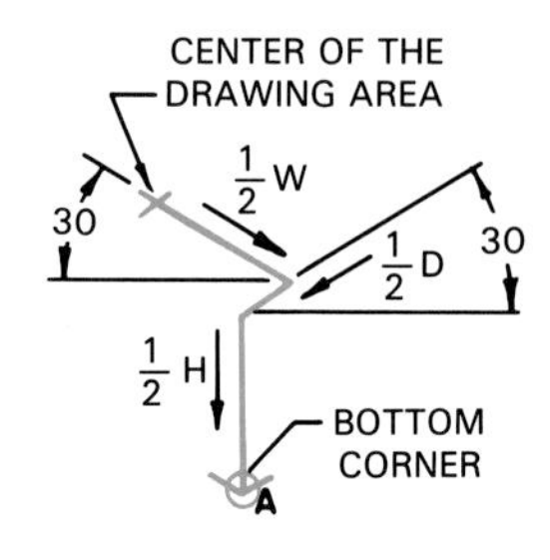

Fig. 13-7. The procedure for centering an isometric drawing is shown here.

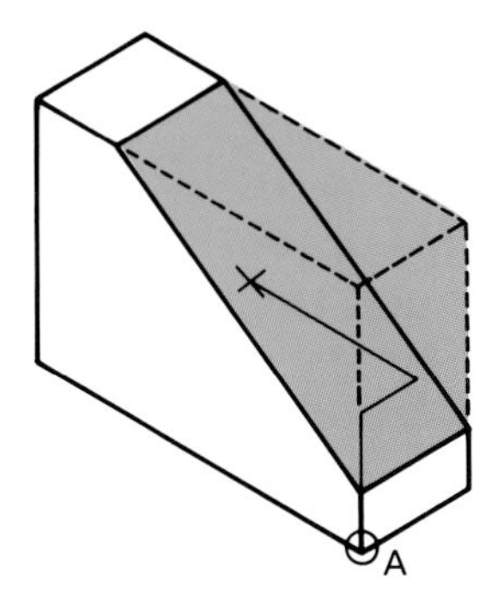

Fig. 13-8. Using the box construction method, an isometric box is drawn and then the object is drawn inside the box. The circled area (A) indicates where the drawing was started.

as in an isometric drawing. The angle for these faces is usually 45°, although any angle between 0° and 90° can be used. The object can recede back to the left or the right, Fig. 13-9.

There are three types of oblique drawings, Fig. 13-10. These include:

- Cavalier.
- Cabinet.
- General.

The **cavalier oblique** has depth dimensions drawn full length. The **cabinet oblique** shows depth as half size. The **general oblique** has a depth dimension between half and full length.

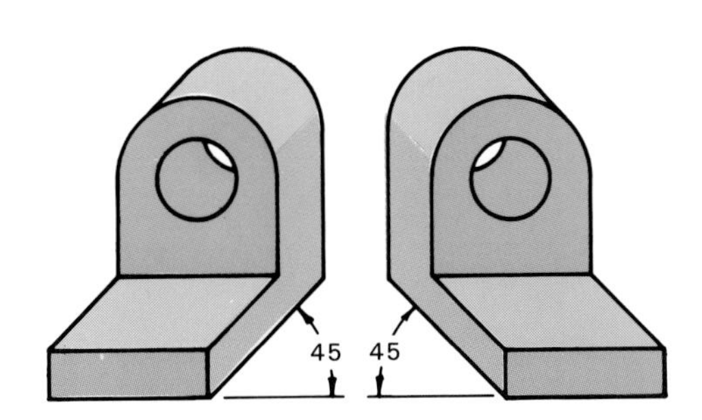

Fig. 13-9. The front face of an oblique is true shape and size. The top and sides recede back at an angle, which is usually 45 °. The oblique can be drawn to make the left or right side visible.

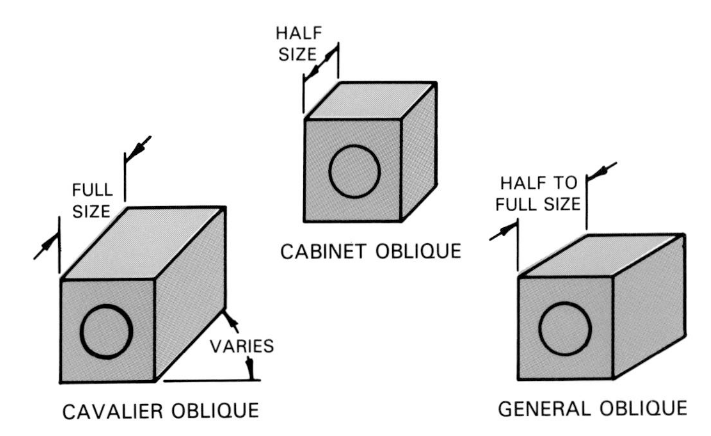

Fig. 13-10. The three types of oblique drawings are shown here.

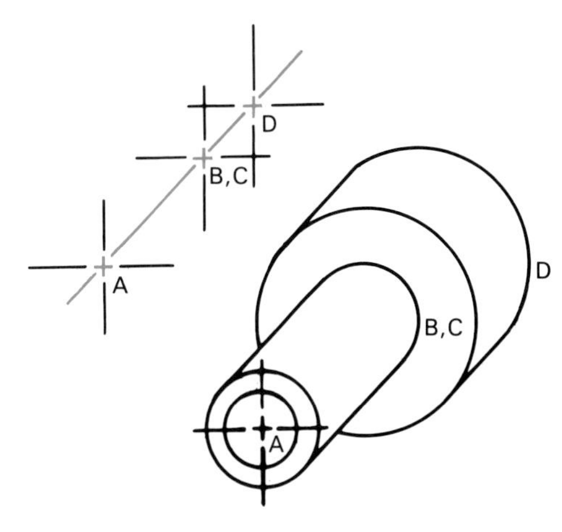

Fig. 13-11. When cylindrical objects are drawn in oblique, a longitudinal center line is first drawn (upper left). Points are then located along this line for drawing the circular features.

The oblique is a very good choice for circular objects, or objects with circular features on the front. For circular objects, such as a cylinder, the oblique can be drawn along a center line. See Fig. 13-11.

PRODUCING AN OBLIQUE DRAWING

The box construction method is often used in producing an oblique drawing. This method is described below.

1. Center the drawing by eye or use the following technique for greater accuracy. See Fig. 13-12.
   a. Locate the drawing space center with diagonal lines.
   b. Draw a horizontal line to the left, half the total width of the object.
   c. Draw a line half the total depth of the object, sloping to the left the same degree as the receding axis.
   d. Draw a vertical line half the total height, which locates the bottom left corner.
2. Draw an oblique box, varying the depth based on the kind of oblique you are drawing.

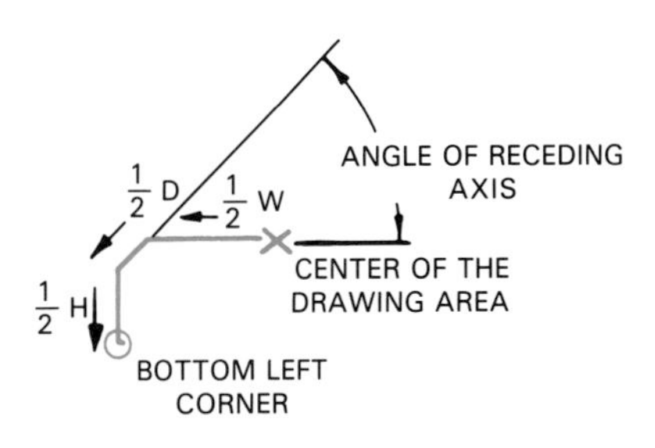

Fig. 13-12. The procedure for centering an oblique drawing is illustrated here.

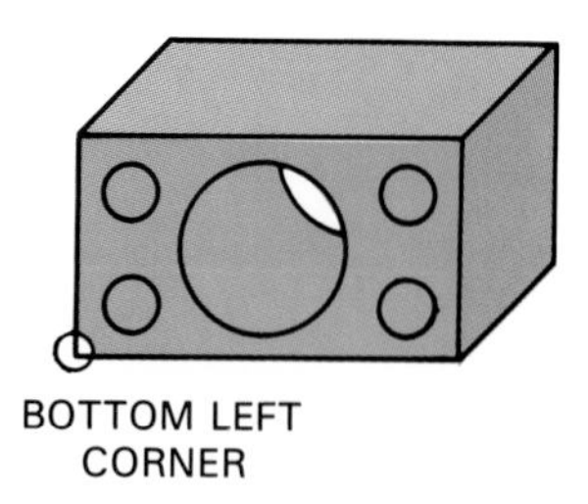

Fig. 13-13. The finished oblique drawing is shown. The circled corner shows where the drawing was started.

3. Layout details within the box starting with lines parallel to the oblique axis. See Fig. 13-13.
4. Darken the visible lines.

PERSPECTIVE DRAWING

A **perspective drawing** is the most realistic pictorial. As in the real world, dimensions on a perspective drawing become smaller as they move away from you. You can see this by looking at a fence, railroad track, or a row of poles.

In perspective drawing, receding dimensions eventually go to a point. This is called a **vanishing point**. Vanishing points are on an imaginary horizon as shown in Fig. 13-14. There are one- point, two-point, and three-point perspectives. This refers to the number of vanishing points, Fig. 13-15. The methods for drawing one- and two-point perspectives will be covered in this section of the text.

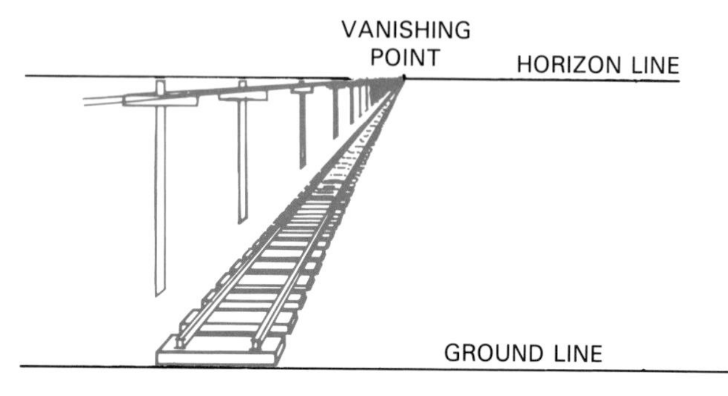

Fig. 13-14. In a perspective, receding dimensions go to a point on the horizon line, as is the case in the real world.

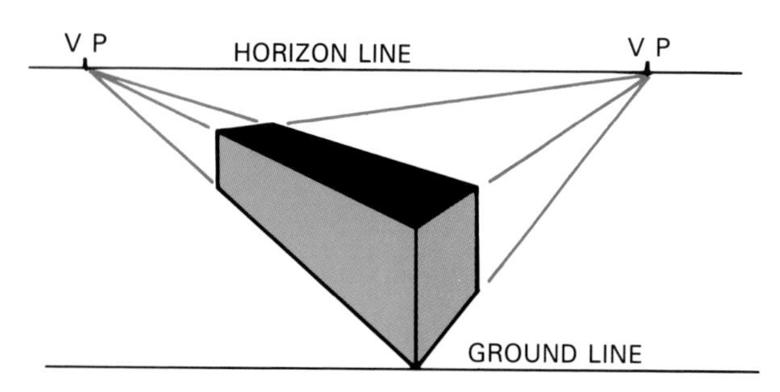

Fig. 13-15. A perspective can be defined by the number of vanishing points. This is a two-point perspective.

ONE-POINT PERSPECTIVE

The **one-point perspective** is like an oblique in that the front face is true shape. It is also known as **parallel perspective** because the front is against or parallel to the projection plane, Fig. 13-16. This is like the side of the glass box in orthographic projection. The rest of the object recedes back toward a single vanishing point.

In one-point perspective, the horizon line can be placed above the object to show the top or below the horizon to show the bottom, Fig. 13-17. This is because the horizon is at eye level. Also, the vanishing point can be located to show either side. For example, if it is to the left, the left side of the object will be shown.

The steps for drawing an approximate one-point perspective are described below and are shown in Fig. 13-18.

1. Draw a light horizon line, which is at eye level, and choose a vanishing point.
2. Lay out the front view. The base of the object is on the ground line.
3. Draw projectors to the vanishing point from the front view.
4. Select a depth dimension that is visually pleasing, but less than full depth.
5. Darken the visible lines on the object.

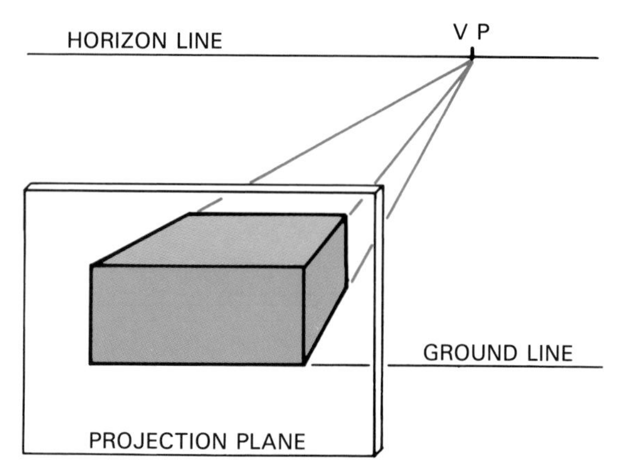

Fig. 13-16. The front face of a one-point perspective is parallel to the projection plane and is drawn true shape.

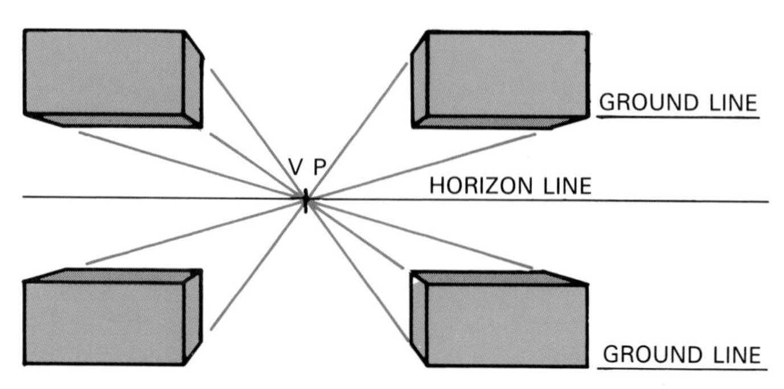

Fig. 13-17. The placement of the horizon line and vanishing point affects the view of the perspective.

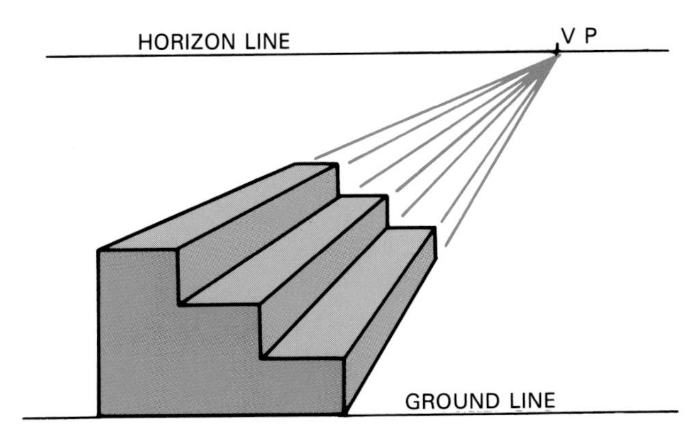

Fig. 13-18. The technique for drawing a one-point perspective is shown here.

TWO-POINT PERSPECTIVE

The **two-point perspective** has two vanishing points. It is also known as **angular perspective** because it is at an angle to the projection plane, Fig. 13-19. The only line that is true length is the one on this plane. Vanishing points recede from this line. Like one-point perspective, the horizon line can be placed above, below, or in the center of the object, depending on the best view, Fig. 13-20. Vanishing points are placed on the horizon line to either side of the object. They can be moved to change the view of the sides of the object, Fig. 13-21.

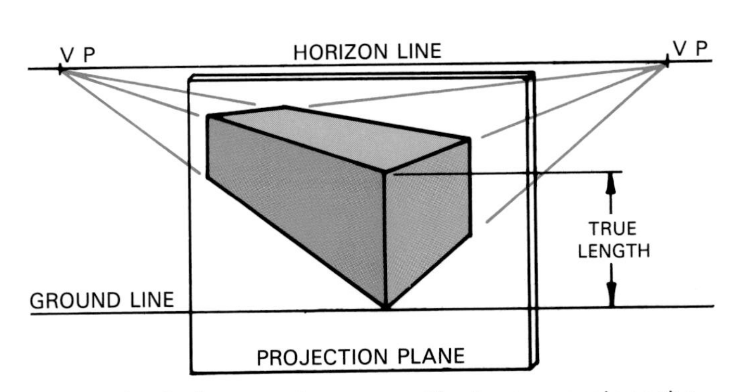

Fig. 13-19. A two-point perspective is at an angle to the projection plane. The only line that is true length on the perspective is the one touching this plane.

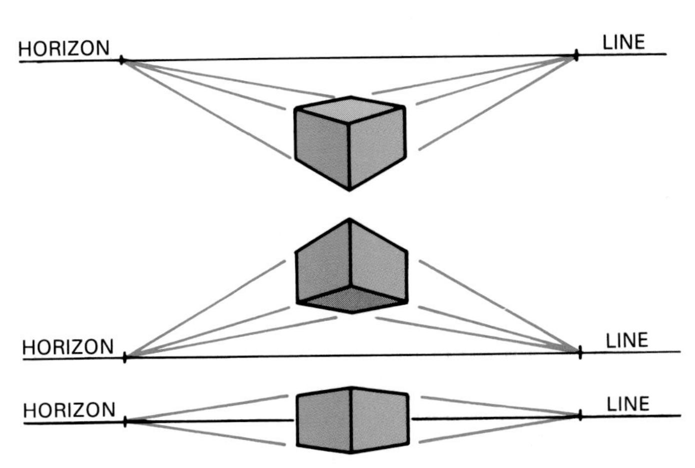

Fig. 13-20. The placement of the horizon line will affect the view of the two-point perspective. Three options are shown.

The steps for drawing an approximate two-point perspective are described below and shown in Fig. 13-22 and Fig. 13-23.

1. Draw the horizon line and place the vanishing points on either side of the object.
2. Draw the front corner, which is true length and is on the *ground line*.
3. Draw projectors from the front corner to the vanishing points.

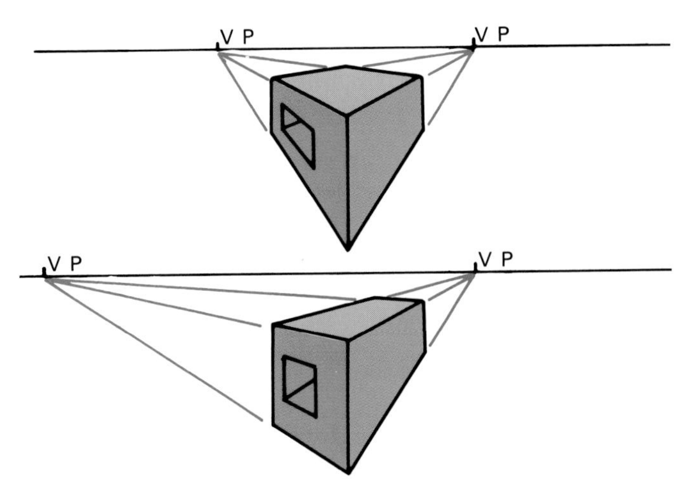

Fig. 13-21. When the vanishing points are moved, the view is changed.

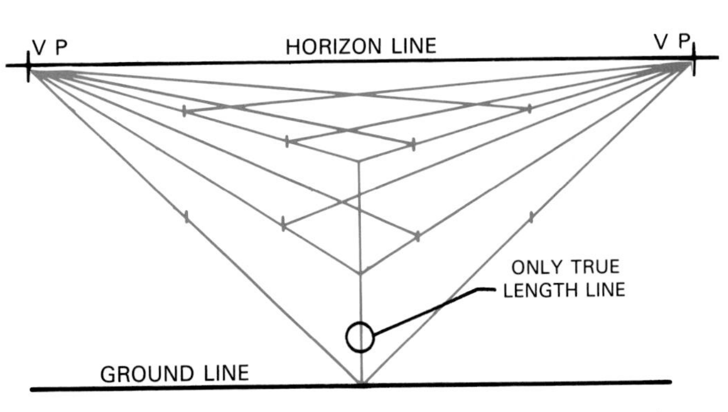

Fig. 13-22. The layout for a two-point perspective is shown here.

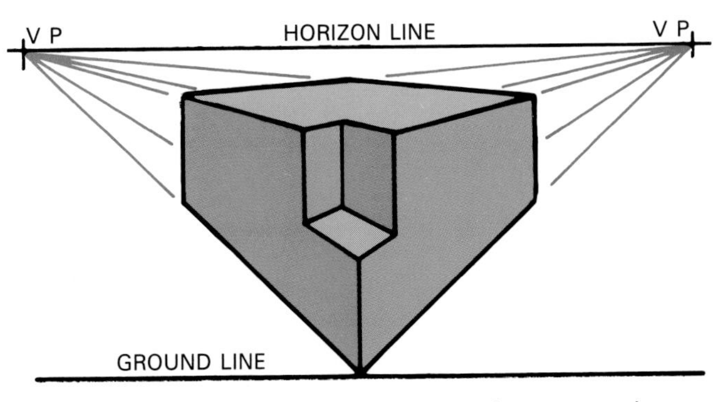

Fig. 13-23. This is the finished two-point perspective.

4. Mark a width and depth that should be proportionately less than the actual dimension. If the vanishing points are placed to emphasize one side, draw this side slightly longer.
5. Draw projectors for the width and depth to the opposite vanishing points.
6. Lay out additional features. Find height dimension on the front edge and vanish this dimension back.
7. Darken the perspective with visible lines, as shown in Fig. 13-23.

DIMENSIONING PICTORIALS

Normally, pictorials are drawn without dimensions. However, this is sometimes done to avoid making multiview drawings. Isometric and oblique drawings are the pictorials most frequently dimensioned.

The easiest method for dimensioning pictorials is the unidirectional system, Fig. 13-24. All dimensions are placed to read from the bottom of the sheet. The extension lines are placed in line with the axis angle on the drawing.

TECHNICAL ILLUSTRATION

Pictorial drawings are often used in technical illustration. This is a broad field that is involved in making technical drawings for publication. These publications include books, magazines, and operation or instruction manuals.

Since the drawings will be printed, they must be checked very carefully. They should also be easy to understand since some readers may not have a technical drawing background. A variety of techniques are used. The drawings are often done in ink, and color and shading are used to provide a more realistic appearance. Computers are also often used to generate technical illustrations.

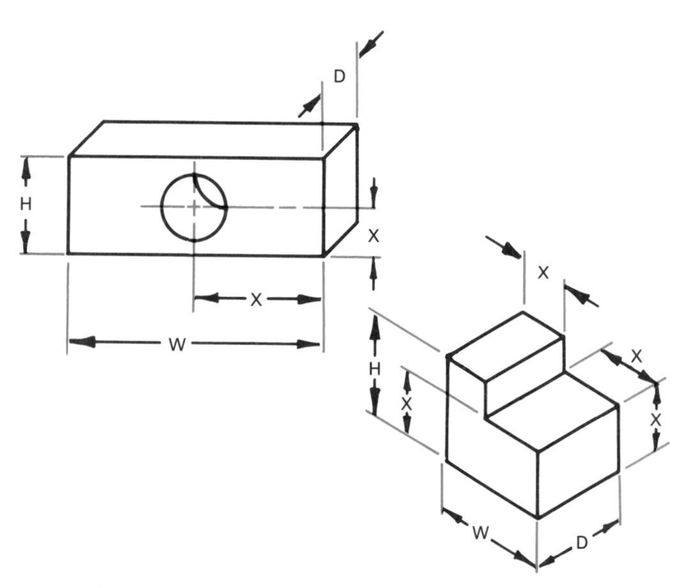

Fig. 13-24. An oblique drawing (left) and isometric drawing (right) can be dimensioned using the unidirectional system. Notice how the extension and dimension lines are at the same angle as the side being dimensioned.

SHADING FOR TECHNICAL ILLUSTRATION

Shading is a technique commonly used in technical illustration. The object appears more natural because it is in light and shadow as it would normally be.

In most instances, light sources are overhead, like the sun or ceiling lights. For shading, you can normally think of the light source as being in the upper left corner of the drawing, shining on the object, Fig. 13-25. Those surfaces facing the light source will have no shading. The surfaces away from the light source will be in shadow. Some surfaces may receive only partial light.

Visible line shading

The **visible line shading** technique is shown in Fig. 13-26. Shading is done by varying the thickness of the

Fig. 13-25. Notice the shadows produced by placing a single light source on the upper left side of these objects.

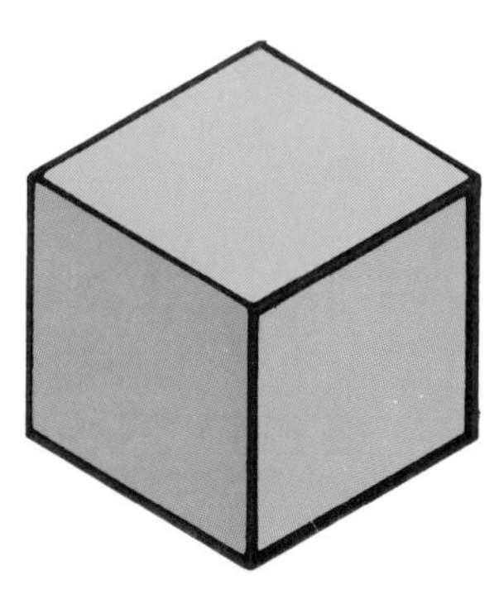

Fig. 13-26. Visible line shading is done by varying the thickness of the visible lines. Since the light is coming from the upper left, the right side and bottom are shaded.

visible lines. Thin, broken lines are used in areas receiving direct light. Shaded areas are bordered with thick lines.

Uniform surface shading

In the **uniform surface shading** technique, whole surfaces are evenly shaded, Fig. 13-27. This can be done with evenly spaced lines, or dots (stippling). Many patterns are available as transfer sheets if a large amount of shading must be done.

Continuous tone shading

The **continuous tone shading** technique provides the most realistic appearance. Patterns on the surface are varied to show the surface gradually becoming shaded, Fig. 13-28. This can be done by placing lines or dots closer together toward the shaded area. Solid black areas can be used for deep shade, with a few lines blending the light and shade areas.

This type of shading is sometimes done with an **airbrush**. An airbrush sprays dots of ink on the drawing, Fig. 13-29. **Transfer sheets** can also be used. These are sheets with a shading pattern on them. The shading is attached to the drawing by burnishing (rubbing) or by using an adhesive.

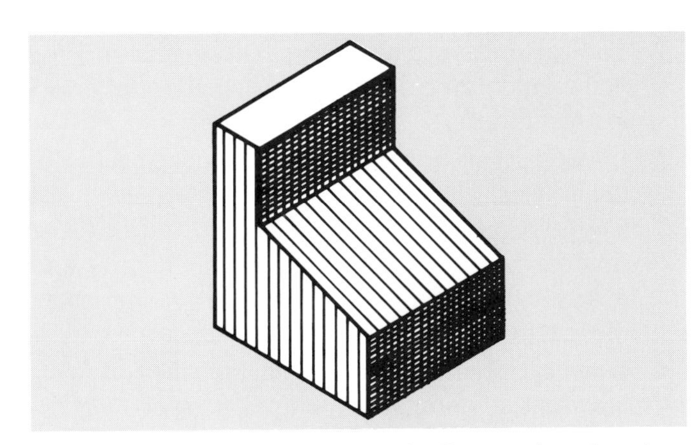

Fig. 13-27. In uniform surface shading, each surface is shaded evenly.

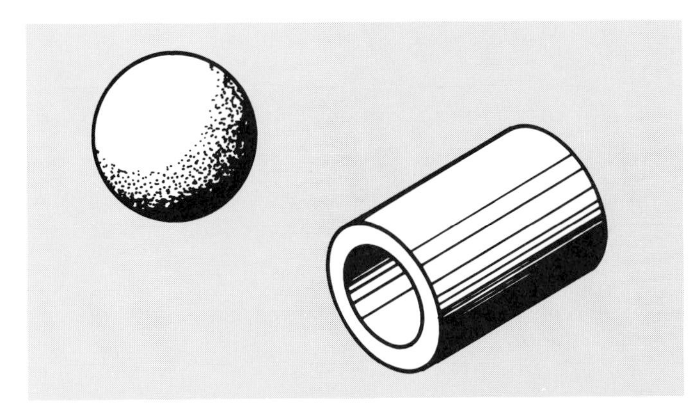

Fig. 13-28. Continuous tone shading is the most realistic. Notice how the shading gradually becomes darker in the shadow areas.

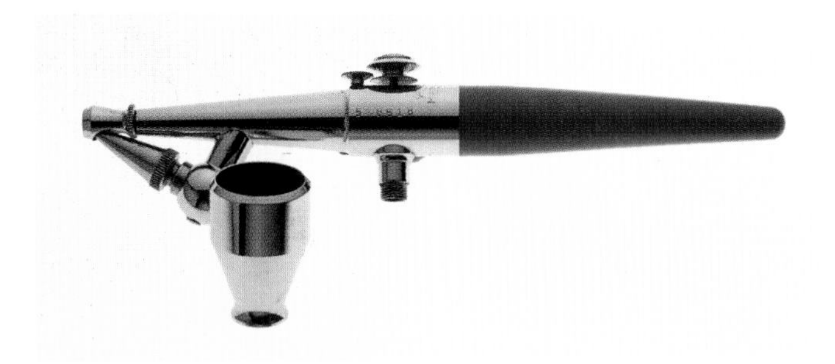

Fig. 13-29. An airbrush can be used for shading. (Paasche Airbrush Co.)

Fig. 13-30. This pictorial was created on a CAD system. Notice the shading technique. (Computervision)

PICTORIALS WITH CAD

This chapter has provided information on pictorials and the way in which they can be drawn with standard drafting equipment. However, CAD can also be used for this purpose. See Fig. 13-30. This will be discussed in Chapter 15.

SUMMARY

There are three types of pictorial drawings. They include axonometric, oblique, and perspective. Axonometric drawings are identified by the angles between the lines joining the front corner or axis. Oblique drawings have a front face that is shown in its true size and shape, like a multiview drawing. In a perspective drawing, receding dimensions eventually go to a vanishing point.

Pictorial drawings are often used in technical illustrations. A common technique used on technical illustrations is shading. Various shading techniques are used to make an object appear more natural.

WORDS TO KNOW

All of the following words have been used in this chapter. Do you know their meanings?

airbrush
angular perspective
axonometric drawings
cabinet oblique
cavalier oblique
continuous tone shading
dimetric drawings
foreshortened
general oblique
isometric drawings
oblique drawing
one-point perspective
parallel perspective
perspective drawing
shading
three-point perspective
transfer sheets
trimetric drawings
two-point perspective
uniform surface shading
vanishing point
visible line shading

REVIEWING YOUR KNOWLEDGE

Please do not write in this text. Write your answers on a separate sheet.

1. Which of the following axonometric drawings is used most often and is easiest to draw?
   a. Trimetric drawings.
   b. Dimetric drawings.
   c. Isometric drawings.
   d. None of the above.
2. Name the three types of oblique drawings.
3. What is the most realistic pictorial drawing?
4. In perspective drawing, receding dimensions eventually go to a point called a ________ point.
5. True or false? The one-point perspective is also known as angular perspective.
6. True or false? The easiest method for dimensioning pictorials is the unidirectional system.
7. Which of the following is true about technical illustrations?
   a. They should be easy to understand.
   b. They are usually done in ink.
   c. Color and shading are often used to provide a more realistic appearance.
   d. All of the above.
8. Which of the following shading techniques provides the most realistic appearance?
   a. Visible line shading.
   b. Uniform surface shading.
   c. Continuous tone shading.
   d. None of the above.

APPLYING YOUR KNOWLEDGE

1. List various types of objects. Discuss the method you would use to draw them using pictorial drawing techniques.
2. Using a set of house plans, draw a perspective of the house.
3. Collect examples of technical illustrations and shading from available magazines. Identify and discuss the techniques used to create these illustrations.
4. Demonstrate how to create each of the following drawings: axonometric, oblique, perspective. Display your work, labeling each one.

14

CHAPTER

PATTERN DEVELOPMENT AND PACKAGING

After studying this chapter, you will be able to:

- *Explain how pattern development applies to packaging and sheet metal products.*
- *Differentiate between the parallel line method of pattern development and the radial line method of pattern development.*
- *Demonstrate pattern development using the parallel line method and the radial line method.*

A **pattern** is a flat plan for a three-dimensional object. The pattern communicates to others how the object will be produced and assembled. Other names for these patterns are *surface developments* and *stretchouts*. Many familiar objects are made from patterns. These include clothes, ducts, automobile bodies, packages, and even paper airplanes, Fig. 14-1. In this chapter, you will explore pattern development applications for packaging and sheet metal products.

PACKAGING

Packaging is one of the largest industries that uses patterns. **Packaging** is a container or a wrapper for a product. Some package examples are shown in Fig. 14-2.

Packages are designed to protect a product as well as to help sell it. In many cases, the container may be all the customer sees before making the purchase, so the package must be an effective marketing tool. This includes both the shape of the package as well as the printed design. An additional consideration for package designers is how the package will stack, both in shipping containers as well as on store shelves. Therefore, the package must be functional as well as communicate effectively.

A pattern for a package is shown in Fig. 14-3. Lines to be folded are usually shown as dashed lines or with an X. **Seams** are often used to hold the package together and are sealed with glue or staples. Various techniques can be used for making a package recloseable, such as a locking tab. Locking tabs allow the customer to open the package and close it back securely. Toothpaste packages often use locking tabs on the ends.

After a pattern is developed, a design is created for the outside. This design is printed on large sheets. Each package is removed from this printed sheet by die-cutting, Fig. 14-4. It is then folded into the finished shape.

A walk through a supermarket reveals a variety of different types of packages, from toothpaste, to oatmeal, to crackers. Take a look at various types of packages and study their designs. At home, before

Fig. 14-1. A paper airplane is a familiar example of a pattern.

Fig. 14-2. These are examples of packages that were developed to hold products. (Sonoco Products Co.)

discarding a package, take it apart along its seam and study its design, Fig. 14-5.

Many packages are made from paperboard, which is a thick paper used more for durability than appearance. A design is printed solid on one side of the paperboard so that the surface is actually hidden from view. Recycled paper is often used to produce paperboard. A special logo is placed on a package if it is made from at least fifty percent recycled material. This logo has three arrows that form a circle placed inside of a circle, Fig. 14-6. If the three arrows are shown alone, without the circle, this indicates that the item is made of recyclable materials (able to be recycled after use).

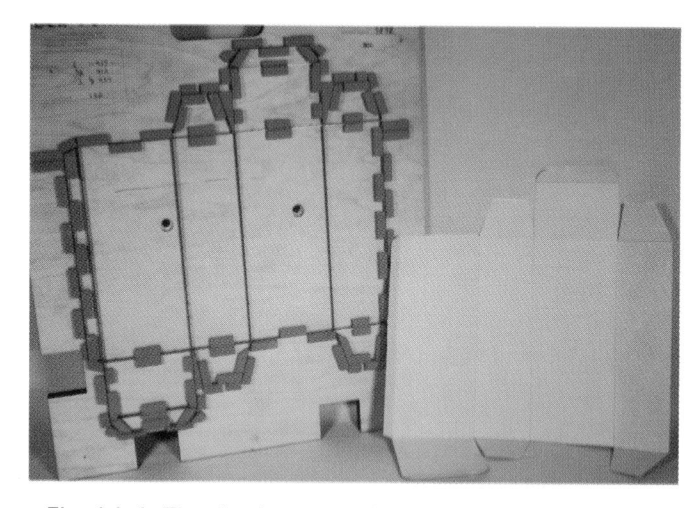

Fig. 14-4. The die shown on the left is used to stamp out the container on the right. The box will then be folded.

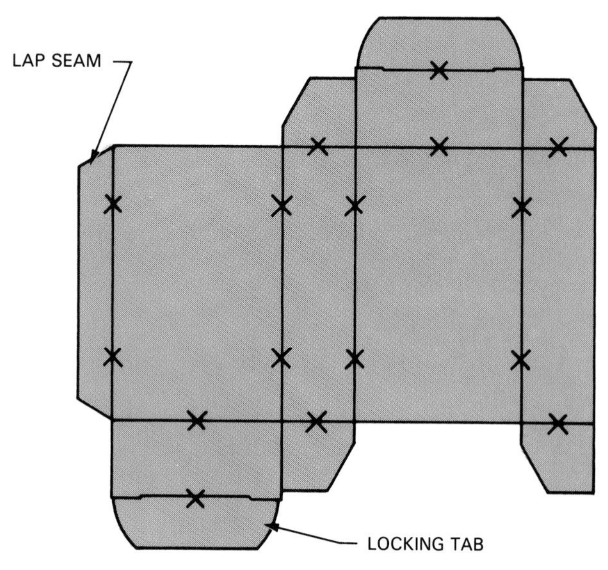

Fig. 14-3. This pattern will be used to make a package. Notice the fold lines (x), seam, and locking tabs.

Fig. 14-5. These packages have been taken apart at the seam so that the patterns can be studied.

Fig. 14-6. This logo is an indication that the package is made from at least fifty percent recycled material.

FOLDING CARTONS

Folding cartons are one of the most common types of packaging. Folded carton design involves several steps. The package dimensions must be determined. See Fig. 14-7. Then paperboard must be selected. Various types of paperboard, their descriptions, and uses are given in Fig. 14-8.

Folding cartons are folded in a variety of ways, depending upon their uses. Some typical folding carton styles are shown in Fig. 14-9.

As with other packages, folding cartons often feature artwork. However, before final artwork can be created, a **die vinyl** must be produced. See Fig. 14-10.

BOARD	DESCRIPTION	USES
Clay Coated Kraftback (''CCKB'')	Heavy duty Kraft board with white top liner (clay coating) used for eye appeal.	For heavier products such as parts or fasteners where strength and a printable surface are required.
Clay Coated Newsback (''CCNB'')	Very smoth white board with good printing surface. Top is white. Back is manila to gray.	Cartons for dry foods, cosmetics, etc., where high quality printing is required.
Solid Bleached Sulfate (''SBS'') Coated One Side	Good printability, uniformity, scoring, and folding. White throughout.	Dry foods, cosmetics, pharmaceuticals, hospital supply. Where quality printing and clean appearance are required.
Kraft	Brown throughout sheets. A coarse board noted for its strength.	Hardware, automotive, mechanical parts, toys, fasteners. Where extra strength is required.
E-Flute	Extra thin corrugated sheet. Corrugations add strength and rigidity.	Retail boxes for toys, hardware. Also point-of-purchase display boxes.

Fig. 14-8. Various types of boards are used in producing folding cartons. (Calumet Carton Co.)

HOW TO DETERMINE THE DIMENSIONS OF A FOLDING CARTON

Place the carton facing you (as it would appear on the retail shelf). The longest open end dimension from left to right – is the LENGTH. The shortest open end dimension front to back – is the WIDTH. The DEPTH is the remaining dimension – top to bottom. **It is important that these dimensions be given in just this sequence – otherwise the carton shape will not be what you had intended.**

L x W x D

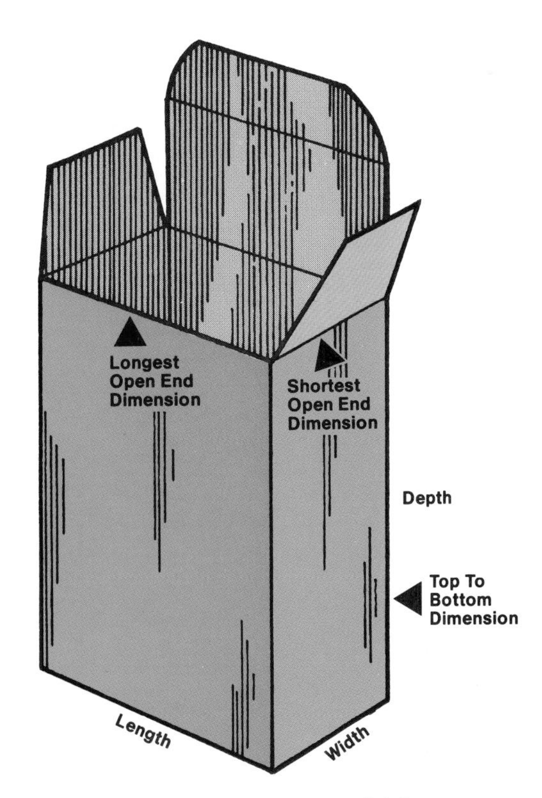

Fig. 14-7. Determining dimensions is the first step in producing a folding carton. (Calumet Carton Co.)

A die vinyl is simply a tracing of the carton outline directly from the cutting die onto a piece of vinyl material. The artist must then work within the outline to assure that the art will fit the carton when it is produced. See Fig. 14-11. The carton is then printed. The final step to complete the finished carton involves applying glue or staples, Fig. 14-12.

SHEET METAL PRODUCTS

Patterns are used for sheet metal products, such as those shown in Fig. 14-13. The sheet metal is cut, shaped, and assembled using special equipment. Seams are used for fastening the object together. Metal seams are soldered, welded, or riveted. For strengthening

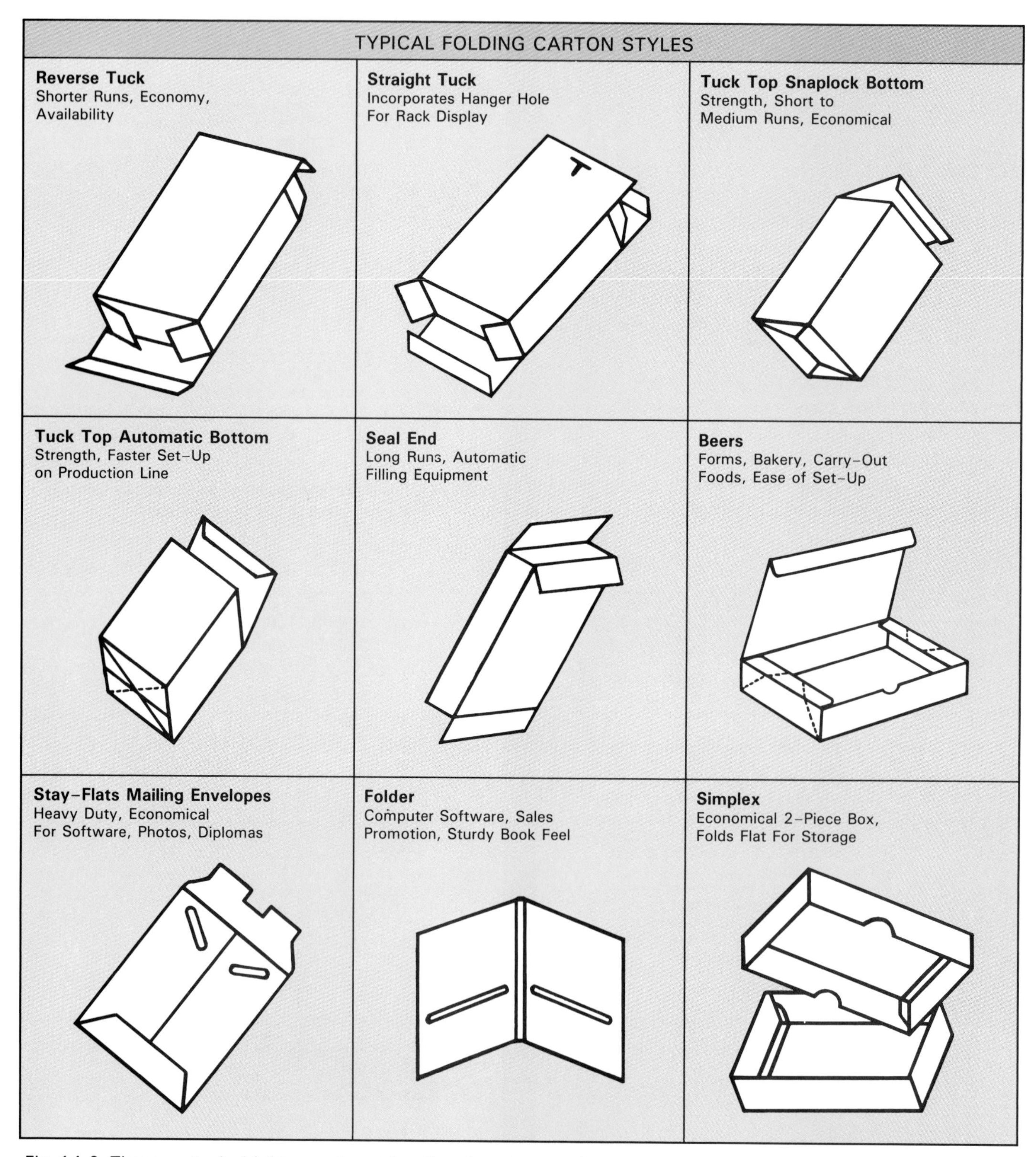

Fig. 14-9. These are typical folding carton styles. Certain styles are often used for specific purposes. (Calumet Carton Co.)

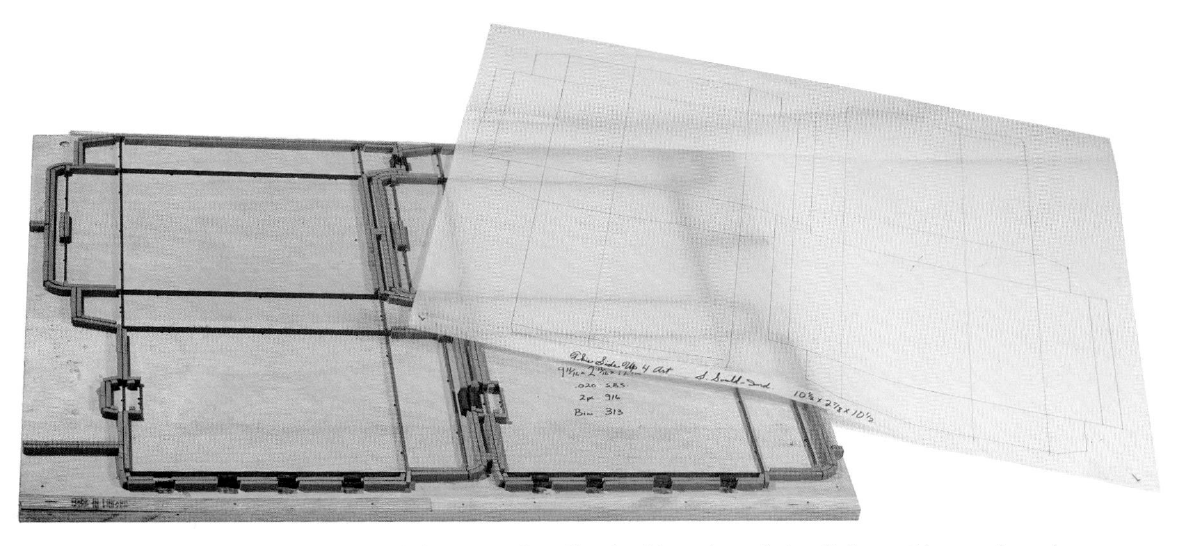

Fig. 14-10. Before artwork is created, a die vinyl (tracing of the die) must be produced. (Calumet Carton Co.)

Fig. 14-11. In producing artwork for a package, the artist must work within the outline of the die vinyl to assure that the art will fit the carton when it is produced. (Calumet Carton Co.)

Fig. 14-13. A pattern was needed for each of these sheet metal products.

Fig. 14-12. One of the final steps in producing a folding carton is the application of glue to the glue flap. (Calumet Carton Co.)

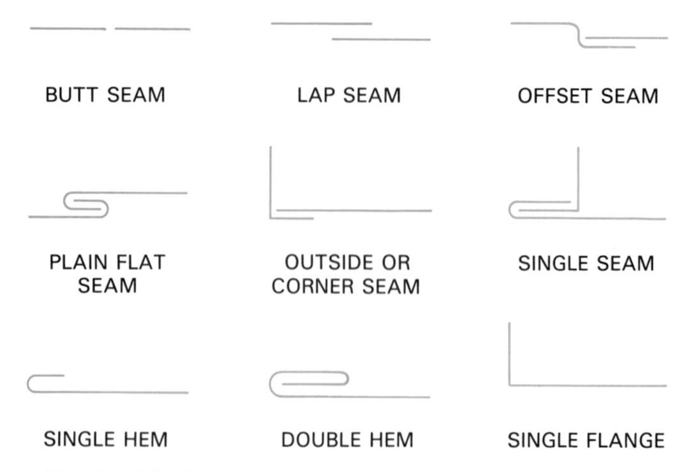

Fig. 14-14. A variety of techniques is used to join sheet metal edges. After the seam or hem is made, it may be soldered, welded, or riveted.

edges, **hems** are often used. A variety of seams and hems are shown in Fig. 14-14.

PARALLEL LINE DEVELOPMENT

Parallel line development is a technique used for making patterns for prisms and cylinders. When unfolded, these objects are square or rectangular in shape. The sides have parallel edges.

PRISM

The steps for drawing a pattern for a prism are described below and shown in Fig. 14-15.

1. Draw a front and top view on the left side of your drawing substrate.
2. Project construction lines over from the front view. Either of these can be the **stretch-out line,** which is a line with a length equal to the sum of the sides.
3. Starting near the front view, lay out the width of each side on the stretch-out line. Find these measurements on the top view. Draw vertical lines for the sides.
4. Lay out the top and bottom of the prism.
5. Add seams, hems, or locking tabs, if needed.
6. Darken the pattern with visible lines. Use dashed lines for folds or mark the side with an X.

TRUNCATED PRISM

The steps for drawing a pattern for a truncated prism are described below and shown in Fig. 14-16. When a geometric solid is truncated, this means that an end has been cut off at an angle to the opposite end.

1. Draw the front and top views. The front view should be in profile.

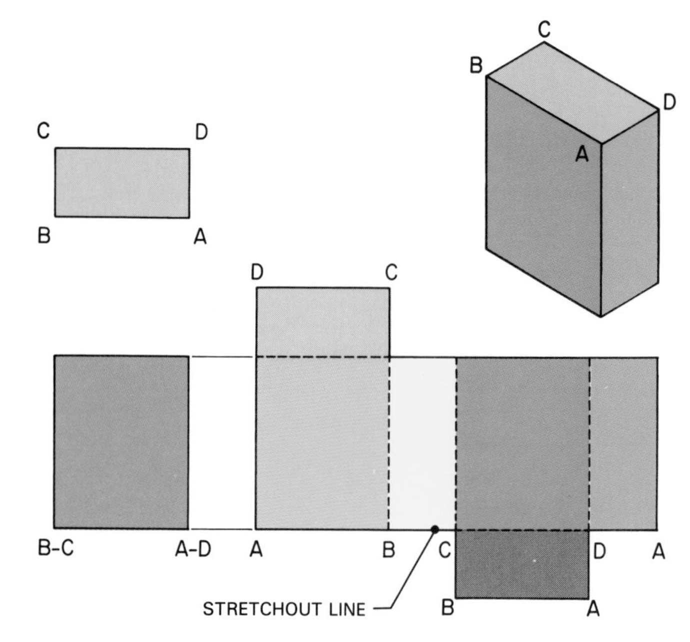

Fig. 14-15. A pattern for a prism is shown. Begin by drawing a front and top view, as shown on the left side.

2. Project three construction lines over from the front view. In this case, the bottom line is the stretch-out line.
3. Starting near the front view, lay off the width of each side on the stretch-out line. Begin with the front side.
4. Lay out the rest of the front side on the pattern. Moving counterclockwise, lay out the remaining sides.
5. Lay out the top and bottom. Add seams, hems, or tabs, if needed.
6. Darken the pattern using visible lines and fold lines.

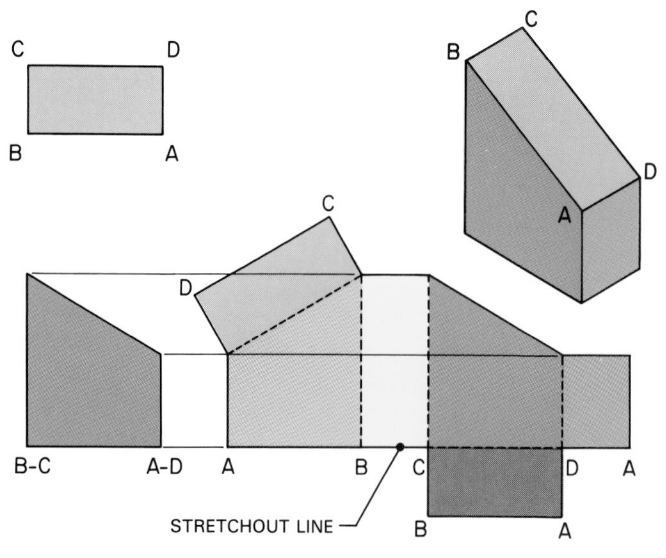

Fig. 14-16. The pattern for a truncated prism is made by first projecting construction lines over from the front view.

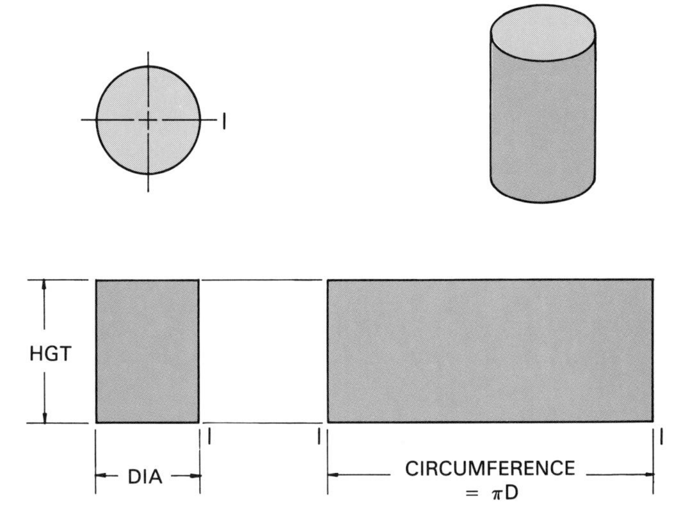

Fig. 14-17. A pattern for a cylinder is shown in the lower right corner. The formula for circumference is given.

CYLINDER

The steps for drawing a pattern for a cylinder are described below and shown in Fig. 14-17.

1. Draw the front and top views.
2. Find the circumference of the cylinder. Multiply 3.1416 (pi) x cylinder diameter.
3. Draw construction lines over from the front view.
4. Mark the circumference of the cylinder on a stretch-out line, and draw vertical lines to complete the rectangle.
5. Add seams, hems, or tabs.
6. Darken with visible lines.

TRUNCATED CYLINDER

The steps for drawing a pattern for a truncated cylinder are described below and shown in Fig. 14-18.

1. Draw the front view (in profile) and the top view.
2. Project the top and bottom over from the front view. In this case, the bottom line will be the stretch-out line.
3. Divide the top view into equal angles, such as 30°. A 30°-60° triangle works well for this. Number the divisions starting with 1.
4. Project down from the top view to find these points on the front view. Record the projection numbers in the front view.
5. Using dividers, transfer the distance between numbers on the circumference to the stretch-out line. Draw vertical construction lines at these points. Number the points.
6. Transfer height dimensions for the front view to the pattern. Start with the lowest side, which is point 1. Also, end with point 1, since they will be the same when connected.
7. Connect the points with an irregular curve.
8. Add seams, hems, or tabs.
9. Darken with visible lines.

RADIAL LINE DEVELOPMENT

In **radial line development**, edges are not parallel as they are for prisms and cylinders. Instead, lines "radiate" out from a point. Cones and pyramids are drawn using this technique.

CONE

The steps for drawing a pattern for a cone are described below and shown in Fig. 14-19.

1. Draw the front and top view of the cone.
2. Divide the circular view into equal angles, such as 30°. Number the divisions.

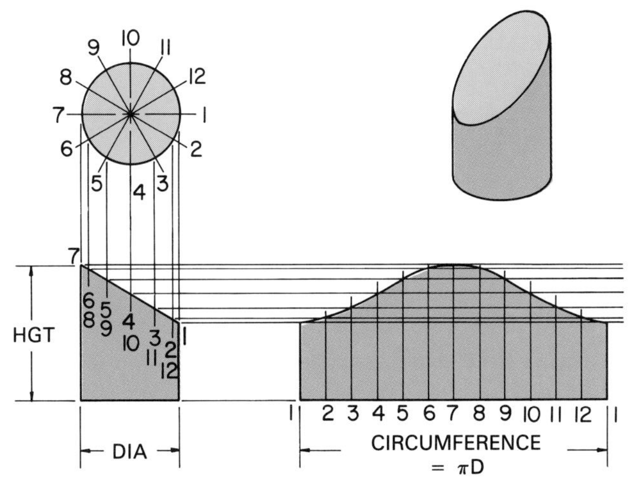

Fig. 14-18. When developing a truncated cylinder, the top view must first be divided into equal angles and numbered.

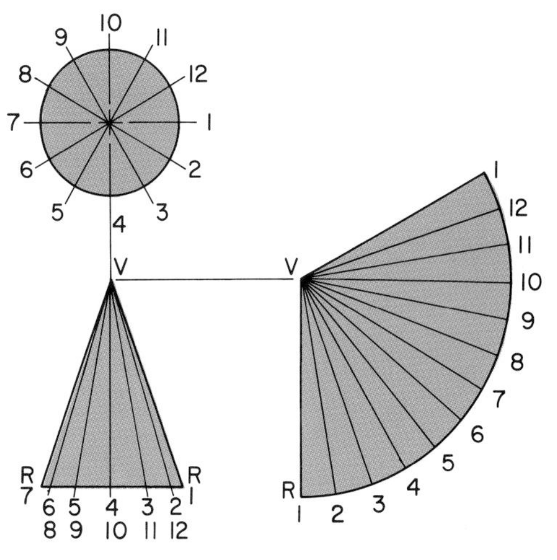

Fig. 14-19. The radial line method is used for developing a cone.

3. Using the side of the cone as a radius, swing an arc for the pattern. Draw a vertical line (VR) through the radius.
4. Using dividers, transfer the divisions from the circular view to the arc. Begin and end with number 1.
5. Connect points on the arc with V (vertex).
6. Add seams, if needed.
7. Darken the pattern with visible lines.

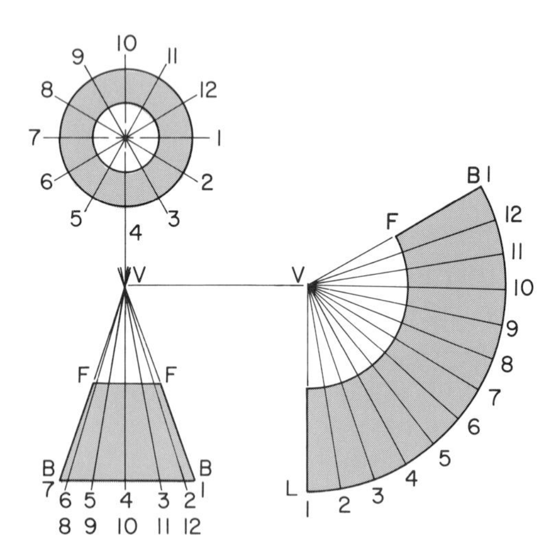

Fig. 14-20. The development of a frustum of a cone is shown here.

FRUSTUM OF A CONE

The steps for drawing a frustum of a cone are described below and shown in Fig. 14-20. The cone is called a frustum because the apex (pointed end) has been cut off parallel with the base.

1. Draw the front and top view.
2. Using construction lines, extend the sides of the frustum until they intersect at the vertex (V).
3. Swing two concentric arcs from point V, one with radius VF, and one with radius VB. Draw a straight line (VL) from the vertex through the arcs to begin the pattern.
4. Equally divide the circular top view, as before, and number the divisions.
5. Transfer the divisions with dividers along the outside arc, beginning at L. Begin and end with the number 1.
6. Connect points on the arc with the vertex on the pattern.
7. Add seams, if needed.
8. Darken with visible lines.

SUMMARY

Patterns are used for products you use every day. Packages and sheet metal products often use patterns. Parallel line development and radial line development are two methods used in producing patterns.

WORDS TO KNOW

All of the following words have been used in this chapter. Do you know their meanings?

die vinyl
packaging
parallel line development
pattern
radial line development
seams
stretch-out line

REVIEWING YOUR KNOWLEDGE

Please do not write in this text. Write your answers on a separate sheet.

1. A _________ is a flat plan for an object that will be produced and assembled.
2. True or false? Packaging is one of the largest industries that uses patterns.
3. Which of the following is used to join parts of a package together?
   a. Tab.
   b. Seam.
   c. Stapler.
   d. All of the above.
4. Which of the following is a method used for fastening sheet metal seams?
   a. Soldering.
   b. Welding.
   c. Riveting.
   d. All of the above.
5. _________ line development is a technique used for making patterns for prisms and cylinders.
6. _________ line development is a technique used for making patterns for cones and pyramids.

APPLYING YOUR KNOWLEDGE

1. Prepare a display of various types of packages and sheet metal items that were produced using a pattern. Label each of them and write brief descriptions.
2. Collect some small packages. Take them apart and study their designs.
3. Design a package for a small object of your choice. Think about seams and locking tabs. Place a design on the outside.
4. Draw patterns for solids and cut them out to see if they are correct.

15

CHAPTER

COMPUTER-AIDED DRAFTING

After studying this chapter, you will be able to:
- *Describe the advantages and disadvantages of a CAD system.*
- *Identify the components of a computer drafting system.*
- *Describe the coordinate systems used in CAD.*
- *Explain the general steps involved in producing and editing a computer drawing.*
- *Demonstrate how to create a drawing with the aid of a computer.*
- *Identify the major kinds of three-dimensional drawings that can be created with CAD.*
- *Explain how a CAD/CAM system operates.*

Computer-aided drafting (CAD) is the process of using special software and a computer system to produce technical drawings. When design is a major part of the process, the system is sometimes called **computer-aided drafting and design (CADD)**.

CAD has a variety of applications in business and industry. It is used for drawing product plans, such as automobiles, appliances, and machinery. Architectural plans are often drawn with CAD. Electronic circuit boards and circuit layouts are drawn with CAD. Even landscaping plans can be drawn in this manner.

In order to make the best use of computer drafting equipment, you must understand how the computer system and software operate, Fig. 15-1. In addition, you must know about the type of drawing to be done. For example, mechanical drawings require an understanding of multiview drawings and dimensioning. Although discussed briefly in this chapter, computers, software, and drafting techniques are covered more extensively in other chapters of this text.

ADVANTAGES AND DISADVANTAGES OF CAD

There are several advantages of CAD that are important to consider. These include:
- Drawings can be generated more quickly using a CAD system than by other means. For example, dimensioning can be done automatically, saving a great deal of drawing time.
- Editing functions, such as erasures, are much simpler when done electronically.
- A design may be modified many times before the best one is chosen. A computer simplifies this task by allowing for quick modifications.
- Quality of drawings is better than by using other means. Lines drawn electronically will be consistent in width and density. Text and dimensioning will also be consistent. Human factors, such as the ability to draw a straight, dense line, are not a consideration with CAD systems.
- Drawings can be stored on magnetic media. Much less space is needed and new drawings can be plotted whenever necessary.

Fig. 15-1. CAD requires an understanding of both computer hardware and software.

- Drawings can be transmitted as data to other locations. If two designers are working on the same product in two different states, they can easily send their ideas to one another via computer.

Although there are many advantages of CAD, there are disadvantages as well. These include:

- A computer, software, and output device are needed in order to draw.
- People must know or learn how to operate the CAD system.

COMPUTER DRAFTING SYSTEM

A **computer drafting system** consists of the hardware and software necessary to create technical drawings. A typical system is shown in Fig. 15-2.

Fig. 15-2. This CAD system has a keyboard and mouse for input. A plotter (right) is being used as an output device. (Hewlett-Packard)

INPUT DEVICES

Input devices are used for placing a drawing in the computer. See Fig. 15-3. This process is sometimes called **digitizing**. For example, if a point is drawn, this is known as digitizing the point.

The **keyboard** is an input device used with most CAD systems. It can be used for entering text on the drawing. In addition, drawing commands can usually be made using the keyboard.

A **mouse** is one of the most common devices used for input. Movement of the mouse results in a corresponding movement of the cursor on the screen. Points are digitized on the screen by using the button on the mouse. The **joystick**, another input device, can be used in a similar manner to the mouse.

The **graphic tablet** is another type of input device. It has a flat surface on which to draw, with an electrical grid underneath. A **puck** or **stylus** is moved over the surface of the tablet as a locator device. A button is pressed on the puck for digitizing. The stylus digitizes by pressing it against the tablet.

The graphic tablet can also be used to digitize existing drawings by placing the drawings on the tablet and tracing them. Larger digitizing tables are also used for this purpose. Devices that allow for input of drawings that have already been created are known as **digitizers**.

CAD COMPUTER

The computer used for CAD must be compatible with the software. The most sophisticated programs require that a hard drive be used and additional **random**

Fig. 15-3. Notice the different input devices being used with these computer systems. A graphics tablet and stylus are shown on the right. (Hewlett-Packard)

access memory (RAM)**. In addition, because of the large number of mathematical calculations needed for computer graphics, a second processor is often needed. This is called a **math co-processor**.

OUTPUT DEVICES

Output devices are used to present the results of CAD. One of the most common output devices is the plotter, Fig. 15-4. A **plotter** is known as a hardcopy device since a drawing can be permanently stored on paper. There are a variety of plotters from which to choose. These include the drum plotter, flatbed plotter, and microgrip plotter. The differences in these are explained in Chapter 31 of this text.

Plotters draw with pens. Most plotters provide a choice of pens, so different colors and line widths can be used as necessary.

CAD SOFTWARE

CAD software varies considerably in capabilities and ease of use. Some programs are easy to learn, but are limited in capabilities. Others may be difficult to learn, but have very advanced features. In general, the decision on which software package to use is based on factors such as:

- Cost.
- Ease of use.
- Advanced features.

CAD software produces **vector graphics**, which are also known as object-oriented graphics. There are several advantages to this type of graphic software. First, objects are defined mathematically so they print or plot very precisely, even though they may show "jaggies" on the computer monitor. Second, since drawings are made of objects instead of dots on the screen, they can be manipulated separately. For example, a circle can be trimmed or moved on the screen as a separate entity.

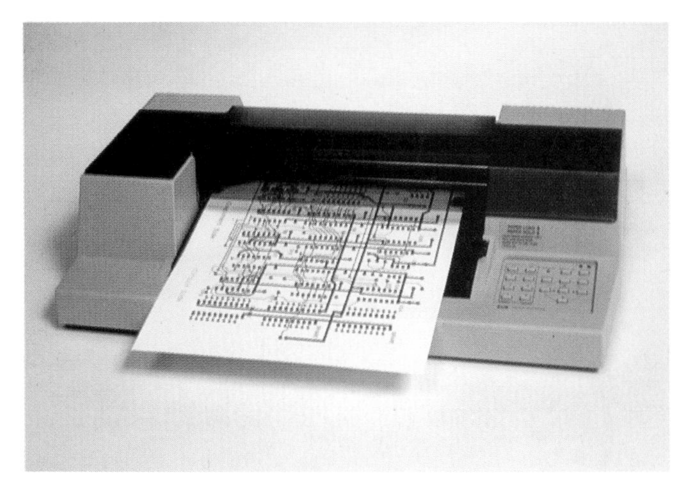

Fig. 15-4. The plotter is the most common output device used with CAD systems. (Hewlett-Packard)

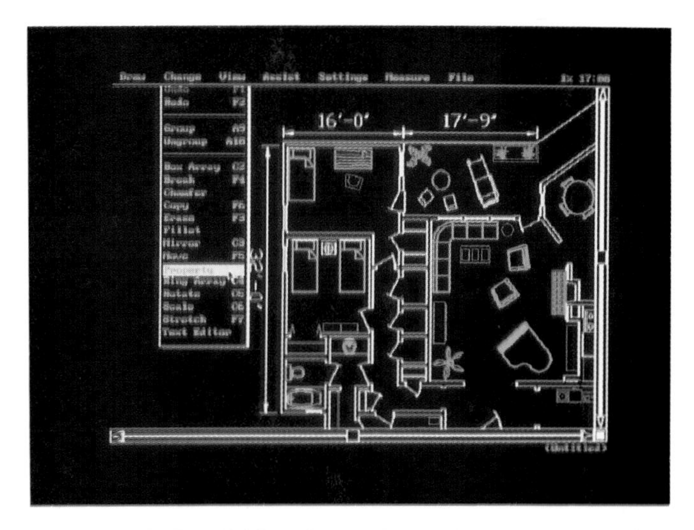

Fig. 15-5. This CAD software has pull-down menus. A command is being selected, as shown by the highlighted area. (Autodesk, Inc.)

CAD software is usually either menu-driven, command-driven, or a combination. In **menu-driven systems**, a menu of command options is available on the drawing screen, or on a tablet menu in the case of the graphic tablet. Menus are easy to use, since little memorization of commands is required.

A common form of screen menu is the **pull-down menu**, Fig. 15-5. When an item is selected from the main menu, other command options appear in a column on the screen.

A **command-driven system** uses commands typed in from the keyboard. This requires either memorization or a reference chart.

Most sophisticated CAD packages provide the combination option of using the keyboard or menu for entering commands. Experienced users usually prefer this combination since some commands are more quickly entered with the keyboard.

COORDINATE SYSTEMS

A **coordinate system** is used in CAD to precisely place or locate entities (objects) on a drawing. For example, there is a dot in the drawing of the display screen in Fig. 15-6. How would you tell someone where the dot is located? You could measure and say it is 2 in. from the left side, but how far is it from the bottom? If you say it is 2 in. from the right side and 2 in. up, then someone can locate it. These horizontal and vertical dimensions are known as *X and Y coordinates.*

The **Cartesian coordinate system** provides an accurate method for locating points on the screen. The X

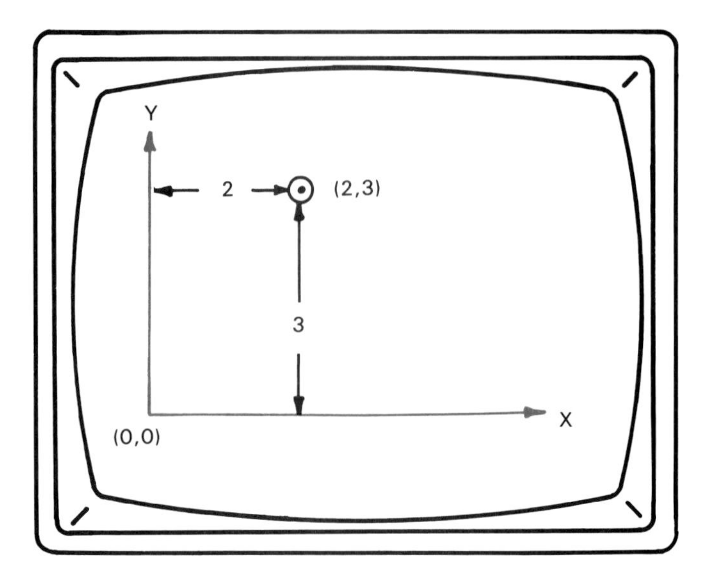

Fig. 15-6. X and Y coordinates are used to locate the position of the dot on the screen.

axis is used for locating points horizontally, and the Y axis is used for locating points vertically. There are four quadrants in this system as shown in Fig. 15-7. However, for many drawings, only the first quadrant is placed on the screen with the origin at the lower left corner. In this way, only positive numbers are needed.

There are three ways that Cartesian coordinates are used for making drawings. These include:

- Absolute coordinates.
- Relative coordinates.
- Polar coordinates.

ABSOLUTE COORDINATES

With **absolute coordinates**, all XY coordinates are specified in relation to the origin (0,0). For example, X3, Y4 means three units over and four units up from the origin.

RELATIVE COORDINATES

With **relative coordinates**, the origin moves to the last point drawn, Fig. 15-8. For instance, a point is placed on the screen at X2, Y2 using absolute coordinates. Now, changing to relative coordinates, this becomes X0, Y0. A point six units to the right is located by X6, Y0. Now this point becomes the new origin. Using this system, the origin is relative, which means it moves dependent upon what was last drawn.

POLAR COORDINATES

Polar coordinates are helpful in drawing round objects or any object where a specific angle is needed. Using this system, a distance and angle are specified to locate points on the drawing. The origin is in the center of concentric circles, with circles having a value of one unit, two units, etc. See Fig. 15-9. Angles are specified on the circles from 0° to 360°. Using this system, the ends of a line eight units long and slanted at a 38° angle would be specified as X0, Y0, (first point) and 8, 38.

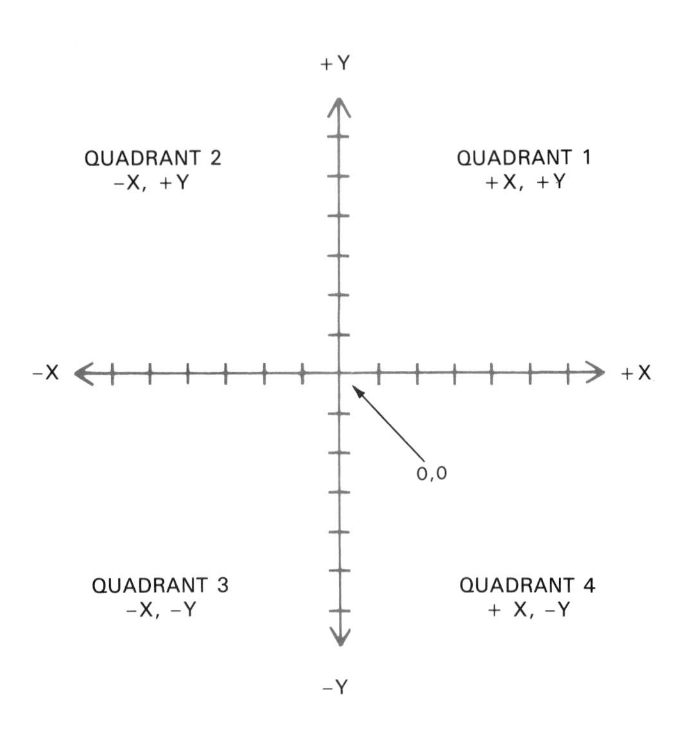

Fig. 15-7. Although the Cartesian coordinate system has four quadrants, the first quadrant is often used for drawing. The quadrants share the same origin (0,0).

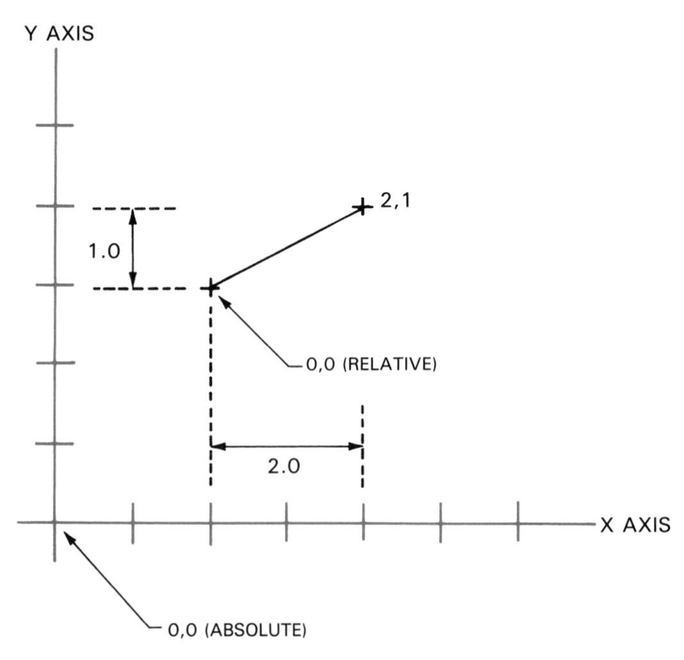

Fig. 15-8. Drawing a line with relative coordinates is shown here. The starting point of the line becomes the new origin, or 0,0. The other end of the line is then specified as 2,1, since it is being drawn from the new origin.

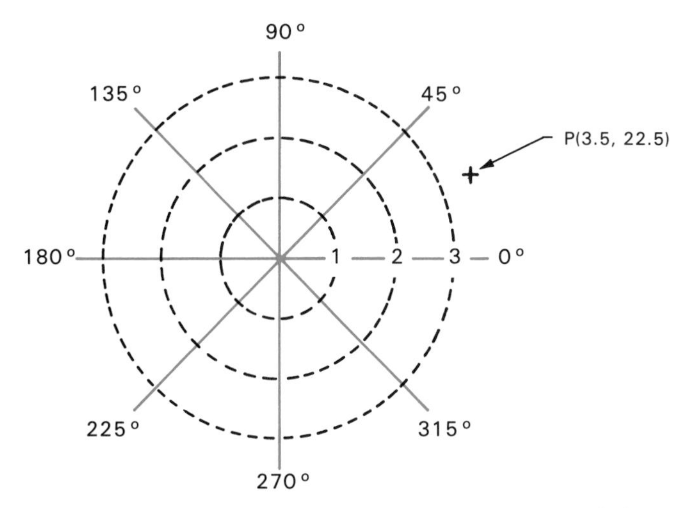

Fig. 15-9. The polar coordinate system is shown. Points are located by the distance from the origin and angle. In this example, the point is located at 3.5 units from the origin (center) and at a 22.5° angle.

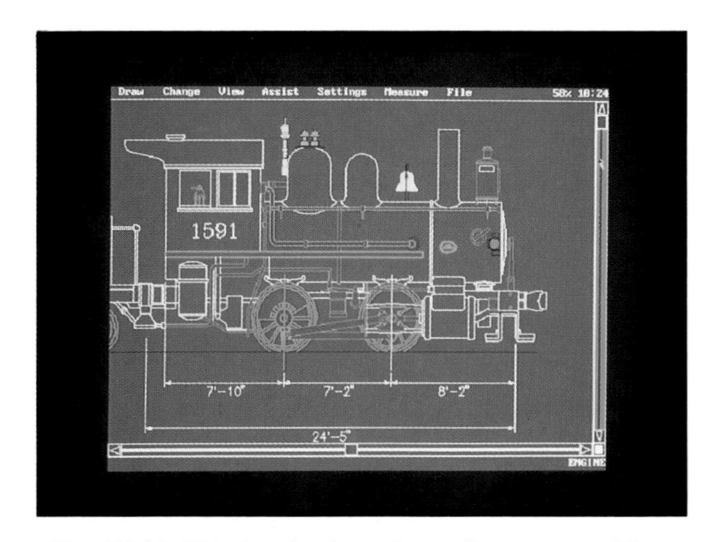

Fig. 15-10. This drawing is made up of separate entities, such as lines and circles, that have been placed together to make a picture. (Autodesk, Inc.)

The 8 specifies length and the 38 specifies the degree of slant.

UNIT MEASUREMENT

Units are a measure of distance on a drawing. They are the same as the coordinate numbers. For example X2, Y3 means to move two units to the right and three units up.

The term "units" is used because it can be used to represent different measurement systems. For example, the user might define a unit as 1 in. for a drawing, and then as 1 mm for another drawing.

Many CAD systems allow the operator to change the format of the measurement. For example, fractional inches could be specified in fraction form or decimal form. The coordinate will then be displayed in this same way as will the dimensions.

ENTITIES

Entities are the elements that make up a drawing, Fig. 15-10. These are sometimes called **primitives**. These include points, lines, arcs, and circles. Entities are combined to make the drawing.

Entities are located on the drawing by using the Cartesian coordinate system. A point can be specified with an XY coordinate, such as X2, Y4. A line is defined by its end points, such as X2, Y4, and X3, Y6.

Modifiers further define how an entity is drawn. For example, there are several ways a circle might be drawn using modifiers. In one method, three points can be digitized on the screen and a circle is drawn whose circumference touches the three points. In another method, the location of the center is digitized, and the diameter is entered from the keyboard. In CAD systems, most entities have modifiers that can be used with them.

DIMENSIONING

Most CAD systems are capable of automatic dimensioning, Fig. 15-11. However, it is up to the CAD operator to specify where these dimensions are located. This is why the CAD user should understand correct dimensioning procedures before dimensioning the drawing.

There are several methods that may be used when dimensioning a drawing, depending upon what is being dimensioned. These include:

- Linear dimensioning.
- Angular dimensioning.
- Diameter dimensioning.
- Radius dimensioning.

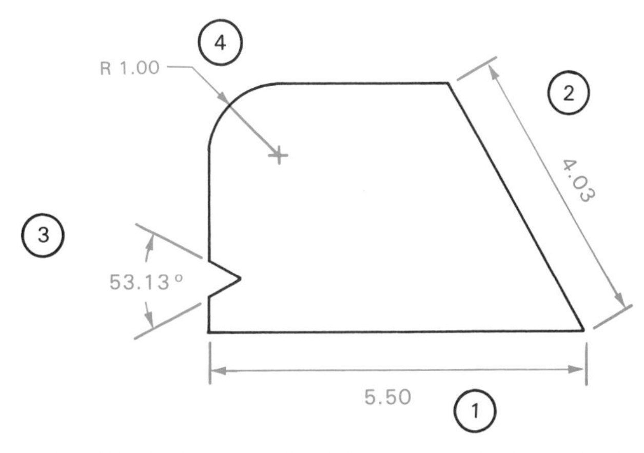

Fig. 15-11. Different methods for automatic dimensioning are shown. These include: (1) horizontal dimensioning, (2) aligned dimensioning, (3) angular dimensioning, and (4) radius dimensioning.

LINEAR DIMENSIONING

Linear dimensioning is used to place measurements parallel to straight lines on a drawing. Modifiers can be used to specify whether the line being dimensioned is horizontal, vertical, or slanted. The modifier for slanted lines is called "aligned" in some programs and "point to point" in others.

ANGULAR DIMENSIONING

The **angular dimensioning** method specifies angles in degrees. The dimension line is drawn as an arc with the angle specified.

DIAMETER DIMENSIONING AND RADIUS DIMENSIONING

Diameter dimensioning and **radius dimensioning** are used for dimensioning round features such as circles and arcs. Diameter dimensioning is typically used for circles, while radius dimensioning is used for arcs. For large circles and arcs, the dimension can be placed inside the feature. For small circles and arcs, the dimension is placed outside the feature with a leader.

DRAWING AIDS

Drawing aids are features in CAD programs that simplify the drawing process. Some of the most common of these include:

- Grid.
- Snap grid.
- Panning.
- Zoom.
- Layering.

GRID

A **grid** is a set of reference dots on the screen in rows and columns. The user can specify how far apart these dots should be in units. Entities, such as lines, can be drawn using the grid points to determine location. The grid is only used as a reference and does not plot out with the drawing.

GRID SNAP

A function that is often used with the grid is **grid snap**. Grid snap ensures that what is drawn goes to a point on the grid. This is very useful since it is sometimes difficult to be sure that the cursor is directly on a grid dot.

ZOOM

The **zoom** command allows the user to enlarge or reduce a portion of the drawing, Fig. 15-12. In some programs, this is known as **windowing**. Zooming is necessary when working on small details on the drawing that are not clearly shown on the monitor. *Zoom in* refers to enlargement of a portion of a drawing. The amount of enlargement can be specified as part of the zoom command.

Zoom out or window out is used to reduce the drawing size on the screen. In order to fit the entire drawing on the screen, zoom origin, zoom full, or window full, are used, depending on the program.

PANNING

Another useful drawing aid is **panning**. Panning allows the user to move the drawing on the screen. For

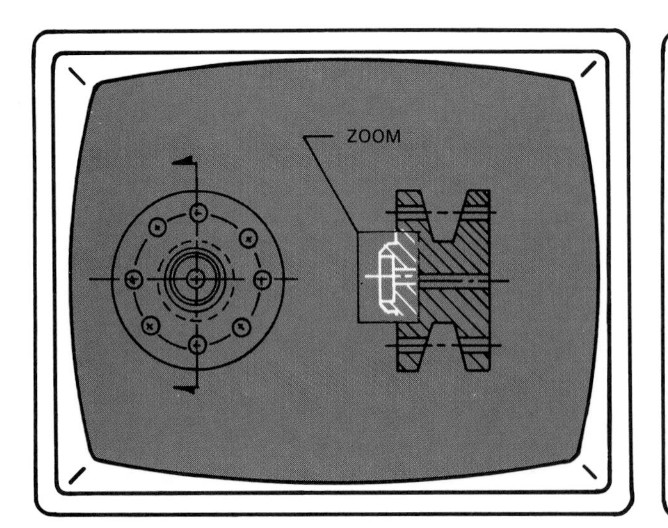

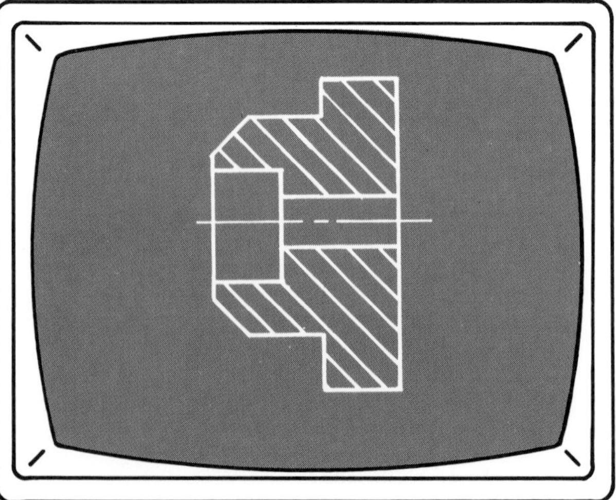

Fig. 15-12. In order to enlarge a portion of the drawing on the left, the zoom command is used. The results are shown on the right.

example, you may zoom in and find that the part of the drawing you wish to work on is half off the screen. Panning is used to center the magnified view.

Panning is accomplished by digitizing at two points on the screen. The second point denotes which direction and how far you wish to move. In some programs, commands can also be used for this, such as pan left or pan down.

LAYERING

Layering occurs when drawings are made that are meant to overlay one another. Most CAD systems are capable of layering. A useful example of layering is in architectural drawing, Fig. 15-13. A floor plan can be drawn as a first layer. Utilities can then be added to other layers, such as electrical on the second layer and plumbing on the third layer. By plotting layers 1 and 2 together, the floor plan is drawn with the electrical details. Layers 1 and 3 provide plumbing details on the floor plan. This technique allows using the floor plan for several drawings without redrawing it. Layers can be plotted separately or together as needed.

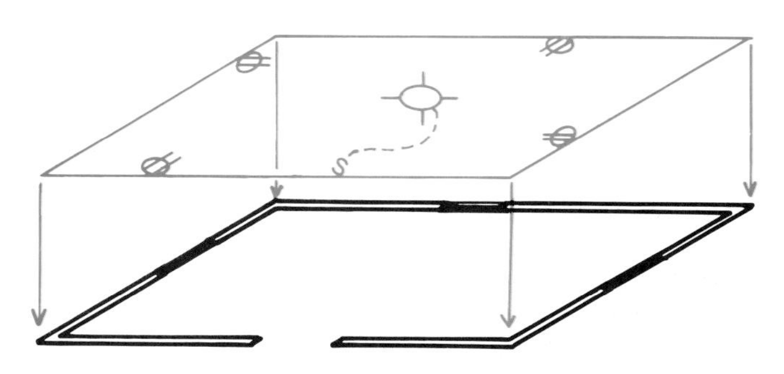

Fig. 15-13. Two layers of a simple floor plan are shown. The first layer shows basic size and shape. The second layer contains electrical details. The arrows show how the layers fit together.

EDITING

Editing involves changing something that has been previously drawn. One of the most common editing commands is **erase**. Erase allows you to remove one or more entities from the drawing. When several entities are to be removed, most CAD programs allow you to create a window around those entities to identify the ones to be removed.

In some instances, only part of an entity needs to be removed. In this case, the **trim** command is used. For example, a line or arc can be shortened when this command is used.

The **move** command allows you to move one or more entities to another location on the drawing. The **rotate** command is similar to the move command, but in this case movement takes place at an angle around a point. The amount of rotation can be specified in several ways, such as digitizing points on the screen or inputting degree of rotation on the keyboard. A group of entities can be rotated by first placing a window around them.

The **copy** command makes a duplicate of one or more entities that can then be placed elsewhere on the drawing, Fig. 15-14. This is a handy command to use to avoid redrawing detailed objects, such as screws.

The **mirror** command copies an entity, or group of entities, in reverse position from the original, like a mirror image, Fig. 15-15. A mirror line is defined and the mirror copy is created on the opposite side of the line. The mirror command is very helpful when a drawing is symmetrical, or similar on both sides. You can draw

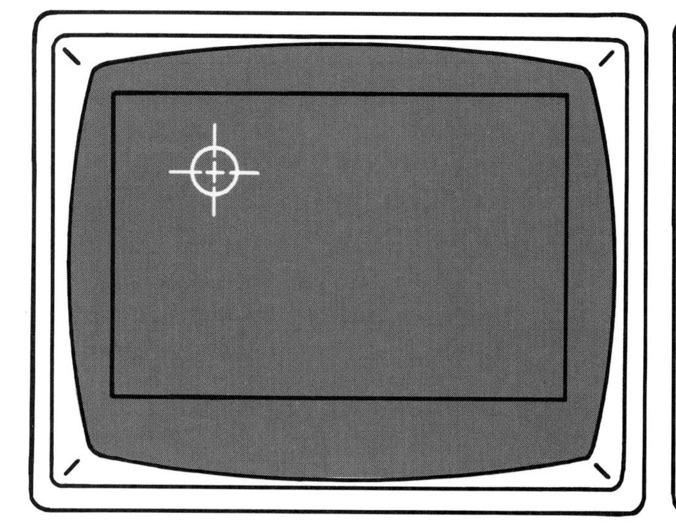

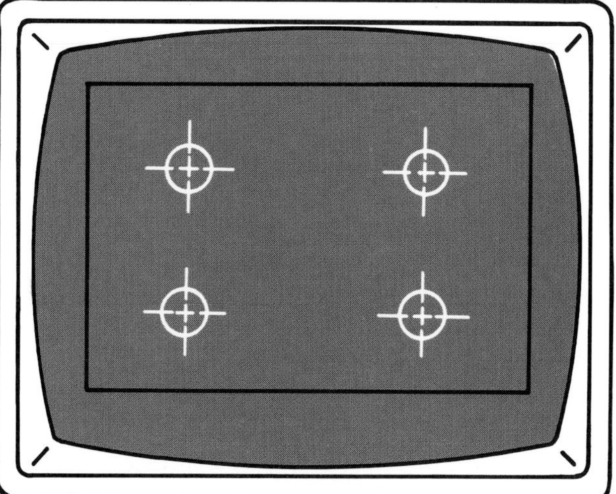

Fig. 15-14. By using the copy command, a single entity (left) can be copied to other locations, as shown on the right.

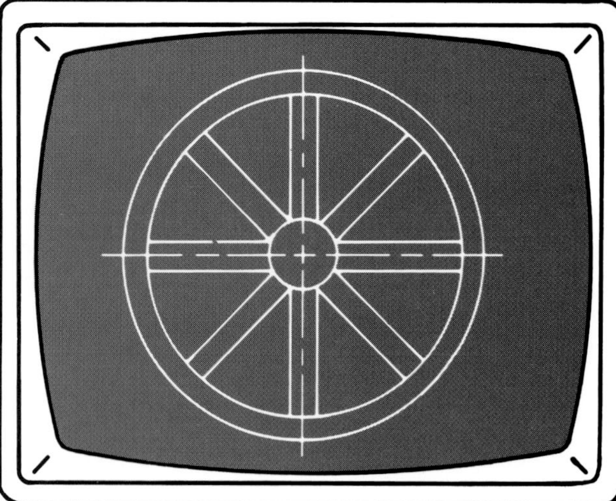

Fig. 15-15. After drawing half of this object, the mirror command is used to draw the other half.

half the drawing and then use the mirror command to create the other half.

Most CAD systems have additional editing features, but some of the more important commands have been explained here. You will learn other editing techniques as you learn about specific CAD systems.

PRODUCING A DRAWING BY COMPUTER

While steps may vary according to the software package and the type of drawing, there is usually a general sequence of events that is followed when producing a drawing by computer. A suggested sequence of events is the following:

1. Make a sketch of what is to be drawn, either on grid paper or by computer. This helps in visualizing the drawing and reduces mistakes.
2. Using the necessary procedures, boot the CAD program.
3. As needed, set up parameters for the drawing. These include such aspects as unit of measurement, substrate size for plotting, and drawing scale.
4. To help you in creating the drawing, set up a grid on the screen to a convenient size. Use snap grid if needed.
5. Draw the entities using absolute, relative, or polar coordinates. Use editing commands for corrections and to manipulate the objects on the screen. Drawing aids, such as zoom and pan, are used when creating detailed portions of the drawing.
6. Dimension the drawing. Use linear, angular, radius, and diameter dimensioning as needed.
7. Place text on the drawing, such as notes.
8. Save the drawing. In fact, you may wish to save the drawing intermittently throughout the process.
9. Plot the drawing.
10. Review, edit, and plot the drawing, as needed, until it is correct.

PICTORIAL OR THREE-DIMENSIONAL DRAWING

Three-dimensional CAD drawings show height, width, and depth in one view. At the simplest level, the objects can be drawn in the same way that pictorial drawings are drawn manually, with the illusion of depth. Isometric and oblique drawings can be made in this way by computer. Some systems have special grids that can be used to assist in creating the drawings.

Some CAD systems allow the creation of models. These are true three-dimensional drawings in electronic form. Models can be rotated on the screen because the computer "thinks" of them as three-dimensional. The three major types of models include:

- Wireframe model.
- Surface model.
- Solid model.

WIREFRAME MODEL

In the **wireframe model**, all edges of the object are shown as lines, including edges that would normally be out of sight to the observer, Fig. 15-16. This can be confusing, so a "hide" command is usually available that removes some of the unnecessary lines.

SURFACE MODEL

In the **surface model**, the object is defined as a series of surfaces. For example, a cube would be stored by computer as six square surfaces joined at the edges.

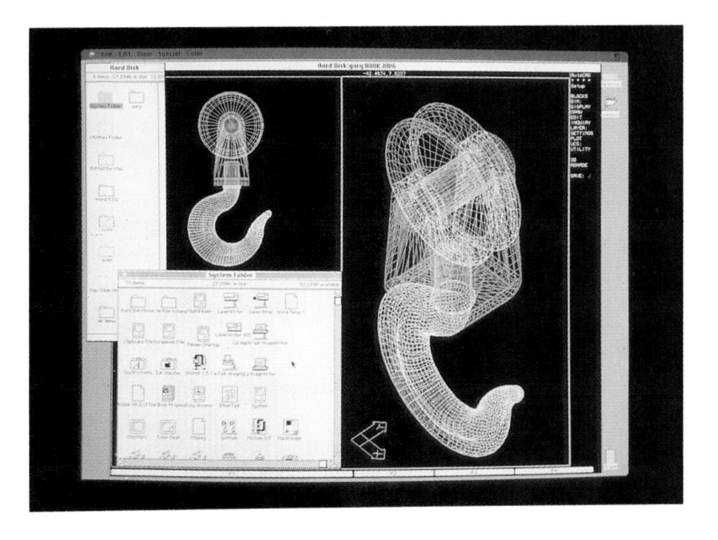

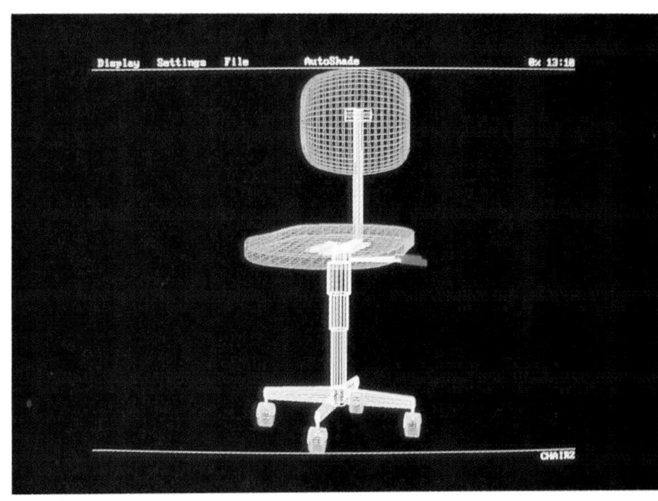

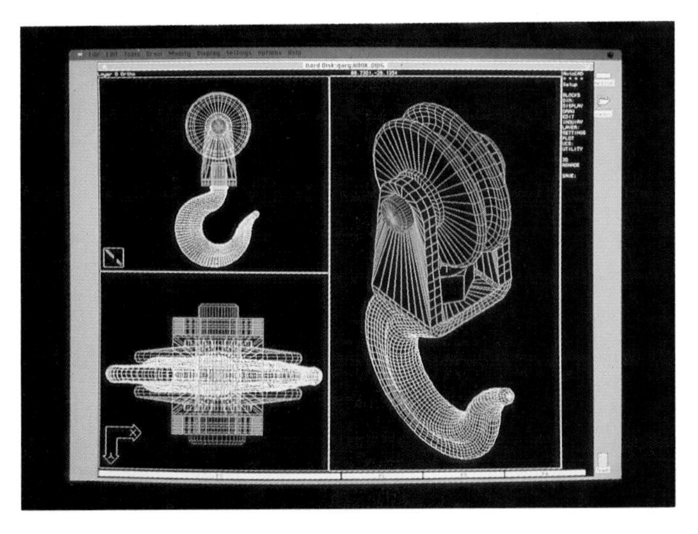

Fig. 15-16. The drawing of the large hook on the top is a wireframe model. In the second example (bottom), lines that are behind the object have been removed. (Autodesk, Inc.)

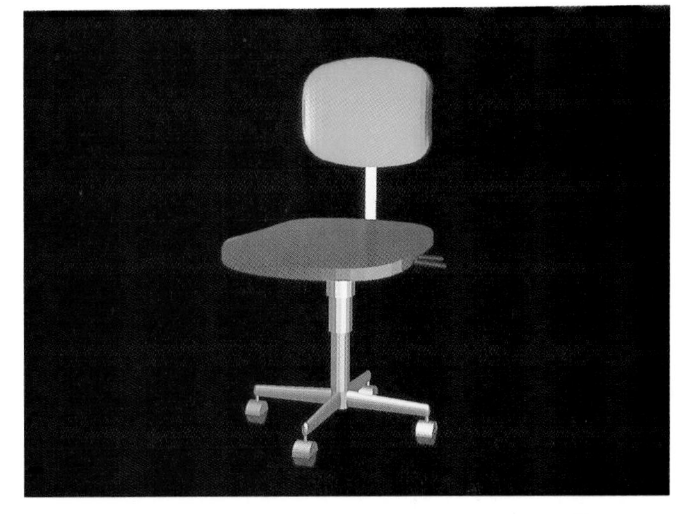

Fig. 15-17. The wireframe model shown on the top is the first step in producing the surface model shown on the bottom. Notice how the surface of the chair has been shaded for a more realistic look. (Autodesk, Inc.)

Colors and shades may be used on the surfaces for a more realistic effect.

A surface model is constructed by first drawing a wireframe model. The surfaces are then defined, which is like putting a cover over the wireframe, Fig. 15-17.

SOLID MODEL

As the name implies, a **solid model** is stored in the computer as an object with mass or volume, Fig. 15-18. This is the most sophisticated modeling technique and is often used for testing parts by computer. For example, a simulation of heat being applied at one side of an object would be seen to actually move through the solid object. Stress on parts can also be tested in this manner.

Another advantage to solid modeling is the capability of seeing section views. For instance, if you had a

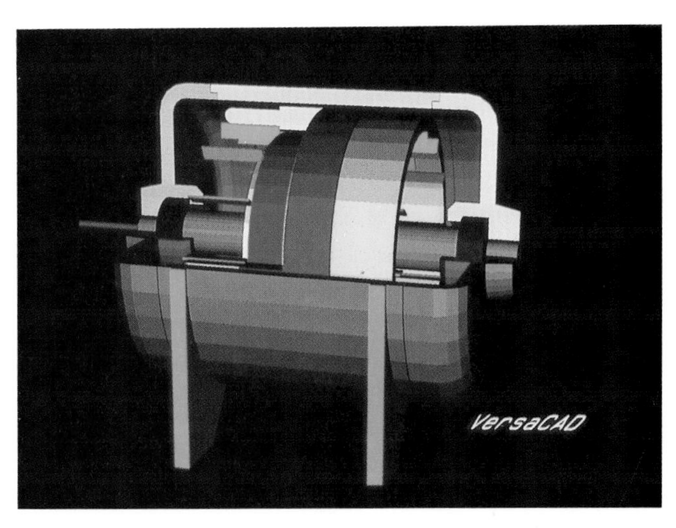

Fig. 15-18. Solid models can be sectioned to show internal details. (Computervision)

drawing of an internal combustion engine drawn as a solid model and then sectioned, you would be able to see all the internal details at the cut line.

When models are drawn, a third axis is used in addition to X and Y. This is known as the *Z axis*. The Z axis is at right angles to both the X and Y axis, and allows the creation of three dimensions for an object on the screen, Fig. 15-19.

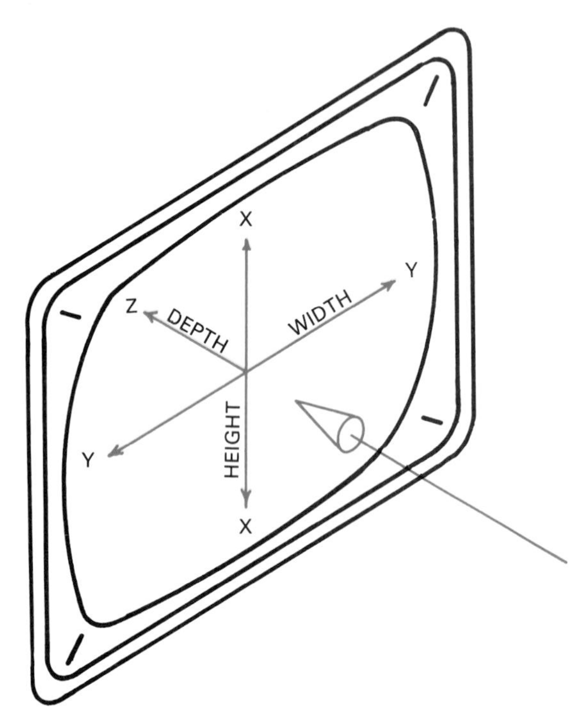

Fig. 15-19. Since three dimensions are needed when constructing models, three axes are used. The third axis is the Z axis.

CAD/CAM

Using a computer to operate machines in industry is known as **computer-aided manufacturing (CAM)**. These computer-operated machines are known as **computer numerical control (CNC)** machines. Tasks that these tools can perform include drilling, machining, welding, and cutting.

In CAD/CAM systems, the CAD system is first used to create drawings of the part to be manufactured. After the drawings are created, they are stored on magnetic tape or a disk that can be used as input to the CNC machine. The parts are then manufactured using the drawings originally created with the CAD system.

SUMMARY

CAD is an important tool for drafting in today's world. This chapter has provided a general introduction to CAD. Although systems will vary, once you have a working knowledge of one program, the information should be transferable as you learn other CAD programs.

WORDS TO KNOW

All of the following words have been used in this chapter. Do you know their meanings?

absolute coordinates
angular dimensioning
Cartesian coordinate system
command-driven system
computer drafting system
computer numerical control (CNC)
computer-aided drafting and design (CADD)
computer-aided drafting (CAD)
computer-aided manufacturing (CAM)
coordinate system
copy
diameter dimensioning
digitizers
digitizing
drawing aids
editing
entities
erase
graphic tablet
grid
grid snap
input devices
joystick
keyboard
layering
linear dimensioning
math co-processor
menu-driven systems
mirror
mouse
move
output devices
panning
plotter
polar coordinates
primitives
puck
pull-down menu
radius dimensioning
random access memory (RAM)
relative coordinates
rotate
solid model
stylus
surface model

trim
units
vector graphics
windowing
wireframe model
zoom

REVIEWING YOUR KNOWLEDGE

Please do not write in this text. Write your answers on a separate sheet.

1. Give several examples of how CAD is used in business and industry.
2. List six advantages and two disadvantages of using CAD.
3. Which of the following can be used as an input device for placing a drawing in a computer?
   a. Puck or stylus.
   b. Joystick.
   c. Mouse.
   d. All of the above.
4. True or false? A plotter is a CAD output device.
5. Name the advantages of vector graphics.
6. True or false? A coordinate system is used in CAD to precisely place or locate entities on a drawing.
7. The elements that make up a CAD drawing, such as points, arcs, and circles, are ________.
8. Name and describe four types of dimensioning that can be done on most CAD systems.
9. Which of the following is a command that allows the user to move the drawing on the screen?
   a. Windowing.
   b. Panning.
   c. Layering.
   d. Editing.
10. Erase, move, rotate, copy, and mirror are common ________ commands.
11. List the general sequence of events in producing a drawing by computer.
12. Name and describe the three major types of three-dimensional drawings.

APPLYING YOUR KNOWLEDGE

1. Design a flowchart that describes the steps necessary to produce a computer drawing.
2. Discuss and list the differences between conventional drafting and computer-aided drafting.
3. Visit local industries to determine how computer drafting is being used. Prepare a report to share with the class.
4. Investigate and compare the features of available computer drawing systems. Write a report outlining the results of your investigation.

Section 3 ACTIVITIES — GRAPHIC COMMUNICATION–DRAFTING AND CAD

BASIC ACTIVITY: Drawing Shapes and Sizes

Introduction:

Most of the objects in the room around you were drawings at one time. Those drawings were probably discussed and changed before they resulted in products. Multiview drawings probably provided the specific shape and size details, while a pictorial drawing likely showed what the object would look like in more realistic form. Let's try to recreate those original drawings for a simple object in the room.

Guidelines:

- The multiview drawing should contain as many views as necessary to fully describe the object. Scale the drawing as necessary.
- The measurement and dimensioning system should be decided upon in advance.
- Use isometric or oblique for the pictorial drawing.

Materials and equipment:

Basic drawing tools or a CAD system.

Procedure:

1. Select a simple object. Some objects may be overly difficult to draw, so obtain your instructor's approval of the object you select.
2. Begin the multiview drawing by completing a sketch. After the sketch appears to be correct, have your instructor review it and initial it if it is satisfactory.
3. If needed, lay out a title block and border for the drawing. Then draft the views, being sure to leave enough space for the dimensions. Use an appropriate scale for the drawings so your views fit nicely on the substrate.
4. Lay out the dimensions on your sketch. Then place the dimensions on the multiview drawing.
5. Decide which pictorial drawing method is best suited to your object and make a sketch. Have the sketch approved by your instructor and initialed before proceeding.
6. Complete the pictorial drawing.

Evaluation:

Considerations for evaluation are:

- Procedure followed, including sketches.
- Correct form for drawings, including number and placement of views, scale used, and correct dimensions used.
- Overall quality of the drawings.

ADVANCED ACTIVITY: Packaging Technology

Introduction:

The goal of packaging is to protect and promote a product. The structure of the package should provide the best protection for the contents at the least cost. The design of the package should help attract potential customers. This activity will involve you in researching package designs and designing one of your own.

Guidelines:

- This is a small group or individual activity.
- The package that you design should be for a commonly found item, but it must be different in design than the typical packaging for the item.
- The package should have at least one seam and, if totally enclosed, locking tabs.
- Like professionally produced packages, the outside design should be on most sides of the package.

Materials and equipment:

- Several empty packages for research. See Section Fig. 3-1.
- Drawing tools or CAD system for pattern and outside design.
- Elements for the outside design, such as headlines and artwork.

Procedure:

1. Select a product that requires a package. Identify the intended audience. Examples of products include a tube of toothpaste, perfume bottle, videotape, or grocery item aimed at a young adult audience. Find similar packages. Take them apart and study their designs.
2. Make thumbnail sketches of different patterns. If necessary, make layouts of several of the patterns

Section Fig. 3-1. This is an example of a package you might decide to study. (Flying Color Graphics)

and try them out. Draw the final pattern using correct layout technique. Do not fold the final pattern since you will be using it for your design.
3. Trace the pattern onto the appropriate substrate for the package. Cut it out, fold it, and glue it. Heavier materials, such as chipboard, may require scoring before they are folded. (See the chapter on finishing in this book.) At minimum, cut out and fold a paper pattern.
4. Design the outside of the package, beginning with thumbnail sketches.
5. Place the final design on the paper pattern for the package. Do not fold it.

Evaluation:

Considerations for evaluation are:

- Group or individual effort.
- Evidence of research, such as different packages and good pattern design.
- Correct pattern layout.
- Design quality and appropriateness for the intended audience.
- Functionality of the pattern.

ABOVE AND BEYOND ACTIVITY:
CAD Comparison

Introduction:

A wise consumer knows that before buying a product, careful research is necessary. Suppose you are interested in purchasing a CAD program. It's a good idea to decide what features are most important and then compare the available products. See Section Fig. 3-2. Your challenge is to find the best CAD program for your needs.

Section Fig. 3-2. In comparing CAD systems, it is helpful to actually try out the system. (CADKEY, Inc.)

Guidelines:

- If necessary, decide on the computer and peripherals that will be used with the CAD software before beginning the activity.
- Work in small groups.
- Find the best system for your needs at the lowest cost.
- Review a minimum of four systems.

Materials and equipment:

- Literature on products that will be compared.
- Magazines, such as Consumer Reports, that will provide ideas on how to compare products.

Procedure:

1. Study the basics of using a CAD system before beginning. Also decide on the specific computer that will be used. You will also need to decide who will be using the computer. For example, will it be for a classroom situation or an architect?
2. Brainstorm on what features the package should include. Does it need to be relatively easy to learn? What capabilities does the program need for the type of work you would like to do? List the features.
3. Begin investigating various systems in and outside of class. Try out at least two systems, but obtain literature on at least four. Modify your list of features as needed, being sure to add only those features you actually need.
4. Make a chart comparing each system with your list of features. Also, make an overall recommendation.
5. Write a short article similar to those found in leading consumer magazines comparing the CAD packages. Mention in the article which systems you actually had an opportunity to try.

Evaluation:

Considerations for evaluation are:

- Group and individual effort.
- Depth of research on the topic, as shown by the comparison chart and article.
- Quality of the chart and article.

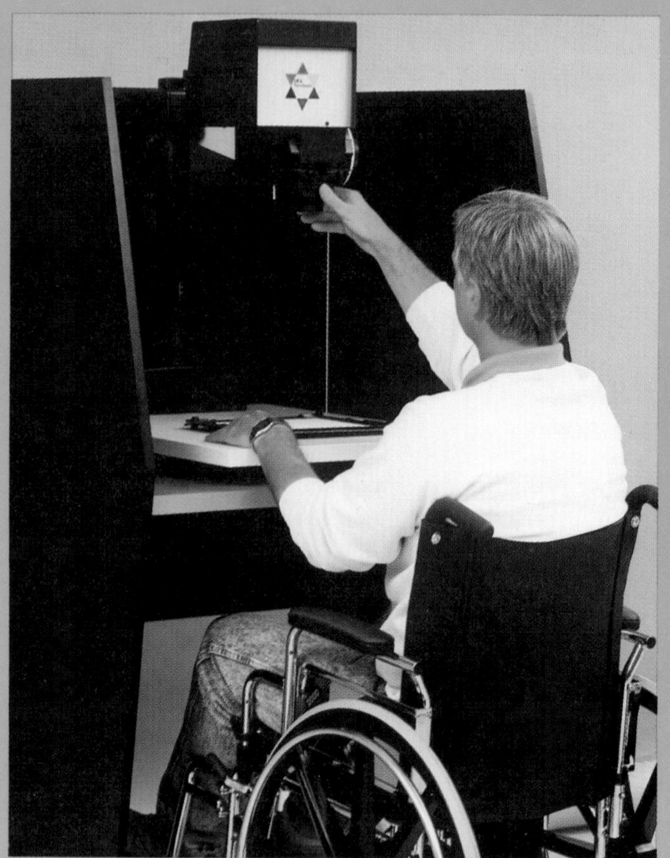

KREONITE, INC.

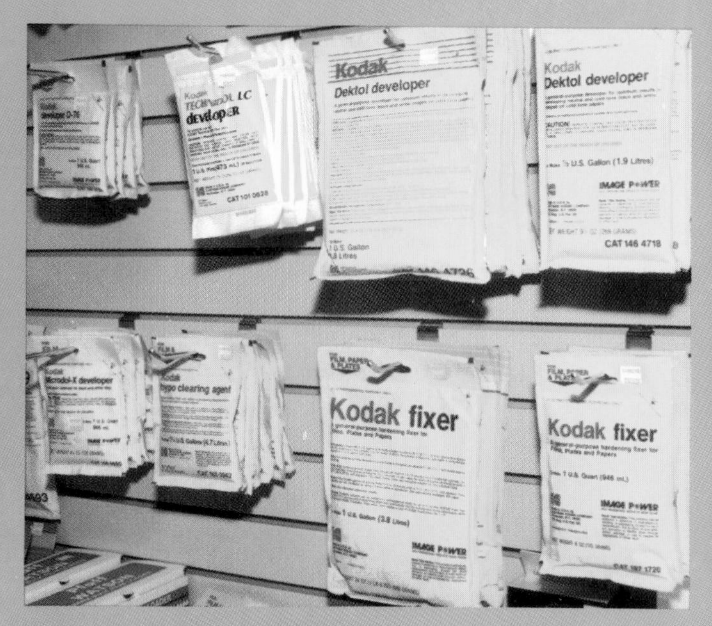

NIKON CORP.

KODAK

section 4

Graphic Communication–Photography

POLAROID CORP.

Understanding how a camera operates is an important first step in taking good pictures. (Polaroid Corp.)

16

CHAPTER

PHOTOGRAPHY BASICS

After studying this chapter, you will be able to:
- *Explain why photographs are continuous tone images.*
- *Describe various types of film and discuss their characteristics.*
- *Identify the types of cameras.*
- *Identify the parts of a camera.*
- *Explain how a camera lens works.*

Photographs are a vital part of communication today. A photograph can communicate information and even feelings. The photographs used throughout this textbook are an example of how photographs can convey ideas and facts. Photographs can record worldwide events in history books as well as preserve individual family histories in photo albums.

Photography is defined as the art or process of producing images on a light-sensitive surface, such as film. A photograph is one type of **continuous tone image**. It is called a continuous tone because a variety of tones are present in a photograph. For example, in a black and white photo, there are various shades (tones) of gray between black and white. See Fig. 16-1. A color photo is also continuous tone. Each color has tones (tints and shades of a color), such as various tones of blue.

Producing a continuous tone photograph requires several steps. First, a camera is used to expose continuous tone film to light. The film is then processed with

Fig. 16-1. A gray scale, placed over a black and white photo, shows that the photo consists of shades of gray.

chemicals. This causes the recorded image to become visible, as a film negative. The negative is placed on an enlarger that projects light through it onto light-sensitive paper. This paper is processed to produce the enlarged photo.

In this chapter, you will explore the basics of photography as you study film, cameras, and lenses. Further chapters in this section will explore taking pictures, processing film, making prints, and color photography.

CONTINUOUS TONE FILM

Light can make objects visible to the human eye. Light can come from primary light sources, such as the sun or a lamp. Light can also come from secondary sources, such as from reflections. In Fig. 16-2, light from a primary light source, the sun coming through the window, is reflected from various objects in the room, which are secondary light sources.

Your eyes convey light waves to your brain as images. Likewise, film, which has a light-sensitive emulsion, can be used to permanently record an image.

Most of the film that is used for continuous tone photography is **panchromatic film**. It is called panchromatic film because it "sees" or is sensitive to about the same range of color as our eyes.

Black and white pan films record images in black and white. The film records very dark shadow areas and bright highlight areas. In addition, tones of gray between these extremes are recorded.

Color film is usually panchromatic also. It has an emulsion layer for each separate color. Filters in the emulsion control the color each emulsion will record. This film is continuous tone because each color is recorded as a gradation of tones. Color films, which will be discussed in more detail in Chapter 19, produce color images.

Photographic film is a strip or sheet of thin plastic coated with a light-sensitive emulsion. A cross-section of black and white film is shown in Fig. 16-3. A plastic film base is coated on both sides. An emulsion layer, containing silver halide crystals, is on one side. An antihalation coating is on the other side. This prevents light rays from reflecting back to the emulsion and producing a double image.

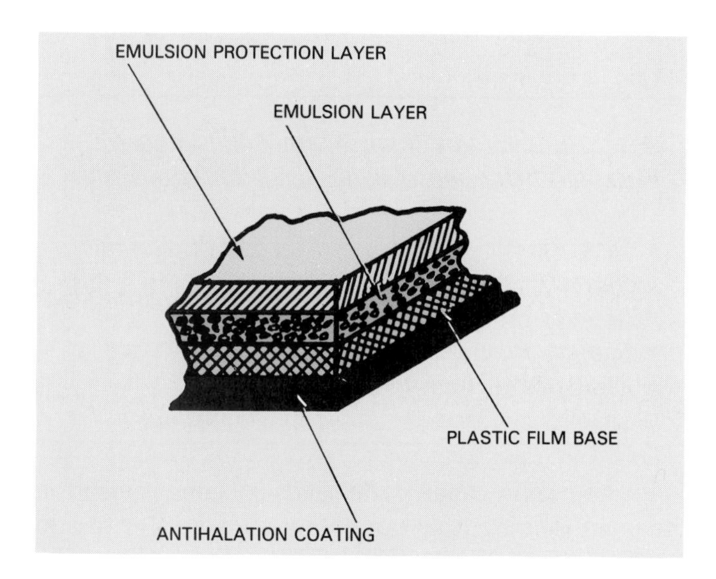

Fig. 16-3. This cross-section shows the various layers that make up black and white film.

Fig. 16-2. When objects in a room reflect light, they become secondary light sources.

TYPES OF PHOTOGRAPHIC FILM

The two basic types of film are negative and positive. Negative film records an image as a reverse of what we see. Film used to make photographic prints is negative film. Positive film records a scene just as our eyes see it. A slide is an example of positive film.

There are many types of film. Films vary primarily in size and speed.

FILM FORMATS

Film is available in different formats. **Format** refers to a certain size of film and kind of packaging. For example, 126 film and 110 film are both rolls of film packaged in a cartridge. However, the 110 format is a smaller film.

Many format numbers, such as 126 film, do not directly correspond to a film size. An exception is 35mm, which has a negative size of 35 mm. See Fig. 16-4. Another exception is sheet film, which is often described by the actual size of the sheet. An example is 4 in. by 5 in. (10.2 cm by 12.7 cm) film. Sheet film for cameras is sometimes called large format, since this is the largest film used in cameras. The sheet is loaded into a special film holder and then placed in the camera.

Most film is packaged as rolls that come on spools, in cartridges, or in canisters (also called "cassettes"). Medium-format (120-size) film is on spools. The 110 and 126 formats are in cartridges, while 35mm and the new Advanced Photo System format are packaged in canisters. See Fig. 16-5.

FILM SPEED

Film speed refers to the sensitivity of the film to light. A "fast" film is highly sensitive to light. A "slow" film is less sensitive and requires more light for a photograph. An **ISO (ASA) number** gives a measure of film speed. As ISO numbers become larger, the film speed increases. For instance, film rated as ISO 400 is twice as fast as ISO 200. See Fig. 16-6.

The ISO stands for International Standards Organization. This organization is responsible for setting up a consistent number system worldwide. The older system that uses the same numbers for film speed is ASA (American Standards Association).

As film speed increases, the crystal size or grain in the emulsion increases. Large grain means less fine detail. Usually, the slowest film is chosen that will work in the light conditions. This reduces the "graininess" of a photo, Fig. 16-7.

Fig. 16-4. Film format sometimes refers to film size, as shown here with 35 mm film.

Fig. 16-5. The APS (at left) and 35 mm canisters appear similar, but cannot be interchanged. The APS film has a smaller frame size.

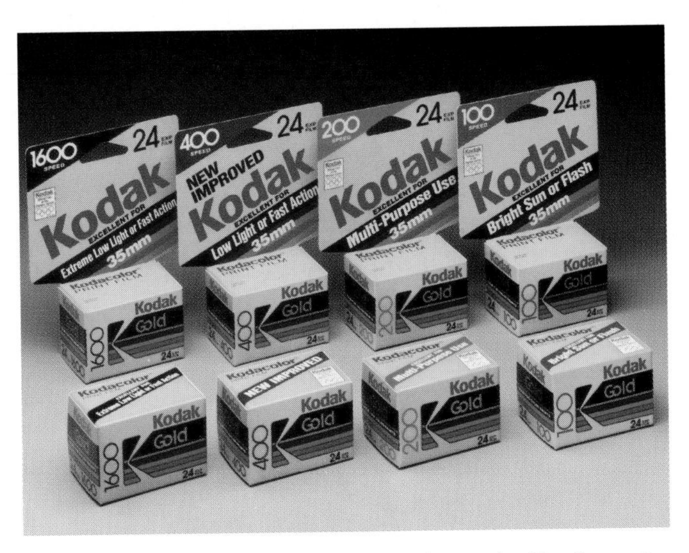

Fig. 16-6. Film comes in a variety of speeds. The fastest film is on the left. (Kodak Corp.)

Fig. 16-7. Note the graininess of this photo.

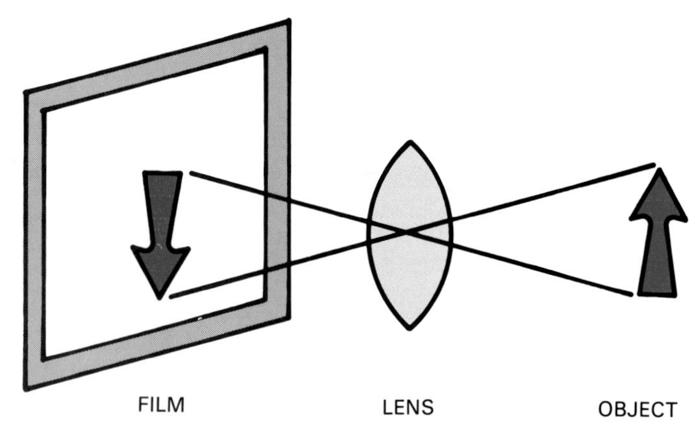

Fig. 16-9. The image is upside down on the film because light travels in a straight line.

THE CAMERA

A camera is made up of a number of parts that work together to make correct exposures on the film, Fig. 16-8. First, the **camera body** is the light-tight box which contains the film. A **viewing device** allows the photographer to see what the film will record. The **shutter release** is pressed to open the shutter and allow light to strike the film. A **lens** is the part of the camera that ensures that the image on the film is in focus. An **aperture** behind the lens controls the size of the hole through which light passes.

In a fraction of a second after the shutter opens, it closes again. The film is now exposed. As shown in Fig. 16-9, the image is upside down on the film. A **film advance system** is used to move unexposed film in position for the next exposure.

TYPES OF CAMERAS

Cameras are often chosen according to the needs of the photographer. Cameras made for the amateur are usually fairly easy to operate. Other types are primarily used by professional photographers or serious amateurs and require more skill.

Pinhole camera

A **pinhole camera** is the simplest type of camera, 16-10. It is basically used in teaching the principles of photography. This is simply a light-tight container with a

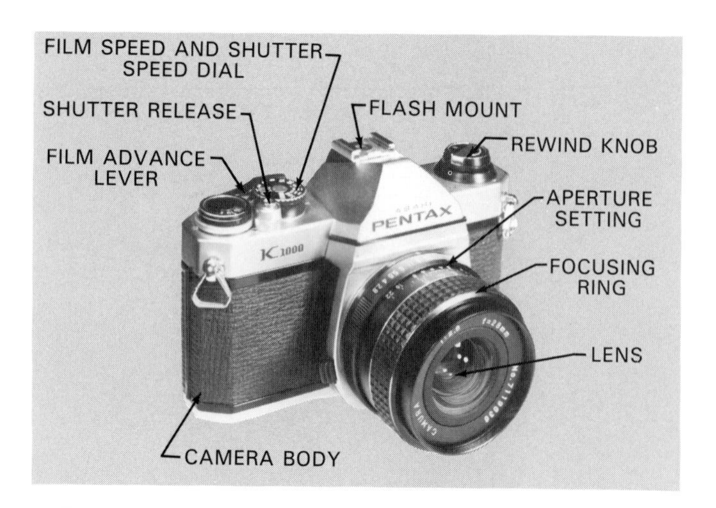

Fig. 16-8. The camera has a number of parts that work together to make correct exposures.

Fig. 16-10. A simple pinhole camera can be made from a can.

pinhole at one end. After the film is loaded, the camera is placed in position for the photo. The pinhole is uncovered to expose the film.

Viewfinder camera

The **viewfinder camera** has a viewing lens through which you look to compose a picture, Fig. 16-11. A separate camera lens is used to focus the scene on the film. The lens is normally a fixed-focus lens that requires no adjustment. This is one of the easiest cameras to operate. Disposable cameras are normally viewfinder cameras, Fig. 16-12.

One disadvantage often encountered with viewfinder cameras is **parallax error**, Fig. 16-13. Since the viewing lens is in a different location, the image seen is slightly different than the image "seen" by the camera lens. This is usually not a problem, except for close-ups. The closer the subject is to the camera, the greater the chance that parallax error may occur.

Fig. 16-11. A viewfinder camera has a fixed focus lens. The viewfinder can be seen at the top center of the camera. (Canon U.S.A., Inc.)

Fig. 16-12. This simple viewfinder camera is meant to be used only one time. (Kodak Corp.)

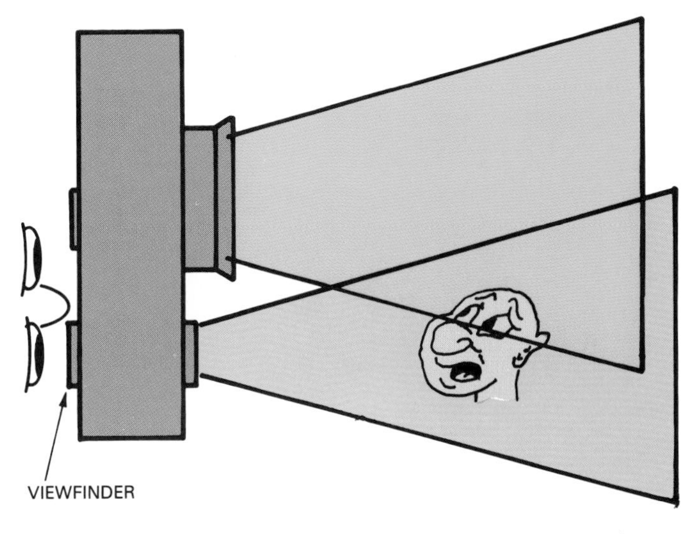

Fig. 16-13. Parallax error occurs because the viewing lens is in a different location from the camera lens. This is usually only a problem with close-ups.

Rangefinder camera

The **rangefinder camera** is an adjustable focus type of viewfinder camera. On the manual type, the camera lens is linked to the viewfinder lens so that both are focused together as the lens is turned, Fig. 16-14. As with the viewfinder camera, parallax error can occur with close-ups. Most new rangefinder cameras have automatic focus.

Single lens reflex camera

The **single lens reflex camera (SLR)** allows the photographer to look directly through the lens, Fig. 16-15. This is done with a mirror at an angle in the camera, Fig. 16-16. The mirror swings up when an exposure is made. One advantage of this system is there is no parallax error.

Most SLR cameras are 35 mm. The 35 mm SLR is very popular and a wide variety of attachments are

Fig. 16-14. This rangefinder camera has automatic focus. (Olympus Corp.)

Fig. 16-15. This SLR camera allows the photographer to look directly through the lens. (Nikon Corp.)

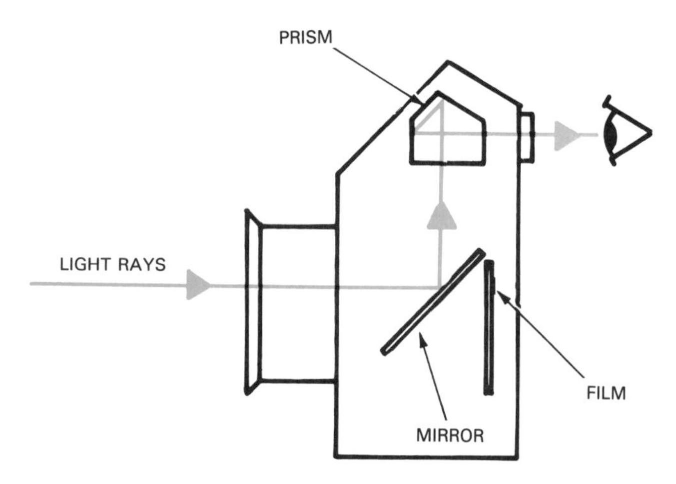

Fig. 16-16. This cross-section of an SLR shows how you are able to look directly through the lens.

available for it. Many new SLR cameras have automatic focus.

Twin lens reflex camera

The **twin lens reflex camera (TLR)** has two lenses that are focused together. See Fig. 16-17. The top lens is used for viewing a subject. A mirror reflects the lens image onto a ground glass. The photographer looks at the image on the ground glass to compose the picture and to focus. The second lens is used for recording the picture on film.

Since the viewing system is separate, parallax error may occur with the twin lens reflex camera. However, the advantages of this camera are a large film format and few moving parts. For example, the mirror remains in position for pictures, instead of swinging away. The TLR is often used in studios by professional photographers.

View camera

The **view camera** is used by professional photographers to eliminate distortion in a picture. See Fig. 16-18.

Fig. 16-17. The twin-lens reflex camera has two lenses that are focused together. One lens is used by the photographer to compose the picture.

It is used most often for pictures of landscapes or architecture. For example, when you look up at tall buildings, they may appear to lean toward each other. The view camera can be adjusted so that buildings appear to be straighter.

The view camera has a lens system that is separate from the camera back. These parts are connected by a rail. Bellows keep the camera light-tight. Each part of the camera can be adjusted along the rail or tilted.

Fig. 16-18. The view camera has a lens that is connected to the camera back with bellows. The flexible bellows allow the camera to be adjusted along the rail at the base of the camera. (Sinar Bron, Inc.)

Like the TLR, the view camera has a ground glass for viewing the picture. When the camera is set, sheet film is placed in the camera for the exposure.

Instant picture camera

There are two kinds of **instant picture cameras** that allow quick viewing of pictures. One type uses an instant film, Fig. 16-19. After the photo is taken, the film is ejected from the camera. At this point, processing chemicals are activated within the layers of the film. During processing, the negative layer transfers an image to the positive print layer. Processing continues for several minutes until the print is ready. The negative layer is then peeled off and thrown away.

Another instant picture camera is the **digital camera,** Fig. 16-20. Instead of film, the image is recorded electronically on a removable memory card. With an adapter, the pictures stored on the card can be copied to a computer's hard drive. They can then be viewed, changed or corrected using special software, and printed out. The electronic files can also be transferred over telephone lines to other computers. *Still video cameras* are also able to capture pictures electronically.

Fig. 16-19. This instant picture camera uses film that is processed automatically when it is ejected from the camera. (Polaroid Corp.)

Fig. 16-20. This digital camera allows the user to preview the picture on a large electronic "viewfinder." It has a zoom lens and can also record sound along with the picture. (Kodak Digital Science)

THE CAMERA LENS

A **camera lens** is a transparent object, usually made of glass or plastic, that has at least one curved surface. It is used to focus light on film.

As you may recall, a lens is not needed to take a picture. For instance, a pinhole camera makes an exposure through a small hole in the front of a light-tight container. However, a pinhole requires longer exposures, and the image is not very sharp. A lens can overcome both of these problems.

If the pinhole is enlarged to allow in more light, a blurred image will occur. This is because light waves of slightly different angles can come from the same point in the scene. If the pinhole is enlarged, all of these will be recorded on the film, creating the blur.

The purpose of a lens is to bend light rays, Fig. 16-21. This is known as **refraction**. If light rays strike clear glass at an angle, they will bend. A lens is shaped to bend these rays exactly to produce a sharp image. In this way, a large opening can be used for recording the image.

A simple lens is a single piece of glass or plastic. A lens that spreads light rays is known as a negative or **divergent lens.** A lens that bends light inward is a positive or **convergent lens**. In a **compound lens,** individual lenses, known as elements, are used together. Compound lenses help in correcting lens defects, known as **aberrations**.

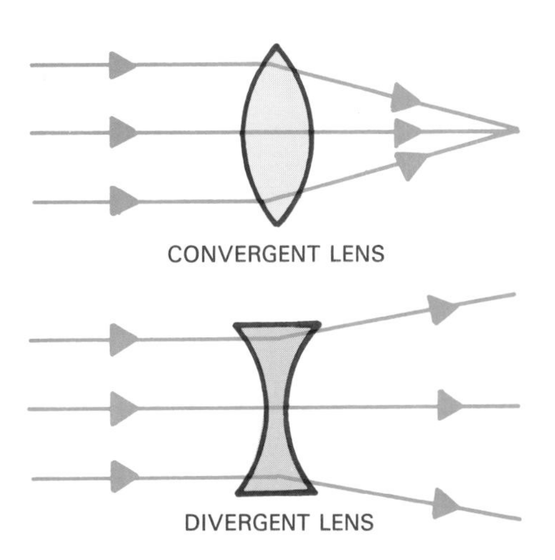

Fig. 16-21. A convergent lens bends light inward, while a divergent lens spreads the rays.

Camera lenses have a **focal length**. This is the distance between the lens and a focused image, when the lens is set at infinity. See Fig. 16-22. The distance between the film and camera lens is set based on this focal length. This length is usually shown in millimeters on the front of the camera lens.

As focal length changes, both angle of view and image size change, Fig. 16-23. For example, a short focal length provides a wide angle of view. This wide view makes the subject in the picture appear small. If a lens with a long focal length is used for the same picture, the angle of view is more narrow. The subject in the picture will now appear large, in comparison to the size of the picture.

A normal lens has about the same angle of view as the human eye. This lens will have a focal length about equal to the diagonal measure of the film format. For example, 35 mm film has a diagonal measure of 44 mm. A normal lens will have a focal length of approximately 50 mm.

A wide angle lens has a short focal length, and a wide angle of view. It also has a greater depth of field since this increases as focal length decreases, Fig. 16-24. A telephoto lens has a long focal length and a narrow angle of view. A zoom lens allows you to change the focal length of the lens, Fig. 16-25. For example, you might be able to vary focal length from 80 mm to 205 mm, until the most pleasing angle of view is found.

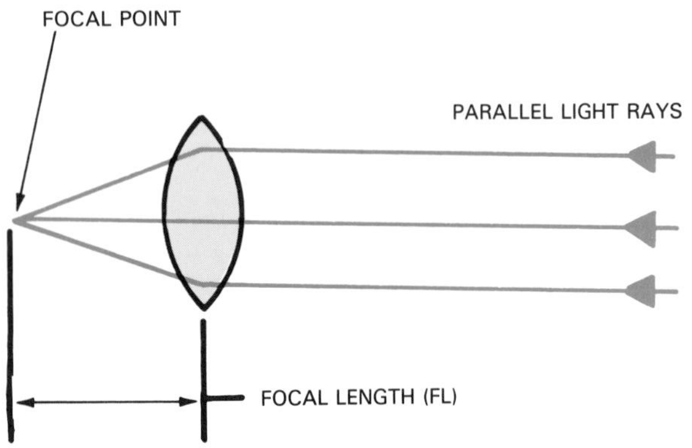

FOCAL POINT

PARALLEL LIGHT RAYS

FOCAL LENGTH (FL)

Fig. 16-22. Lenses have a focal length. This is the distance between the lens and film (focal point) when the lens is set at infinity.

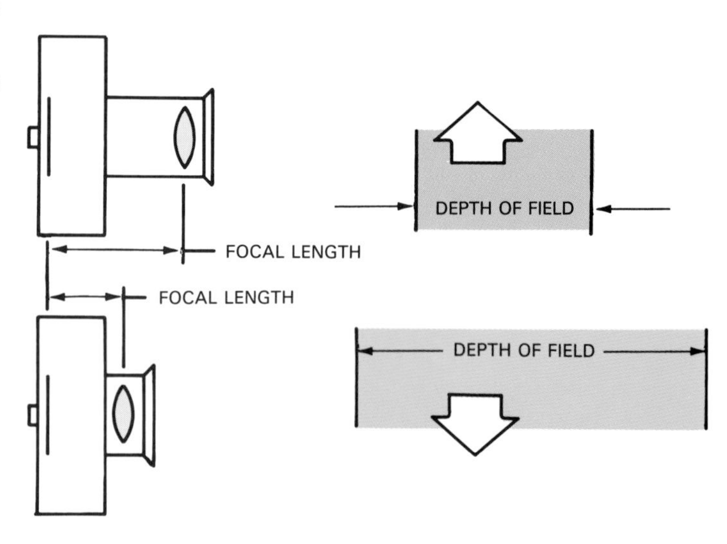

Fig. 16-24. Depth of field increases as the focal length decreases. Using an arrow as the subject, two examples are shown. Depth of field is the area in focus in front and behind the subject.

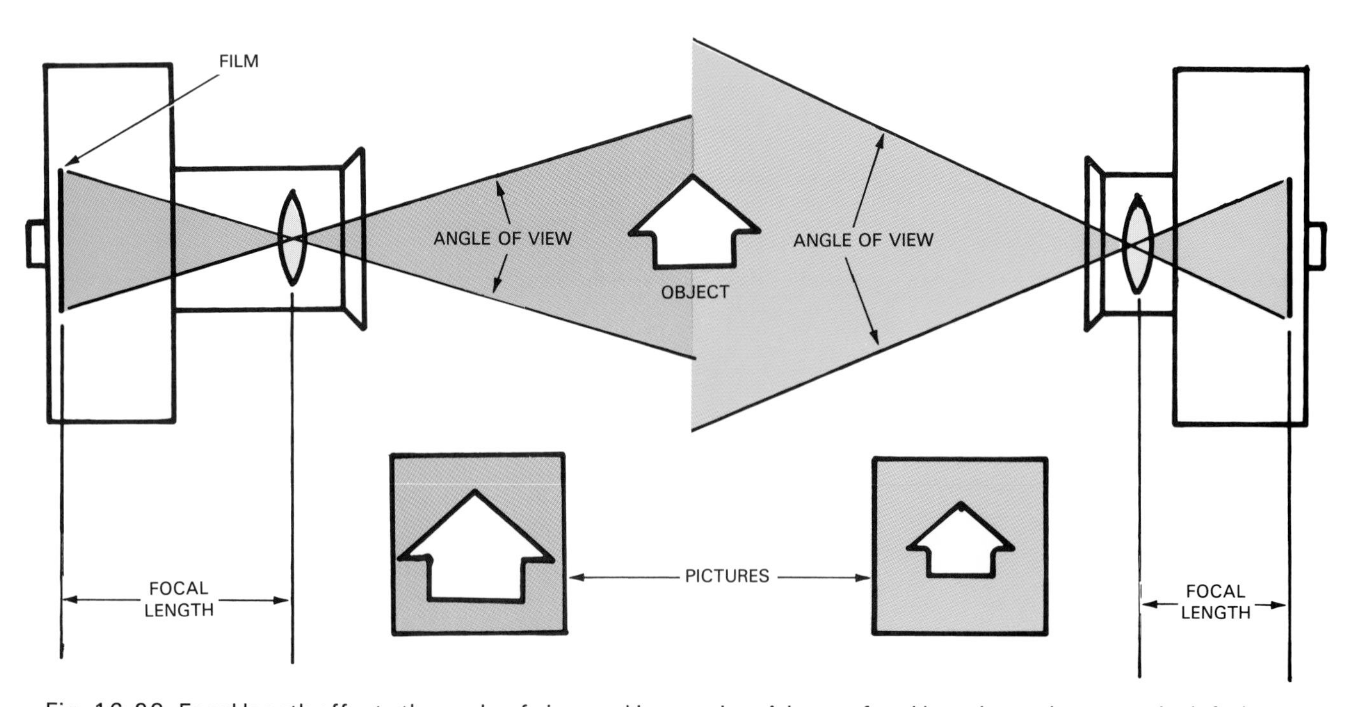

Fig. 16-23. Focal length affects the angle of view and image size. A longer focal length, as shown on the left, has a narrower angle of view.

Fig. 16-25. Several zoom lenses are shown that can be attached to the camera. (Nikon Corp.)

Fig. 16-26. This camera has a zoom lens that is an integral part of the camera. (Olympus Corp.)

Zoom lenses are often built into the camera body, Fig. 16-26.

SUMMARY

Photography is a powerful communication tool. In this chapter, you were introduced to continuous tone photography. Information about photographic films, cameras and camera lenses was also presented.

WORDS TO KNOW

All of the following words have been used in this chapter. Do you know their meanings?

aberrations
aperture
camera body
compound lens
continuous tone image
convergent lens
divergent lens
film advance system
film speed
focal length
format
instant picture camera
lens
panchromatic film
parallax error
photographic film
photography
pinhole camera
rangefinder camera
refraction
shutter release
single lens reflex camera (SLR)
still video camera
twin lens reflex camera (TLR)
view camera
viewfinder camera
viewing device

REVIEWING YOUR KNOWLEDGE

1. A photograph is called a continuous tone image because:
   a. A variety of tones are present in a photograph.
   b. One tone runs continuously throughout a photograph.
   c. Only black and white are shown in a photograph.
   d. None of the above.
2. Photographic _________ is a strip or sheet of thin plastic coated with a light-sensitive emulsion
3. True or false? A slide is produced from negative film.
4. Which film rating indicates the fastest film?
   a. ISO 100.
   b. ISO 200.
   c. ISO 400.
   d. All are the same speed.
5. Which of the following controls the size of the hole through which light passes into a camera?
   a. Shutter.
   b. Aperture.
   c. Lens.
   d. None of the above.
6. True or false? The closer a subject is to a viewfinder camera, the greater the chance that parallax error may occur.
7. Which type of camera allows the photographer to look directly through the lens?
   a. Viewfinder camera.
   b. Single lens reflex camera (SLR).
   c. Twin lens reflex camera (TLR).
   d. None of the above.
8. Which of the following cameras is used most often by professional photographers to eliminate distortion in a picture?
   a. Viewfinder camera.
   b. Instant picture camera.
   c. View camera.
   d. Twin lens reflex camera (TLR).
9. True or false? A divergent lens bends light inward.
10. True or false? A short focal length provides a wide angle of view.

APPLYING YOUR KNOWLEDGE

1. Visit a camera store. Investigate new film and camera technology. Prepare a report describing your findings.
2. Identify the parts of a camera and describe their function.
3. Interview a professional photographer. Ask him or her to discuss various types of cameras and their advantages and disadvantages.
4. Photograph the same scene with different lenses or with a zoom lens set at various focal lengths. Compare perspective and depth of field.

17

CHAPTER

TAKING PICTURES

After studying this chapter, you will be able to:
- *Demonstrate the proper way to use a camera.*
- *Describe camera focusing systems.*
- *Discuss exposure controls commonly found on cameras.*
- *List different types of lighting options often used in photography.*
- *Identify various photo composition techniques.*

Anyone can take pictures. However, to take quality pictures that communicate what you want them to communicate takes skill and an understanding of some basic techniques. See Fig. 17-1.

This chapter provides information on how to operate a camera. In addition, photo composition is covered. At the end of this chapter, a summary of picture-taking steps is presented.

Fig. 17-1. Sometimes, taking quality photographs can be a challenge.

LOADING AND UNLOADING FILM

The first step in taking pictures is to load the film into the camera body. Since there are different types of film and cameras, it is best to read directions before loading. In general, load and unload film in subdued light. This will help to avoid **fogging** (accidentally exposing) the film. Also, be sure to adjust the camera for the film speed (ISO). Some cameras automatically set the film speed by reading a DX code on the film package.

Different film formats require different loading procedures. The four most common formats include:
- Cartridge film.
- Sheet film.
- Roll film.
- Canister film.

CARTRIDGE FILM

Cartridge film is the easiest to load. The camera back is opened and the cartridge is placed in position, Fig. 17-2. The back is closed. Then the film is advanced for the first picture. After the film has been exposed, the cartridge is simply removed from the camera.

SHEET FILM

Sheet film is loaded into a light-tight film holder. After the holder is placed in the camera, the cover is removed so the film can be exposed. The cover is then replaced so the film can be taken to the darkroom for processing.

ROLL FILM

Roll film must be threaded onto a take-up spool, Fig. 17-3. To load roll film, open the camera and place the

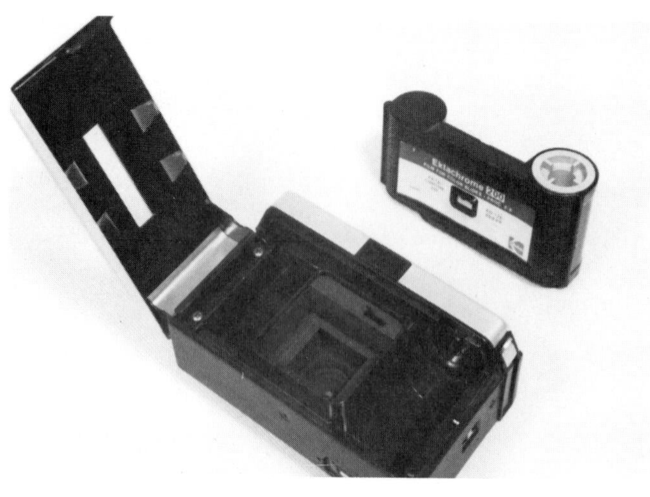

Fig. 17-2. To load cartridge film into a camera, you simply place it into the back of the camera.

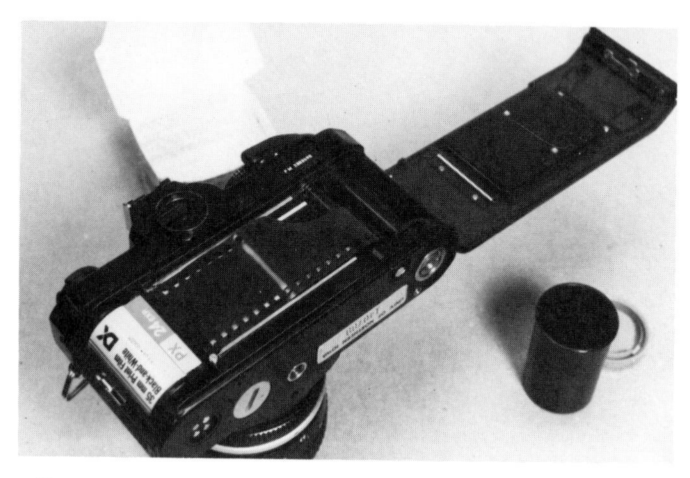

Fig. 17-4. Canister film is loaded by attaching one end of the film to the take-up spool. The film perforations are aligned on the sprocket roller.

Fig. 17-3. To load roll film into a camera, you thread it onto a take-up spool.

Fig. 17-5. When using canister film, once the film has been exposed, the film must be rewound back into the canister.

spool in the take-up position. Attach the roll film and unroll a small amount with the paper side up. Attach the film end to the take-up spool. Close the back of the camera. Then advance the film for the first picture. After the exposures are made, the film is completely rolled onto the take-up spool before removal.

CANISTER FILM

Canister film, like roll film, must be threaded onto a take-up spool. Install the canister and unroll a small amount of film. Attach the loose end to the permanent take-up spool. If the film is aligned properly, the film sprocket roller and film perforations will be aligned, Fig. 17-4. Close the back of the camera. Then advance the film for the first picture. The rewind knob should turn as you advance the film. This shows that film is being pulled from the canister. After exposure, this type of film is rewound into the canister, Fig. 17-5. A button is pressed that disengages the film advance. The rewind knob is then turned in the direction shown on the knob. Rewind the film completely and remove the canister.

Some cameras that use canister film offer automatic loading. The film leader is pulled out to a mark on the camera. When the camera back is closed, the film is threaded automatically. The new APS cameras, Fig. 17-6, use a simple "drop-in" loading method.

EXPOSURE CONTROL

The main exposure control devices in a camera are shutter speed and f-stop. **Shutter speed** is the length of time the film is exposed to light. The **f-stop**, or lens aperture, controls the size of the hole where light enters the camera body. Shutter speeds and f-stops can be fixed (one setting only) or adjustable.

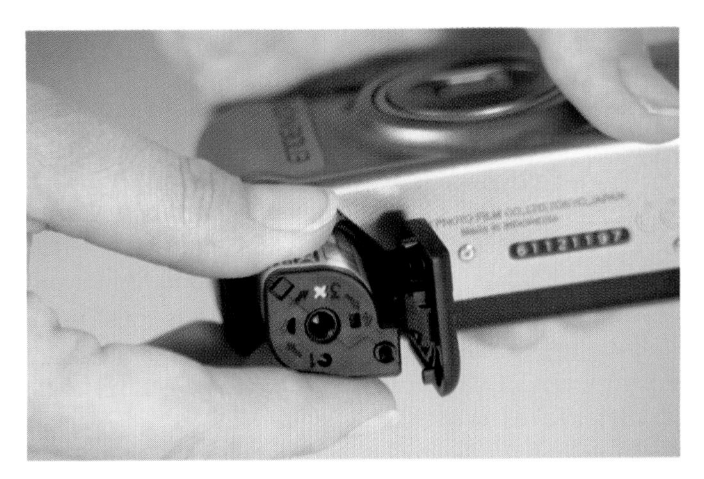

Fig. 17-6. All Advanced Photo System cameras are designed for "drop-in" loading. The canister is slipped in place and the door closed. Frame advance and rewinding are fully automatic. (Jack Klasey)

Most pictures are taken with a shutter speed of less than one second. A movement of one click on the shutter speed dial either halves or doubles the speed. For example, if the camera is set at 1/125 second, moving to 1/60 will allow the shutter to remain open twice as long. Moving the other direction, from 1/125 to 1/500, will allow the shutter to remain open half as long.

Fast shutter speeds can be used to stop the action in pictures, Fig. 17-7. Slow shutter speeds are helpful in low light situations. Shutter speeds slower than 1/30 second usually require a tripod.

The f-stops of a lens are also important for exposure control. As f-stops become larger, the lens opening becomes smaller, Fig. 17-8. Like shutter speed, moving one f-stop will halve or double the light coming through the lens. In addition to controlling light, the f-stop also affects depth of field. This is the distance in focus, in

Fig. 17-8. The lens opening becomes smaller as the f-stop number becomes larger.

front and back of the subject. The smaller the lens opening, the greater the depth of field will be, Fig. 17-9.

EXPOSURE ADJUSTMENT

Cameras that require exposure adjustment can be adjusted in several ways. These include:

- Referring to the chart that often comes with film, Fig. 17-10. This chart gives an approximate f-stop and shutter speed for different light conditions.
- Using a hand-held exposure meter as shown in Fig. 17-11. These meters can be reflective or incidental, Fig. 17-12. Reflective meters are aimed at the subject for a light reading. They measure light reflected from the subject. **Incident meters** measure the light at the subject location. This is normally done by standing where the subject will be and taking a meter reading while pointing the meter towards the camera. This type is often used to set up lights for use in studio photography.

Fig. 17-7. A fast shutter speed was used here to stop the running water.

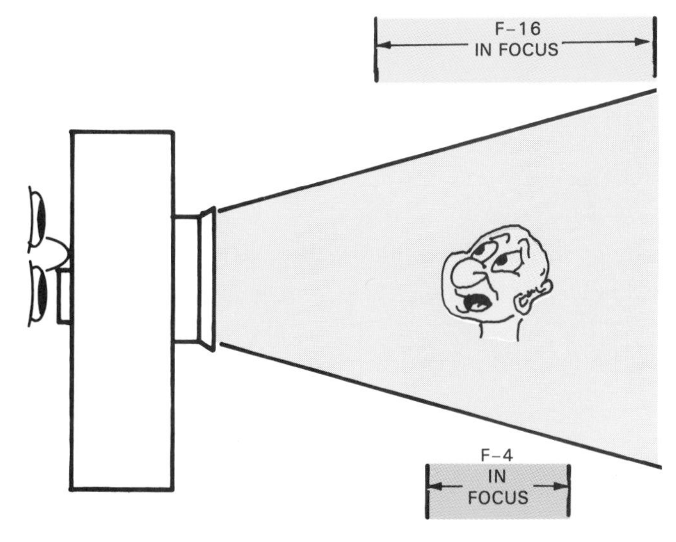

Fig. 17-9. As the lens opening decreases in size, the depth of field increases.

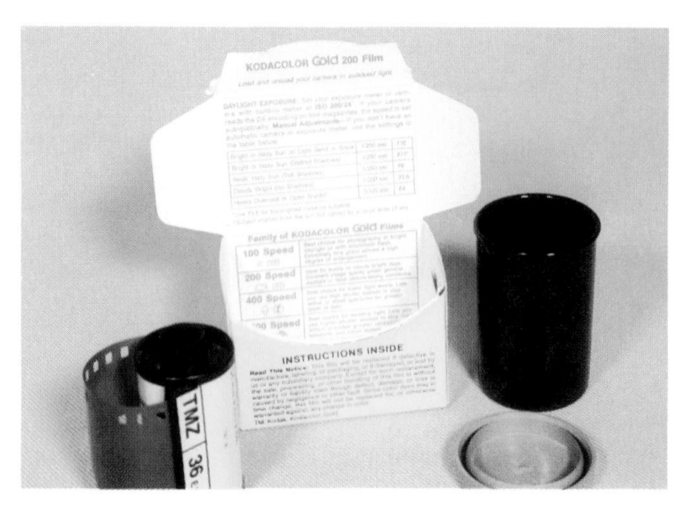

Fig. 17-10. A chart is packed with most types of film. It will give suggested settings for correct exposures.

Fig. 17-11. This hand-held light meter is used for determining the camera f-stop and shutter speed needed for a correct exposure.

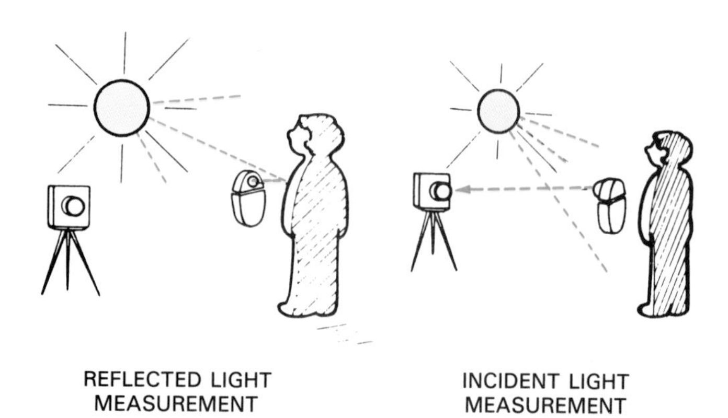

Fig. 17-12. Reflective meters measure light reflecting from the subject (left), while incident meters measure the light striking the subject (right).

- Using the built-in meter found in many cameras. Many cameras have built-in exposure meters. The ISO, or film speed, must always be adjusted on the camera before reading the meter. This adjusts the meter for the speed of the film. Built-in light meters are *reflective*. They measure the light coming from the subject. See Fig. 17-13. A **spot meter** is a reflective meter that measures the reflected light in a very small area in the picture. A **center-weighted meter** measures all the light in the picture but places more emphasis on the center of the picture. An **averaging meter** takes an average of all the light in the scene.

Built-in light meters can be manual, automatic, or a combination. The manual exposure meter is adjusted by first setting either the f-stop or shutter speed to a desired setting. The viewfinder display then provides information on what the other setting should be, Fig. 17-14. A light or needle movement is often used to show correct exposure. Based on the viewfinder information, the other setting (f-stop or shutter speed) is then set manually by the camera operator.

The law of reciprocity

A variety of f-stop and shutter speed combinations can be used for different exposure needs. This is known

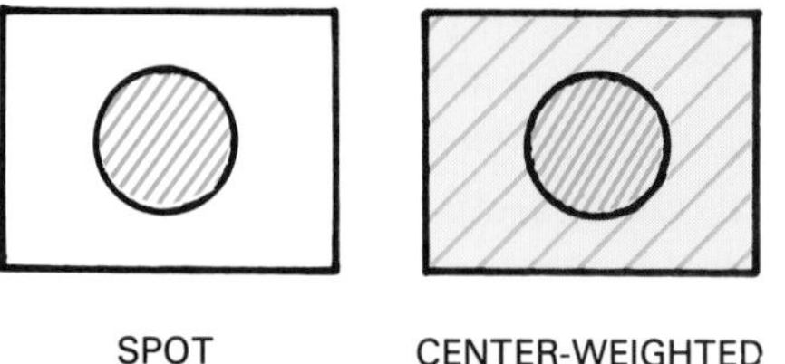

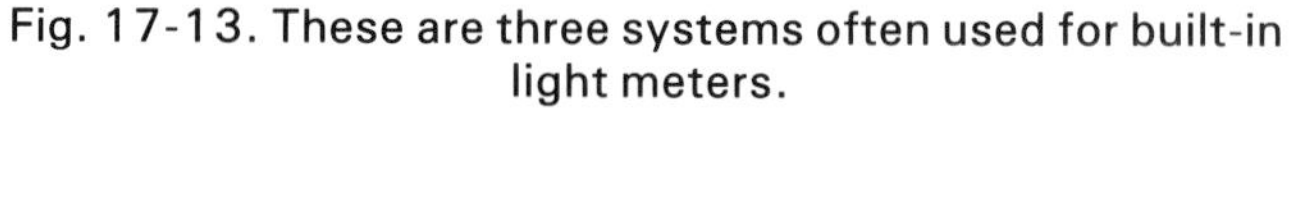

Fig. 17-13. These are three systems often used for built-in light meters.

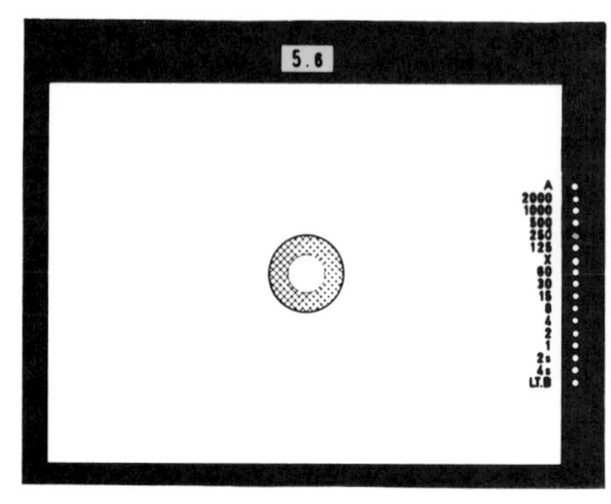

Fig. 17-14. This viewfinder display shows the camera is set at f/5.6. The pointer on the right shows the correct shutter speed, which is 1/250 second. (Pentax Corp.)

as the **law of reciprocity**. For example, a camera is set to make a correct exposure at f/11 and 1/250 second. However, if you are photographing a sports event and want to stop the action of the players. You can do this by moving to 1/500 second. Now you have halved the amount of light for a correct exposure. By moving to f/8, you are now back to a correct exposure. Other combinations could also be used to achieve a desired effect.

AUTOMATIC EXPOSURE SYSTEMS

Some cameras require no exposure adjustment. Automatic exposure units automatically set the camera for correct exposure. In *aperture priority systems*, you preset the aperture. The correct shutter speed is automatically chosen. In *shutter priority systems*, you preset the shutter speed and the f-stop is chosen automatically. *Programmed cameras* can automatically choose both shutter speed and f-stop.

HOLDING THE CAMERA

A camera should normally be held very still when taking a picture. See Fig. 17-15 and Fig. 17-16. Camera movement will be recorded as a blur. Cameras can be held by hand with a shutter speed as slow as 1/60 second. A general rule is to use the lens focal length as a fraction denominator and 1 as the numerator. This fraction will be the slowest speed for holding the camera. For example, a 100 mm lens can be safely held at 1/100 second. A shutter speed is selected which is closest to this, which is 1/125 second. A 50 mm normal lens can be hand held at 1/50 or 1/60 second.

In some cases, a slower shutter speed is needed. For example, low light conditions may require a longer exposure. Leaning against a wall, or setting the camera on a surface will help for slower speeds, such as 1/15 or 1/30 second, Fig. 17-17. However, a **tripod** is the best method for steadying the camera. A **cable release** is also useful. See Fig. 17-18. This tool allows you to press the shutter release without the danger of moving the camera.

Fig. 17-16. This is the correct way to hold the camera for a vertical format picture.

SUBJECT LIGHTING

As was discussed in Chapter 16, **primary light sources** emit light. These sources can be natural, like the sun, or artificial, like a lamp. **Secondary light sources** are those that reflect light emitted from a primary source. The reason we see most objects is because they act as secondary light sources. They reflect light back to our eyes.

Available light is the combination of the light sources you have for a photograph. For example, in a room,

Fig. 17-15. This is the correct way to hold a camera for a horizontal format picture.

Fig. 17-17. Leaning against a solid object can help steady the camera when using slow shutter speeds.

Fig. 17-18. A tripod and cable release are useful for avoiding camera movement during longer exposures.

primary light sources might be lamps and sun shining through a window. In addition, the surfaces in the room are secondary light sources, which reflect the light.

TYPES OF LIGHTING

Many outdoor photos are taken with only natural light from the sun. Other photographs use other light sources.

Frontlighting is the easiest method to use. In this method, the photographer stands between the subject and the sun or other light source, Fig. 17-19.

Sidelighting can be used to show more surface texture and create light and shadow areas on a subject, Fig. 17-20. In this method, the light is to the side of the subject.

In **backlighting**, the light is behind the subject. This method must be carefully used, since correct exposure for the bright light may underexpose the subject. See Fig. 17-21.

EXTRA LIGHTING

Extra lighting can be used for special effects and to eliminate shadows. Extra lighting can come from two

Fig. 17-19. This is an example of how frontlighting was used.

Fig. 17-20. Sidelighting can be used to show surface texture.

Fig. 17-21. When backlighting is used, the subject might be underexposed.

types of light sources. These include a continuous light source or a flash.

Continuous light source

A **continuous light source** such as a floodlamp, gives constant lighting. Continuous light is often used by professionals in a studio so that the effect of the light can be seen before the picture is taken.

Flash

A **flash** provides a short burst of light. Older flashes used a disposable flashbulb. Newer flash units, known as electronic flash units, have a permanent tube powered by batteries, Fig. 17-22.

Some electronic flash units are built into the camera, Fig. 17-23. Others must have an electrical attachment to the camera. This is done with a "hot-shoe" or cord. See Fig. 17-24. For cord attachment, an input marked "X" on the camera is used for electronic flash. The electricity is provided by batteries.

After the flash is attached, it is adjusted for the film speed (ISO). See Fig. 17-25. A calculator on the back of the flash provides information on which f-stop to use for the subject to camera distance. The camera is also set for the shutter speed recommended for electronic flash.

Most flash units are automatic. They measure light reflected back to a sensor. Refer again to Fig. 17-22. This sensor turns the flash off when the subject is properly exposed.

Fig. 17-23. This camera has a built-in flash. (Sigma Corp.)

Fig. 17-24. An electronic flash can be attached to a camera with a power cord.

Fig. 17-22. This electronic flash is mounted to a ''hot-shoe'' on the camera. Then sensor reads reflected light. (Nikon, Inc.)

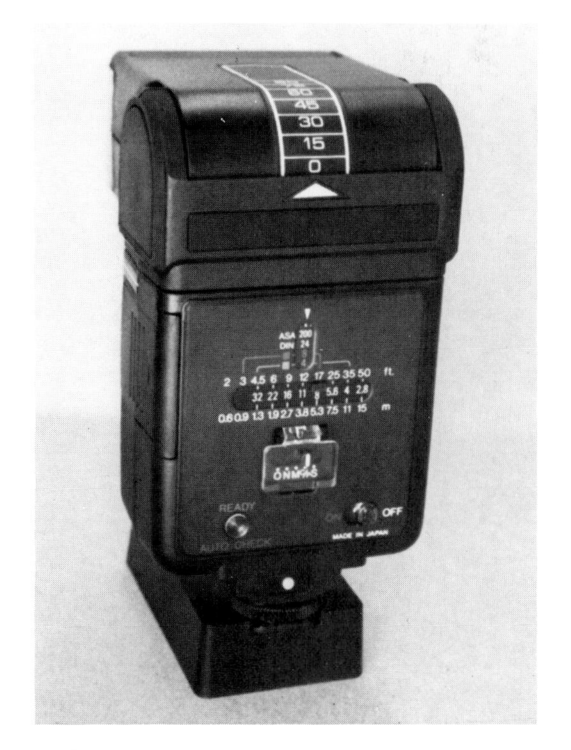

Fig. 17-25. An electronic flash must be set for exposure.

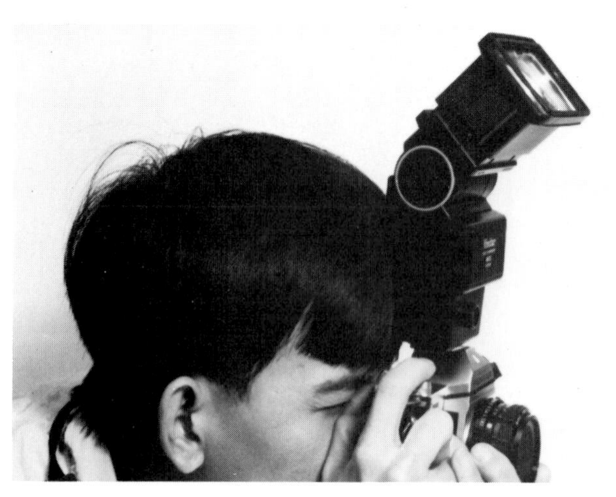

Fig. 17-26. This flash unit has been set in the bounce position to avoid ''flattening'' the subject.

A **dedicated flash** works only with a particular camera. It is easiest to use because separate settings do not have to be made on the flash. For example, when the film speed is set for the camera, this information is used by the flash also.

A common problem with a flash is the reflection from eyes, which show up as red spots in a photograph. Moving the flash to the side with a bracket will help eliminate this. Another solution is to have the subject look slightly away from the camera.

Another problem with flash is that it often eliminates shadows on a subject. This "flattens" the subject. Sometimes the flash is bounced off the ceiling or held to the side of the camera to help eliminate this problem, Fig. 17-26.

CAMERA FOCUS

Some cameras have a fixed lens that requires no focusing. Others have automatic focusing that takes care of the focusing for you. For cameras that require manual focusing, one of three focusing screen systems are normally used, Fig. 17-27.

- **Superimposed image**: With this system, the camera is focused so that two images of the scene come together into one.
- **Split image spot**: With this system, a small circle in the center shows a broken image unless the camera is in focus.
- **Microprism**: With this system, a small circle in the center shows a dot pattern until focused properly.

In low light conditions it is sometimes difficult to focus the camera. In this case, the distance numbers on the focusing ring can be used for approximate focusing, Fig. 17-28.

Autofocus cameras either have an active or passive focusing system. The **active focusing system** emits an ultrasonic or infrared beam, Fig. 17-29. This beam is

Fig. 17-28. At f/8, everything should be in focus from 3 ft. (3.5 m) to 30 ft. (10 m).

Fig. 17-29. This camera utilizes an active focusing system. (Canon U.S.A.,Inc.)

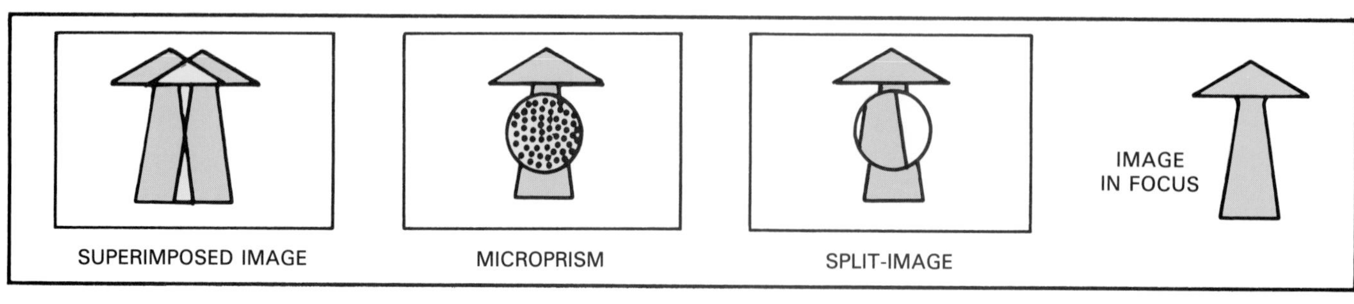

Fig. 17-27. These three systems are often used to focus an image.

reflected back to a camera sensor from the subject. A processor in the camera calculates the distance and a motor moves the lens to achieve focus. The **passive focusing system** uses the normal light reflected from the subject to determine correct focus. The subject image, after entering the camera, is split by two small lenses and projected onto a sensor. A motor moves the lens until the two images are focused with one another.

PHOTO COMPOSITION

Composition is the design phase of photography. In graphic arts, elements are composed or produced for a paste-up. This might include type matter and illustrations. A photographer also composes by arranging the elements in the scene for the best photo, Fig. 17-30. Design principles, as discussed in Chapter 4, are helpful in composing photos. However, some additional suggestions will be presented in this section. Knowing some rules of composition does not guarantee that each photo will be a masterpiece. However, with practice, these rules of composition will help you to be a better photographer.

Fig. 17-30. The photographer must arrange elements in the scene for the best photograph. This is especially true in the field of food photography. (National Pork Producers Council)

CENTER OF INTEREST

A **center of interest (COI)** or main subject should be chosen for a photo. Then all the other elements in the photo can be placed to complement (look pleasing with) the main subject. Centered subjects are usually less interesting because they do not suggest movement. An alternative is to divide the scene into thirds like a tic-tac-toe board, Fig. 17-31. This is the **rule of thirds**. Try placing the subject at one corner of the central box. Which corner do you choose? Pick a corner that shows the subject looking or moving toward the center of the picture. Also, pick a spot where the other picture elements complement the main subject. If the film format

Fig. 17-31. The rule of thirds can help make pictures more interesting.

Fig. 17-32. The pilings provide lines that "lead you into" this photograph.

is rectangular, then you have two options. Width or length can be the longest dimension. Look at the scene in both ways before deciding.

Watch for lines in the scene as you compose. Use them to complement the subject. Lines should lead into the picture, Fig. 17-32.

PERSPECTIVE

Perspective, or point of view, is important in composition. Usually, you should move around the subject to find the best point of view. Your hands can be used to quickly frame the scene in different locations. Move close enough to the subject to eliminate unwanted elements. Moving closer allows your subject to be larger in the scene. Also, consider camera angle. You may choose to shoot photos from above or below the subject to create different effects.

BACKGROUND

Choose a background that is uncluttered. Distracting objects in a photograph can take interest away from the main subject of the photo.

As you compose a photo, watch for elements in the background that appear to be part of the main subject. For instance, you don't want something to appear to be "growing" out of a person's head. See Fig. 17-33. For safety, never back away from the subject without looking back.

FRAMING

Framing is a technique of using elements in the foreground to provide a pleasing border for the subject. For example, a tree limb in the foreground can be a natural frame. Framing adds a feeling of depth to a photograph. See Fig. 17-34.

SELECTIVE FOCUSING

Selective focusing involves using depth of field in composing the photo. For instance, a large lens aperture shortens the depth of field. The camera focus can be adjusted so the foreground, background, or both are fuzzy and out of focus. Thus, attention is drawn to the main object of the photograph.

PANNING

When the subject is moving, composition must be done quickly, Fig. 17-35. Try to show movement into the picture, as with still shots. A fast shutter speed will "freeze" or stop the motion. **Panning** (moving the camera with the subject) will blur the background, but stop

Fig. 17-33. Use care when choosing backgrounds. Sometimes, the background can be distracting or, as in the case of this antenna, appear to be a part of the subject.

Fig. 17-34. This is an example of how framing can be used in composing a photograph.

Fig. 17-35. Shots featuring moving subjects must be composed quickly.

the motion of the subject, Fig. 17-36. A slower shutter speed can be used to show subject movement by slightly blurring the subject.

PUTTING IT TOGETHER

A great deal of information has been presented in this chapter on how to take pictures. A list of 10 steps for taking pictures is provided here:

1. Load the camera in subdued light and advance the film for the first picture. Remember the number of exposures available.
2. Set the camera for film speed (ISO), if needed.
3. Compose the picture.
4. Meter the scene. Use extra light (flash or lamps), if needed.
5. Set the shutter speed and f-stop.
6. Focus.

Fig. 17-36. Using the panning technique will stop the motion of the subject in the photograph, but it will blur the background.

7. Hold the camera steady and depress the shutter release smoothly.
8. Advance the film and continue taking pictures.
9. After the roll is exposed, rewind or forward wind the film.
10. Remove the film in subdued light for processing.

SUMMARY

Photography is a skill. You can develop your skills in photography by learning how to use your camera properly. This involves regulating exposure, focusing, and lighting. Using various composition techniques allows you to compose a photo that will communicate what you want it to say.

WORDS TO KNOW

All of the following words have been used in this chapter. Do you know their meanings?

active focusing system
averaging meter
backlighting
cable release
canister film
cartridge film
center of interest (COI)
center-weighted meter
composition
continuous light source
dedicated flash
flash
fogging
framing
frontlighting
f-stop
incident meter
law of reciprocity
microprism
panning
passive focusing system
perspective
primary light sources
rule of thirds
secondary light sources
selective focusing
sheet film
shutter speed
sidelighting
split image spot
spot meter
superimposed image
tripod

REVIEWING YOUR KNOWLEDGE

Please do not write in this text. Write your answers on a separate sheet.

1. Name the four most common film formats.
2. True or false? Shutter speed controls the size of the hole where light enters the camera body.
3. True or false? A fast shutter speed can freeze motion on the film.
4. True or false? The smaller the lens opening, the greater the depth of field will be.
5. Which type of light meter measures the reflected light in a very small area in the picture?
   a. Spot meter.
   b. Center-weighted meter.
   c. Averaging meter.
   d. None of the above.
6. Suppose your camera is set to make a correct exposure at f/11 and 1/250 second. However, you are photographing a bike race and want to stop the action in the photograph. If you moved the shutter speed to 1/500 second, what f-stop would be needed?
7. When shooting at slower speeds, such as 1/15 or 1/30 second, what is the best method of steadying the camera?
8. When the photographer stands between the subject and the light source, the type of lighting is:
   a. Sidelighting.
   b. Frontlighting.
   c. Backlighting.
   d. None of the above.
9. Describe how to eliminate the reflection from eyes that shows up as red spots in the photograph when using a flash.
10. Name three camera focusing systems that are normally used.
11. __________ is the design phase of photography.
12. Which of the following involves using a shallow depth of field?
    a. Framing.
    b. Selective focusing.
    c. Panning.
    d. Perspective.

APPLYING YOUR KNOWLEDGE

1. Investigate composition by taking a variety of photos using the composition techniques discussed in this chapter.
2. Find examples of good composition in magazines. Share them with the class and explain what types of composition techniques helped to make these good photos.
3. Visit a photography studio and discuss the effects of lighting on a subject. Ask the photographer to share lighting technique hints with you. Prepare a report to share with the class.
4. In small groups, discuss the steps involved in taking good pictures.
5. Demonstrate how to properly use a camera. Take some pictures. Then analyze them in class, citing good points and describing how you think they could be improved.

18

CHAPTER

PROCESSING FILM AND MAKING PRINTS

After studying this chapter, you will be able to:
- *Describe the steps involved in processing black and white film.*
- *Explain the printing process.*
- *Describe the purpose of a contact print.*
- *Identify techniques for improving and mounting prints.*

Once you have taken your pictures, you will probably be curious about how they will look, Fig. 18-1. Several steps are involved in producing a finished print from exposed film. First, the film is processed in chemicals to make the latent image visible. Next, an enlarger is used to enlarge and expose the negative to light-sensitive paper. The paper is then processed, and the finished print is ready. This chapter will discuss these steps.

PROCESSING FILM

Processing changes the latent film image into a visible and permanent image. Three steps are required. These include:
- Chemical preparation.
- Film loading.
- Actual processing.

CHEMICAL PREPARATION

The three main chemicals used in film processing are developer, stop bath, and fixer. The **developer** makes the latent image on the film or paper visible. It reacts with the emulsion to change exposed silver halide crystals into black metallic silver. The **stop bath** halts the developing process. The **fixer** removes the unexposed emulsion and makes the image permanent.

Chemicals can be purchased in liquid or powder forms. Chemicals must be measured accurately. See Fig. 18-2. Chemical goggles and adequate ventilation are necessary when mixing and using chemicals to avoid splashes and fumes. Use gloves to avoid getting chemicals on your hands. Always pour an acid, such as stop bath, into the water instead of the water into acid. Remember to read all warnings on labels, Fig. 18-3. Place the mixed chemicals in labeled, opaque, plastic containers, such as those in Fig. 18-4.

Chemicals that are mixed and ready to use are called *working solutions*. Developer is often mixed first as a

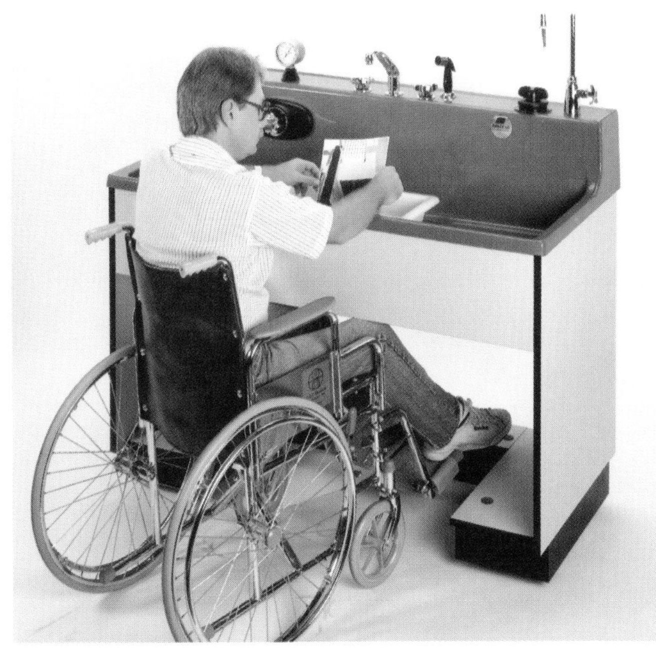

Fig. 18-1. It's fun to process film and produce prints. Special darkroom equipment is available for physically challenged people. (Kreonite, Inc.)

full strength *stock solution* and placed in a separate jug. As needed, this is further diluted with water to make a working solution.

The chemicals should be 68°F for processing. See Fig. 18-5. Containers of chemicals can be placed in a water bath to bring them to the correct temperature. For a quick temperature adjustment, the container can be placed under hot or cold running water. The temperature should constantly be checked with a thermometer to maintain proper temperature.

LOADING THE FILM

Roll film is placed on a developing reel for processing. Since this is done in complete darkness, you should practice with scrap film first. As shown in Fig. 18-6, the

Fig. 18-2. A graduate is used to ensure that chemicals are measured correctly.

Fig. 18-3. Since stop bath is a strong acid, when mixing it with water, always pour the acid into the measured water. Be sure you read and understand all label warnings and instructions.

Fig. 18-4. Chemical solutions should be kept in clearly labeled opaque containers.

Fig. 18-5. Chemicals should be about 68 °F for processing.

Fig. 18-6. These are examples of both ratchet and spiral wire developing reels.

two common types of developing reels are the *spiral wire reel* and the *ratchet reel*. On both, the emulsion side of the film is placed toward the center of the reel. The film curls toward the emulsion side.

Film is fed into the ratchet reel starting from the outside, Fig. 18-7. Advance the film by holding it with your thumb and rotating the reels. The spiral wire reel is loaded starting in the center. Cup the film and orient it so it is facing the spiral ends. Now attach the film in the center of the reel, Fig. 18-8. Feed the film as you turn the reel. This can be done by rolling it on a flat surface. If loaded properly, the film should follow the spiral grooves. If the film surfaces touch, "burn marks" can result, which will ruin the film in those particular spots. After practicing in a lighted area and then in the dark using scrap film, you will be ready to load the exposed film.

Prepare the materials needed for loading the film. For 35 mm film, this includes a developing tank and reel, the film, scissors, and a bottle opener or film canister opener, Fig. 18-9. Turn the lights off and open the film canister. This is done from the end without the protruding spool. See Fig. 18-10. Remove the film and square the end with scissors. Handling the film by its edges, place it on the reel with the emulsion toward the center. Near the end, cut the spool from the film and finish winding the film on the reel.

Place the reel in the developing tank and cover it. Some covers screw on clockwise. Since developing tanks are light-tight, the light can now be turned on.

Fig. 18-7. A ratchet reel is loaded with film starting from the outside of the reel.

Fig. 18-9. Assemble all materials you will need for loading film. Once the lights are out, it may be difficult to locate a needed item.

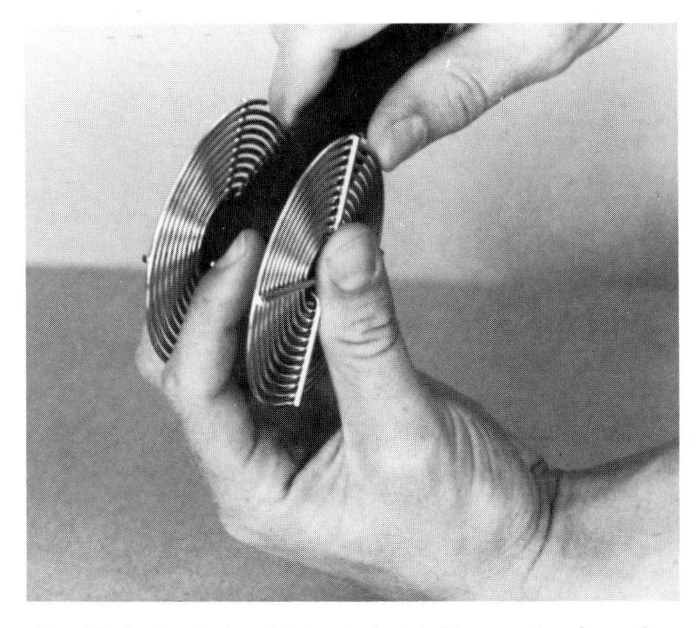

Fig. 18-8. A spiral reel is loaded with film starting from the inside of the reel.

Fig. 18-10. This device is useful for opening film canisters, although a bottle opener can also be used.

FILM PROCESSING

A time and temperature method is used for film processing. Check the directions for the film and developer used. You will need to know how long to develop, stop, and fix the film at 68°F.

Pour the developer in the tank and cap the opening, if needed. Begin timing. Tap the bottom of the tank several times on a flat surface to dislodge air bubbles from the film. Agitate constantly for 30 seconds and then 5 seconds during each 30 seconds after that. For tanks with a thermometer, rotate the thermometer for agitation. See Fig. 18-11. For others, hold the cap in place and rotate 45° to each side. Agitate smoothly and gently. About 15 seconds before development is complete, remove the cap (NOT the lid of the container) and begin pouring the developer out of the container.

Pour stop bath in the tank. Agitate it for the required length of time. Use a funnel to return the stop bath to its container.

Pour fixer into the tank and agitate as you did with the developer. If a range of fixing times is given, choose the longer one to ensure a permanent image. For example, if two to four minutes is recommended, fix for four minutes. Pour out the fixer.

Open the tank and wash the film, Fig. 18-12. The water should be about 68°F and flow at a low pressure. This helps to prevent film damage. A sudden change in temperature can cause **reticulation**, which is a cracking of the emulsion. If a cleaning agent is used, washing time can be reduced considerably.

After washing the film, a **wetting agent** is normally used to prevent water spots on the film. Leave the wetting agent in the tank for about 30 seconds, with some agitation. This is usually discarded after use.

Fig. 18-12. Film must be washed after fixing to remove chemicals.

Remove the film from the reel. Using a clean sponge or squeegee, gently remove the excess water, Fig. 18-13. The emulsion will still be soft, so be careful. Attach a film clip at one end (so that the film will hang straight and not curl) and hang the negatives in a dust-free location to dry. See Fig. 18-14.

After the film is dry, carefully cut between frames to form strips. Still handling the film by the edges, slide these strips into negative sleeves for protection. See Fig. 18-15.

MAKING A PRINT

Printing is done by projecting light through a film negative onto light-sensitive paper. Because a negative is used, the image on the print will be a reversed, posi-

Fig. 18-11. This tank has a thermometer that can be turned to provide agitation.

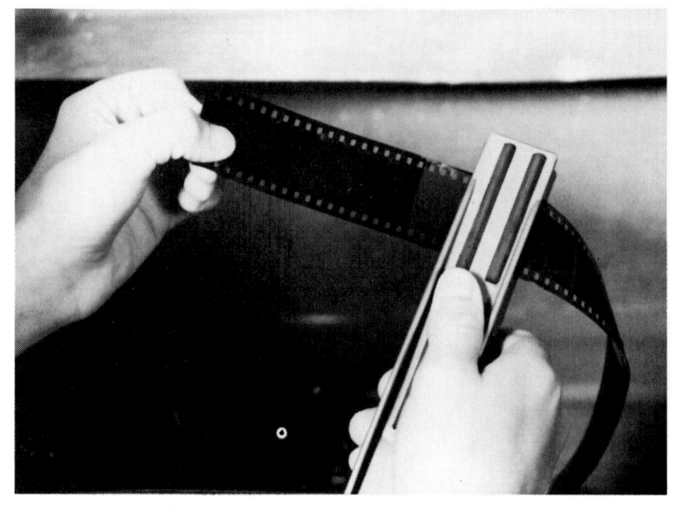
Fig. 18-13. This film squeegee is used to remove excess water from the film.

Fig. 18-14. Film clips are used to hang processed film to dry.

Fig. 18-15. Once negatives are dry, they are cut into smaller strips and placed in protective sleeves.

tive image. This image is formed with a light-sensitive emulsion similar to film emulsion. However, this emulsion is not sensitive to the entire visible light spectrum. Therefore, a **safelight** can be used without exposing the emulsion. This is normally amber (yellow) for panchromatic black and white film.

Unlike film that has a plastic base, print paper has a paper base. See Fig. 18-16. **Fiber base paper** has four layers. The brightness coating reflects light, which improves the appearance of the finished print. **Resin coated paper** has an additional plastic coating on either side of the paper base. Since the plastic prevents absorption of chemicals, washing time is reduced. In addition, resin coated (RC) paper is easier to dry.

Print paper can be purchased with different surface finishes. For example, a glossy surface is used for pictures that will be printed in books and magazines. A

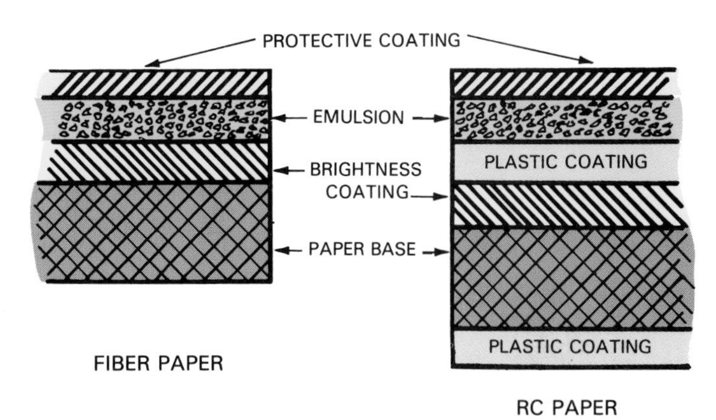

Fig. 18-16. This diagram illustrates the differences between fiber base paper and resin coated (RC) paper.

matte (non-glossy) surface might be used for exhibiting large photos. This surface reduces reflection that might be distracting to the observer.

Many print papers are available in a series of contrast grades, from 0 to 5. These are used to make good prints from problem negatives. A No. 2 paper is used for a negative with a normal density range. However, if you have a negative with a short density range (flat), it will reproduce as a flat, low contrast print. This print can be improved by choosing a higher contrast paper, such as a No. 3 or No. 4. If the negative has too much contrast, then a grade less than No. 2 can be used to reduce the contrast on the print.

Another way of controlling density is by using variable-contrast paper, Fig. 18-17. One paper is used with a set of filters to control contrast, Fig. 18-18. The paper has several emulsion layers that are selectively exposed, depending on the filter used. The filters are numbered like graded print papers. A No. 2 filter is used for a normal negative. The filters are numbered in half steps, such as 1, 1 1/2, 2, 2 1/2, 3, 3 1/2, or 4.

Fig. 18-17. Several brands of variable contrast paper are available.

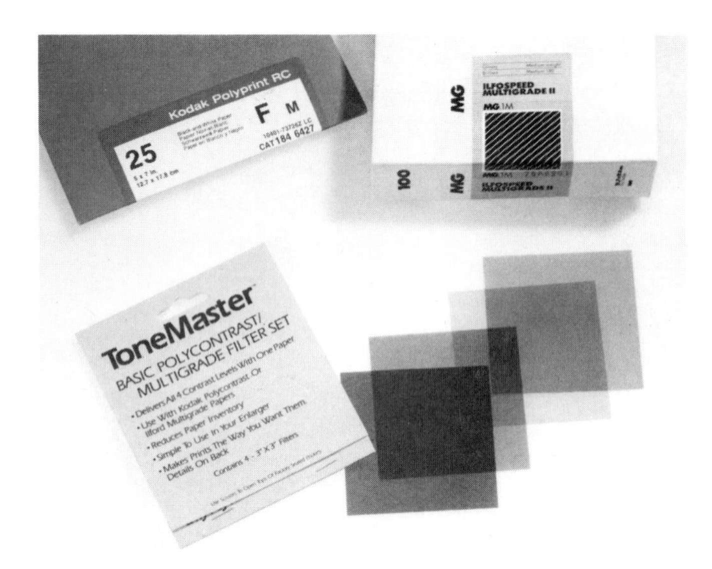

Fig. 18-18. Using filters such as these with variable contrast paper can enable you to adjust the print density.

Fig. 18-19. Chemical trays for print processing should be arranged in the order in which they will be used.

PRINT CHEMICAL PREPARATION

Before making a contact print, processing chemicals should be prepared. Choose three trays slightly larger than the print or contact sheet. Mix enough chemicals to fill these to a depth of at least 1/2 in. A developer, stop bath, and fixer are used, Fig. 18-19. The same stop bath and fixer for film processing can be used in making prints. However, the dilution may be different. As with film processing, printing chemicals should be about 68°F. The processing trays should be arranged in order from developing to fixing and then washing.

MAKING A CONTACT PRINT

The best way to evaluate negatives is to make a contact print from them. A **contact print**, often called a proof sheet, produces a positive image that is much easier to view than looking at the negative.

A contact printing frame and enlarger are normally used for making the contact print, as shown in Fig. 18-20. Clean the glass surface, if needed, with a glass cleaner. Dust and dirt can cause spotted prints.

Place an empty negative carrier in the enlarger. See Fig. 18-21. Adjust to the largest lens opening. Turn on the enlarger and raise the head so the rectangle of light is larger than the paper to be used. Then turn off the enlarger.

Turn off the white light in the darkroom and turn on the amber safelights. With clean hands, remove one sheet of normal contrast print paper by the edge. Close the box. Place the paper in the printing frame with the emulsion side facing up. (The paper often curls toward the emulsion side.) Remove the negatives from the sleeves and place them emulsion side down on the

Fig. 18-20. A glass sheet is placed over print paper and negatives to keep them in place as a contact print is made.

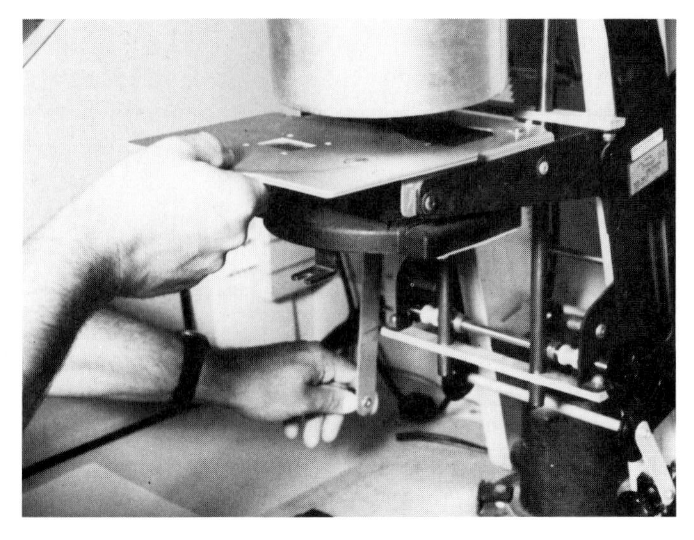

Fig. 18-21. When making a contact print, an empty negative carrier is placed in the enlarger.

paper. (The negatives curl toward the emulsion side.) Then close the frame.

Make a **test strip** to determine the correct exposure time for a contact print, Fig. 18-22. Using a black card, cover all but about 1 1/2 in. on the end of the paper. Make a three-second exposure. Continue moving up the paper and making three-second exposures. Remove the paper.

Hold the paper by its edge. With the emulsion up, slide it into the developer, Fig. 18-23. Develop for the recommended time for the developer being used. Agitate the tray by gently rocking it from each corner. About five seconds before developing is complete, lift the sheet with the tongs, let it drip for a few seconds. Now slide it into the stop bath without letting the tongs touch the stop bath. Agitate the print in the stop bath. Using a second set of tongs, lift the print. Let it drip a

Fig. 18-22. When making a test strip to determine the correct exposure time for a contact print, expose sections of the paper in three-second intervals.

Fig. 18-23. Exposed print paper is slid into the developer emulsion side up.

Fig. 18-24. A print washer is convenient for rinsing prints.

few seconds. Then slide it into the fixer. Fix and wash the print for the recommended length of time, Fig. 18-24. After washing, place the print on a flat surface and remove excess water, Fig. 18-25.

Now you can turn on the lights. Inspect the print and choose an exposure time. The end with the least exposure (lightest) had a three-second exposure. The next strip is a six-second exposure, then a nine-second exposure, and so on. Record the best time and use it for making contact prints.

Make a second contact sheet, using the time chosen. Process as before and dry the print. Inspect the sheet and choose those negatives you wish to use for prints. A magnifying glass is helpful for this. A grease marker can be used for marking on the contact sheet.

Fig. 18-25. Excess water is being removed from a contact sheet using a print chamois.

MAKING ENLARGEMENTS

The final print is made with an enlarger. Place the negative in the negative carrier with the emulsion side down, Fig. 18-26. If needed, clean the negative with a special brush. Place the negative carrier in the enlarger head.

Adjust a print easel for the print size desired and place it under the enlarger, Fig. 18-27. Adjust the lens for the largest aperture, Fig. 18-28. Switch off the white lights and turn on the enlarger light. Raise or lower the head on the column until the negative image is slightly larger than the print size on the easel. Focus the image with the knob that raises and lowers the lens. A special magnifier can be used to check the focus, Fig. 18-29. After focusing, the lens aperture is changed to about f/11. As with a camera, this provides more depth of field (range of distance from lens to easel in focus). This eliminates most small focusing errors.

Fig. 18-26. When making an enlargement, the negative is placed in the negative carrier emulsion side down.

Making a test print

A **test print** is made to determine proper exposure time, Fig. 18-30. Place a piece of print paper in the easel, emulsion up. Make a series of five-second exposures using a black card as a mask. Process the print as described for contact printing. Choose the best exposure time. With experience, you will be able to choose a time based on the contrast of the negative.

A test print can also be made with a *print scale* as shown in Fig. 18-31. The scale is placed over the print paper and one exposure is made. You choose one section that is properly exposed. Set the enlarger timer for the exposure time printed on the section.

Fig. 18-27. When making an enlargement, a print easel is placed in position under the enlarger.

Making the final print

After the exposure time is determined, you can make the final print. Process the print following the steps used for making a contact print and a test print, Fig. 18-32. The print can be inspected with the lights on after one minute in the fixer. Slight adjustments in exposure time may be necessary. If needed, make another print. Then dry the prints, Fig. 18-33.

Some negatives may require a change in contrast. In this case, use a different grade of paper or a different filter with variable contrast papers.

IMPROVING PRINTS

There are several methods of improving prints. These include:

- Changing the contrast.
- Cropping.
- Burning-in and dodging.
- Spotting.

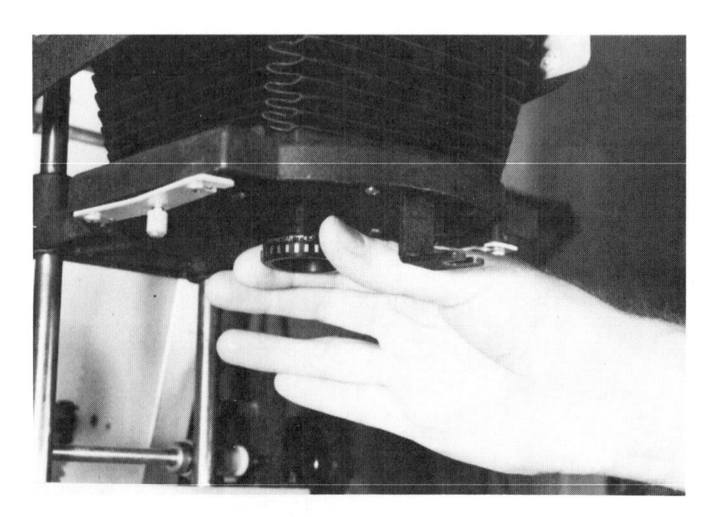

Fig. 18-28. Before focusing, the aperture on the enlarger should be adjusted to the largest aperture setting. After focusing, the aperture setting is changed to about f/11.

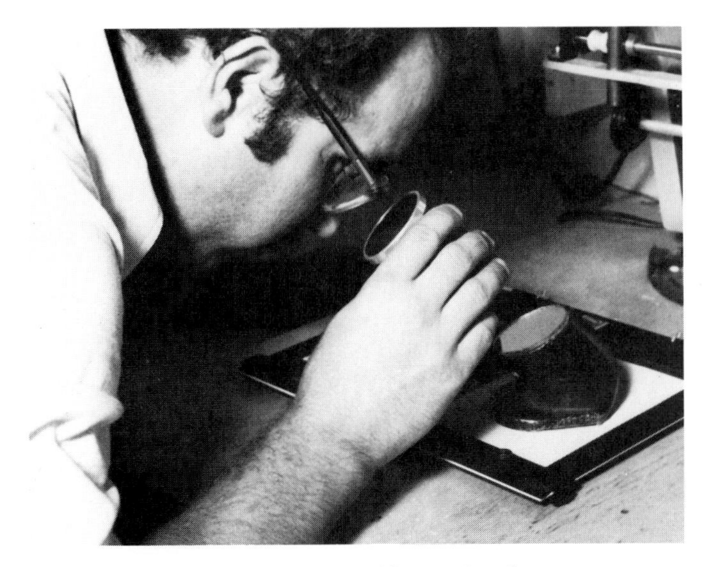

Fig. 18-29. This print magnifier assists in accurate focusing.

Fig. 18-30. An exposure time can be selected from this test print.

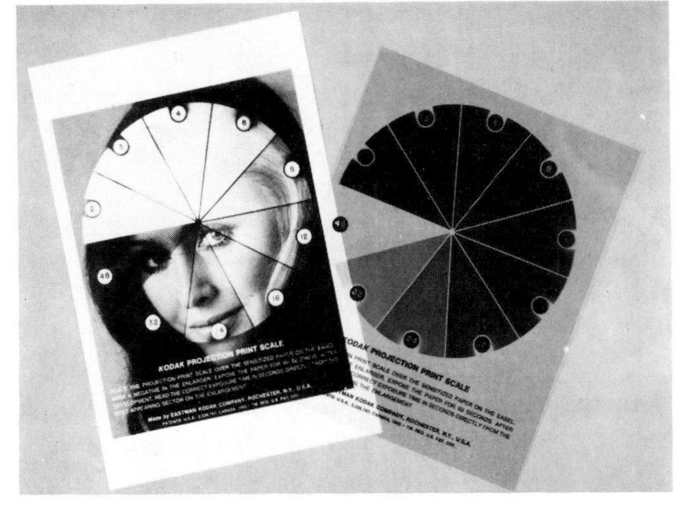

Fig. 18-31. A print scale can be used to determine correct exposure.

Fig. 18-32. The steps for processing a final print are similar to those used in processing a contact print.

Fig. 18-33. After excess water is removed, prints are placed in a film dryer.

Changing the contrast

One method for improving prints is to change the contrast. This is done when the contrast of the negative is not correct for a normal contrast print. In this case, a different paper grade is used. However, in the case of variable contrast paper, a different filter is used.

Cropping

Cropping is another method for improving a print. It is done by selecting only part of the negative to print. This is accomplished on the enlarger by further enlarging the image. The easel is then moved to frame the desired scene. **Cropping L's** can be used to visualize how a crop will look, Fig. 18-34. These can be placed on the proof sheet or print. Be careful about enlarging the negative too much. This can produce a "grainy" image.

Fig. 18-34. Cropping L's help you visualize how a cropped photo will look.

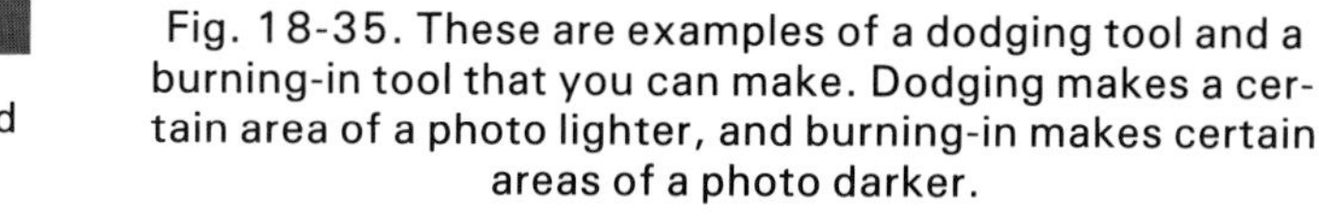

Fig. 18-35. These are examples of a dodging tool and a burning-in tool that you can make. Dodging makes a certain area of a photo lighter, and burning-in makes certain areas of a photo darker.

Dodging and burning-in

Dodging and burning-in are methods used to improve the quality of a photograph. Typical, homemade dodging and burning-in tools are shown in Fig. 18-35.

Dodging is used to lighten an area of a print by giving it less exposure. Dodging tools can be made or purchased for this purpose. During exposure, move the dodging tools in small circles several inches above the area to be lightened. The movement blends the light and dark areas on the print.

Burning-in is used to darken an area of a print by giving it more exposure. A tool for this purpose can be made from a large card with a hole in it. To burn-in, make the initial exposure. Then make a second exposure with the burning tool in position above the print. Keep the card moving to blend the burned area with lighter areas.

Fig. 18-36. Special dyes are available to cover spots on prints.

Spotting

Prints with spots can be touched up using a process called **spotting**. A special dye is mixed so that it matches the spot, Fig. 18-36. It is applied with a brush by lightly touching the spot with the brush end. Keeping negatives clean will greatly reduce the need for spotting.

MOUNTING A PRINT

You may wish to mount your photographs so that they can be exhibited. Mount board can be purchased in various colors and sizes for this purpose. The board should be cut so that it is larger than the photo.

Spray mounting can be done using spray mounting adhesives. To do this, make very light corner marks on the board where the photo will be positioned. It is usually placed slightly above center. Place the photo face down on a covered surface, such as cardboard, and spray the back lightly with adhesive. See Fig. 18-37. Let it dry a short time, as directed. Then position the print on the board. Place a scrap print or piece of paper over the top and use a hard, rubber roller or a smooth, plastic squeegee to smooth the surface. See Fig. 18-38 and Fig. 18-39.

Dry mounting is a more precise method of mounting prints. A dry mount tissue is used that has an adhesive on both sides. This is placed between the board and print. A dry mount press is used to adhere the layers, Fig. 18-40.

Fig. 18-37. When using the spray mounting technique, the back of the photo is sprayed lightly with adhesive.

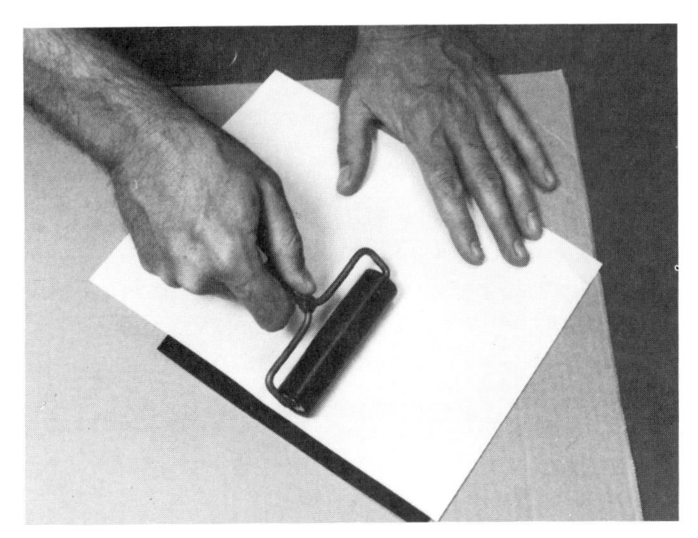

Fig. 18-38. A rubber roller is being used to attach the photo to the mount board. Scrap paper prevents the roller from contacting the print.

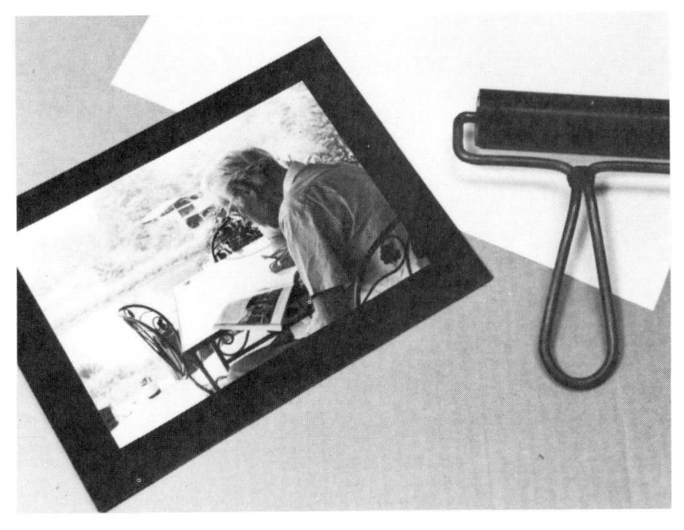

Fig. 18-39. A mounted print gives a photograph a professional look.

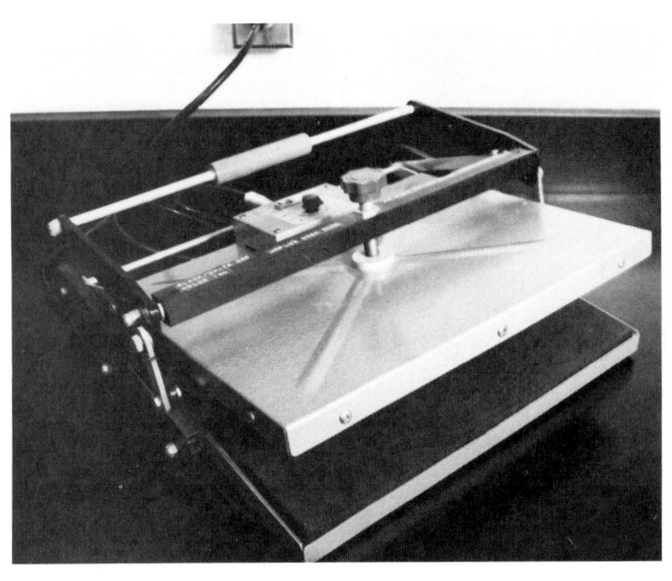

Fig. 18-40. When mounting a photo using the dry mounting technique, a dry mount press is used to apply heat and pressure to the photo, dry mount adhesive tissue, and mounting board.

SUMMARY

This chapter has introduced you to black-and-white film and print processing. Additional techniques have been presented for improving the print. In industry, automatic film and print processors are used for high production. However, the techniques described in this chapter are still used by professionals for high- quality prints.

WORDS TO KNOW

All of the following words have been used in this chapter. Do you know their meanings?

burning-in
contact print
cropping L's
developer
dodging
dry mounting
fiber base paper
fixer
printing
processing
resin coated paper
reticulation
safelight
spotting
spray mounting
stop bath
test print
test strip
wetting agent

REVIEWING YOUR KNOWLEDGE

Please do not write in this text. Write your answers on a separate sheet.

1. Which of the following chemicals makes the latent image on the film or paper visible?
   a. Developer.
   b. Stop bath.
   c. Fixer.
   d. None of the above.
2. True or false? Chemicals that are mixed and ready to use are called stock solutions.
3. For processing, chemicals should be at:
   a. 32°F.
   b. 68°F.
   c. 100°F.
   d. 120°F.
4. True or false? When loading film on a reel, the emulsion side of the film is placed toward the center of the reel.
5. The steps involved in processing black and white film are:
   a. Developer, fixer, wash, stop bath.
   b. Fixer, developer, wash, stop bath.
   c. Developer, stop bath, fixer, wash.
   d. Developer, fixer, stop bath, wash.
6. Which contrast grade of paper would you choose to use with a negative with a normal density range?
   a. 0.
   b. 2.
   c. 4.
   d. 5.
7. A _________ _________, often called a proof sheet, produces a positive image that is much easier to view than looking at the negative.
8. When focusing an image for an enlargement, the enlarger aperture is set at:
   a. The smallest aperture.
   b. f/11.
   c. The largest aperture.
   d. None of the above.
9. Name four methods of improving prints.
10. Enlarging a negative too much can produce a "_________" image.
11. Which of the following is used to darken an area of a print by giving it more exposure?
    a. Dodging.
    b. Burning-in.
    c. Spotting.
    d. Cropping.
12. True or false? A dry mount press is needed when spray mounting.

APPLYING YOUR KNOWLEDGE

1. Tour a local film processing facility. Prepare a written report about the experience.
2. Collect literature on photo processing and develop a floor plan for a home or school darkroom.
3. Demonstrate various techniques often used to improve prints.
4. Using a central theme, develop an exhibit of photographs you have taken, processed, printed, and mounted.

19

CHAPTER

COLOR PHOTOGRAPHY

After studying this chapter, you will be able to:
- *Describe how colors of light are seen by the human eye.*
- *Compare additive primary colors and subtractive primary colors.*
- *Recognize types of color film.*
- *Describe color temperature and its relationship to light and film.*

Color photography is a way of recording the world around us in full color, just as our eyes see it. Special film and processing techniques are necessary to do this. This chapter will explore the world of color photography. First, however, let's find out how our eyes see color.

HOW WE SEE COLOR

The structure of the human eye is shown in Fig. 19-1. When you look at something, the image is focused on the retina. The retina has thousands of light-sensitive cells known as rods and cones. The rods send information to the brain about how bright the light is. The cones send information to the brain about the color of the light. These cones are sensitive to the primary colors of light, which are red, green, and blue.

LIGHT AND COLOR

Light can be thought of as a wave that travels across distance. It is part of the **electromagnetic spectrum**, Fig. 19-2. This spectrum shows waves of different wavelengths. The left end contains the long radio waves. A radio receives this type of wave. In the middle, there are light waves. At the right side, there are very short waves such as x-rays.

Most light sources produce a combination of wavelengths. For example, a prism in sunlight will divide the light into a rainbow of colors, Fig. 19-3. These various wavelengths are all in the sunlight.

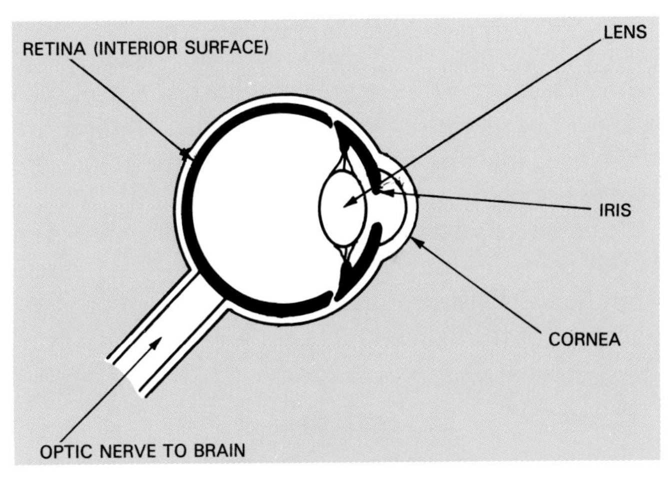

Fig. 19-1. This is the basic structure of the human eye.

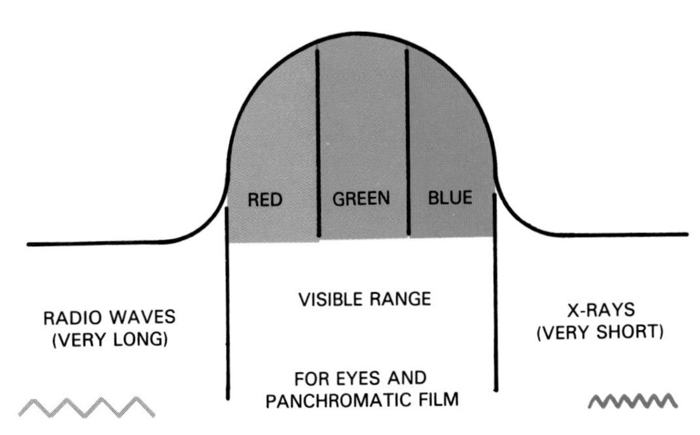

Fig. 19-2. This simplified electromagnetic spectrum shows various types of wavelengths.

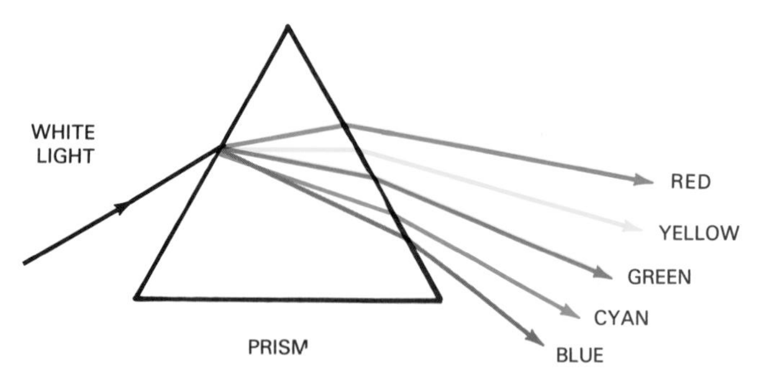

Fig. 19-3. A prism divides white light into its various colors giving a rainbow effect.

When red, green, and blue wavelengths are combined, they produce white light, as shown in Fig. 19-4. They are called **additive primaries** because they are *added* together to form white light. They can also be added in varying amounts to create other visible colors. For example, red and green light will combine to produce yellow. The additive primaries are used in color television to form all the colors you see on the screen.

Look at Fig. 19-5. You will notice that the three additive primaries are overlapped. This combination of two primary colors produces yellow, magenta, or cyan. These colors are known as the **subtractive primaries** because they are used to *subtract* colors from white light. When two of these are combined, they can subtract all but one additive primary. When all three are combined, black is produced. Subtractive colors are used as ink pigments or dyes.

Subtractive primaries are used for color photos and in full color printing in graphic arts. In both cases, white light strikes a colored surface. These colors subtract part of the light and reflect the rest, Fig. 19-6. In various combinations, we see these reflections as full color.

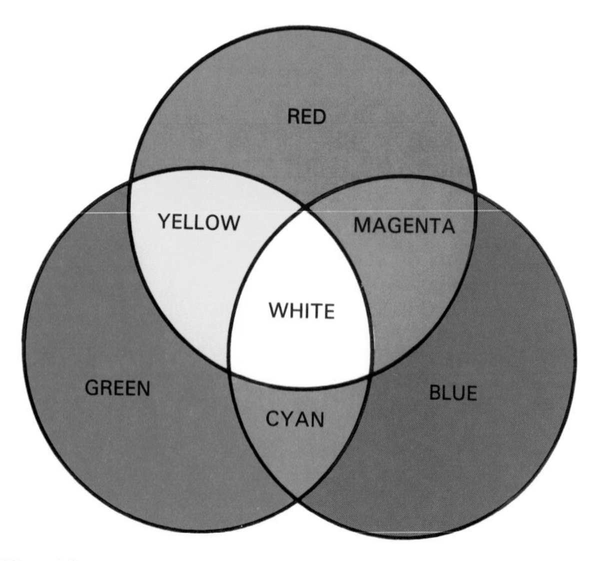

Fig. 19-4. The additive primaries of red, green, and blue produce white light. Color television is an example of where additive primaries are used.

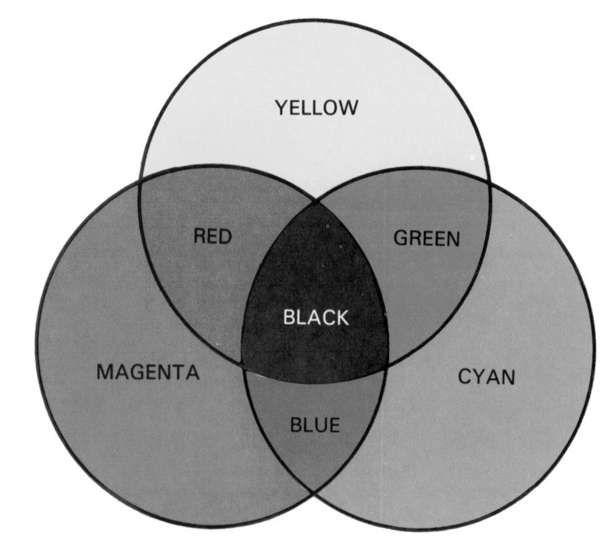

Fig. 19-5. The subtractive primaries of yellow, magenta, and cyan produce black. Ink pigments or dyes are examples of where subtractive primaries are used.

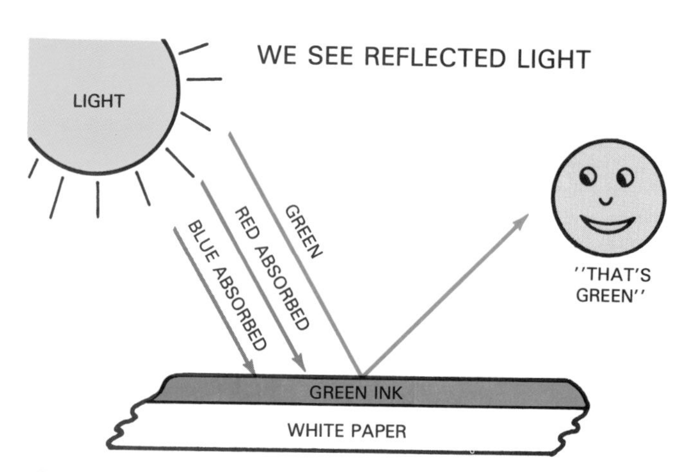

Fig. 19-6. In this example, red and blue are subtracted or absorbed. Green is reflected. Thus, green is the color that is seen.

COLOR FILM AND COLOR PRINT PAPER

Color film and color print paper typically have three emulsion layers. Each emulsion layer is sensitive to one color of the spectrum–red, green, or blue. When it is processed, the layers are dyed as cyan, magenta, and yellow. Refer again to Fig. 19-5. Various combinations of these subtractive colors produce any color needed. For example, magenta (red + blue) and cyan (green + blue) filter out all but blue.

COLOR FILM

Color film is made in the same sizes as black and white film. Like black and white film, color film has a

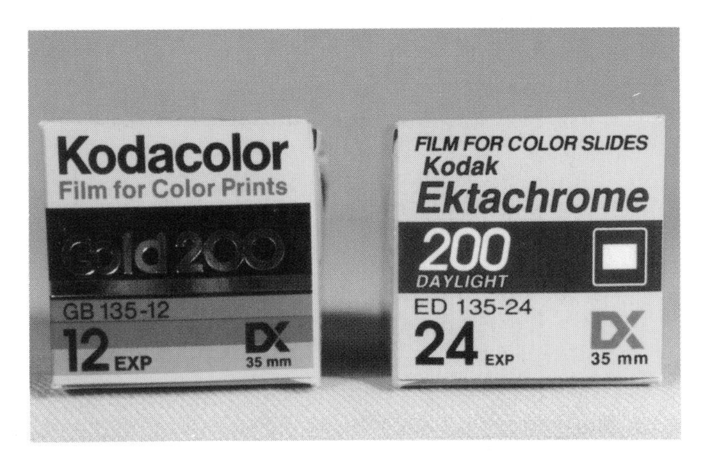

Fig. 19-7. Color negative film is shown at left and positive film on the right. Positive film is used for slides.

plastic backing and an antihalation layer. Color film can be negative or positive, Fig. 19-7.

Color positive film

Color positive film (reversal film) is used for slides. Light is projected through the film for viewing on a screen. This film usually has the suffix, "chrome," in its name. Ektachrome and Agfachrome are two examples.

Color negative film

Color negative film is used for color prints. The colors are recorded on the film as an opposite or complementary color, Fig. 19-8. For example, blue would be recorded as yellow, which is a combination of red and green. When a color print is made from the negative, the colors will be correct.

COLOR PRINT PAPER

Color print papers have layers of emulsion similar to those that make up color film. However, instead of a plastic backing, a paper base is used. Color print paper is available in the same sizes as black and white paper. However, it is not produced in different contrast grades.

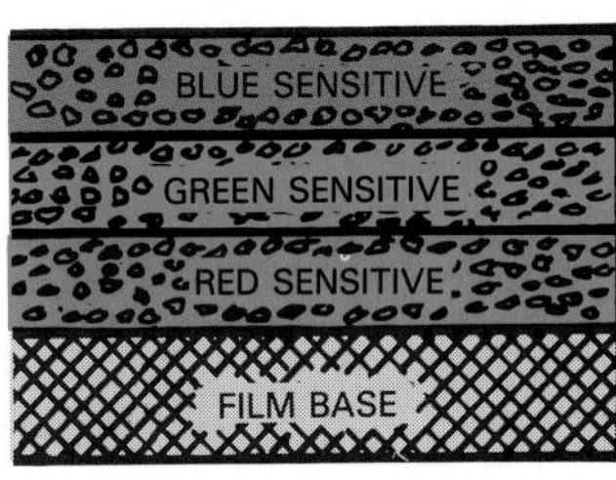

Fig. 19-8. Color negative film records the additive primaries on three emulsion layers that are dyed subtractive primary colors during processing.

COLORING THE EMULSION

One of three systems is used to color the emulsion layers of film or paper. These are:
- Chromogenic system.
- Dye-bleach system.
- Diffusion transfer system.

Chromogenic system

In the **chromogenic system**, the emulsion layers have color added during processing. For one type of chromogenic film, dye is added during processing to each of the emulsion layers. Kodachrome is an example. This is a complex process, so it is normally done by a commercial processor.

A newer technique uses special chemicals (couplers) in the emulsion that combine with the developer to form the dye. This is a common system used for film and print paper, Fig. 19-9. It has the advantage of being fairly easy to process. Films that use this process include Ektachrome, Kodacolor, and Fujichrome.

Dye-bleach system

The **dye-bleach system** starts with color in the emulsion layers. After exposure and processing, colors are removed as needed. This is a very permanent image. Cibachrome was the first film of this type, Fig. 19-10.

Diffusion transfer system

The **diffusion transfer system** can be used for print making in the darkroom. A negative is exposed and placed in contact with a receiver sheet. These are then

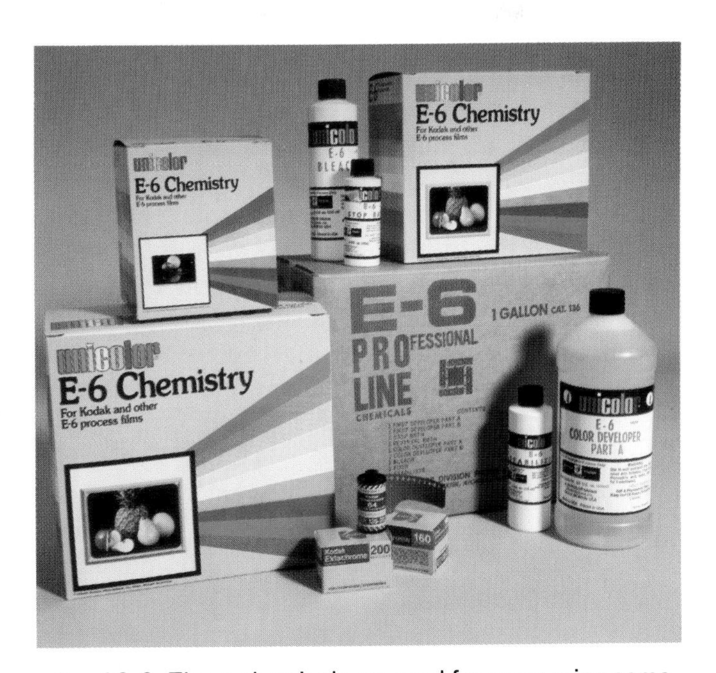

Fig. 19-9. These chemicals are used for processing some chromogenic type slide films. (Photo Systems, Inc.)

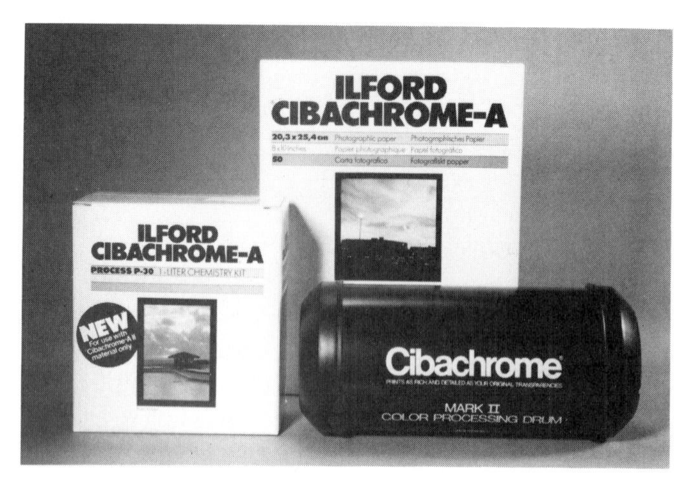

Fig. 19-10. This kit produces a dye bleach type of color print in the print drum shown here.

sent through a processor. After a few minutes, the sheets are separated. The negative color image has now transferred as a positive color image on the receiver. Ektaflex and Agfachrome-Speed are two types of diffusion transfer print materials.

Instant camera print film uses a diffusion transfer technique. The film layers and chemicals are made together. As the material exits the camera after exposure, the pressure activates the chemicals and causes the print to be processed.

MATCHING LIGHT AND FILM

Most white light sources are made up of a combination of light wavelengths. The sum of these wavelengths is known as **color temperature**. This is measured on a Kelvin Scale as shown in Fig. 19-11. Unlike other measures of temperature, the Kelvin Scale does not represent the amount of heat in the light source. Instead the temperature is a measure of the colors in the light.

Low temperature light sources, such as a candle and house lighting, have more red wavelengths in comparison to blue. High temperature light sources, such as daylight and electronic flash, have more blue wavelengths as compared to red. Red is considered to be a warm color, and blue is considered to be a cool color. Therefore, light with excess red is considered to be warm, and light with excess blue is considered to be cool. On the Kelvin Scale, a high temperature is a cool light source.

Color film is balanced for different light sources. The two most common are daylight (5500° K) and tungsten. Daylight film is balanced for pictures outside as well as electronic flash. Tungsten film is needed when photofloods are used to light the scene. Tungsten film is available as Type A (3400° K) or Type B (3200° K) depending on which light is used. Tungsten film is balanced to receive more red wavelengths as compared to daylight film.

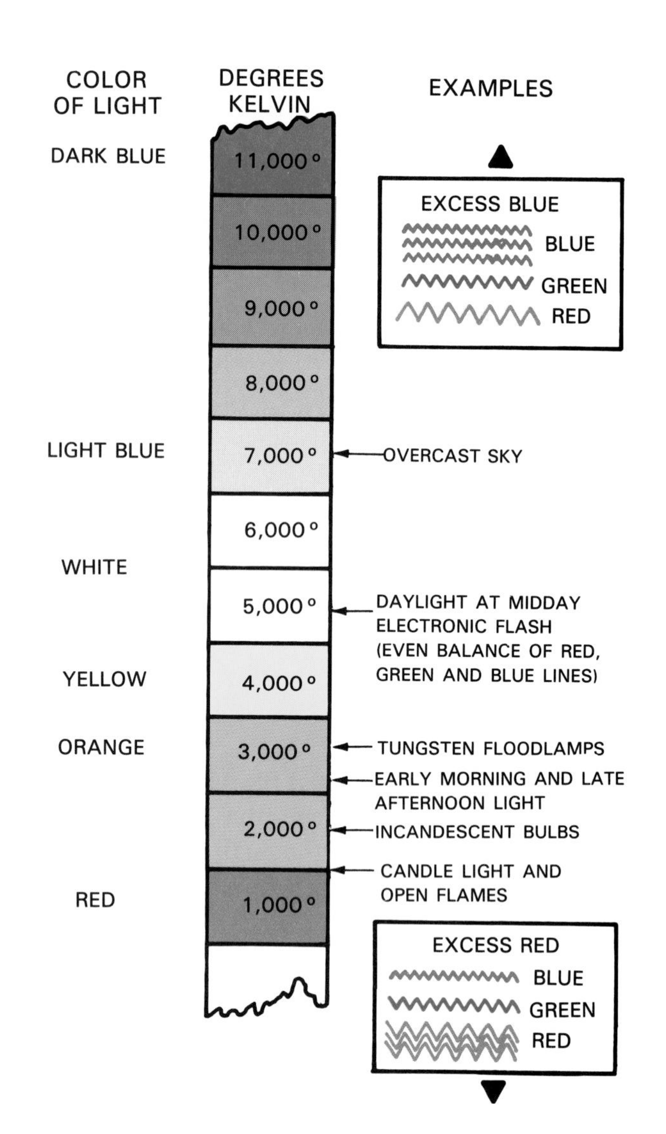

Fig. 19-11. Color temperature is measured on a Kelvin scale. Lower temperature light has more red wavelengths and higher temperature light has more blue wavelengths.

Film can be used with the wrong light source if a filter is used. The filter color usually matches the color "expected" by the film. For example, if daylight film is used with red-dominant tungsten lights or household bulbs, the print will have too much yellow in it. A special blue filter will balance the film and light by blocking red and adding more blue wavelengths. See Fig. 19-12. If tungsten film is used with a flash or in blue-dominant daylight, the print will have too much blue in it. A special yellow filter will balance the film and light by blocking some of the blue, while also allowing more red to pass through the filter, Fig. 19-13.

Some light sources have a very uneven distribution of colors. See Fig. 19-14. For example, fluorescent light has a large amount of green, so the resulting photo has too much green. A special filter can be used that will absorb or block these excessive green wavelengths.

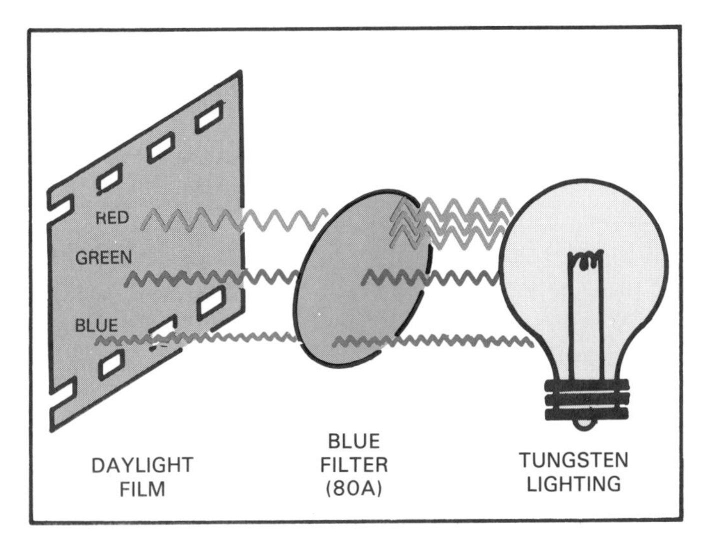

Fig. 19-12. By using a blue filter, tungsten light can be balanced for daylight film.

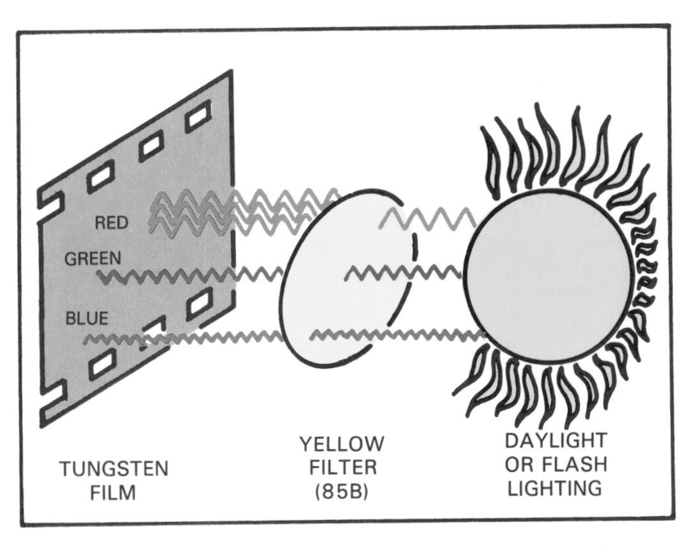

Fig. 19-13. By using a yellow filter, daylight can be balanced for tungsten film.

COLOR AND COMPOSITION

The composition rules for black and white photography apply to color photography as well. However, you will also need to pay attention to color in the scene. Just as highlight areas attract your attention in a black and white photo, bright colors attract your attention in a color photo. Therefore, a bright color works well as the center of interest, with subdued colors elsewhere in the scene. Too many bright colors can be distracting in a photo. Looking at photos by professional photographers is a good way to learn about using color.

COLOR PROCESSING

Color film can be processed in a darkroom set up for black and white photography. Processing kits can be purchased that contain the additional chemicals needed. See Fig. 19-15. Color film is usually processed at a higher temperature than black and white film. Gloves and adequate ventilation should be used when working with color processing chemicals.

When color slide film is processed, it can be mounted in frames and used without additional steps, Fig. 19-16. For this reason, it is the most popular type of color film for home and school use.

Color prints require an exposure for each emulsion layer in the print paper. This is done with color filters in an enlarger. Yellow, magenta and cyan filters are used to make the exposures. Color corrections are made in the print by varying the density of the three filters.

Color enlargers have the filters built into the color head, Fig. 19-17. Color correction can be adjusted directly on the enlarger. Filters can also be purchased for use with black and white enlargers. The set contains the

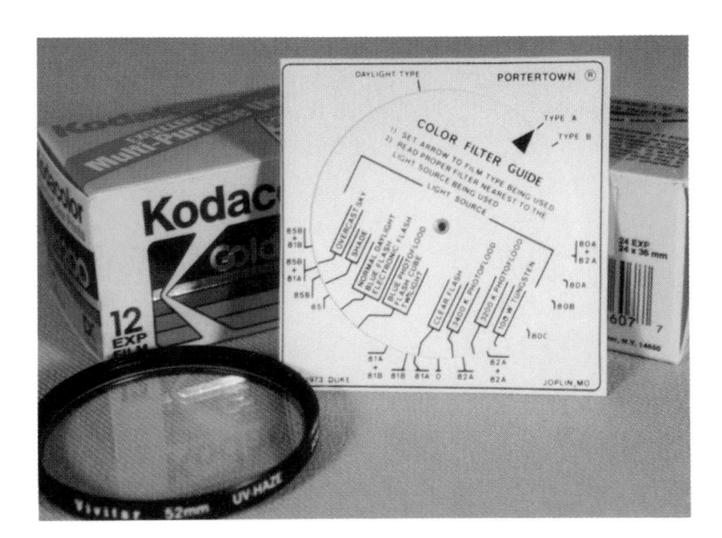

Fig. 19-14. This filter guide helps in choosing the right filter to balance the film and light source. (Porter's Camera Store, Inc.)

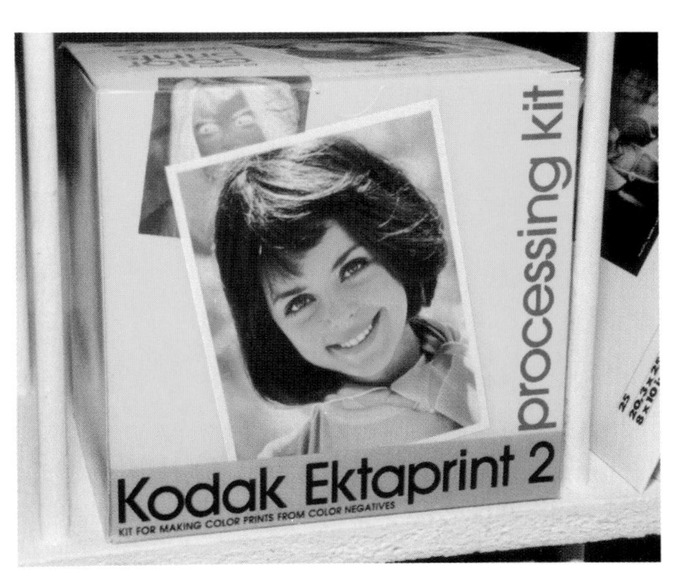

Fig. 19-15. This processing kit is used for color negatives.

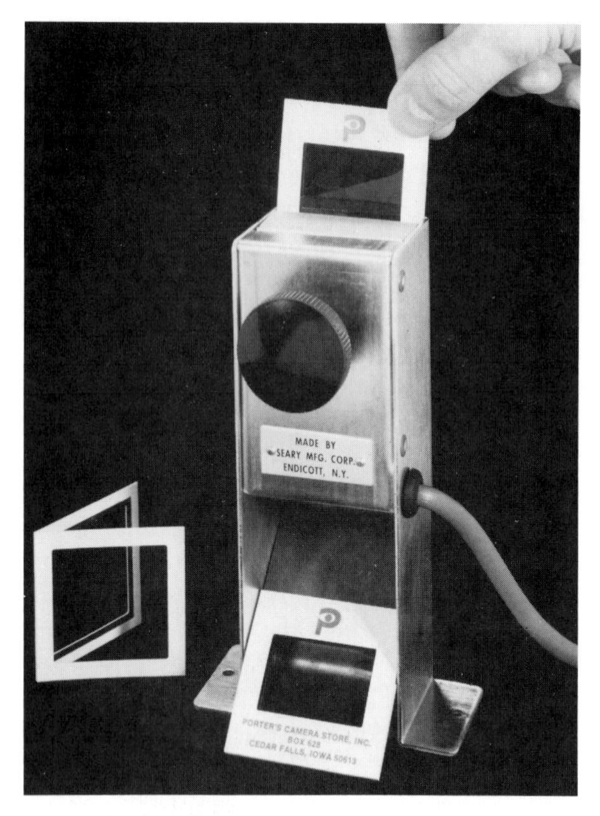

Fig. 19-16. This slide mounting unit automatically seals the slide in a cardboard mount.

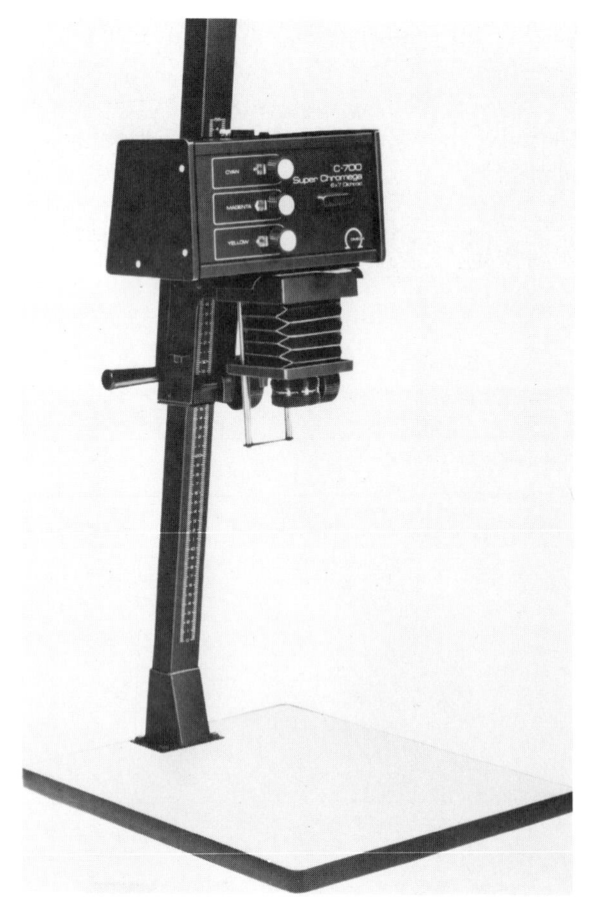

Fig. 19-17. A color enlarger is used to expose each emulsion layer in color print paper.

three filter colors in different densities. Tests are made to find the correct filter combination. An ultraviolet filter is also used because the light source in a black and white enlarger must be corrected for color printing.

Color prints can be processed in trays. However, since the processing must be done in the dark, tubes or drums are often used. Refer again to Fig. 19-10. After the print paper is loaded in the drum, the lights can be turned on. After chemicals are added, the tube is rolled for agitation. Again, adequate ventilation and gloves should be used with color processing chemicals.Diffusion transfer print papers are very easy to process. Kodak Ektaflex is one example. After exposure, the negative and receiver are placed together. These sheets are sent through a processor with an activator chemical. When the sheets are separated, the color image will be transferred to the receiver sheet. This system is usually slightly more expensive per print.

COMMERCIAL PROCESSING

In recent years, color film and print processing have become faster and less expensive. Processing equipment has automated the processing steps. See Fig. 19-18. For this reason, some establishments can process photos in one hour or less.

Although commercial processing is done by machine, the company personnel must still understand how color film works, Fig. 19-19. They must adjust the equipment as needed for quality color film and prints.

Many professional photographers process their own color film. This allows them to have more control over the final product.

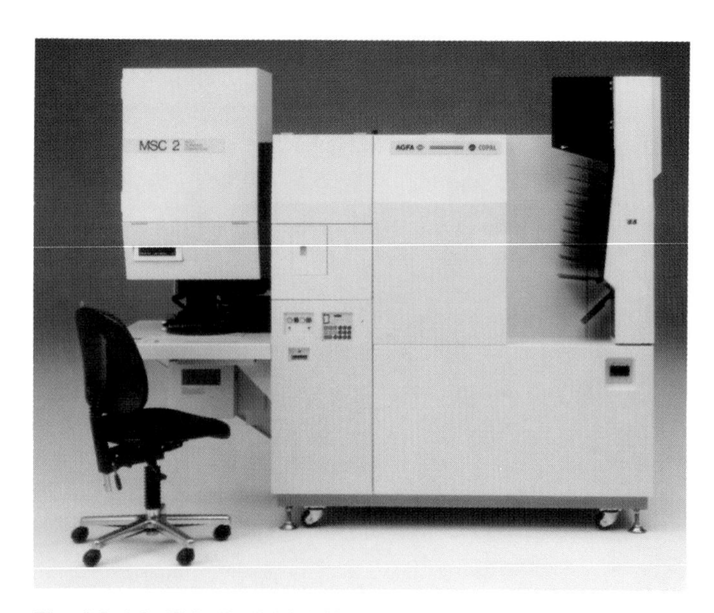

Fig. 19-18. This "mini-lab" uses computer control to automatically process color film and produce prints in less than one hour. (Agfa Copal Inc.)

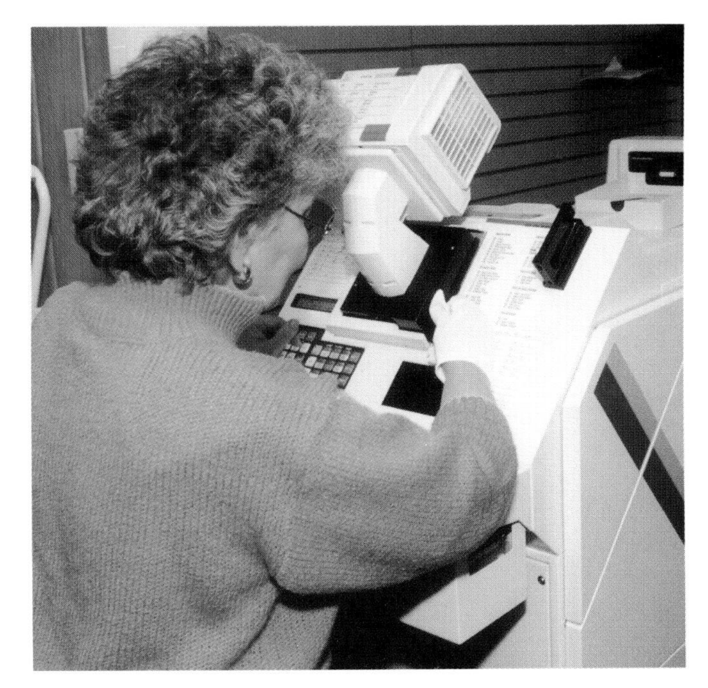

Fig. 19-19. This person is preparing to make prints from the negatives.

PRODUCING A SLIDE SHOW

A slide show is a very effective method of presenting a message to a large group. It can be easy and inexpensive to produce. In addition, sound can be added if you wish.

The steps involved in producing a slide show are similar to the steps involved in producing a video production, which will be discussed in Chapter 36.

First, decide on the audience for the production and the purpose. Now make a rough outline of the presentation. An overview in the beginning and summary at the end are helpful.

After making a rough outline, storyboard cards are made. An example is shown in Fig. 19-20. Each card is for one slide. A quick sketch of the scene is made on one end of the card. The narration that goes with the slide is summarized next to the sketch. These cards can be rearranged as needed until you are satisfied with the sequence.

The next step is to produce the photos. Be sure the film you use is balanced for the type of light used. Photos under fluorescent light require a filter or electronic flash. Try to use a horizontal format for most of the slides.

If title slides are to be made, a copystand is helpful. See Fig. 19-21. The title should be composed on heavy paper. An example is shown in Fig. 19-22. A one-inch border is needed around the title for handling. A good, quality image is needed for the title. Dry transfer letters or computer-generated letters work well. If the image is

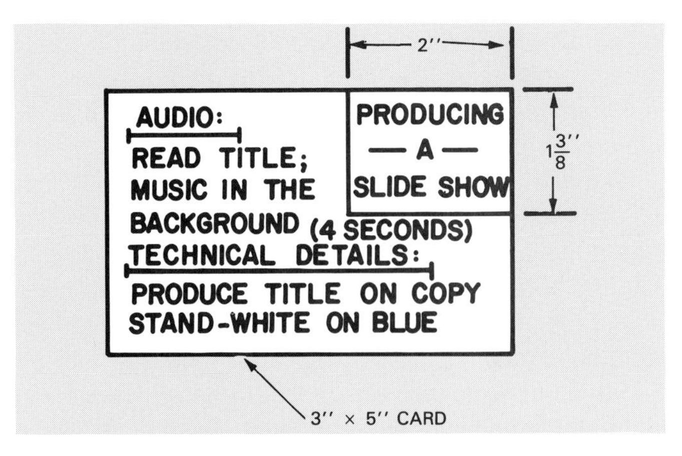

Fig. 19-20. This is an example of a typical storyboard card.

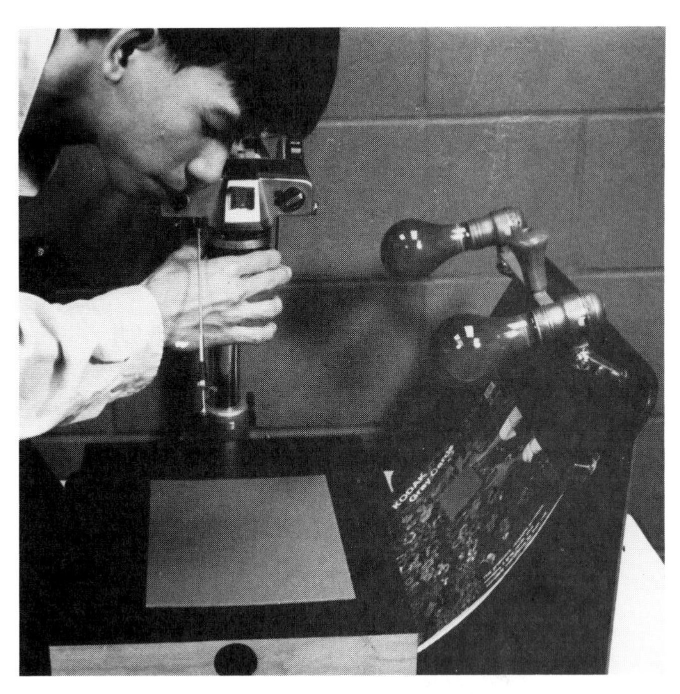

Fig. 19-21. Correct exposure may be obtained on a copystand by first taking a light reading from a neutral gray card while the copystand lights are on.

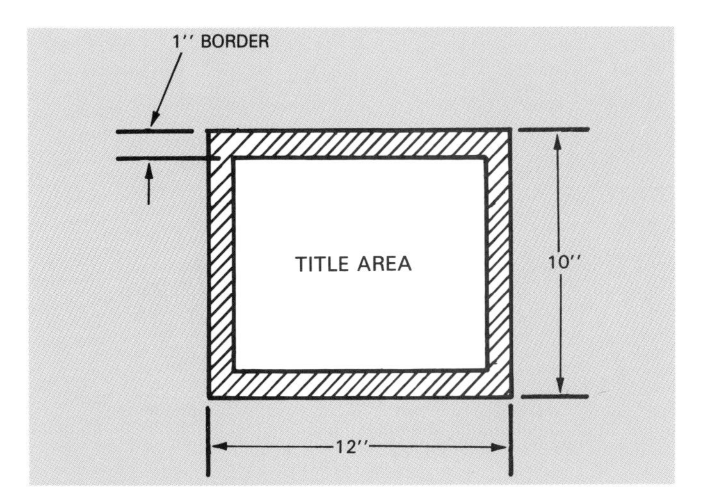

Fig. 19-22. Title slides are created by photographing a title board like this one.

small, a close-up (macro) lens will be needed. Plastic or ceramic letters are also available that can be placed directly on a copystand. A straightedge can be used to align each row of letters.

When a copystand is used a neutral (18%) gray card is used to set the exposure. Place the card on the copystand and turn on the lights. Determine the correct exposure setting for your camera. A cable-release is best with copystand work so that you can be sure the camera does not move.

After the slides are processed, they are placed on a light table and arranged in order. Make sure the slides are "right-reading." In this position, lightly mark the lower left corner of the slides. Now turn the slides over and number them at the mark. The slides are loaded in the slide tray by first orienting the first slide for correct projection. The other slides are placed in the tray with the numbers in the same position. See Fig. 19-23. When the slides are removed from the tray, they should be placed in a protective sleeve or container.

After the slides are ready, the narration should be developed. If the slides will accompany a talk, note cards may be all that is necessary. However, if an audio tape is to be used, a script should be developed. This is done by viewing the slides and writing out the narration. Indicate slide changes on the script. Once the script is ready, record it on a tape recorder in a quiet location. See Fig. 19-24. Pause between slides and use an audible tone such as snapping fingers or a musical note. Make sure the sound is the same each time and not too distracting. It may be necessary to record the script several times before it is correct. Play the tape while viewing the slides to make sure they are synchronized.

Fig. 19-23. Slides should be numbered so that the number is in the upper right corner when standing behind the projector as shown. (Porter's Camera Store, Inc.)

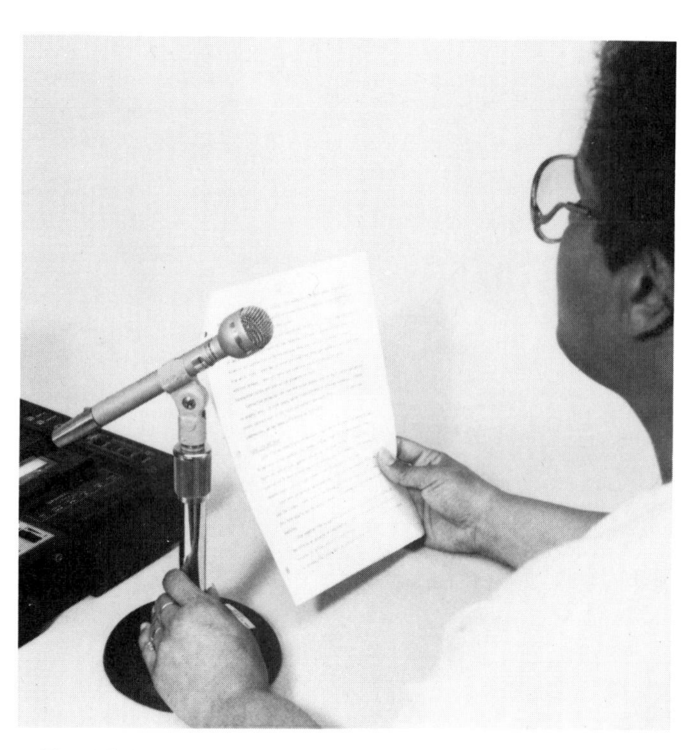
Fig. 19-24. You can record a script to accompany your slides. A signal can also be placed on the audio tape to indicate slide changes.

If a synchronizing recorder is available, an inaudible signal can be placed on the tape for automatic slide changes. The script is recorded as before, with or without an audible signal. The tape is then replayed and an inaudible signal is placed on the second track of the tape. The slide projector is then connected with a cord to the tape player for automatic slide changing. Since a second track on the tape is used, only one side can be used for recording.

A second tape player can be used to add music to the production. The music can fade in (become louder) and fade out (become softer) by adjusting the volume. For example, music might be used with the title slide and then fade out as the narration begins. Music could fade back in at the end. With additional equipment, the music can be mixed more precisely. However, the method described works well for the beginning production.

SUMMARY

This chapter has presented the basics of color photography. How the human eye sees color was discussed. Then various types of film and processing methods were described. Finally, steps for producing a slide show were presented. A slide show is an exciting application of color photography.

WORDS TO KNOW

All of the following words have been used in this chapter. Do you know their meanings?

additive primaries
chromogenic system
color negative film
color positive film
color temperature
diffusion transfer system
dye-bleach system
electromagnetic spectrum
subtractive primaries

REVIEWING YOUR KNOWLEDGE

Please do not write in this text. Write your answers on a separate sheet.

1. When red, green and blue wavelengths are combined, they produce:
   a. White light.
   b. Black.
   c. Red, green, and blue light.
   d. None of the above.
2. Yellow, magenta, and cyan are ____________ (additive, subtractive) primaries.
3. True or false? Each emulsion layer in color film and color print paper is sensitive to one color of the spectrum–red, green, or blue.
4. True or false? Color negative film is used for color slides.
5. True or false? Film with the suffix, "chrome," is usually color positive film.
6. In which of the following systems are the emulsion layers colored during processing?
   a. Chromogenic.
   b. Dye-bleach.
   c. Diffusion transfer.
   d. None of the above.
7. If daylight film is used with red-dominant tungsten lights or household bulbs, what filter color would balance the film and light?
   a. Red.
   b. Green.
   c. Blue.
   d. None of the above.
8. True or false? Color film is usually processed at a higher temperature than black and white film.
9. Which color filters are used in making exposures for color prints?
   a. Yellow.
   b. Cyan.
   c. Magenta.
   d. All of the above.
10. The correct sequence of events for producing a slide show is:
    a. Storyboard, slides, outline, script.
    b. Outline, storyboard, slides, script.
    c. Storyboard, outline, slides, script.
    d. Outline, slides, storyboard, script.

APPLYING YOUR KNOWLEDGE

1. List examples of technology that use color theory. Investigate these examples further.
2. Experiment with color temperature by taking photos in warm and cool light. Compare the photos.
3. Process color slides.
4. Develop a slide/tape presentation. Show it to the class.

Section 4 ACTIVITIES GRAPHIC COMMUNICATION-PHOTOGRAPHY

BASIC ACTIVITY: Cameras and Composition

Introduction:

After reading about photography, you are probably eager and ready to take some pictures. An excellent way to begin is to acquaint yourself with the camera and experiment with a roll of film.

Guidelines:

- Complete this activity individually or in small groups.
- Use a medium speed (for example, 200 ASA) color print film.
- Film should be 24 exposures minimum.
- Camera should have a flash.

Materials and equipment:

- Camera, flash, and film.
- Index cards.
- Exposure meter if available.
- Tripod, if available.

Procedure:

1. Begin by becoming acquainted with the camera equipment. Check to see that the camera and flash have fresh batteries. If the flash is one that mounts on the camera, check on the correct camera and flash settings. Check to see that the camera lens is clean. See Section Fig. 4-1.
2. Load the camera and prepare to take pictures. Keep track of what was done for each shot on the cards.
3. Begin with outdoor shots. Try shooting each of the following:
   a. Still subject with various lighting techniques (front, back, and side).
   b. Framing and selective focusing techniques.
   c. Different backgrounds and perspective for the same subject.
   d. Different composition for a moving subject, including stopping the subject, blurring the subject, and blurring the background while stopping the subject.
   e. Photo using the rule of thirds.
   f. Indoor pictures, with and without flash.
4. When finished, unload the camera and have the film processed.
5. Mount and number the photos. Write a brief report explaining the technique used for each photo as well as any other findings from the photo study.

Section Fig. 4-1. Part of knowing how to use the camera is understanding basic maintenance, such as lens cleaning.

Evaluation:

Considerations for evaluation are:

- Overall quality of the photos.
- Correct shots.
- Quality of the report and mounting for photos.

ADVANCED ACTIVITY: Black and White Processing

Introduction:

Using black and white film has several advantages for the beginning photographer. First, it allows you to work on composition without having to be concerned with color. Instead, you can concentrate on the highlights, shadows, and middle tones in the scene. Another advantage is that black and white print film is easier to develop and print than its color counterpart.

Guidelines:

- This activity can be completed individually or in a small group.

Materials and equipment:

- Camera and black and white film.
- Processing chemicals.
- Standard darkroom equipment.

Procedure:

1. Practice composition techniques by taking some photos.
2. In a darkroom, load the film on a developing reel and place it in the developing tank. Place the lid on the tank, and turn on the lights.
3. Process the film, being sure to use the correct chemicals for the film that is being processed. See Section Fig. 4-2.
4. Remove the film from the reel and squeegee off the excess water. Hang the negatives to dry.
5. Carefully cut the negatives into strips and place them in sleeves.
6. Prepare the chemicals for making prints.
7. Make a contact print of the negatives. Evaluate your results.
8. Make enlargements of several of the negatives, beginning with a test print if necessary. Crop as needed to improve the prints.

Section Fig. 4-2. Be sure that you use the correct processing chemicals for the film being used. (Photo Systems, Inc.)

Evaluation:

Considerations for evaluation are:

- Processing skill, for both negatives and prints.
- Proficiency in cutting and storing the negatives.
- Quality of the finished prints.

ABOVE AND BEYOND ACTIVITY:
Slide Presentation

Introduction:

A slide presentation is an effective way to present information to a group of people. This activity will give you experience in putting together a slide program.

Guidelines:

- Small group activity.
- Presentation should be about five minutes in length.

Materials and equipment:

- Camera equipment.
- Medium speed (around 200 ASA) "chrome" slide film. Examples are Ektachrome, Fujichrome, etc.
- Plastic sleeve for the slides.
- Slide projector and slide tray.
- Audio tape recorder and tape.

Procedure:

1. Select a topic for a presentation and the intended audience. Demonstrations of concepts learned in class are one possibility. Develop an outline and storyboard cards for the presentation.
2. Based on the storyboard, decide on specific shots needed and take the pictures. When in doubt, try several shots of the same scene. Bracketing, which is the technique of trying the same shot with slightly different exposures, can also help.
3. Process the film. (If you are not able to do your own processing have it done. Often, you can find one-day processing for this type of film.)
4. Review the finished slides and decide on the sequence. Number the slides.
5. Develop a script for the presentation and record it onto an audio tape. Use an audible signal for the slide changes, or if a synchronizing recorder is available, use an inaudible signal. Review the presentation and make changes if necessary.
6. Present the show to a group. Be sure to check out the equipment and have everything set up before the show is to begin.

Evaluation:

Considerations for evaluation are:

- Quality of the slides.
- Correct procedure followed in developing the slide show.
- Overall quality of the slide show.

IBM CORP.

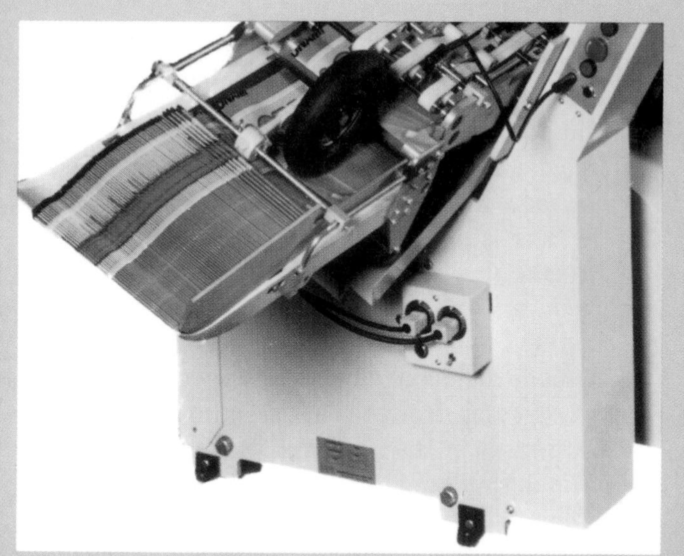

F. P. ROSBACK CO.

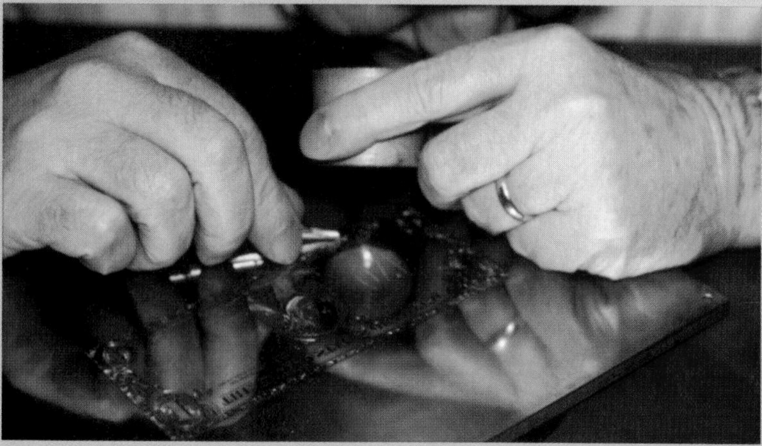

U. S. DEPT. OF TREASURY

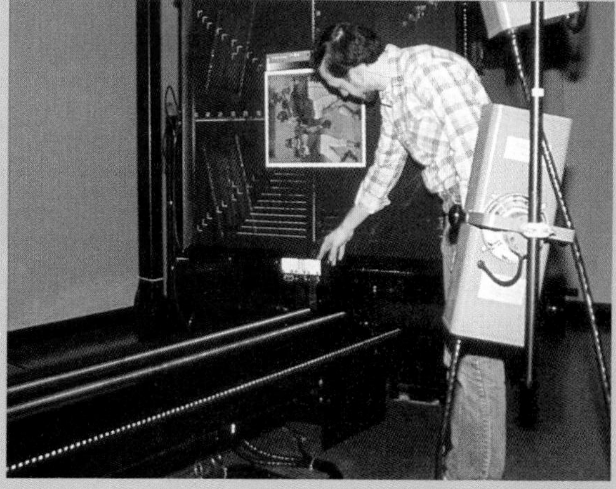

FLYING COLOR GRAPHICS

An Involved and Responsive Readership

FLYING COLOR GRAPHICS

section 5

Graphic Communication–Graphic Arts and Desktop Publishing

In this chapter, you will explore how a graphic arts product is developed.
(Flying Color Graphics)

20

CHAPTER

DESIGNING A GRAPHIC ARTS PRODUCT

After studying this chapter, you will be able to:
- *List the steps necessary in producing a printed product.*
- *Determine the best image transfer method for a designed product.*
- *Describe how color harmony can be achieved in a design.*
- *Identify various type styles and describe the kinds of products on which you would use them.*
- *Demonstrate how to mark-up a rough layout.*

Just think about the printed products you see every day. The books and magazines you read were designed by graphic artists. Everything from a cereal box to a garment label involves some type of graphic design. Graphic arts is a broad field that involves printing. Since printed materials are often sent to a large number of people, this is one form of mass communication.

Before you begin to design a graphic arts product, you need some information about this field. The next few chapters will provide information about the exciting area of graphic arts.

STEPS IN PRODUCING A GRAPHIC ARTS PRODUCT

Designing a graphic arts product can be divided into stages or phases of production. See Fig. 20-1. These stages are:

1. Image design.
2. Image generation and assembly.
3. Image carrier preparation.
4. Image transfer.
5. Product finishing.

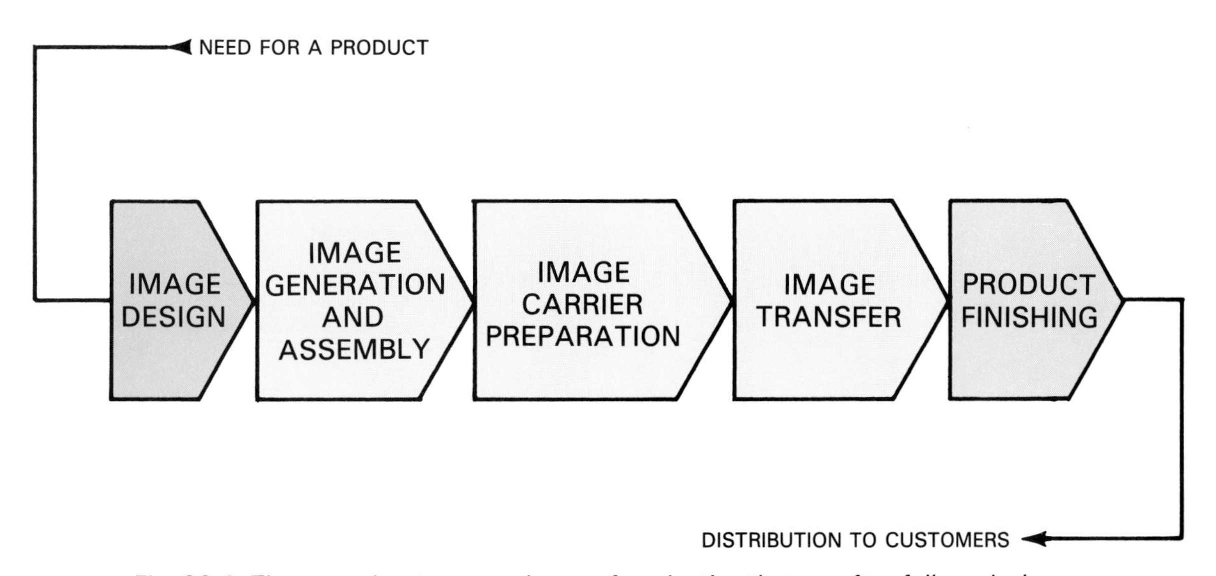

Fig. 20-1. These are the stages or phases of production that are often followed when a graphic arts product is produced.

GRAPHIC IMAGE DESIGN

Graphic image design follows a sequence of steps. First, small thumbnail sketches are made. A full-size, rough layout is then developed. If needed, a comprehensive layout can be produced. This layout is very exact and is used for expensive products that must be approved before printing.

Types of images

The part of a product that is printed is known as the **image**. The background, which is not printed, is the **non-image area**. Images are also known as **copy**.

The two kinds of copy are line and continuous tone images. The type that you are now reading is a line image. **Line copy** has a uniformly dark, or dense image. This is the only type of image that can be printed with ink. This is because presses place a uniform layer of ink in the image area.

Continuous tone images have variations in density. For example, a black and white photograph has shades of gray, from black to white. Since presses cannot print continuous tone images, they must be converted to line images. This is done by breaking the image into different sizes of dots to give the appearance of a continuous tone. This type of line image is known as a **halftone**, Fig. 20-2.

Line copy can be changed for special effects. A **tint** is produced when the image is broken into dots of equal size. A line image can be printed as a **reverse**, Fig. 20-3. In this case, the image is an area that is not printed. A **surprint** involves printing over an area that is already printed, Fig. 20-4. This is often used to place lettering on the lighter part (highlight) of a halftone.

Fig. 20-2. This continuous tone photo was converted into a halftone for printing in this book. (Polaroid Corp.)

Fig. 20-3. This is an example of a reverse line image.

Fig. 20-4. This surprint was made by placing a line image over a tinted background.

Color

All printed products have at least two colors, which are the paper color and ink color. When a second ink color is used for some of the image on a page, this is known as **spot color**.

Colors on the product should be in *harmony*. This means they should look good together. There are various types of color harmony. These include:

- Monochromatic harmony.
- Analogous harmony.
- Complimentary harmony.
- Triadic color harmony.

Monochromatic harmony is achieved by designing in various shades or tints of a single color, Fig. 20-5. A *shade* is produced by adding black to the primary color. A *tint* is produced by adding white to the primary color.

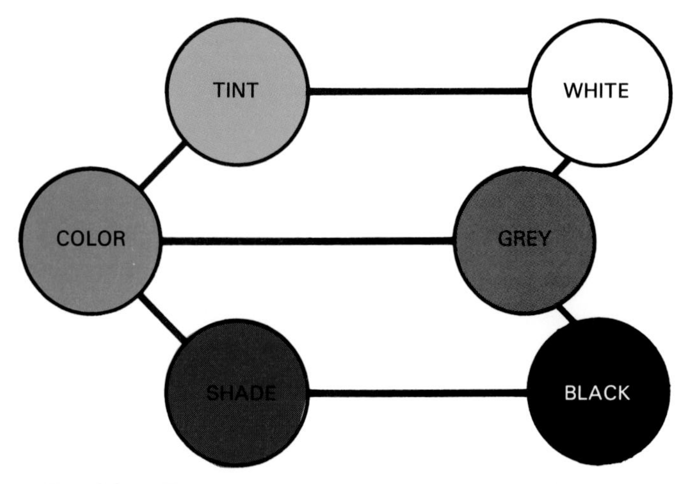

Fig. 20-5. This diagram illustrates the effect of adding black or white to a color.

Other types of harmony can be obtained by using a color wheel, Fig. 20-6. **Analogous harmony** involves using colors that are next to one another on the color wheel. **Complementary harmony** involves using colors directly across from each other on the color wheel. **Triadic harmony** involves using colors that divide the color wheel into equal thirds.

Colored inks can be opaque or transparent. **Opaque inks** cover what is underneath when printed and stay the same color. **Transparent inks** combine with the color underneath when printed.

Process color printing is printing that is used to reproduce colored, continuous tone pictures and photographs. Four colors are used for printing, which are yellow, magenta, cyan, and black, Fig. 20-7.

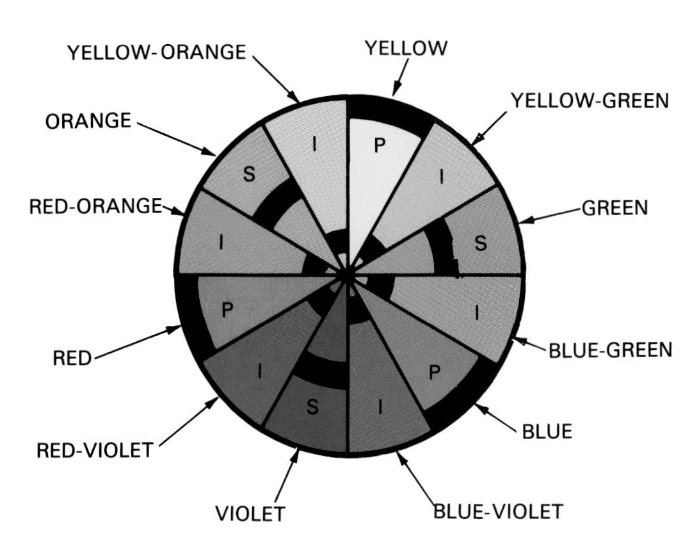

Fig. 20-6. This color wheel shows primary colors (P), secondary colors (S), and intermediate colors (I). When two primaries are mixed, a secondary color is produced, and when a secondary and primary are mixed, an intermediate is formed.

Fig. 20-7. This color picture was printed with four colors. The colors are yellow, magenta, cyan, and black.

Fig. 20-8. These ink samples are numbered according to a color system. (Van Son Holland Ink Corp. of America)

Color systems, discussed in Chapter 4, provide a method for clearly communicating what color is desired, Fig. 20-8. By using a color system, the hue that is specified by the designer should closely match the hue on the final product.

Typography

Typography is the art of effectively using type to convey a message. A person who is highly skilled in this technique is called a *typographer*.

Special terms are used for letters in graphic arts. Capital letters are also known as *upper case* letters. Small letters are known as *lower case* letters. Large letter sizes that are used to attract attention are called headlines or **display type**. Small letter sizes are called text or **body copy**.

Typeface classifications. A **typeface** is a specific type design. There are thousands of typefaces that can be placed in six general classifications, Fig. 20-9. These include:

- Serif.
- Sans serif.
- Square serif.
- Script.
- Old English.
- Novelty.

Serif type is also known as *Roman*. This type has serifs that are thin strokes, usually at the ends of letters. This type is very legible, and it is used for a variety of publications, including books and magazines. It can be used as a headline or text.

Sans serif type has a clean design and is also very legible. The word, "sans," is French for "without." Thus, sans serif means a style without serifs. This style is used for a variety of publications. It works well as a headline or text.

Square serif type is like serif. However, the serifs are blocks. Square serif can capture attention, but it is not as easy to read as the serif or sans serif type styles. This is why the square serif type style is primarily used in headlines and in advertisements.

Script type resembles handwriting. This gives it a warm, personal image. Some script styles have joining letters and some do not. It is almost never set in all upper case letters. This style is popular for announcements and advertising.

Old English type is also known as *Text*. The style, which is very ornate, is a copy of the style used by scribes to write books. The letters have extra strokes for ornamentation. Old English should never be used in all

TYPEFACE CLASSIFICATION

SERIF	ABCDEFGHIJKLMNOPQRSTUVWXYZ abcdefghijklmnopqrstuvwxyz 1234567890 1234567890 1234567890 $¢.,:;!?& " "( )[ ]%/— -*§†£flfi
SANS SERIF	ABCDEFGHIJKLMNOPQRSTUVWXYZ abcdefghijklmnopqrstuvwxyz 1234567890 1234567890 1234567890 $¢.,:;!?& " "( )[ ]%/— -*§†£¢$
SQUARE SERIF	ABCDEFGHIJKLMNOPQRSTUVWXYZ abcdefghijklmnopqrstuvwxyz 1234567890 1234567890 1234567890 $¢.,:;!?& " "( )[ ]%/— -*§†£¢$
SCRIPT	ABCDEFGHIJKLMNOPQRSTUVWXYZ abcdefghijklmnopqrstuvwxyz 1234567890 1234567890 1234567890 $¢.,:;!?& " "( )[ ]%/— -*§†£¢$
OLD ENGLISH	ABCDEFGHIJKLMNOPQRSTUVWXYZ abcdefghijklmnopqrstuvwxyz 1234567890$$£¢¢%/ 1234567890 &.,:;' "" !?--•( ) *
NOVELTY	ABCCDEEFGHHIJKLLMNOPQRSTUVWXYYZ 12334556 7890 ¢¢%/.,# 12334556 7890 &.,:;'"""!?--•( )*$$£

Fig. 20-9. These are six common typeface classifications. (Type House of Iowa, Inc.)

upper case because it is difficult to read. This style provides a formal appearance to the printed product. It is used for invitations, announcements, and advertising.

Novelty type is a category for those typefaces that do not fit in the other five categories. These are unusual and decorative type designs. Large companies often have a novelty typeface developed to represent themselves or their products and services. Advertising is the major use for these typefaces.

After a general typeface category is determined to be best for the message, a specific typeface must be chosen. For example, Times Roman and English Times are both serif type, but each is a different typeface.

Typefaces are further defined by style. Typical styles include italic and bold. All of the variations of one typeface are known as a family.

Font may have several meanings. It was originally used to mean a specific typeface, style, and size. An example would be 10 point Bodoni Bold. However, in electronic publishing systems, font sometimes means the same thing as typeface. In this case, you might have a Times Roman font. When you hear this term, be sure to find out which way it is being used.

Type size. When a typeface has been chosen for a printed piece, then a specific size is determined. For example, you might choose 10 point type, Fig. 20-10. In addition, for text matter, you must decide on the extra space between lines. The total line spacing is known as **leading**. It includes the type size and extra space, Fig. 20-11. For example, 10/12 Univers means 10 point Univers is set with 12 point leading. If no extra space is used between lines, this is called **set solid**.

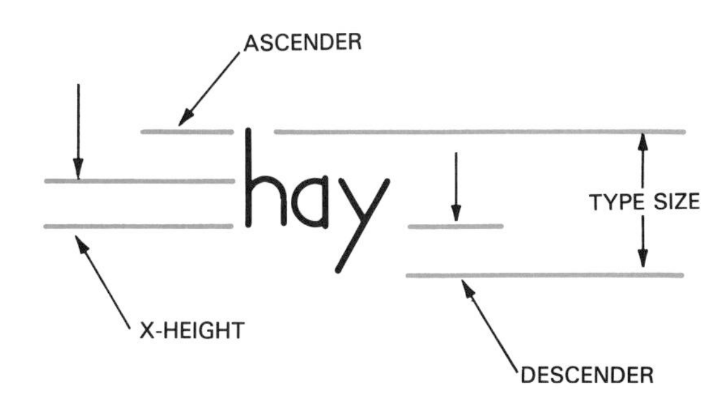

Fig. 20-10. Type is measured from an ascender to a descender.

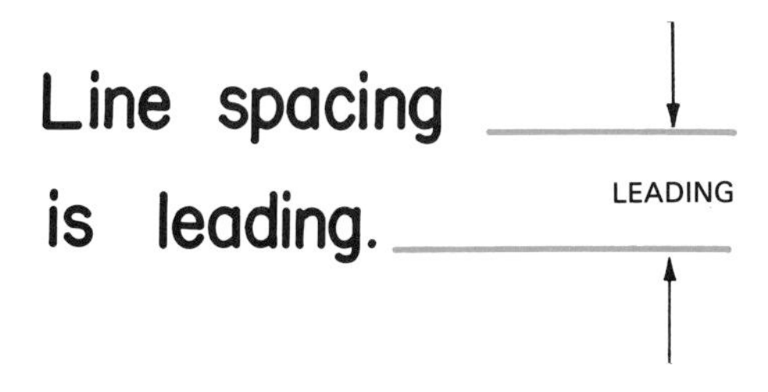

Fig. 20-11. Leading is the same as line spacing.

Product type and page layout specifications

There is a variety of formats for printed products. These include brochures, books, pamphlets, newspapers, etc. Part of the design process is choosing the best format for the message that is to be sent. Budget can also play a role in the decision. If the budget is limited, a less expensive format may need to be chosen.

After the type of product is chosen, page layout specifications may be needed. This book, for example, has two columns. A margin was chosen so the page will be optically centered.

For folded or complicated products, a **dummy** can be made. This is a quickly made representation of the finished product. For example, multiple pages are sometimes printed on both sides of a single sheet and then folded. This is called a **signature**, Fig. 20-12. The dummy is often made to ensure that the pages are in the correct order. Placing pages in the correct position for printing is known as **imposition**.

Image design steps

Before layout of a printed product can begin, a decision must be made about the type of product and the printing method to be used. (Printing methods are described later in this chapter.) A dummy can be made at this stage, if needed. Layout can then proceed, which involves designing the pages of the product.

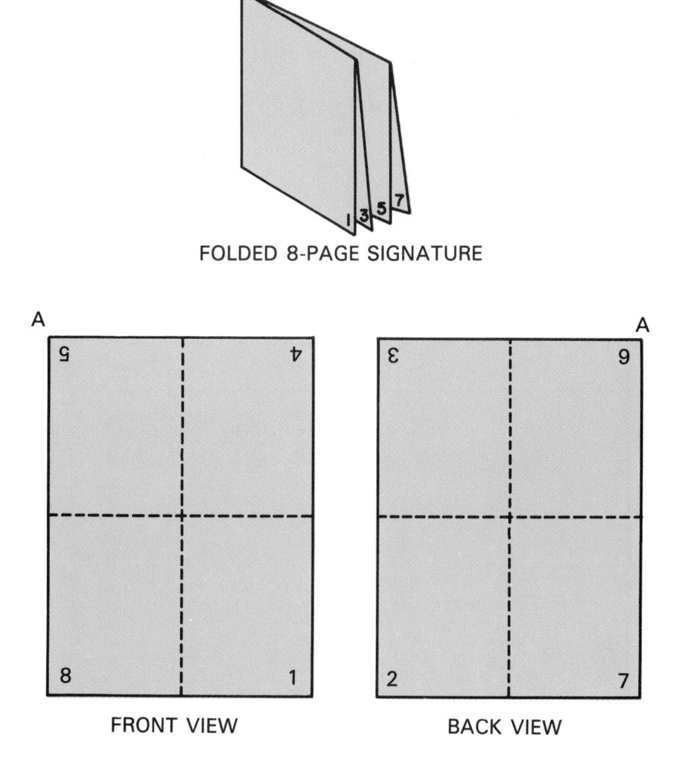

Fig. 20-12. A signature is a single sheet that is printed on both sides and is then folded. A dummy signature is often made to ensure that pages are in correct order.

Creating thumbnail sketches is the first layout step, Fig. 20-13. These are often drawn on grid paper and should be proportionately correct. If both width and height of the final size are divided in half, proportionately correct spaces can be made for thumbnail sketches. At least four thumbnail sketches should be drawn. Lines or blocks can be used for both headlines and text. Quickly sketch illustrations. Decide on a thumbnail sketch or a combination of thumbnail sketches for the rough layout.

The rough layout is often made full-size on grid paper, Fig. 20-14. Headlines, illustrations, and small quantities of text are sketched in to approximate size. Large amounts of text can be shown with a rectangle or lines. Text can then be written or typed on a separate sheet. Photos should have a rectangle around the rough illustration to indicate size.

The rough layout contains **mark-up information.** This is information that is needed to complete the product. The following areas are normally included:

- Width of margins.
- Color and type of paper, ink color, and location of colors.
- Typeface, size, leading, and column width.
- Conversion information for halftones and special effects such as tints.
- Scaling information for reductions and enlargements of the product.

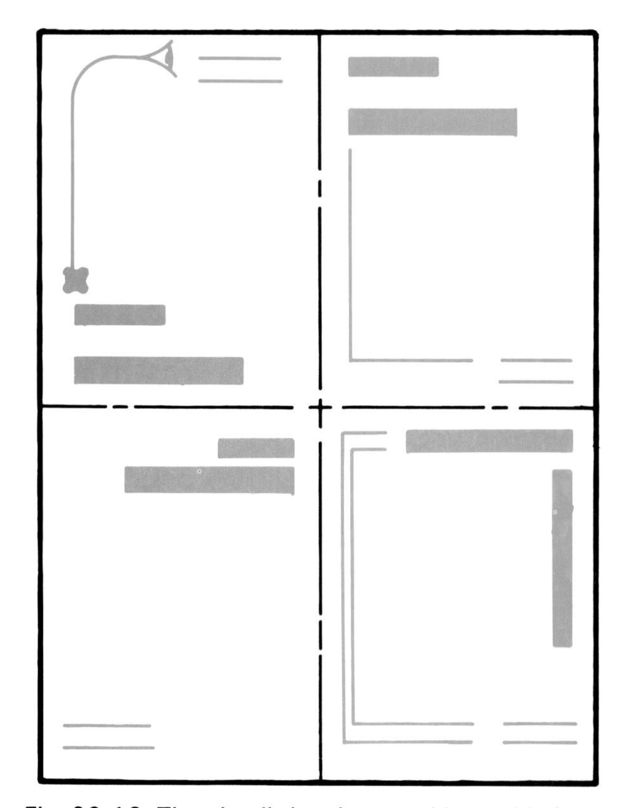

Fig. 20-13. Thumbnail sketches provide a guide for the rough layout.

Fig. 20-14. A rough layout includes headlines, illustrations, and small quantities of text sketched to approximate size.

- Finishing and binding information, including fold and cut lines.

In many instances, the rough layout is the final layout for approval. However, as mentioned, a *comprehensive layout* may also be completed in industry. This is a very exact, full color representation of the final product.

GRAPHIC IMAGE GENERATION AND ASSEMBLY

Image generation involves producing the images as described on the rough layout. When these are made, they are often assembled in the form of a **pasteup.** Graphic image generation and assembly will be discussed in Chapter 21 of this text.

GRAPHIC IMAGE CARRIER PREPARATION

Graphic image carrier preparation involves producing an image carrier that is used for printing. An image carrier "carries" the image during printing. A printing plate is an example.

For some printing processes, the pasteup is converted into a film image. This film image is then used for producing the image carrier.

GRAPHIC IMAGE TRANSFER

Graphic image transfer is the process of transferring an image from the image carrier to the substrate (the surface on which an image is printed). There are four

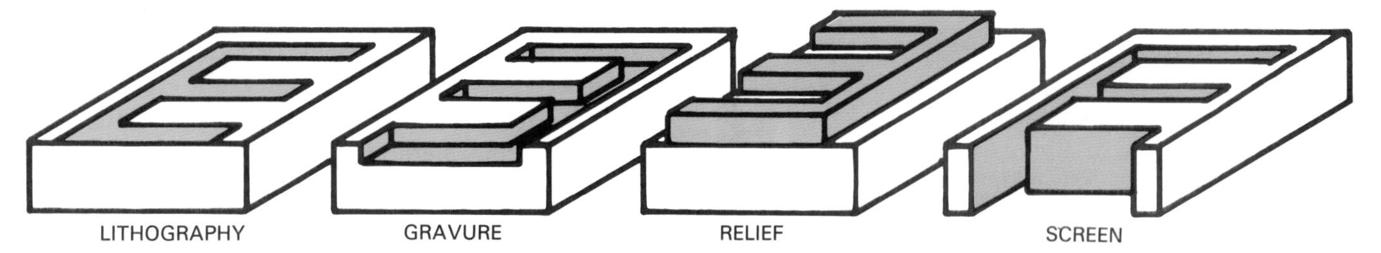

Fig. 20-15. These are four common methods of printing in industry.

common methods of printing in industry, Fig. 20-15. These are:

- Relief printing.
- Lithography.
- Gravure printing.
- Screen printing.

There is also a variety of duplicating and copying methods used for printing a small number of copies. These will be discussed in a later chapter.

Relief printing

Relief printing is printing from a raised surface. A rubber stamp is a simple example of this technique. Some product examples of relief printing are shown in Fig. 20-16. One type of relief printing is *letterpress*. This is printing from metal type or plastic (photopolymer) plates. Although this method is used less than in the past, it is still used for items such as fancy stationery and numbered tickets.

The modern form of relief printing for high volume work is *rotary relief printing*. In this method, an image carrier is wrapped around a cylinder for printing. *Flexography* is the most popular type of rotary relief printing. A rubber or plastic (photopolymer) plate is used as the image carrier. This type of printing is used to print on different kinds of packaging materials. These packaging materials include thin plastic and paperboard, and even some newspapers.

Gravure printing

Gravure printing is printing from a recessed image. The image carrier for this process is expensive, but very durable. Thus, gravure is normally used for high volume printing only. Examples include stamps, money, candy and gum wrappers, and magazines, Fig. 20-17.

Lithography

Lithography is printing done from a smooth surface. This is the most widely used form of printing. Printed products include forms, magazines, newspapers, and books, Fig. 20-18. Small lithographic printing units are known as duplicators.

Screen printing

Screen printing is done with a stencil on a screen. Ink passes through the screen onto the product. Screen printing is popular because of the ability to print on a variety of materials and shapes. Printed products include clothes, posters, wallpaper, and outdoor signs, Fig. 20-19.

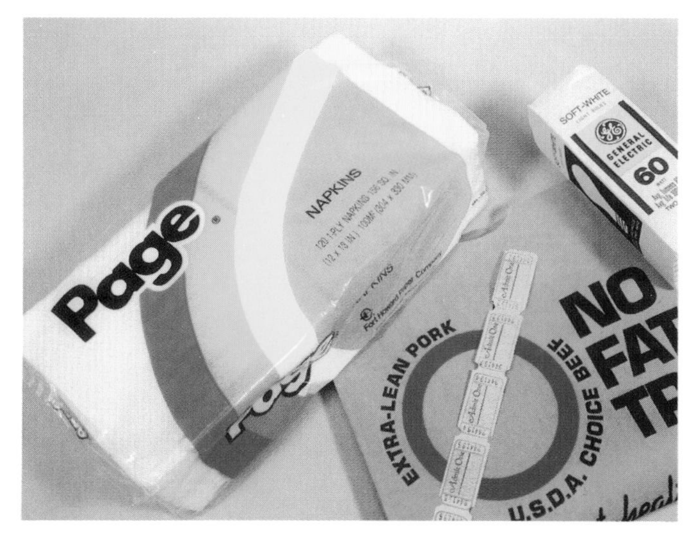

Fig. 20-16. These are examples of products that use relief printing.

Fig. 20-17. These are typical examples of products that are produced using gravure printing.

Fig. 20-18. Lithography was used to print these products.

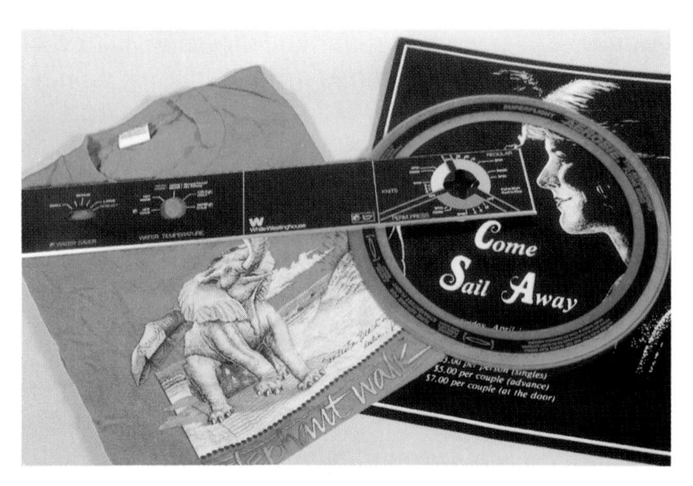

Fig. 20-19. These products were screen-printed.

GRAPHIC PRODUCT FINISHING

Graphic product finishing methods are used to convert the printed matter into the final product. Finishing includes processes such as:

- Cutting.
- Folding.
- Trimming.
- Binding.

Binding methods are used to hold parts of products together, such as books and magazines. This textbook, for instance, has been bound.

When finishing operations are completed, the product is distributed. **Distribution** is the delivery of the printed matter to the intended audience, which is the customer.

SUMMARY

Designing graphic arts products involves several steps. First, the graphic image is designed, taking into consideration color and type style and size. Next, graphic image generation and assembly takes place. This involves laying out the product. The graphic image carrier is then produced, and the image is transferred using a printing method. Graphic finishing methods then convert the printed matter into the final product that will be distributed.

WORDS TO KNOW

All of the following words have been used in this chapter. Do you know their meanings?

analogous harmony
body copy
complementary harmony
continuous tone images
copy
display type
distribution
dummy
font
gravure printing
halftone
image
imposition
leading
line copy
lithography
mark-up information
monochromatic harmony
non-image area
novelty type
Old English type
opaque inks
pasteup
process color printing
relief printing
reverse
sans serif type
screen printing
script type
serif type
set solid
signature
spot color
square serif type
surprint
tint
transparent inks
triadic harmony
typeface
typography

REVIEWING YOUR KNOWLEDGE

Please do not write in this text. Write your answers on a separate sheet.

1. List the five stages of production when designing a graphic arts product.
2. True or false? Since presses cannot print continuous tone images, they must be converted to line images.
3. An image that is broken into different sizes of dots to give the appearance of a continuous tone is:
   a. Line copy.
   b. A halftone.
   c. A surprint.
   d. None of the above.
4. Which of the following is achieved by designing in various shades or tints of a single color?
   a. Monochromatic harmony.
   b. Analogous harmony.
   c. Complimentary harmony.
   d. Triadic harmony.
5. Name the four colors used in process color printing to reproduce continuous tone photographs.
6. Choose one of the typeface categories discussed in this chapter. Describe its unique features and describe graphic arts products on which this typeface would be used.
7. Mark-up information includes:
   a. Width of margins.
   b. Color and type of paper, ink color, and location of colors.
   c. Conversion information for halftones.
   d. All of the above.
8. Which of the following is the most widely used form of printing?
   a. Gravure printing.
   b. Relief printing.
   c. Lithography.
   d. Screen printing.
9. List three examples of product finishing methods that would have been necessary in producing this textbook.
10. _________ is the delivery of printed matter to the intended audience.

APPLYING YOUR KNOWLEDGE

1. Choose any graphic arts product. Develop a poster that depicts the phases of how the product was produced.
2. Using a magnifying glass, find examples of the various printing methods.
3. Design a graphic arts product through the rough layout stage.

This person is designing a publication with the help of a computer. (Xerox Corp.)

21

CHAPTER

IMAGE GENERATION AND ASSEMBLY

After studying this chapter, you will be able to:
- *Discuss the process involved in generating various kinds of images.*
- *Identify common proofreading marks.*
- *List the steps involved in image assembly.*
- *Demonstrate simple copyfitting.*

The actual production of images is called ***image generation*** or ***composition.*** Just as composing music is a creative process, so is image generation, Fig. 21-1. Most of the methods discussed in this chapter have been replaced in recent years by computer-based composition and layout.

HOT AND COLD COMPOSITION

Image generation is sometimes divided into hot composition and cold composition. **Hot composition** refers to metal type that is cast in molds, which is the older of the two methods. **Cold composition** refers to other methods that do not use metal type. Hot type composition will be covered in Chapter 26 of this text.

Fig. 21-1. Composing images and composing music are both creative processes that require patience and skill.

There are five methods of cold composition. These methods are:
- Manual image generation.
- Impact composition.
- Computer composition.
- Photographic composition.
- Electronic publishing system.

MANUAL IMAGE GENERATION

Manual image generation refers to producing images without the use of machines. The main kinds of manual images include:
- Inked illustrations.
- Clip art.
- Transfer images.

Inked illustrations

Freehand illustrations are normally done in pencil first. When complete, a translucent sheet, such as vellum, is attached over the sketch on a light table. The drawing is then traced in ink, Fig. 21-2. For inked lettering, a template and stylus is used.

Clip art

Copyright-free illustrations that are purchased in book or disk form are referred to as **clip art**. See Fig. 21-3. Since the drawings are copyright-free, they can be reproduced. To use clip art from a book, carefully cut out the desired image. Stay away from the image when cutting and be careful not to cut through other layers of clip art in the book.

Fig. 21-2. This inked illustration is being prepared to be used in an architecture textbook. (Jack Klasey)

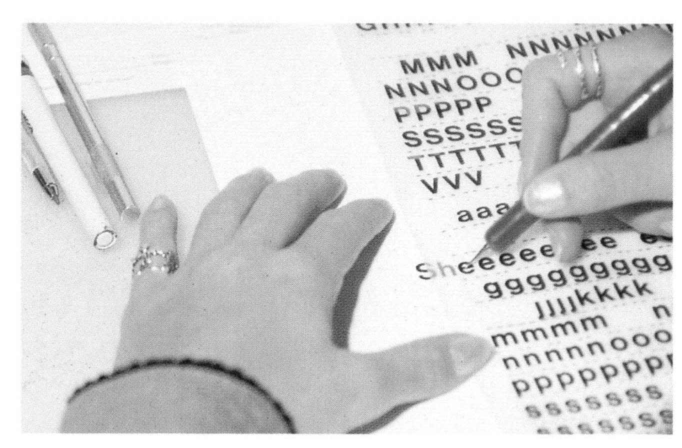

Fig. 21-4. Transfer images can be easily applied using a round-tipped object such as a pencil or pen. (Jack Klasey)

It may be desirable to save clip art so that it can be used again. In this case, make a copy rather than cut it out. Diffusion transfer is used for the best copy quality.

Transfer images

Transfer images are carbon or adhesive images. Carbon image transfers are available in sheet form with instructions on how to use them. To apply them, begin by placing light blue guidelines on vellum with drafting instruments. Align the letter or image and burnish it in place. This is done by rubbing over the surface with a round-tipped object such as a pencil or pen, Fig. 21-4. After the composition is completed, it is checked for alignment with a T-square or triangle. Then it is burnished again with the backing sheet in the package placed over the image. In order to remove a defective image, place masking tape on it and carefully lift.

Adhesive images are available in sheet and roll form, Fig. 21-5. As with carbon images, use light blue guidelines for image placement. When applying adhesive borders, they are overlapped and cut at a 45° angle with an artist knife and metal straight edge, Fig. 21-6.

Fig. 21-3. An assortment of clip art books is shown here. (Graphic Products Corp.)

IMPACT COMPOSITION

Impact composition is a method of creating an image by striking a substrate such as paper, Fig. 21-7. Another name for this is *strike-on*. A typewriter is used to produce impact composition. A dot matrix printer, often used for computer output, is also a strike-on machine.

Fig. 21-5. Adhesive borders are purchased in roll form. (Graphic Products Corp.)

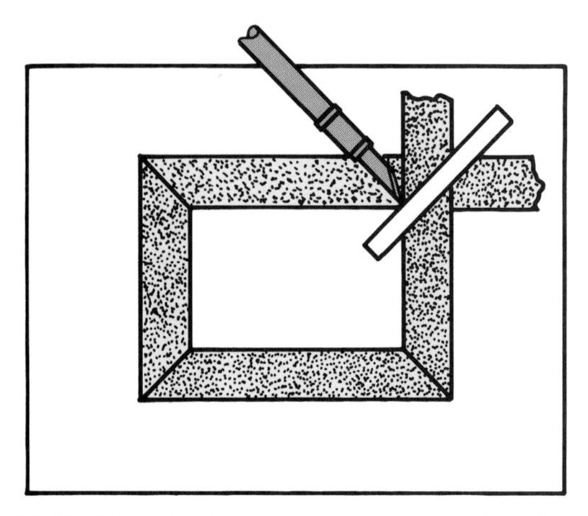

Fig. 21-6. Adhesive borders are overlapped and cut at a 45° angle.

Fig. 21-7. This is an example of an impact composition unit.

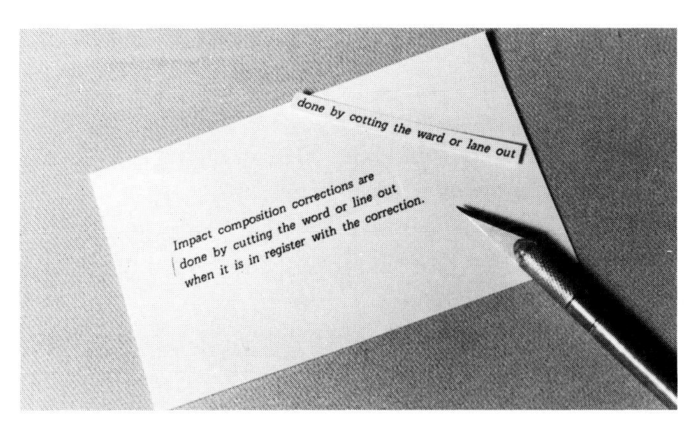

Fig. 21-8. Correcting impact composition is done by cutting the incorrect line or word out, resetting it correctly, and replacing the correction.

However, it is most often used for proof copies since the image is lower resolution than some other composition methods.

Impact composition for a pasteup should be of very high quality. A carbon ribbon is normally used. This produces images similar to transfer images by placing the carbon on the paper upon impact. A white, coated paper is used for best results.

Fonts can be changed on many impact units. If proportional letterspacing (different space for each letter) is used, then the image spacing will be more pleasing to the eye.

If a standard typewriter is used for impact composition, a ragged, uneven right margin results. More sophisticated machines can **justify** the margins automatically. This means that the line lengths are even, like in this book.

Correcting impact composition requires cutting out the mistake, Fig. 21-8. It is preferable to replace a line, but a word can also be replaced. Place the correct line or word on the light table and tape it in position. Align the mistake over the top and tape it down. Using the artist knife, carefully cut out the mistake. Knife cuts should overlap on the corners. Untape the copy and correction. Then tape the correction in place from behind with transparent tape.

COMPUTERS AND COMPOSITION

Personal computers, with the proper software, provide an effective way to compose images, Fig. 21-9. Word processing programs can be used for text, and graphic programs can be used for illustrations. A pasteup can then be made using these images. Page layout software allows you to do computer layout on the computer.

A dot matrix printer can be used for proof output, although the resolution is not good enough for high quality work. A laser printer or ink jet printer are desirable if available.

Microcomputers can be used as part of an electronic publishing system. This will be discussed later in this chapter.

Fig. 21-9. Computers are a common device used for image composition. (Bowling Green State University, Visual Communication Technology Program)

PHOTOGRAPHIC COMPOSITION

Photographic composition uses light to expose a light-sensitive paper or film. One category of photographic machines is for headlines only (display type). The font is a negative image through which light is projected, Fig. 21-10. Two common font types are a film strip and a plastic disk.

Most headline units require that each letter be positioned and exposed individually. Spacing can be automatic or manual. For machines that are spaced manually, marks on the font show approximate letterspacing. However, you should also space "by eye" to be sure the spacing is visually pleasing. The letter "0" can be used for wordspacing.

Some display units can change the font with lenses and contact screens. This includes scaling (enlarging or reducing), distorting the letters, or tinting the letters.

Photographic composition for text and headlines

Most photocomposition machines that produce text will make headlines as well. These are sometimes called **phototypesetters,** Fig. 21-11. Some phototypesetters work only on an output device and must be connected to a computer. In other instances, the typesetter is a stand-alone unit that will do composition and printer output.

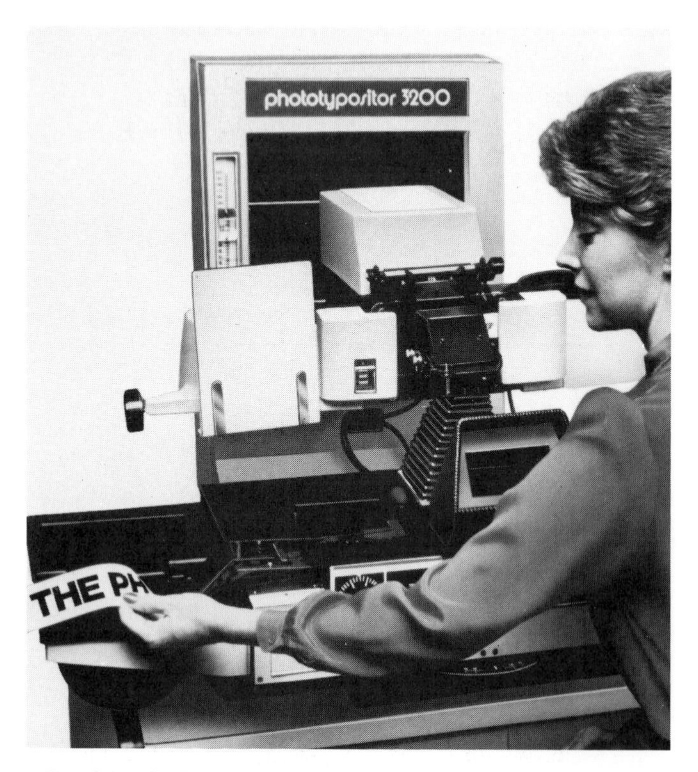

Fig. 21-10. This unit uses a film font for producing headlines. (Visual Graphics Corp.)

Fig. 21-11. A phototypesetter is being used here. (Jack Klasey)

Phototypesetters contain fonts that are stored as digital information in magnetic memory. Using a computer, this digital information is converted to electronic pulses that form letters and numbers. For output, a CRT (cathode ray tube) or laser is used to expose photographic paper or film. A CRT is like the picture tube in a television. The image that forms on the screen is used for exposure. A laser beam is the most current method for exposure. This beam scans the photo paper to make exposures.

Typesetters produce high quality images. Image quality, or resolution, can be described as the number of dots per inch in the image. Higher quality images have more dots per inch. Therefore, they are said to have higher resolution. For example, 2400 dpi would be a high resolution image.

Stand alone typesetters have a variety of settings that must be learned. Settings should be made for type style and size, line length, and leading. Most of these units offer automatic hyphenation and justification (h & j). In order to make every line the same length (justification), words have to be divided between lines and hyphenated. In automatic systems, an electronic dictionary is part of the software and automatically checks words for correct hyphenation.

Photographic composition must be processed to bring out the image. For stabilization papers, two chemicals are used. An activator causes the image to darken, and a stabilizer stops the activator. This type of image will darken with age, so it should be used within a few months. Conventional photographic papers use three steps for processing. The image passes through a developer, fixer, and water. This method produces a more permanent image.

The newest generation of phototypesetters has the capability of imaging both text and graphics. For this reason, they are referred to as **imagesetters**, Fig. 21-12. Imagesetters can output complete pages, thus eliminating the need for pasteup. Since they are primarily an output device, a separate computer is used for composition. Imagesetters produce very high resolution images and are often used as part of an electronic publishing system.

ELECTRONIC PUBLISHING SYSTEMS

An **electronic publishing system** is one that has page make-up capabilities, Fig. 21-13. This means that all text and illustrations can be composed in page form on a computer screen. This capability of seeing the page on the monitor as it is to print is known as *WYSIWYG* (*What You See Is What You Get*). These systems are also known as computer-aided publishing systems or automated composition systems.

Fig. 21-12. Several types of imagesetters are shown here. (Agfa-Compugraphic)

Electronic publishing systems vary greatly. The most sophisticated systems, normally used in graphic arts, are capable of high quality output. In addition, color systems can be used to generate color pages, including color photographs. These systems are often used by companies and organizations that produce high quality

Electronic publishing systems are often networked so that different devices can communicate with one another. Often, several computers will be networked to an imagesetter for output.

The desktop publishing system is a less sophisticated version of the electronic publishing system. See Fig. 21-14. A personal computer, laser printer, and special software are used to produce page layouts. This system is often used for publications such as newsletters and reports. Desktop publishing will be discussed further in Chapter 22 of this text.

PROOFREADING

Text and headlines that have not been checked for errors are known as **proof copy**. Checking for errors is known as **proofreading**. Special proofreading marks are used to indicate corrections that need to be made, Fig. 21-15. These are placed in the text and margins.

After proofreading, another proof is made with the corrections. If no mistakes are found after proofreading this second proof, it is used as the final copy.

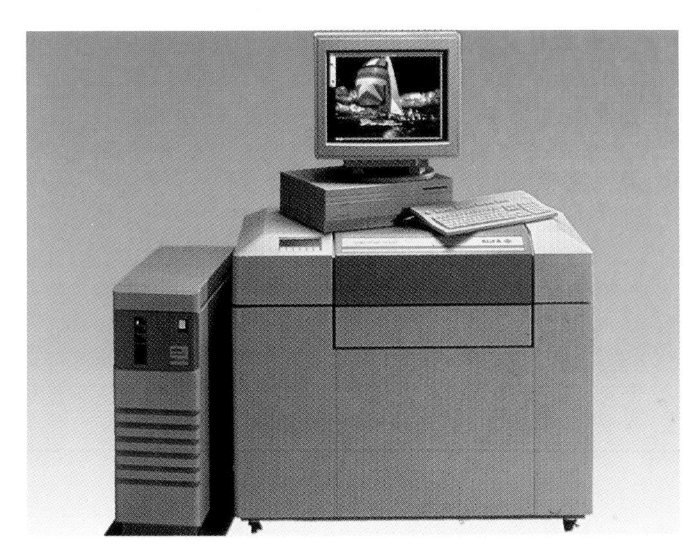

Fig. 21-13. This is an electronic publishing system. (Agfa-Compugraphic)

Fig. 21-14. This desktop publishing system includes a personal computer (right), laser printer (left), and mouse. (IBM Corp.)

PROOFREADING MARKS

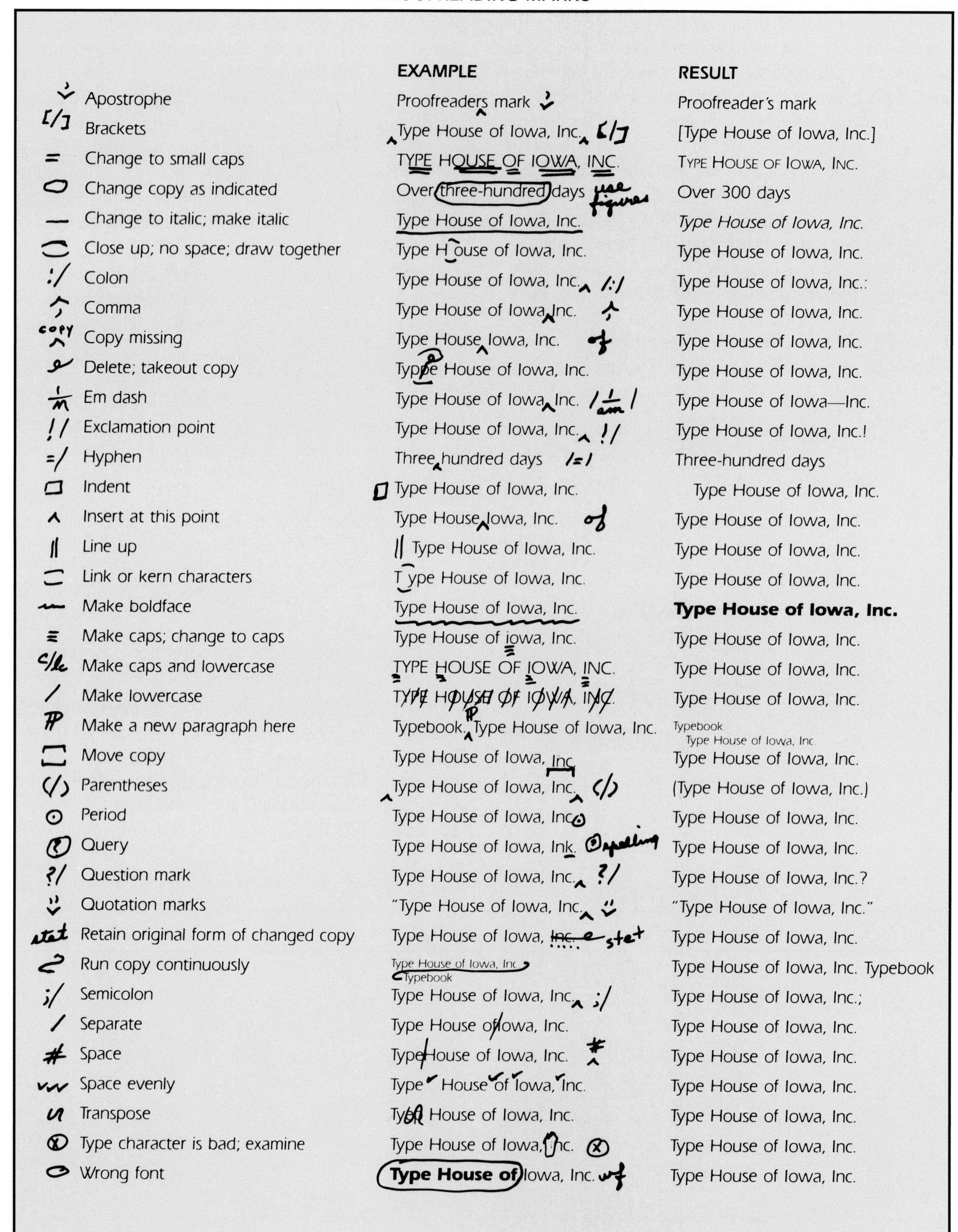

Mark	EXAMPLE	RESULT
Apostrophe	Proofreaders mark	Proofreader's mark
Brackets	Type House of Iowa, Inc.	[Type House of Iowa, Inc.]
Change to small caps	TYPE HOUSE OF IOWA, INC.	TYPE HOUSE OF IOWA, INC.
Change copy as indicated	Over three-hundred days use figures	Over 300 days
Change to italic; make italic	Type House of Iowa, Inc.	*Type House of Iowa, Inc.*
Close up; no space; draw together	Type H ouse of Iowa, Inc.	Type House of Iowa, Inc.
Colon	Type House of Iowa, Inc.	Type House of Iowa, Inc.:
Comma	Type House of Iowa Inc.	Type House of Iowa, Inc.
Copy missing	Type House Iowa, Inc. of	Type House of Iowa, Inc.
Delete; takeout copy	Typpe House of Iowa, Inc.	Type House of Iowa, Inc.
Em dash	Type House of Iowa Inc.	Type House of Iowa—Inc.
Exclamation point	Type House of Iowa, Inc.	Type House of Iowa, Inc.!
Hyphen	Three hundred days	Three-hundred days
Indent	Type House of Iowa, Inc.	Type House of Iowa, Inc.
Insert at this point	Type House Iowa, Inc. of	Type House of Iowa, Inc.
Line up	Type House of Iowa, Inc.	Type House of Iowa, Inc.
Link or kern characters	T ype House of Iowa, Inc.	Type House of Iowa, Inc.
Make boldface	Type House of Iowa, Inc.	**Type House of Iowa, Inc.**
Make caps; change to caps	Type House of iowa, Inc.	Type House of Iowa, Inc.
Make caps and lowercase	TYPE HOUSE OF IOWA, INC.	Type House of Iowa, Inc.
Make lowercase	TYPE HOUSE OF IOWA, INC.	Type House of Iowa, Inc.
Make a new paragraph here	Typebook. Type House of Iowa, Inc.	Typebook. Type House of Iowa, Inc.
Move copy	Type House of Iowa, Inc.	Type House of Iowa, Inc.
Parentheses	Type House of Iowa, Inc.	(Type House of Iowa, Inc.)
Period	Type House of Iowa, Inc	Type House of Iowa, Inc.
Query	Type House of Iowa, Ink. spelling	Type House of Iowa, Inc.
Question mark	Type House of Iowa, Inc.	Type House of Iowa, Inc.?
Quotation marks	"Type House of Iowa, Inc.	"Type House of Iowa, Inc."
Retain original form of changed copy	Type House of Iowa, Inc. stet	Type House of Iowa, Inc.
Run copy continuously	Type House of Iowa, Inc. Typebook	Type House of Iowa, Inc. Typebook
Semicolon	Type House of Iowa, Inc.	Type House of Iowa, Inc.;
Separate	Type House ofIowa, Inc.	Type House of Iowa, Inc.
Space	TypeHouse of Iowa, Inc.	Type House of Iowa, Inc.
Space evenly	Type House of Iowa, Inc.	Type House of Iowa, Inc.
Transpose	Tyep House of Iowa, Inc.	Type House of Iowa, Inc.
Type character is bad; examine	Type House of Iowa, Inc.	Type House of Iowa, Inc.
Wrong font	**Type House of** Iowa, Inc. wf	Type House of Iowa, Inc.

Fig. 21-15. Some common proofreading marks are shown, along with examples of their uses. (Type House of Iowa, Inc.)

CROPPING AND SCALING

Images are sometimes cropped and scaled. **Cropping** is done when an unwanted portion is removed from an image. As shown in Fig. 21-16, cropping L's can help in this decision. Crop marks are placed in the border in ink, or on an overlay. If cropping and scaling must both be done, cropping is done first because this may change the proportion of the illustration.

Scaling copy is enlarging or reducing the copy proportionately, Fig. 21-17. The *proportion wheel* is most often used for this, Fig. 21-18. Using the height or width, decide what the new dimension should be when scaled. On the proportion wheel, align the original size on the inner disk with the reproduction size on the outer disk. For example, a 10-inch original width can be aligned with a 1-inch reproduction width. Without moving the wheel, check to see what reproduction size is aligned with the original height. This will be the new height dimension. Also, record the percentage in the window. This will be used for scaling the copy on the camera.

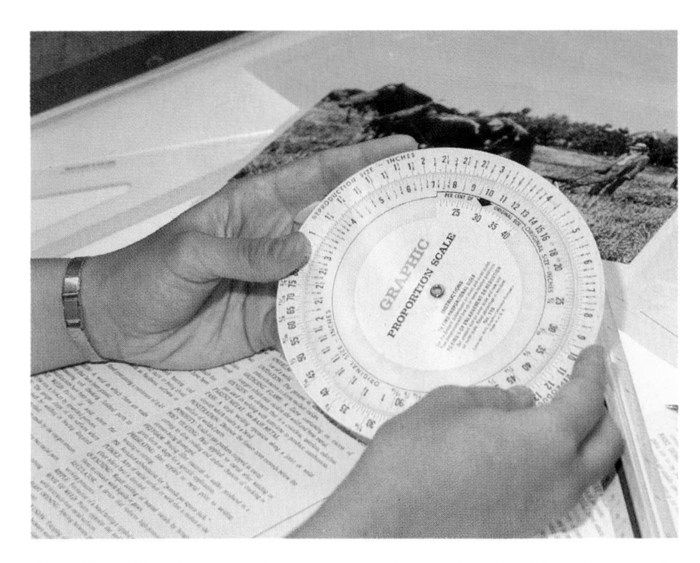

Fig. 21-18. When using a proportion wheel, align the original size on the inner disk with the reproduction size on the outer disk. (Jack Klasey)

Fig. 21-16. These cropping L's help in visualizing how a cropped image will appear. (Jack Klasey)

Fig. 21-17. This large clip art was reduced as shown to fit a layout. (Jack Klasey)

IMAGE ASSEMBLY

When images are assembled into the final layout, this is known as a **pasteup** or **mechanical**. Images are placed on a pasteup board made of heavy white paper such as index or offset. See Fig. 21-19. The pasteup board is normally cut 1 in. to 2 in. larger than the sheet to be printed. Using drafting tools and a non-photo blue pencil,

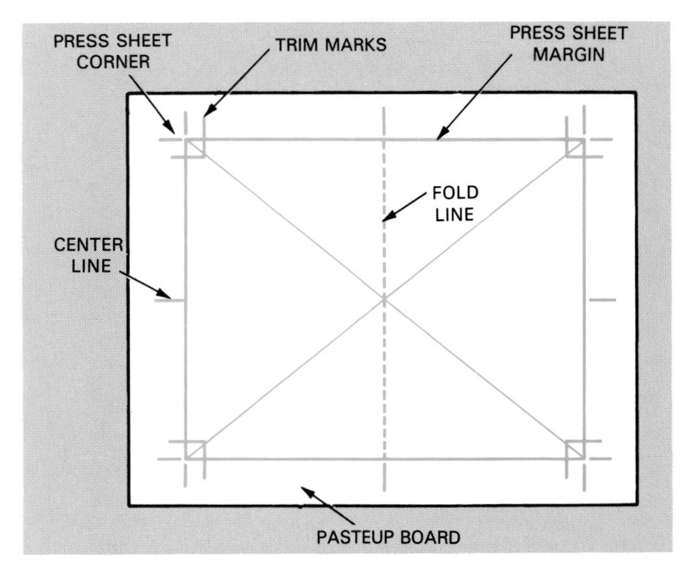

Fig. 21-19. This is a typical layout for a pasteup board.

diagonal lines are drawn to find the center. Draw the press sheet dimensions measuring out from the center. Fold lines are drawn as a dashed line. All layout lines to this point are in non-photo blue. Other layout lines are first made with blue pencil, and then traced with black ink. Press sheet corners and trim marks are shown as right angle lines. Fold marks are drawn in ink outside the press sheet border. Register marks are used if more than one ink color is to be printed, Fig. 21-20.

Draw non-photo blue lines on the pasteup for aligning the images. These will show through the image on the light table. For text, draw this line at the bottom of the X-height.

ATTACHING IMAGES

Leave a 1/8 in. margin around the images when trimming. Miter the corners (45°) since the corners may tend to lift up, Fig. 21-21.

Wax is normally used as an adhesive for the images because it allows repositioning. This is applied with a waxer, Fig. 21-22. Because smaller images can fall into the wax pool of table-model waxers, these images may be left untrimmed until after waxing.

Other methods for attaching images include using rubber cement, glue, or tape. Like wax, the adhesive is placed on the back of the image before it is attached to the pasteup. Rubber cement allows repositioning. Glue and double-surface tape are permanent, so there is no chance for repositioning. Special tape is available that can be repositioned. It is used on the edge of the image.

After attaching the images, check one more time for alignment, Fig. 21-23. If the images are aligned correctly, attach a cover sheet with tape on one side. With

Fig. 21-21. When trimming an image, corners should be mitered to prevent them from lifting up. (Jack Klasey)

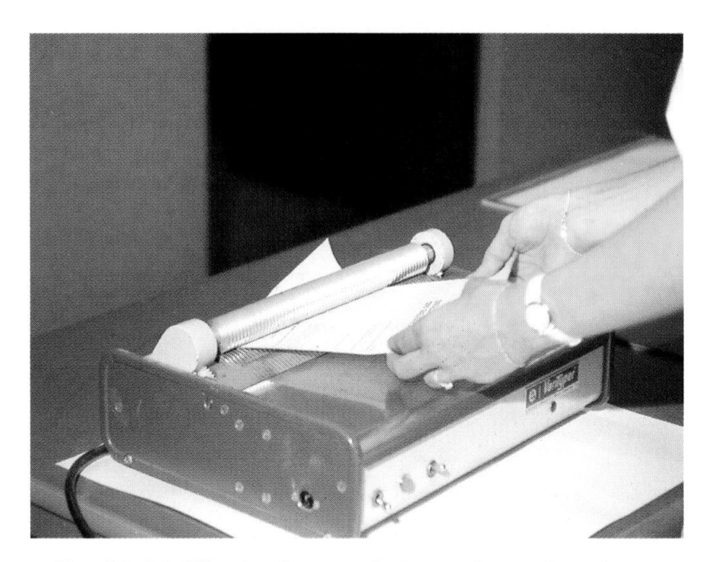

Fig. 21-22. Wax is often applied to an image in order to attach it to a pasteup. (Jack Klasey)

Fig. 21-20. Register marks are used for multiple colors. (Jack Klasey)

Fig. 21-23. All images on a pasteup must be checked for alignment. This can be done with a T-square and triangle or with an overlay as shown here. (Graphic Products Corp.)

the cover sheet over the images, burnish the images down with a roller.

CONTINUOUS TONE COPY

If continuous tone copy has been converted into a halftone positive on paper, it can be attached to the pasteup like other copy. If a halftone film negative is to be used, a black or red rectangle is placed on the pasteup. This will leave a clear window when the pasteup is photographed, Fig. 21-24. The film halftone is then attached behind the window.

IMAGE ASSEMBLY FOR MULTIPLE COLORS

Several methods can be used for preparing the pasteup if more than one ink color is used. If ink colors will not touch or overlap, a **masking** technique can be used. This requires labeling the colors for each image in blue pencil. Registration marks are also needed inside the press sheet area. Each color is covered or "masked" at a later stage for making image carriers. This is explained in a later chapter.

When different colors will touch or overlap on the pasteup, an overlay is used, Fig. 21-25. Images for one color are attached to the pasteup board, and registration marks are added in the borders. A clear acetate sheet is taped by one edge over the pasteup. Images for the second color are attached to the acetate. In this method, one negative is made from the base pasteup or **key**. A second negative is made with white paper between the base and acetate. The white paper does not cover the register marks so they will also be visible on the second negative.

Fig. 21-24. The pasteup on the left was photographed to produce the film negative on the right. (Jack Klasey)

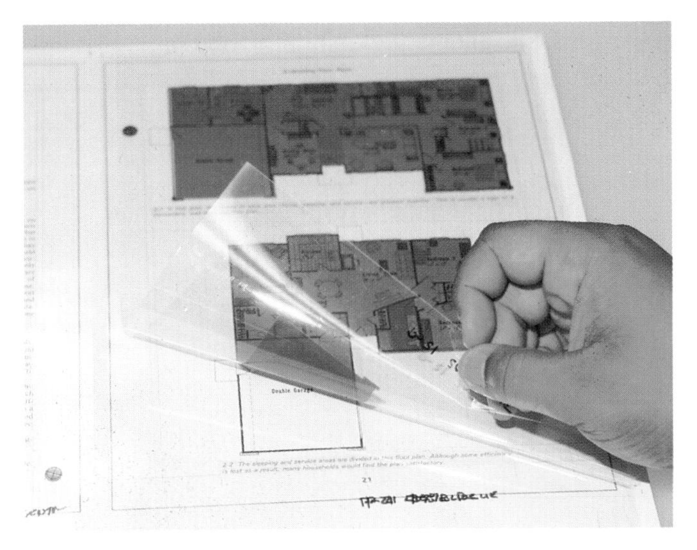

Fig. 21-25. When an overlay is used, registration marks are needed. (Jack Klasey)

COPYFITTING

Copyfitting involves fitting a certain amount of copy into a limited amount of space. When text is needed for a publication, it is useful to know how much space the rough copy will require. The character count method is an accurate method for determining copyfit. The steps are as follows:

1. Count the characters in five full lines of rough copy. Spaces and punctuation marks are counted as characters. Divide by 5 to find the average number of characters per line.
2. Multiply the number of rough copy lines by the characters per line. This is the total number of characters.
3. Find the characters per unit of measure (pica or inch) for your type style and size. Charts are available for this purpose. Multiply this number by the column width that will be used. This is the characters per line in the finished copy. If a chart is not available, simply print out some text in the column width desired and use the character count method as described above.
4. Divide the total number of characters in the rough by the characters per line for the finished copy. This is the total number of typeset lines.
5. Multiply the number of lines by the leading (linespacing) per line for the column depth.

SUMMARY

Image generation and assembly is a stage in production that requires patience, skill, and creativity. If a mistake in spelling or alignment is made at this stage and it is not corrected, all the printed copies will have the mistake on them. Make sure the final layout is correct before proceeding.

WORDS TO KNOW

All of the following words have been used in this chapter. Do you know their meanings?

clip art
cold composition
composition
copyfitting
cropping
electronic publishing system
hot composition
image generation
imagesetters
impact composition
justify
key
manual image generation
masking
mechanical
pasteup
photographic composition
phototypesetter
proof copy
proofreading
scaling
transfer images

REVIEWING YOUR KNOWLEDGE

Please do not write in this text. Write your answers on a separate sheet.

1. Which of the following is NOT a cold composition image generation method?
   a. Manual.
   b. Metal type cast in molds.
   c. Impact.
   d. Photographic.
2. True or false? When cutting out clip art, it is a good idea to use an artist knife.
3. Name two types of transfer images.
4. True or false? Phototypesetters contain fonts that are stored as digital information in magnetic memory that become letters and numbers.
5. In a desktop publishing system, which of the following would be necessary to produce page layouts?
   a. A microcomputer.
   b. A laser printer.
   c. Special software.
   d. All of the above.
6. True or false? Electronic publishing systems are often networked so that different devices can communicate with one another.
7. True or false? Text and headlines that have been checked for errors are known as proof copy.
8. Enlarging or reducing copy proportionately is called:
   a. Proofreading.
   b. Cropping.
   c. Scaling.
   d. None of the above.
9. When drawing lines on a pasteup or mechanical for aligning images, which of the following should be used?
   a. Lead pencil.
   b. Non-photo blue pencil.
   c. Black ink.
   d. None of the above.
10. True or false? If continuous tone copy has been converted into a halftone positive on paper, it can be attached to the pasteup like other copy.

APPLYING YOUR KNOWLEDGE

1. Design and produce a page layout utilizing available composition methods.
2. Research and compare the features included on current electronic publishing systems.
3. Practice placing proofreading marks on rough copy.
4. Inquire about copyfitting methods used by local graphic arts businesses. Report your findings to the class.

22

CHAPTER

DESKTOP/ELECTRONIC PUBLISHING

After studying this chapter, you will be able to:
- *Define desktop publishing.*
- *Identify hardware and software needed for a desktop publishing system.*
- *Explain the procedure to follow when creating a product with desktop publishing.*
- *Utilize page design considerations.*

Desktop/electronic publishing has had a great impact on graphic communications. Desktop publishing enables you to produce professional quality graphics quickly and easily. The terms electronic publishing and desktop publishing are often used interchangeably.

In Chapter 21, you explored various methods of image generation and assembly. In this chapter, you will become acquainted with desktop/electronic publishing. You will find out how camera-ready images are generated and assembled electronically.

WHAT IS DESKTOP PUBLISHING?

Desktop publishing (DTP) involves using a computer, specialized software, and a laser printer as a system for page layout, Fig. 22-1. Other terms sometimes used for this process are electronic publishing or computer-aided publishing. However, **electronic publishing** is a broader term that includes the sophisticated page layout systems used in the graphic arts industry.

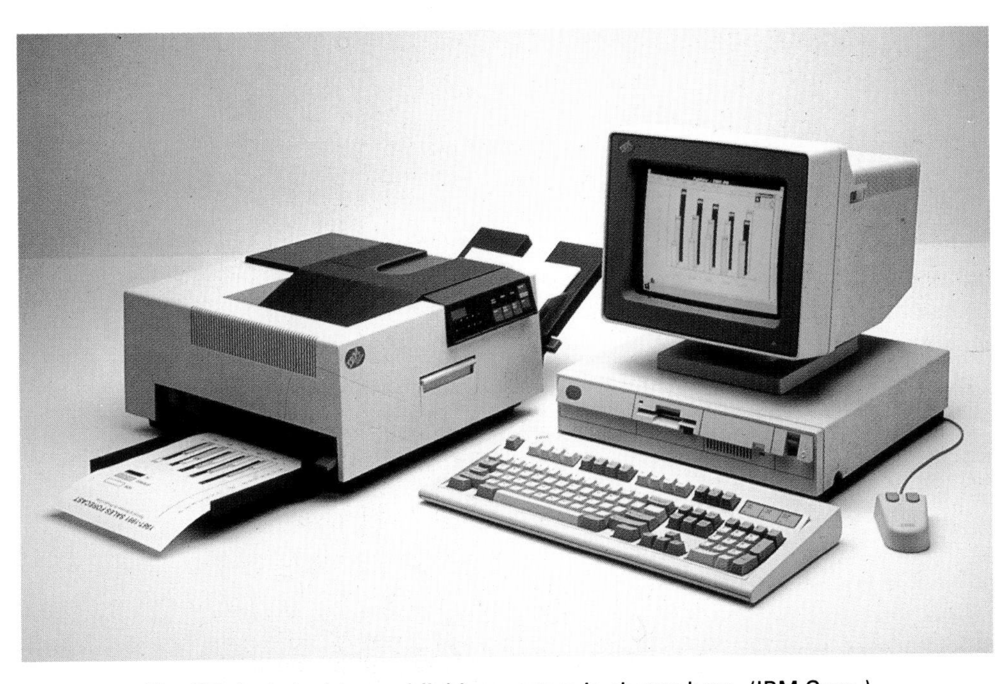

Fig. 22-1. A desktop publishing system is shown here. (IBM Corp.)

These systems are usually characterized by greater complexity, cost, and higher quality output. In contrast, the **desktop publishing system** is relatively inexpensive, smaller in size, and easier to operate.

The desktop publishing system relies upon page layout software, which allows you to place both text and graphics on a page. This is normally in *WYSIWYG* form (*What You See Is What You Get*).

In traditional page composition, text is produced with a typesetter and attached to a pasteup along with illustrations. The illustrations might be drawn by hand, or clip art might be used. In desktop publishing, text is produced on a word processing program or with the DTP software. Images are either drawn with graphics software, converted to digital form with a scanner, or chosen from electronic clip art. The DTP software allows you to then merge the text and graphics on a page on the computer monitor and ultimately print out the completed page, Fig. 22-2.

Fig. 22-2. A desktop publishing system allows you to merge text and graphics electronically. (Bowling Green State University, Visual Communication Technology Program)

EQUIPMENT FOR DESKTOP PUBLISHING

Each of the items of equipment in a desktop publishing system performs one or more of the functions of the system: input, process, and output. Since the computer is the most important part of the system, this is a good place to begin.

COMPUTER

The kind of computer you choose for desktop publishing depends upon the software you will be using. High quality publications require sophisticated software and more expensive computer systems. For newsletters, brochures and other DTP tasks that many people want to do, simpler software that will run on a less expensive computer will usually be sufficient.

Software for desktop publishing is available in versions that will run on either the Macintosh computer platform or the Windows (IBM-compatible platform). See Fig. 22-3. The simpler forms of desktop publishing can be done using full-featured word processing programs. For more complex tasks, a page layout program is usually chosen.

Computer capabilities have grown rapidly in recent years, so that even most "entry level" models available today are capable of using even the most complex page layout software. During the 1990s, typical computers progressed from clock (operating) speeds of 66Mhz or less to 200Mhz or more. Hard disk drives for program and file storage saw even more dramatic change, from approximately 20 megabytes (20 million bytes) capacity to gigabytes (billions of bytes) of storage space Drives with 6 Gb or even higher capacities are common.

The amount of **RAM (Random Access Memory)** in the computer is important, because it must temporarily store data and instructions while processing is being done. An insufficient amount of RAM can make a program run slowly, or even refuse to run. Just as clock speeds and hard drive capacities have increased dramatically, the amount of RAM supplied with a typical computer has quickly increased from 4 Mb or 8 Mb to 32 Mb, 64 Mb, or even 128 Mb.

Fig. 22-3. Adobe PageMaker is a page composition program that is widely used on the Macintosh platform. It allows the operator to combine text and graphics into finished pages, ready for output. (Apple Computer, Inc.)

A high-resolution color monitor allows you to clearly view the text and graphics you are combining while using a desktop publishing program. The higher the monitor resolution, the easier it is to see fine details. For faster display of material on the monitor screen, the graphics cards installed in today's computers include several megabytes of memory (separate from the computer's RAM). Some systems, like the one shown in Fig. 22-4, use dual monitors. This allows the operator to "zoom in" on one screen for a detailed view of changes being made on the other screen.

INPUT DEVICES

Many devices are available for assisting in creation and arrangement of text and graphics. These include the computer keyboard, the mouse, the graphics tablet, and the scanner. The *keyboard* is used for text input and program commands. The **mouse** is a point and select device that allows movement of objects around the screen and selection of commands with the click of a button. The **graphics tablet**, Fig. 22-5, can be used for drawing or tracing an image onto the screen. A mouse-like "puck" used with the tablet can select commands, as well.

When an existing drawing or photograph must be converted to digital form for use in a DTP program, scanning is done. The **electronic scanner**, Fig. 22-6, converts the illustration into electronic signals that can be stored as a file in the computer's memory. The images are given filenames with special extensions to identify them as graphics.

Fig. 22-4. For graphics work, a high-resolution monitor is a vital tool. This system has dual monitors, which lets the artist working with an image manipulation program compare versions of an image, or get an enlarged view of changes being made. (Intergraph Corporation)

Fig. 22-5. A graphics tablet is a very precise input device. (Summagraphics)

OUTPUT DEVICES

Output devices include the dot matrix, ink jet, and laser printers. The **dot matrix printer** produces a low-quality output, and is seldom used except for proofing or draft printing of output from word processing programs. **Ink jet printers** have gained in popularity in recent years, and are widely used for the output of some DTP projects. They are well-suited for newsletters, fliers, and similar jobs. With special paper, they can produce prints of photographic quality.

Fig. 22-6 A scanner can be used to input illustrations (art or photos). Some scanners can be used only with artwork or photographic prints; others can handle photo negatives or transparencies (slides). This model can do both. (Microtek)

Fig. 22-7. A laser printer and sample print are shown here. (Hewlett-Packard Co.)

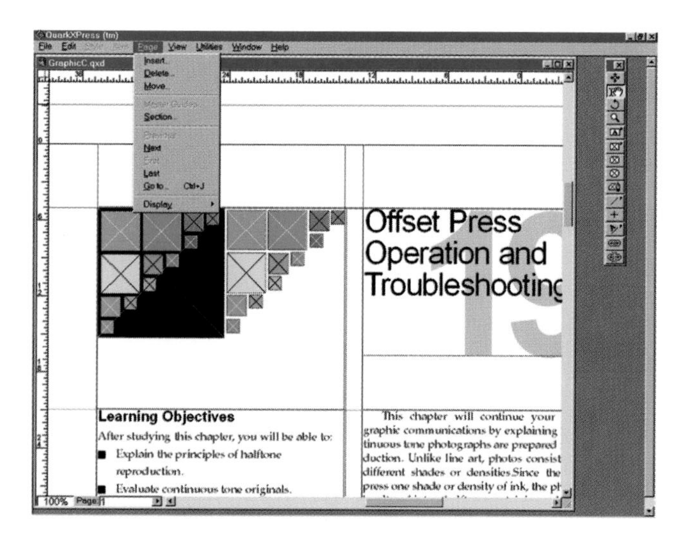

Fig. 22-8. Desktop publishing software is an important part of a desktop publishing system.

The **laser printer** is the most common output device used in desktop publishing systems, Fig. 22-7. It combines small size with reasonably high resolution. Typical resolution for this printer is about 300 dpi. At this resolution, letters on a page appear to have smooth edges, and reasonably good halftones can be placed on a page. This is adequate resolution for many publications, such as newsletters.

When extremely high resolution is needed, an imagesetter is used as the output device. Typical resolutions for this system are 1200 dpi and 2400 dpi. This resolution is needed for high quality publications, such as magazines or books.

SOFTWARE

Normally, several software packages are needed for desktop publishing. These include:

- Desktop publishing software.
- Word processing.
- Graphics programs.
- Electronic clip art.
- Font software.

DESKTOP PUBLISHING SOFTWARE

The most important program for page layout is the desktop publishing software, Fig. 22-8. This program allows you to electronically merge text and graphics on a page. Most programs can also create limited amounts of text and simple graphics. However, larger amounts of text and sophisticated graphics are usually created in other programs and then later imported.

As with pasteup, nonprinting guidelines are first placed on the page to be created. These include page margins, columns and any other guidelines, such as the location for a heading. This layout, often called a **grid**, can be saved and duplicated, if needed, for multiple pages. Next, text and graphics are placed on the page. The desktop publishing software allows you to manipulate both of these. For example, different type styles and sizes can be used, and the heading may be changed. Columns can be justified or not. Graphics can be scaled and cropped if needed.

Completed pages are saved and changed as needed. Normally, pages may be printed and edited several times. This is done, in part, because the screen image is very similar, but not exactly like the printed page.

WORD PROCESSING PROGRAMS

Word processing software is the most common program used with desktop publishing programs. Although text can be created within the DTP program, a separate computer with word processing software is often used. In this way, several people can be working on text for a publication at the same time.

In word processing, the emphasis is on writing quality, and several drafts may be created to ensure that grammar and spelling are correct. When done, the text files are imported to the DTP program. Special filters are a part of the desktop publishing program, so the text from different word processing programs can be imported correctly.

When word processing is done on a separate microcomputer, it is saved on disk. The disk can be placed on the DTP system and the text can be placed on the page being created.

Some word processing programs have features similar to desktop publishing programs. They can set type in columns, use different fonts, and import graphics. However, they are usually limited for complicated layout projects.

GRAPHIC PROGRAMS

As is the case with word processing, simple graphics can be created with the DTP software. This includes circles, lines, rectangles, and other simple illustrations. However, creating sophisticated graphics requires a separate program. Some drawing programs, such as CAD, are best for technical drawings. When a more artistic (freehand) drawing is needed, illustration or paint programs are normally used, Fig. 22-9. These programs provide a greater variety of lines, patterns, tones, and even colors that can be used for illustrations.

Color image editing software is used to make changes in full color images such as photographs. After an image is converted to electronic form using a scanner, the software is used to make changes and even produce color separations for printing.

ELECTRONIC CLIP ART

Most desktop programs have a limited library of graphics that can be used. In order to expand these options, electronic clip art can be purchased. This is simply a large collection, or library, of illustrations in digital form. When an illustration is chosen, it can be scaled and cropped as needed for the page layout.

Fig. 22-9. This illustration software provides a way to create high quality graphics. (Quantel)

FONTS

The major forms of fonts for desktop publishing are printer fonts and downloadable fonts. The **printer fonts** are actually built into the printer. Some printers have slots that will accept slip-in cartridge fonts.

Downloadable fonts are available that must be sent to the printer from the computer. These software fonts may stay on diskettes or be placed on a hard drive. When needed, they are sent through RAM to the printer. For this reason, downloadable fonts normally require extra time for printing.

THE DESKTOP PUBLISHING PROCEDURE

As discussed, the primary purpose of a desktop publishing program is to create a pasteup electronically. The steps in this process are similar to those used for manual pasteup. Although variations may occur on individual jobs, the following steps are typical for this process:

1. Product and topic selection.
2. Design and layout.
3. Grid development.
4. Preparation of the text and graphics from other programs.
5. Creating the page.
6. Saving the page.
7. Printing the page.

PRODUCT AND TOPIC SELECTION

The product and topic selection is based on the purpose of the publication (for example, to inform or to persuade) and the intended audience. Products might include a newsletter, advertising flyer, or brochure.

DESIGN AND LAYOUT

Design and layout begin by drawing thumbnails and a rough layout for each page design as described in Chapter 20. Both rough and final layouts can be done with the DTP software. Normally, a publication will require that several pages be designed. For example, a newsletter has a first page with a masthead and columns below it while subsequent pages just have columns.

Whether created on paper or the computer, the rough layout will need the following information:

- Page size and margins.
- Number of columns and space between columns.

- Type size and style for text, headlines, and subheads.
- Running heads and feet (headers and footers). This is information at the top or bottom of the page that is repeated on every page, such as page numbers.
- Any other graphic or text elements that are important on the page.

GRID DEVELOPMENT

In desktop publishing, a grid is created for the page. This includes all the nonprinting lines, such as page margins, columns, and other nonprinting guidelines helpful for placing elements on the page.

If the same grid is useful for several pages, it is saved without text and graphics. It is then placed on the screen each time it is to be used. If facing pages (pages that lie next to each other) are needed, a different grid will be needed for each, since the inside and outside margins will vary.

Fig. 22-10. A page consisting of both text and graphics is shown on the monitor. (Apple Computer, Inc.)

PREPARATION OF TEXT AND GRAPHICS FROM OTHER PROGRAMS

If space is limited, it might be helpful to know approximately how many words are needed. Text and graphics that are developed are usually saved with specific file extensions (e.g. TIFF). The DTP program can then identify these files as text or graphics.

CREATING THE PAGE

A page is created by placing the text and graphics on the page, Fig. 22-10. The grid, which was previously developed, is used to assist in placement. Text and graphics, such as mastheads, headlines and subheads, headers and footers, and page numbers, are created with the DTP software. As needed, the text and graphics created from other programs are placed on the page. The graphics are usually placed first and the text can flow around them. Review the entire page when finished and make adjustments as needed.

SAVING THE PAGE

To avoid losing your work, be sure to save the page. Remember, even a slight power failure can cause you to lose the information in RAM.

PRINTING THE PAGE

Print the page and edit as needed, Fig. 22-11. Even in WYSIWYG programs, there are differences between the screen and printed image. One reason for this is that screen fonts are often used on the monitor that only

Fig. 22-11. The printed page must be carefully reviewed so that any needed corrections can be made.

approximate the font that will print. Make corrections and print the page again if necessary.

USING A DESKTOP PUBLISHING PROGRAM

Although programs differ, there are many similarities among DTP programs. An understanding of these basics will simplify learning any of these programs.

USING THE MOUSE

One of the most important tools to master in order to use desktop publishing effectively is the mouse, Fig. 22-12. This device allows you to move around on the screen and to make selections. Specific operations are as follows:

- *Move or point:* Simply move the mouse over the table or mouse pad to move an arrow on the screen.
- *Click:* Tap the main mouse button quickly with your finger to make a selection.
- *Double-click:* Tap the main mouse button twice with your finger. This is often used as a shortcut when you must click on a selection and then also verify the choice by clicking a second selection, such as O.K. The double click does both.
- *Drag:* Press and hold the main mouse button, then move to another location. Release the button when the drag is complete.

USING THE DIALOG BOX

After beginning the desktop publishing program, you will be using **dialog boxes**. These are display screens that allow you to make selections or enter new

Fig. 22-12. The mouse is a pointing and selection device extensively used in DTP.

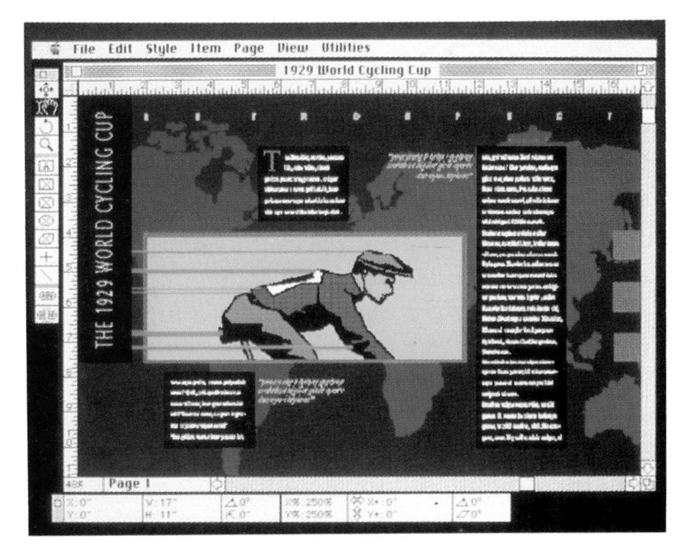

Fig. 22-13. This is a publication window. (Quark, Inc.)

information. Selections are made by clicking with the mouse on your choice. New information is entered from the keyboard.

USING THE PUBLICATION WINDOW

A publication window is shown in Fig. 22-13. The **title bar** along the top shows the name of the publication on the screen. Also, a **menu bar** is along the top of the window. This is a pull-down menu that is activated by the mouse. By pointing to a menu item and holding the mouse button down, the commands under the menu drop down. A command is chosen by dragging down to it with the arrow and then releasing the mouse button. For example, when drawing a line on the screen, commands under the line menu provide options for the width of line.

Usually, the first menu you use when beginning a publication is the **file menu**. This menu contains some of the basic utility routines that are needed. These include *new* (starting a publication), *open* (open a publication previously created), *save*, *close*, and *print*.

The area inside the publication window that holds the page is the **pasteboard**. This area can be used for temporary storage of text or graphics as a publication is being created. Items can be moved between the pasteboard and page.

Rulers are usually placed vertically and horizontally in the publication window, which provides a means of measurement on the page. As the mouse moves the arrow on the page, dashed lines in the ruler show the exact location, both vertically and horizontally. *Ruler guides* or *guidelines* can be placed on the page by clicking on either ruler and dragging a guideline to the desired position.

Page control

The page in the publication window can be changed in several ways. The **page menu** contains commands for enlarging or reducing the page as needed. Scroll bars allow you to move the page around the screen in several ways. Clicking on the arrow moves the page a small amount. You can drag the box on the scroll bar to move the page greater distances.

The page icons in the publication window allow you to move from page to page as needed. Just click on the desired page. Master pages may also be used. These are pages that contain the features that are repeated on every page. For example, margins and column guides are usually placed on master pages. These guidelines then show up on every page without having to be redrawn over again each time.

DRAWING TOOLS

Drawing tools, sometimes known as a **toolbox**, are icons that can be selected by pointing and clicking the mouse, Fig. 22-14. **Icons** are simple pictures that represent a command. These tools, or icons, are available for placing text, creating simple graphics, and cropping illustrations.

PAGE DESIGN CONSIDERATIONS

Good page design is accomplished in several ways. First, you should understand the basic design principles. Second, when a particular type of publication is to be developed, it is a good idea to study designs that others have used. Third, find and use references on designs, such as books and magazine articles. Fourth, remember that design involves using your creativity. Be willing to experiment and try different layouts.

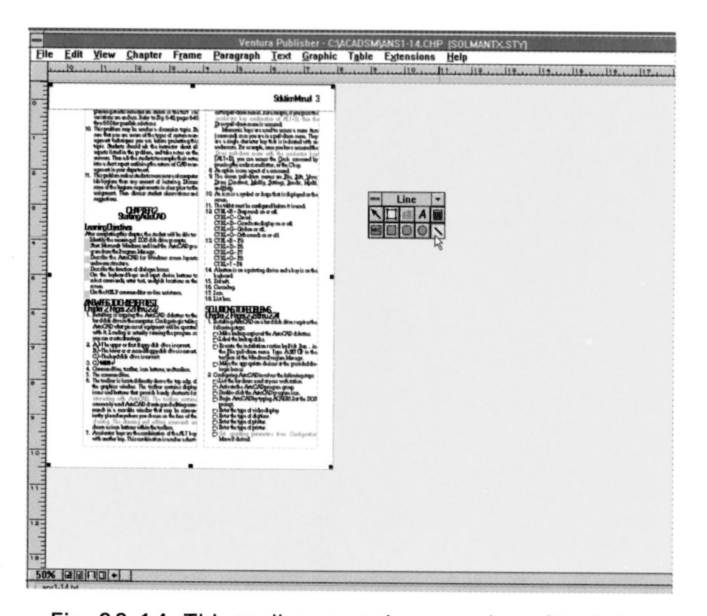

Fig. 22-14. This toolbox contains a number of tools for making simple illustrations, such as this rectangle with rounded corners.

The following is a list of guidelines to help you in laying out pages:

- In a large document, try to be consistent in page layout. Make sure facing pages look good together.
- The minimum page margin is normally 1/2 in. If a publication is to be bound (for example three-hole punched), the inside margin or gutter should be wider than the outside margin.
- Use column widths appropriate for the type size. Larger type requires wider columns. Normally, you should have at least four words on each line within the column. Another rule of thumb is to make the column width no narrower than one and one-half times the type size, measured in picas. For example, 12-point type should not be placed in a column narrower than 18 picas.
- Two or three columns are normally used on a page. Columns do not have to be of an equal width. Space between columns should be at least one pica. If a ruled line is used between the columns, more space is needed.
- Justified columns give a formal appearance. Unjustified (ragged right) columns are fine for an informal look and are easier to read since there are fewer hyphens. Unjustified columns are often used for short lines.
- Think of white space as a design element. Use it carefully and do not crowd the page.
- Serif type, by custom, is normally used for text.
- The average size for text is 9 to 12 points. Older and younger people often find larger type easier to read. The wider the columns are, the larger the type can be.
- If possible, use one type family on a page. For variety, different styles from the family can be used (for example, italics and bold). This is a very efficient procedure because printing is much faster and the entire page can be changed to a different font, if needed, with little difficulty.
- When two typefaces are used on a page, choose contrasting faces. For example, a serif and sans-serif face could be used. Unusual styles should be used sparingly, and for headlines only.
- Be consistent with heads and subheads in regard to style, and use of upper and lower case letters. For example, if the head is all caps, use all caps for the subhead. Heads and subheads often look good in upper and lower case.
- If not automatic, 2-point leading is often used for text.

- Avoid widows and orphans. These are short lines at the top or bottom of the column. **Widows** are at the end of a paragraph or bottom of the column. **Orphans** are at the top of a paragraph or column.
- Use italics instead of underlining for text.
- Use running heads and feet to keep track of page numbers and topics. The outside of the page is normally used.
- Use only one space after periods.

COMBINING DESKTOP PUBLISHING WITH OTHER TECHNIQUES

It is important to remember that standard pasteup procedures may be used with desktop publishing if needed, Fig. 22-15. For example, clip art can simply be placed in position if a scanner is not available. The same technique can also be used for mastheads or logos. When you understand the options available for composition, there are greater design possibilities.

Fig. 22-15. Standard pasteup procedures are sometimes necessary even with desktop publishing. (Xerox Corp.)

SUMMARY

This chapter has provided an overview of desktop/electronic publishing, which is a system for creating page layout with a computer. This technique can speed up the process of producing many publications, such as newsletters and brochures. The information that was presented in this chapter applies to desktop/electronic publishing systems in general. It can prepare you to use a desktop/electronic publishing system and to follow the operating manual when learning to use a specific system.

WORDS TO KNOW

All of the following words have been used in this chapter. Do you know their meanings?

desktop publishing system
dialog boxes
dot matrix printer
downloadable fonts
electronic publishing
electronic scanner
file menu
graphics tablet
grid
icons
laser printer
menu bar
mouse
orphans
page menu
pasteboard
printer fonts
RAM (random access memory)
rulers
scanning
title bar
toolbox
widows

REVIEWING YOUR KNOWLEDGE

Please do not write in this text. Write your answers on a separate sheet.

1. Desktop publishing involves using which of the following for page layout?
   a. A microcomputer.
   b. Specialized software.
   c. A laser printer.
   d. All of the above.
2. True or false? A computer with a large RAM allows faster processing because the disk drives do not have to be accessed as often for additional information.
3. The __________ converts illustrations and text into electronic signals so they can be imported into pages.
4. True or false? The dot matrix printer is the most common output device used in desktop publishing systems.
5. Name five types of software packages that are needed for desktop publishing.
6. List the seven typical steps used in creating a pasteup electronically.

7. Which of the following is a display screen that allows you to make selections?
   a. Title bar.
   b. Dialog box.
   c. Pasteboard.
   d. None of the above.
8. True or false? Justified columns give a formal appearance to a document.
9. Which of the following is NOT a guideline to follow when using a desktop publishing system.
   a. Try to be consistent in page layout.
   b. Use two spaces after periods.
   c. Think of white space as a design element.
   d. Avoid widows and orphans.
10. True or false? Standard pasteup procedures may be used with desktop publishing if needed.

APPLYING YOUR KNOWLEDGE

1. Investigate the various desktop publishing programs and develop a chart that compares their major features.
2. Identify the parts of the publication window for the desktop publishing program you are using.
3. Outline the sequence of events that must be followed when creating a document using desktop publishing.
4. As a class, decide on a design for a calendar that can be produced by desktop publishing. Divide responsibilities and complete the calendar.
5. Investigate professional applications of desktop publishing. Visit a company that uses desktop publishing and write a report describing its uses.

23

CHAPTER

GRAPHIC ARTS PHOTOGRAPHY

After studying this chapter, you will be able to:
- *Compare high contrast and continuous tone film.*
- *List and describe the steps involved in producing a line negative.*
- *Discuss the purpose of a sensitivity guide.*
- *Describe alternative methods for producing a line negative, such as diffusion transfer and contact printing.*
- *Explain the process of producing a halftone.*
- *Define process color photography.*

Changing flat images into photographic images is the function of **graphic arts photography**. See Fig. 23-1. A variety of techniques is used to change flat images into photographic images. This is also known as **image conversion**. This chapter will explore various image conversion techniques. Image conversions that use graphic arts photography procedures include:
- Lithographic image carriers (plates) that are made using a film negative.
- Enlarging or reducing copy.
- Continuous tone images that must be converted to halftone images for printing.

Fig. 23-1. Graphic arts photography involves changing flat images into photographic images.

ORTHOCHROMATIC FILM

The film that is used for snapshots is called *panchromatic film*. As discussed in previous chapters, it records color and shades of color as your eye does. This is called *continuous tone film*. The film most often used in graphic arts is **orthochromatic film**, Fig. 23-2. It is high contrast film. The processed film contains dark areas and light areas with no tones in between. It is called

Fig. 23-2. The label on a box of film will indicate what type it is as well as other types of information.

red-blind film because it is not sensitive to red light. For this reason, a red safelight can be used in the darkroom when using this film.

Orthochromatic film has a base side and emulsion side, Fig. 23-3. The emulsion is the light-sensitive layer. When an exposure is made, the light forms a *latent* image (not yet visible) on the emulsion. The antihilation coating stops the light from reflecting back to make a second exposure or "halo."

When film is exposed with a pasteup, a negative is formed. Special duplicating film allows you to go directly to a positive. Orthochromatic film should only be removed from the box when a safelight is in use. Handle the film by the edges with clean, dry hands. Close the box before proceeding. If the film must be laid down, place it emulsion side up.

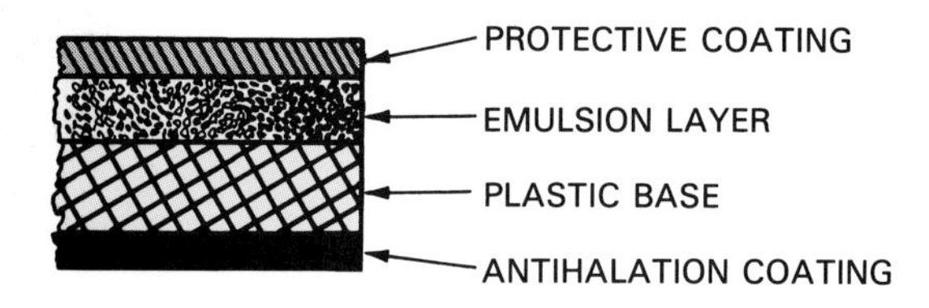

Fig. 23-3. Orthochromatic film layers are shown here.

PRODUCING A LINE NEGATIVE WITH THE PROCESS CAMERA

The **process camera** is used for photographing flat copy. Cameras are either vertical or horizontal. The entire vertical camera is placed in the darkroom, Fig. 23-4. Most horizontal cameras pass through a darkroom wall, Fig. 23-5. A **gallery camera** is entirely outside the darkroom, Fig. 23-6. A light-proof film holder may be used for transporting the film to and from the darkroom. Some gallery cameras have a film dispenser and processor on the camera, so no darkroom is needed.

Fig. 23-4. This is a vertical process camera. (nuArc Co., Inc.)

To set up the camera, first open the copyboard and clean the glass if needed. Using the guidelines, place the line copy on the copyboard. Place a sensitivity guide (gray scale) on the copy, keeping it away from images, Fig. 23-7. Close and latch the copyboard and swing it into position if necessary. Check the lights. They should be at about a 30° angle to the copyboard. It is important to never touch the lamps.

Set the camera to the desired f-stop. See Fig. 23-8. This is usually one to two stops from wide open. This position gives satisfactory resolution (image quality) while keeping exposure time fairly short. If the camera has a *variable diaphragm control*, find the correct f-stop and then set the percentage of reproduction along this scale. This automatically varies the light for enlargements and reductions.

Set the *scaling controls* on the camera, Fig. 23-9. Settings can be for the same size (100 percent) or other percentages. These changes are accomplished by moving the lensboard, which holds the lens, and the copyboard.

Check the image location by swinging the ground glass into position and turning on the exposure lamps.

Fig. 23-5. This is a horizontal process camera. (nuArc Co., Inc.)

See Fig. 23-10. This is not normally needed for same size reproductions. Swing the ground glass back and latch it.

Set the exposure time. In safelight conditions, center the film on the film holder with the emulsion up or facing the lens. If available, turn on the vacuum pump to hold the film, and wait a few seconds for full vacuum. Then close the camera back.

Fig. 23-6. A gallery process camera can be placed outside of the darkroom. (Visual Graphics Corp.)

Fig. 23-9. The handwheels allow scaling of the copy by moving the lensboard and copyboard.

Fig. 23-7. When setting up the camera, place the copy on the copyboard and place a sensitivity guide on the copy.

Fig. 23-8. This variable diaphragm control is used to set the f-stop.

Fig. 23-10. A ground glass can be used to check image size and location.

Make the exposure. Light is reflected from white areas on the copy, exposing the film. The image area, which was black, absorbs the light and does not reflect it back to the film. After the exposure is complete, open the camera, turn the vacuum off, and remove the film for processing.

STEP TEST

If you do not know the exposure time for line copy, a **step test** is necessary. To perform a step test, follow the following procedure:

1. Place normal copy, such as a typewritten page, on the copyboard. Use the f-stop next to the largest lens opening, and set the scaling controls on 100 percent.
2. Cut masking strips using black or red paper, 2 in. to 3 in. larger than the film size.
3. Set the timer for four seconds and place the film on the camera.
4. With the vacuum on, mask all but 2 in. of the film, Fig. 23-11.
5. Make a four-second exposure.
6. Continue making four-second exposures, removing one strip each time.
7. Process the film using the recommended development time (usually 2 minutes and 45 seconds).
8. Choose the strip with the best image quality and record the time, Fig. 23-12.
9. Place a gray scale (sensitivity guide) on the typewritten copy. Make an exposure using the time previously chosen. The gray scale should show a black 4, foggy 5. If not, make slight adjustments in the time.

Fig. 23-11. This is how a step test is performed.

Fig. 23-12. The resulting negative from the step test is shown at left, and a correctly exposed and developed negative at right.

FILM PROCESSING

Film processing changes a latent film image to a visible and permanent image. Processing can be done with trays or by using a machine processor, Fig. 23-13. The two major forms of processing are:

- Sensitivity guide.
- Time and temperature.

When the **sensitivity guide method** is used, development time can be varied as long as the correct step is reached on the gray scale. In the **time and temperature method**, developing time remains constant. Changes in image density are produced by changing the exposure time. The sensitivity guide method is explained here.

For tray processing, the chemicals should be first mixed in the darkroom with lights on for visibility. The room should be well ventilated, and all safety precautions should be followed. If a temperature-controlled

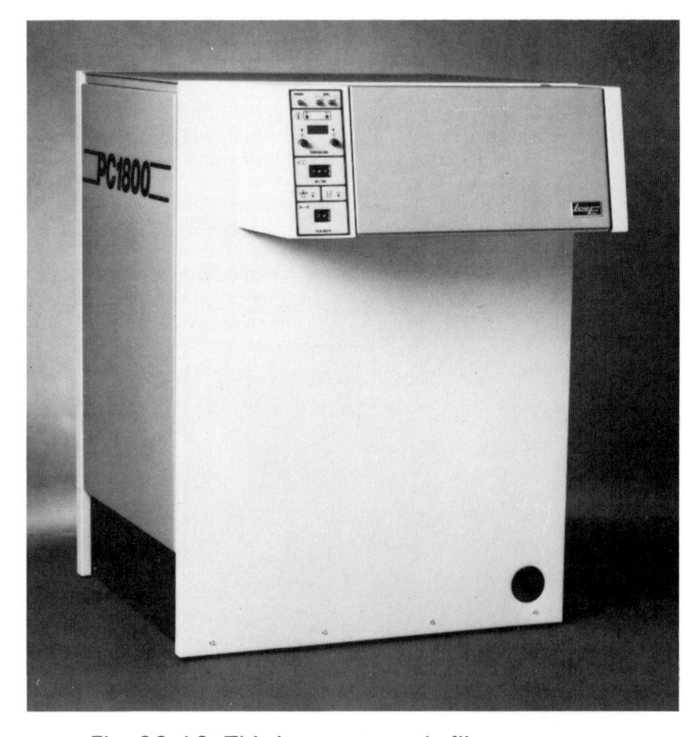

Fig. 23-13. This is an automatic film processor. (LogEtronics, Inc.)

sink is used, fill it with water that is at the temperature appropriate for the chemicals you are using. Use trays slightly larger than the film size. While wearing chemical goggles, mix enough chemicals for a 1/2 in. depth in the tray. The stop bath is an acid. (Remember to always add acid to water.) Place the trays so that, starting from the left, there is developer, stop bath, fixer, and a running water bath, Fig. 23-14.

Process the film emulsion side up to prevent scratching. Starting with the developer, tilt the tray away from you to form a "pool" at one end. Holding the film in a "V" with the fingers, slide it into the chemicals and lower the tray, Fig. 23-15. The chemicals should wash over the film evenly. Agitate the tray gently from alternate corners. This keeps fresh chemicals in contact with the film surface. Develop to the recommended step on the gray scale. This is usually a black 4, and foggy 5, Fig. 23-16. Immediately remove the film and place it in the stop bath with agitation for 10 seconds. This halts the film development.

Next, the film is placed in the fixer. The fixer removes the unexposed emulsion. This step is also known as clearing the film. The fixing process usually requires about three minutes.

The film is then placed in a running water bath for 10 minutes. This removes chemicals that might stain the film. After washing, use a squeegee on the film with the emulsion side down, Fig. 23-17. Dry the film by hanging it or by machine drying. This processed film is now a line negative.

Different orthochromatic film may require different processing. For example, rapid access film, which can be processed more quickly than traditional "lith" films, use different chemicals and higher temperatures. Newer films, often called hybrids, also require different chemicals. Make sure that you are using the correct chemicals for the film type you are using and that all safety precautions are followed.

USING THE SENSITIVITY GUIDE

The **sensitivity guide** or **gray scale** is an important tool in graphic arts. It measures density. **Density** is a

Fig. 23-15. Use this procedure when placing the film in the chemicals.

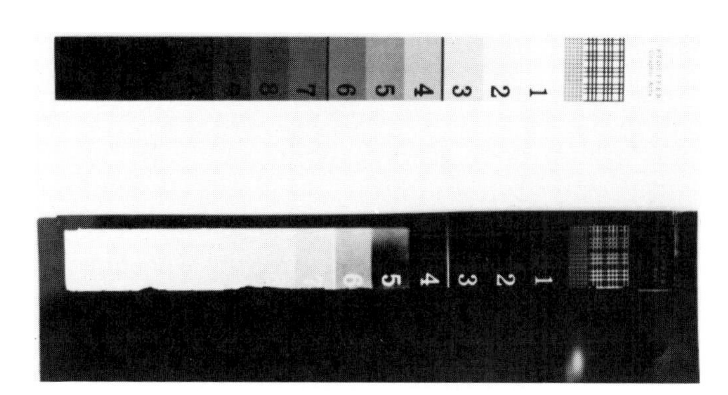

Fig. 23-16. Develop the film to the correct step on the sensitivity scale, as shown on the right.

Fig. 23-17. Remove excess water from the negative with a squeegee.

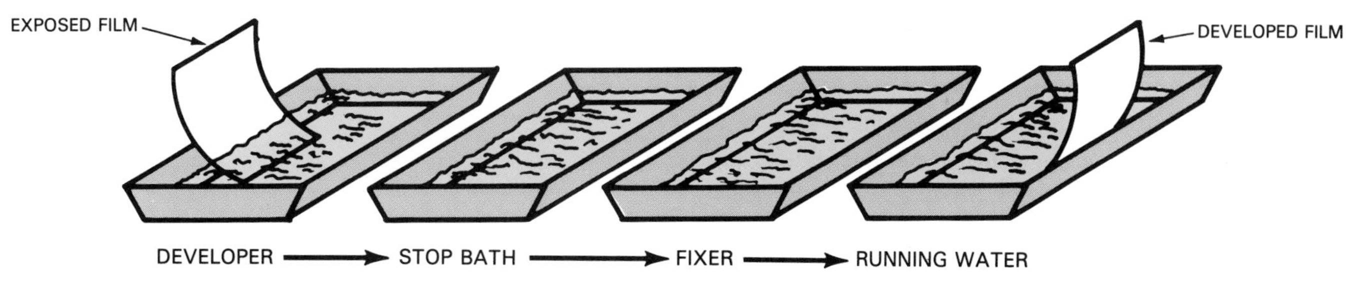

Fig. 23-14. This is the processing sequence that is often followed for many orthochromatic films.

measurement that shows the "darkness" of an image. As an image becomes darker, it is able to absorb more light. Therefore, density is also a measure of the light-absorbing ability of an object.

Density measurements start at 0.0 (white) and continue up the scale to about 3.0 (black). Each step on a gray scale is approximately .15 away from the next step.

When you develop to a step on the gray scale, white areas as dense as that step will be black on the negative. Why did we develop to a step 4 on the gray scale for a line negative? Why not step 1? Step 1 is very white (close to 0.0 density). Smudges and fingerprints on the pasteup are not this white. If you develop to a black step 1, fingerprints and other unwanted marks will be clear on the negative, just like the image areas, Fig. 23-18. When a step 4 is used, this eliminates unwanted, grayish marks.

If development time is constant (time and temperature method), doubling the exposure time will result in a .30 density (2 steps) shift up the gray scale. Half the exposure time will result in a .30 shift down the scale. This technique can be used for correcting exposures for line and continuous tone images.

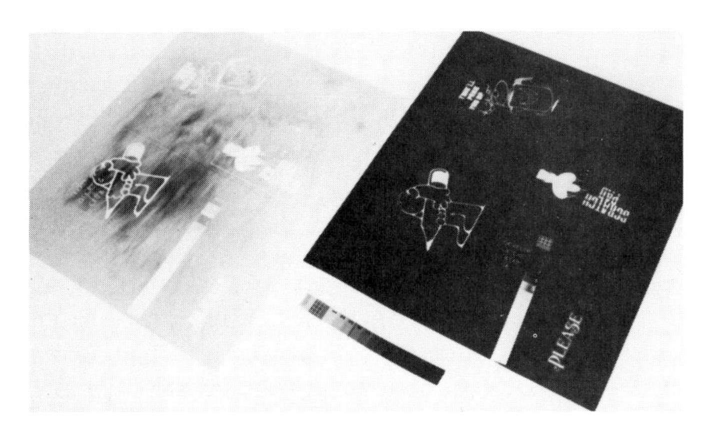

Fig. 23-18. The negative on the left was developed to a step 1 on the gray scale while the negative on the right was correctly developed.

DIFFUSION TRANSFER

Diffusion transfer is used to reproduce copy as a positive instead of a negative, Fig. 23-19. This reproduction can be on paper or film. Diffusion transfer paper copies are often used for enlarging and reducing illustrations that will be used on the pasteup.

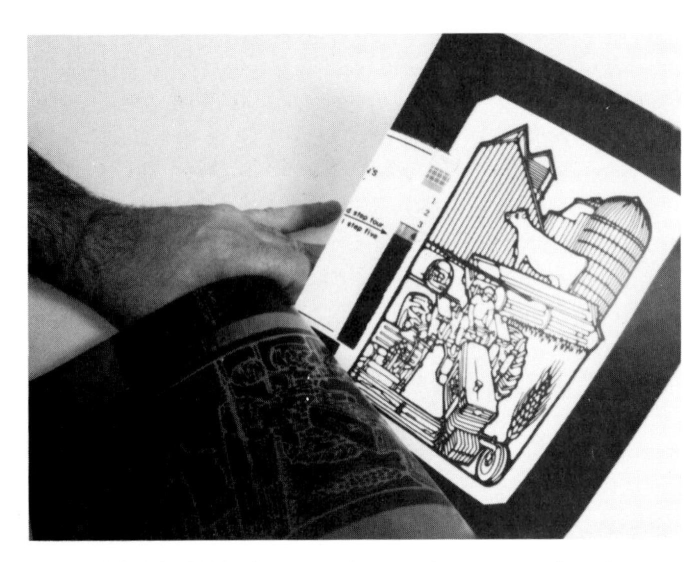

Fig. 23-19. Diffusion transfer requires a negative sheet and receiver sheet.

This process uses a *negative* (donor) sheet and *receiver* sheet. The negative sheet is exposed like a line negative, with the emulsion up. However, exposure time will be different. Use a step test to determine exposure time if necessary. After exposure, the negative is placed emulsion to emulsion with a same size receiver sheet. These sheets are fed into a diffusion transfer processor, Fig. 23-20. An activator liquid causes the image to transfer from the negative to the receiver. Peel the sheets apart after 30 seconds and discard the negative, Fig. 23-19. Wash the receiver sheet for two minutes, squeegee, and dry it. Additional images can be transferred to the receiver sheet for special effects. Follow the same procedure again.

Evaluate the positive print. If it is too light or too dark, change the exposure time since development time is constant. This will be opposite the change that would be made for a line negative. More exposure will result in a lighter image. It may help to remember that more exposure will result in a darker image on the negative which then produces a lighter image on the receiver.

Fig. 23-20. This is a diffusion transfer processor. (nuArc Co., Inc.)

CONTACT PRINTING

Contact printing is a method of producing same size images on light-sensitive film or paper. The reproduction is reversed from the original. For example, a negative is used to make a positive.

The equipment needed for this process is a contact frame and pinpoint light source, Fig. 23-21. To make a film positive, unexposed orthochromatic film is placed in the contact frame, emulsion side up. Place a film negative over the positive, emulsion side down. Close the glass cover and turn on the vacuum.

Set the exposure time. A step test can be used if exposure time is not known. Make the exposure, being careful not to shade the film. Turn off the vacuum and remove the exposed film.

The film is processed using a time and temperature method. For tray processing, a development time of 2 minutes, 45 seconds is normally used.

The contact frame can be used for making a duplicate of a film positive or negative. Diffusion transfer materials can be used for this purpose. A step test will be necessary if exposure time is not known. Duplicating film can also be used for this purpose.

Fig. 23-21. This contact frame and pinpoint light source are placed in a darkroom.

HALFTONES

Continuous tone images must be converted to halftone images for printing. By breaking the tones into various dot sizes, the viewer has the illusion of tones on the printed page. This is because the eye "blends" the dots and background together. Black and white photos are an example of copy that must be converted to a halftone, Fig. 23-22.

Fig. 23-22. This continuous tone image, which has highlights, shadows, and middle tones, must be converted to a halftone for printing.

COPY DENSITY

Before a halftone is made, the copy density must be found in the extreme highlight and shadow areas. Highlights are the lightest areas, and shadows are the darkest areas. A **visual densitometer** can be used for this, Fig. 23-23. This is a card with holes in the center of density patches. Place the card over the highlight area and find the matching density. Repeat this procedure for the shadow area.

A more precise method of finding copy density is by using a **reflection densitometer**, Fig. 23-24. This instrument must be calibrated to a known density before use. In this case, the densitometer reflects light from a small area of the copy and provides a density readout.

The density readings should be recorded for later use. The shadow density minus the highlight density is the **copy density range**. Photos are easiest to use if they have good copy density range or contrast.

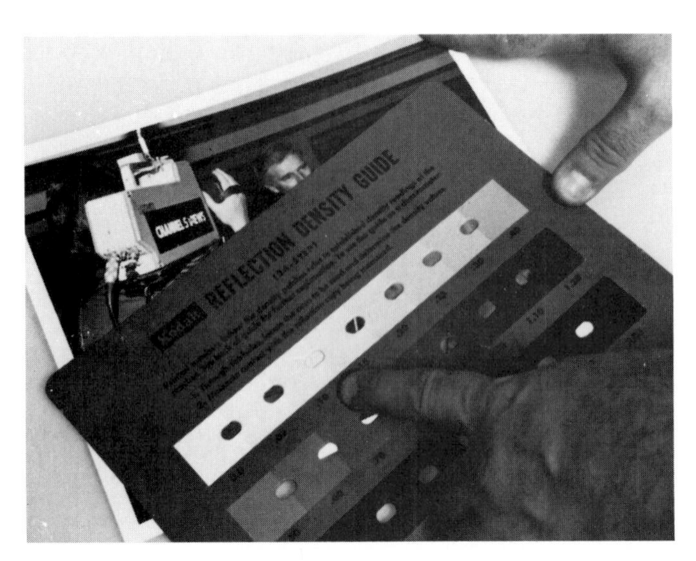

Fig. 23-23. A visual densitometer is being used to find the density of this photo.

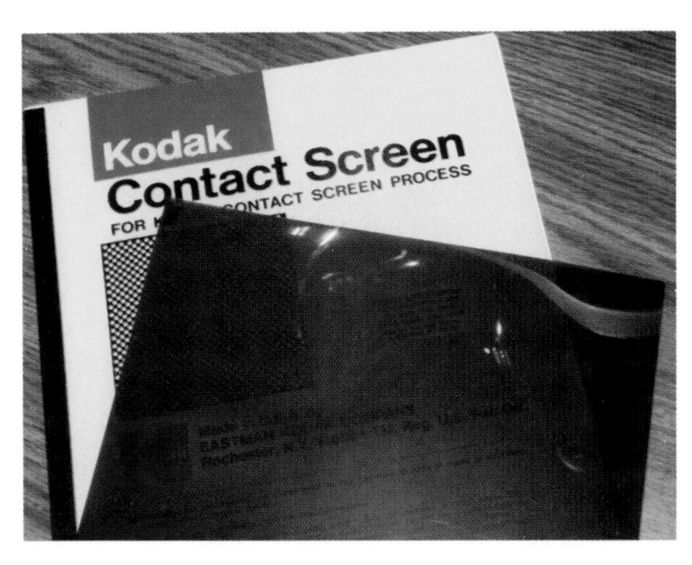

Fig. 23-25. This contact screen should be handled very carefully by its edges and returned to the folder when done.

HALFTONE SCREENS

Halftone screens are used to break the continuous tone image into dots. This is done by placing the screen over the film before exposure. The original screens were glass. Most screens today are plastic and are known as **contact screens**, Fig. 23-25. The most common screen colors are magenta and gray. Magenta screens are for black and white copy only.

Halftone screens have round, square, or elliptical dots. Regardless of shape, the dots are *vignetted*. This means the dot is almost clear in the center and becomes

Fig. 23-24. A reflective densitometer is being used to find the densities on the photograph. (Tobias Associates, Inc.)

progressively darker or more dense near the outside. This dot is the key to making a halftone. A very weak light will only pass through the center of the dot. As a light becomes brighter, it will pass through a larger area of the dot, Fig. 23-26.

Screen ruling is a measure of the screen dots in a linear inch. This is also known as "lines per inch" since the dots are aligned on the screen. Coarse screens are under 100 lines per inch, Fig. 23-27. Coarse screens make larger dots that can be printed on rough papers. For example, newspapers use about a 65-line screen. Very fine screens require smooth or coated paper. The dots are very small and would be lost on rough paper such as newsprint.

Halftone screens have a **basic density range (BDR)**. This is the maximum density range the screen can record on film with a single exposure.

HALFTONE EXPOSURES

There are three exposures that can be used to make a halftone. The exposures needed depend on the copy density range and the screen density range. The exposures are:

- Main exposures.
- Flash exposures.
- Bump exposures.

The **main exposure** is also known as the detail exposure. This exposure is made with the halftone screen over the film, Fig. 23-28. It records most of the highlight, middletone, and shadow detail.

Some copy has a density range that exceeds the screen range. In this case, a **flash exposure** is used, Fig. 23-29. This exposure places detail in the shadow area. It is produced using a yellow filtered lamp. The exposure is also called a **non-image exposure** because the exposure is made with the camera film back open. This weak light only passes through the center of the dot. This enlarges the shadow dots, thus creating more detail.

In most cases, a main and flash exposure will be used. However, if copy has low contrast, it may have less density range than the screen is capable of recording. In this case a **bump exposure** is made without the screen. This enlarges the dots in the highlight area which "compresses" the screen range. The bump is made at the beginning or end of the exposure sequence. It is normally made with a neutral density filter over the lens, which only allows a percentage of the light to pass through. The main exposure time is used for this exposure also.

The exposure sequence should always be the same. A bump-main-flash sequence is recommended.

Determining exposure time

A **halftone calculator** is normally used for determining halftone exposures, Fig. 23-30. These are calibrated

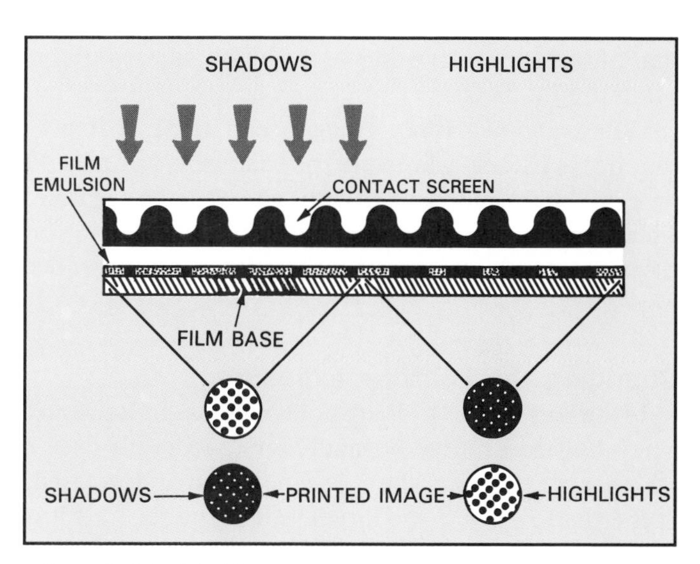

Fig. 23-26. A halftone negative is made by exposing the film through a contact screen. The printed image will be a reverse of this negative image.

Fig. 23-27. An 85-line halftone on the left is compared with a 110-line halftone on the right. (Caprock Developments, Inc.)

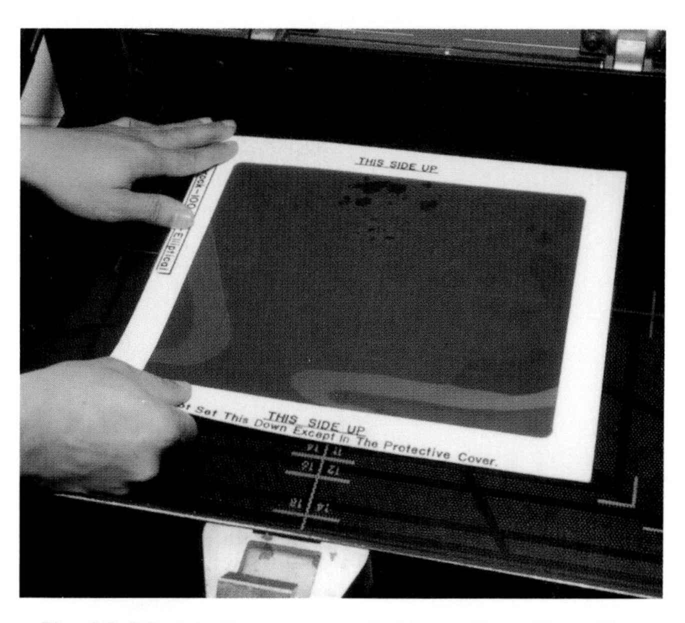

Fig. 23-28. A halftone screen (with cardboard handling frame) is being placed over the film on the process camera.

using tests in your own lab. In order to use the calculator, the copy density ranges are entered. The correct exposure times for your lab can then be found.

Fig. 23-29. The camera is in position for a flash exposure from the lamp shown at top.

Producing the halftone

Camera preparation for a halftone is similar to setup for a line negative. Place the continuous tone copy in the copyboard. Place film in the film holder and turn on the vacuum. Handling the halftone screen by the edges, place it over the film. A brayer can be used to ensure good contact. Make the main exposure. Leaving the vacuum on, lower the camera back and make the flash exposure. If a bump is needed, this is done before the other exposures. A neutral density filter is placed over the lens and the main exposure time is used for the bump.

Process the halftone using the time and temperature method. This is normally 2 minutes, 45 seconds, with agitation.

Halftone dots

Dot size on a halftone refers to the size of the black dot, whether the dot is on the negative or final printed copy. The halftone negative has small dots in the shadow area and large dots in the highlight area. Refer again to Fig. 23-26.

Dots are shown as percentages, Fig. 23-31. A 10 percent dot is usually wanted in the shadow area and a 90 percent dot is wanted in the highlight area of a negative. This percentage is determined by checking the dot on the negative with the dot on a **percentage indicator**. See Fig. 23-32.

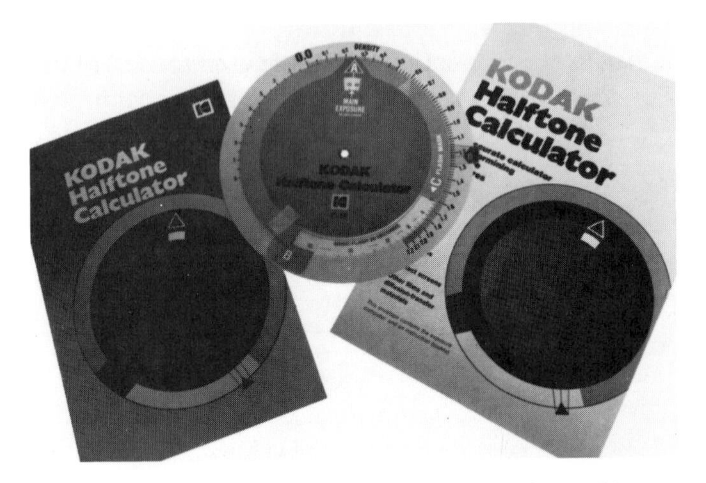

Fig. 23-30. The halftone calculator (center), provides main, flash, and bump exposure times.

Changing the halftone exposure

If the correct dot sizes are not found on the halftone, a new halftone should be made. First, generally decide whether the exposure should be increased or decreased. If the shadow dot is too large, reduce the flash. If it is too small, increase the flash. If the highlight dot is too large, decrease the main exposure. If it is too small, increase the main exposure.

The corrected exposure times can be found more precisely if a gray scale is photographed along with the copy. If a dot size is incorrect on the negative, find this same dot on the gray scale. Now find the correct dot size

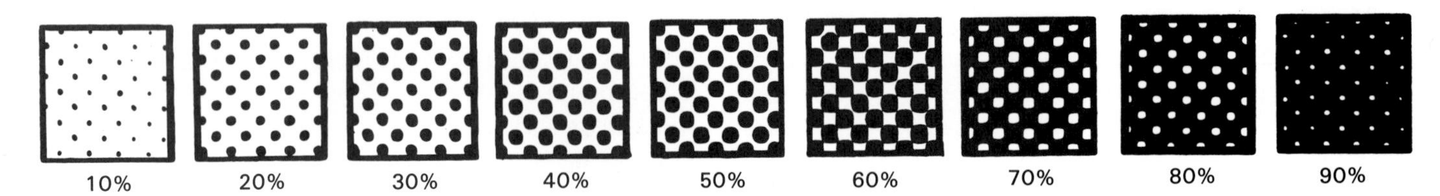

Fig. 23-31. These are dot percentages.

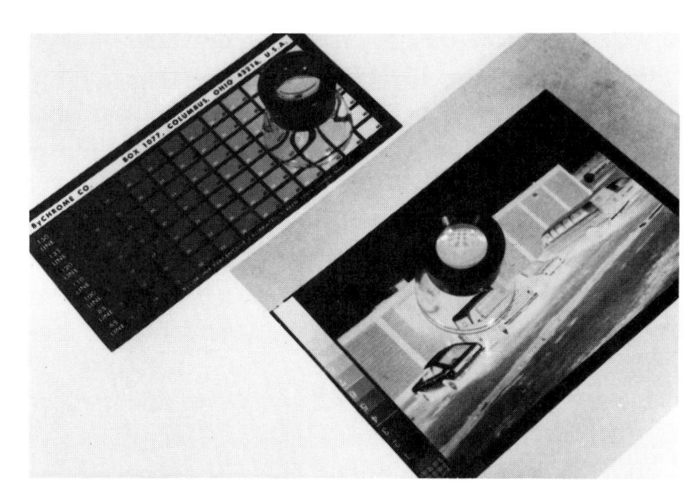

Fig. 23-32. Compare the dots on the halftone negative with a percentage indicator. A film loupe (magnifier) is used to enlarge the dots for viewing.

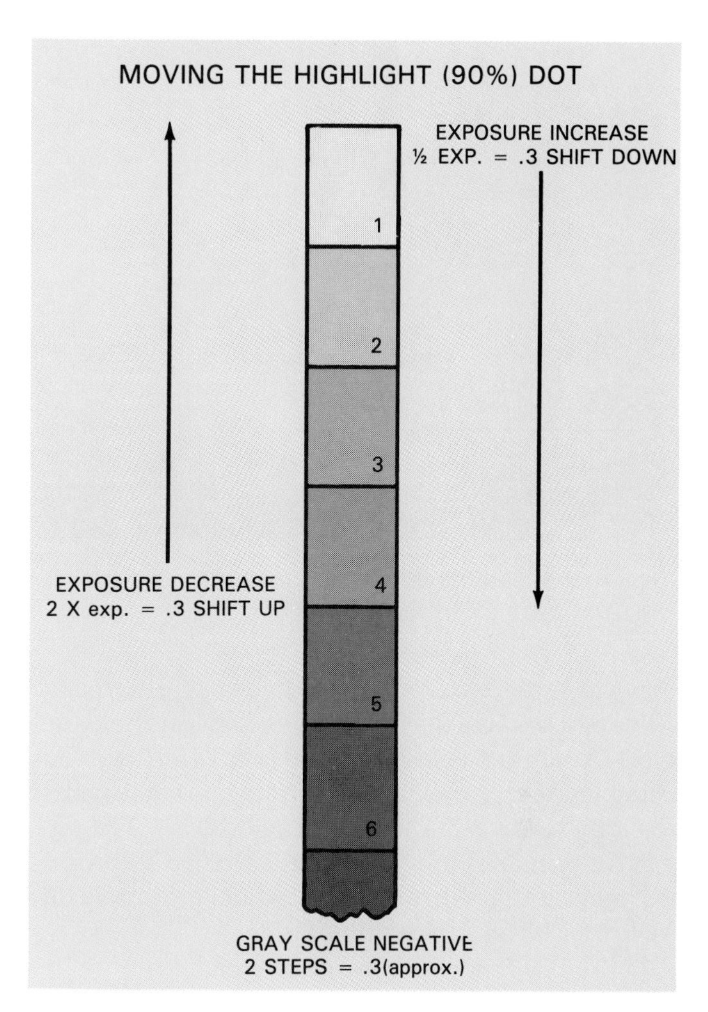

Fig. 23-33. Halftone dot size can be corrected by varying the exposure time as shown.

on the gray scale. Figuring .15 density per step, find the density difference. This difference can usually be "dialed in" to the original calculator settings to revise the times. Another method is to vary exposure time. Doubling the exposure will move a dot size 2 steps (.30) up the scale, Fig. 23-33. Half the exposure time will move the dot 2 steps (.30) down the scale. For example, the dot found in the negative highlight area is too small. This dot is found in step 7 of the gray scale. The correct dot size is found in step 8 of the gray scale. The main exposure time was 40 seconds. Doubling that will cause a .30 density shift. Since a .15 shift is needed, a 60 second exposure is used.

PROCESS COLOR PHOTOGRAPHY

In order to print a full color, continuous tone image, the colors in the picture must be photographically separated. This is necessary because presses print one ink color at the time. Four ink colors are used to produce full or process color.

In **color separation**, the color original is photographed four times using color filters and a halftone screen. This produces one set of color separation negatives, with each having a dot pattern. The filters used are the same as the additive primaries of light, which are red, green, and blue, as discussed in Chapter 19 of this text. Since red, green, and blue are combined to produce white light (the sun, for example), these filters can be used to separate all the colors in the original. In addition, all three filters are used in combination to produce a fourth separation negative. This negative is used because of imperfections in production materials (filter and ink color).

Each separation negative will be used to make a printing plate for one color. This color will be the complimentary (combination of other two primaries) of the filter used originally. For example, the negative produced with the blue filter is the yellow printer. When the negative is made with the blue filter, only blue on the original passes through the filter to expose the film. The unexposed areas, which will be clear on the negative, are the reds, greens, and combinations of reds and greens in the original. Red and green in combination produce yellow. When a plate (positive image) is produced from the negative, the image will be in those areas containing yellow. Therefore, the negative is called the yellow separation negative or the yellow printer.

As mentioned, each colored filter used for separation produces a complementary color printer. The red filter is used to produce its complement, which is the cyan (blue and green) printer. The green filter is used to produce the magenta (red and blue) printer. The filters in combination produce the black printer. This printer will add density in the shadow areas.

After image carriers are produced from the negatives, they are used for printing magenta, cyan, yellow, and black, Fig. 23-34. As was discussed in Chapter 19,

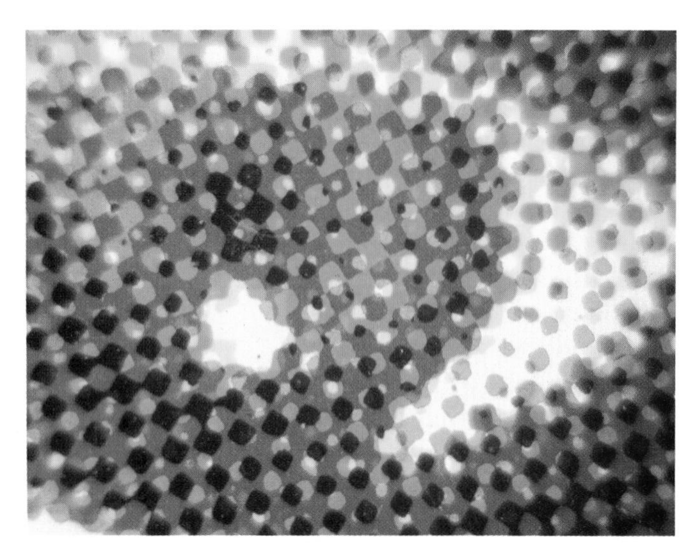

Fig. 23-34. This is a close-up of a printed color picture. (Flying Color Graphics)

magenta, cyan, and yellow are known as subtractive primaries, because they are used as colorants (inks and dyes) to subtract color from the printed surface. For example, when magenta ink is printed on white paper, you only see magenta, not white (all colors). The subtractive primaries are sometimes specified as process red (magenta), process blue (cyan), and process yellow (yellow). The inks are translucent.

When each color is printed, it has a range of dot sizes as with any halftone. In this way, different values of the same color can be placed on the paper. In addition, dots of different colors can overlap to produce other colors. For example, when magenta and cyan overlap, blue is produced.

Color separation negatives are produced by several methods. They can be produced with a process camera using filters for the lens and a halftone screen. However, color separation is most often done in industry by **electronic scanning**, Fig. 23-35. An electronic scanner usually has a drum on which the color original is placed. Film is then placed on the other side of the drum. As the drum rotates, a photocell "reads" or scans the original. This produces an electronic signal that is used to expose the film with light. For color originals, filters are used so scanning can be done for each color. In this way, a set of separations can be generated.

DIGITAL PHOTOGRAPHY

Digital photography is another way images can be converted in graphic arts. In fact, there are digital alternatives for most photographic techniques. For example, color or black and white images can be captured electronically using a scanner and then edited as needed

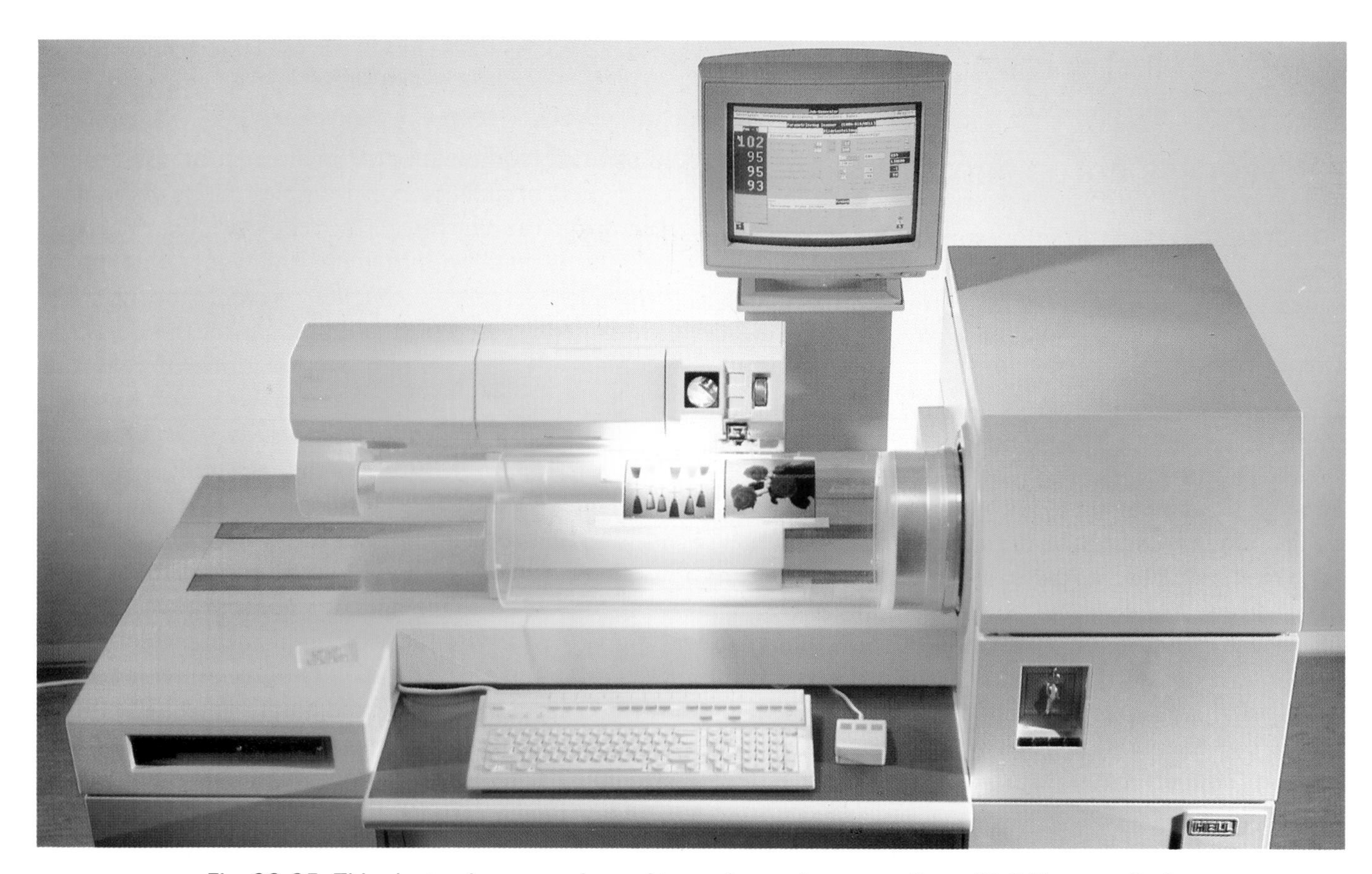

Fig. 23-35. This electronic scanner is used to produce color separations. (Hell-Linotype Co.)

by computer. Electronic publishing systems can make color separations from the full color images. In addition, the images can be printed directly to film using an imagesetter.

Continuous tone images that are converted to halftones electronically are actually different than traditional halftones. A halftone screen creates various dot sizes when an exposure is made. However, digital images are made up of pixel dots of equal sizes. In this case, several pixels may be necessary to be roughly equivalent to a halftone dot. See Fig. 23-36.

Color separations can also be done on electronic publishing systems. Color image editing software is typically used to create the four separations. These can then be printed onto film with an output device such as an imagesetter.

Fig. 23-36. This color image editing software allows you to electronically create a halftone screen.

Since high quality output devices can be expensive, many companies choose to use a service bureau for printing out the final images. Proof copies are done in-house. After corrections, they are sent to the service bureau. Most service bureaus provide several film output options.

SUMMARY

This chapter has provided information on converting line and continuous tone images to photographic images. These techniques can be used to produce image carriers for printing. This topic will be explored in future chapters.

WORDS TO KNOW

All of the following words have been used in this chapter. Do you know their meanings?

basic density range (BDR)
bump exposure
color separation
color separation negatives
contact screens
copy density range
density
diffusion transfer
digital photography
dot size
electronic scanning
flash exposure
gallery camera
graphic arts photography
gray scale
halftone calculator
image conversion
main exposure
non-image exposure
orthochromatic film
percentage indicator
process camera
reflection densitometer
screen ruling
sensitivity guide
sensitivity guide method
step test
time and temperature method
visual densitometer

REVIEWING YOUR KNOWLEDGE

Please do not write in this text. Write your answers on a separate sheet.

1. Which film is used most often in graphic arts?
   a. Continuous tone film.
   b. Orthochromatic film.
   c. Halftone film.
   d. None of the above.
2. True or false? When making an exposure with the process camera, the image area (black) absorbs the light and does not reflect it back to the film.
3. If you do not know the exposure time for line copy:
   a. Guess.
   b. Look it up in a manual.
   c. Perform a step test.
   d. None of the above.
4. Name the two major forms of film processing.
5. The sensitivity guide measures __________.

6. _________ _________ is used to reproduce copy as a positive instead of a negative.
7. Continuous tone images must be converted to which of the following before printing can take place?
   a. Halftone images.
   b. Negative.
   c. Positives.
   d. None of the above.
8. Name the three exposures that can be used to make a halftone.
9. In color separation, the color original is photographed how many times using color filters and a halftone screen:
   a. One.
   b. Two.
   c. Three.
   d. Four.
10. Color separation is most often done in industry by using:
   a. A process camera using filters for the lens and a halftone screen.
   b. A 35 mm camera.
   c. Electronic scanning.
   d. None of the above.

APPLYING YOUR KNOWLEDGE

1. Perform a step test to determine exposure times for your equipment.
2. Calibrate a halftone calculator to produce a quality halftone with shadows and highlights.
3. Investigate the differences between tray and machine processing.
4. Research current trends in color printing. Prepare a report.

24

CHAPTER

SCREEN PRINTING

After studying this chapter, you will be able to:

- *Recognize important elements in screen printing and describe their characteristics.*
- *Select and demonstrate the proper use of the appropriate screen printing stencil for specific applications.*
- *Discuss image carrier preparation and image transfer as it relates to screen printing.*
- *Describe the steps involved in multicolor screen printing.*
- *Describe and demonstrate how screen printing is done on textiles.*

Fig. 24-1. Screen printing can be used to print on a variety of surfaces with inexpensive equipment.

Screen printing, the process of forcing ink through an image stencil, is a popular method of printing, Fig. 24-1. This method is versatile because it can be used to print items that would be difficult to print using other methods. Objects that are commonly screen printed include textiles, bottles, metal signs, and posters. See Fig. 24-2.

The image carrier for screen printing is a **stencil**. The stencil is attached to a screen. Ink passes through this stencil and on to the surface to be printed.

A wide variety of screen printing presses is used in industry. An example is shown in Fig. 24-3.

Fig. 24-2. All of these objects have been screen printed. (John Walker)

BASIC SCREEN PRINTING FRAME

A basic screen printing frame is shown in Fig. 24-4. A screen frame is attached to a base with hinges. Some hinges are adjustable for printing on thick items. The kick stand is used to hold the screen in a raised position.

An object to be printed is placed on the base. The screen frame, with stencil, is lowered to the base. Ink is

Fig. 24-3. This screen printing press prints and dries in multiple colors of ink on a single product. (Advance Process Supply Co.)

Fig. 24-4. A basic screen printing unit is shown here. (Bowling Green State University, Visual Communication Technology Program)

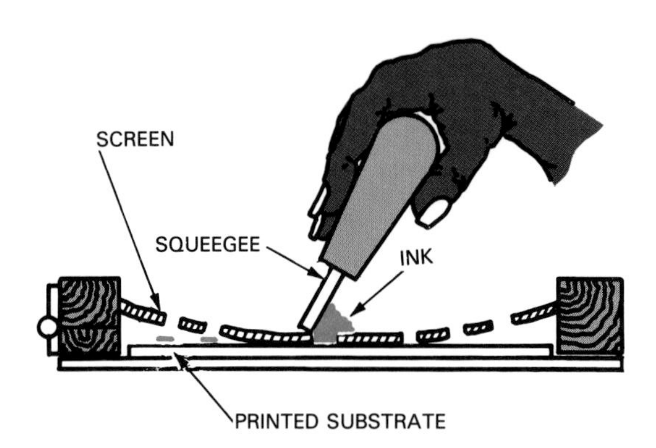

Fig. 24-5. In screen printing, the squeegee forces ink through the screen and onto the substrate.

placed on the frame and a squeegee is used to press ink through the stencil, Fig. 24-5. The frame is lifted and the printed object is removed.

SCREEN FABRICS

The fabric for the screen is made of woven threads called a screen mesh or **screen fabric.** Originally, only natural fibers such as silk were used. Today, natural and synthetic fibers are available. Some examples include silk, nylon, polyester, and stainless steel. Fabric can be made from monofilaments or multifilaments. In monofilament fabrics, a single fiber makes up one thread. In multifilament fabrics, a number of strands are wound together to form each thread, like a rope.

A number system and mesh count are both used to designate how fine a weave is. The fibers must be woven very close together to hold a stencil with fine detail. For example, a halftone with small dots, must be printed with a fine screen mesh.

Multifilament screens are designated by a number system with the prefix "xx." **Screen numbers** range from 0000 to 35. As the number becomes larger, the screen becomes more fine, with more threads per inch. Examples of multifilament fabrics are silk and multifilament polyester.

For most monofilament fabrics, a **mesh count** is used to designate how closely the fibers are woven together. The mesh count is the number of threads in an inch. Examples of monofilament fabrics include nylon, stainless steel, and monofilament polyester.

A comparison of screen numbers and mesh counts is shown in Fig. 24-6. Notice that the screen number becomes larger as the mesh count becomes larger.

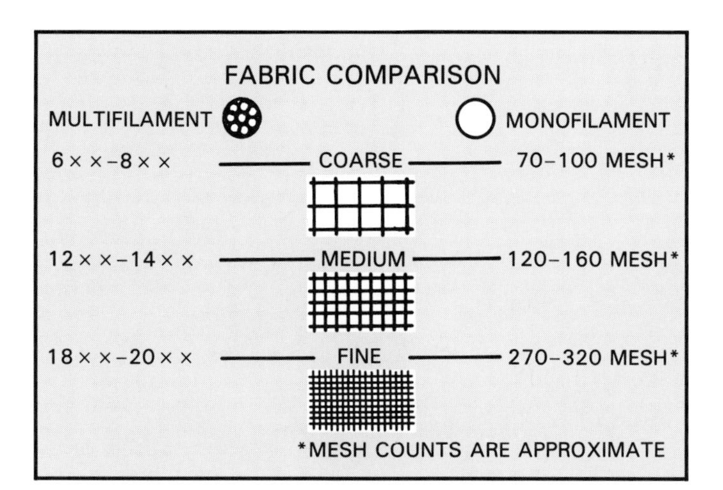

Fig. 24-6. This chart provides a comparison of screen numbers and mesh counts.

STRETCHING THE FABRIC

There are two common methods of attaching fabric to the screen frame. These are the staple method and the cord and groove method. In either method, the fabric should be cut so there is a 1 in. overhang on the frame. After cutting, the fabric is rinsed in warm water to remove the sizing (starch) that keeps the new fabric flat for cutting. The fabric is attached while it is still wet.

Staple method

To use the **staple method**, begin by placing several staples in one corner, Fig. 24-7. Stretch to a near corner and staple along that side. Place the staples at an angle to the screen, about 1/2 in. apart. Pull and staple the center of the opposite side. Pull the fabric diagonally to attach it on each side of center. Pulling straight out, staple the center of a remaining side and staple to each corner. Finish the remaining side in a similar manner.

Fig. 24-7. As shown, staples are being used to attach a screen. Staple tape is being placed over the fabric so the staples can be removed later.

As each side is stapled, a hammer should be used to tap in staples that are not firmly seated.

Two methods can be used to replace the screen. First, the old screen can be ripped or cut off and the old staples hammered flat with the wood. Next, the new screen is restapled in place. The second method involves removing the staples. This is helpful for restretching a loose screen as well. For this method, use short (about 1/4 in.) staples. Staple tape is attached over the fabric as the fabric is stretched. This cloth tape allows staple removal without tearing the fabric.

Cord and groove method

Begin the **cord and groove method** by centering the wet screen on the frame. Starting at one corner, press the cord lightly into the groove on one side. Use a screwdriver or cord tool for this. Continue this procedure for the three remaining sides, Fig. 24-8. Using the cord tool and mallet, tap the cord deeper into the groove around the frame. This is done in stages until the cord is at the

Fig. 24-8. The cord and groove method is being used for stretching a screen.

bottom of the groove. The screen should be very tight. If not, remove the cord with needle-nose pliers or a screwdriver and try again.

PREPARING THE FABRIC FOR PRINTING

The screen fabric should be thoroughly cleaned or degreased when new and then after each use. Any residue in the fabric can create stencil adhering problems.

Using a soft brush, clean the surface with a commercial screen cleaner. Automatic dishwashing powder or trisodium phosphate can also be used for this purpose. For strong cleaners, such as trisodium phosphate, chemical goggles and gloves are necessary.

Some fabrics should be roughened before each use. This is necessary when fibers are smooth, such as monofilaments. A smooth fiber will not hold the stencil.

Roughen or fray the fabric with a fine, silicon carbide grit. Place it on the stencil side of the screen and rub it on the surface with a damp cloth for two minutes. Rinse the screen well and clean the fabric before use.

KNIFE-CUT STENCILS

A **knife-cut stencil** is a stencil on which a knife is used to cut a thin film. The parts that will print are then peeled away from the backing sheet. Knife-cut stencils include the lacquer-soluble type or the water-soluble type. The **lacquer-soluble knife-cut stencils** are attached to the screen with lacquer thinner. The **water-soluble knife-cut stencils** are attached with water. In both cases, the film is simply a layer of gelatin on a backing sheet. The gelatin is cut and removed from the image area to produce the stencil. The basic steps used in producing a knife-cut stencil are shown in Fig. 24-9.

Full-size artwork is used for knife-cut stencils. For the beginner, artwork should not contain fine detail and many lines. Cut the stencil film 3 in. larger than the

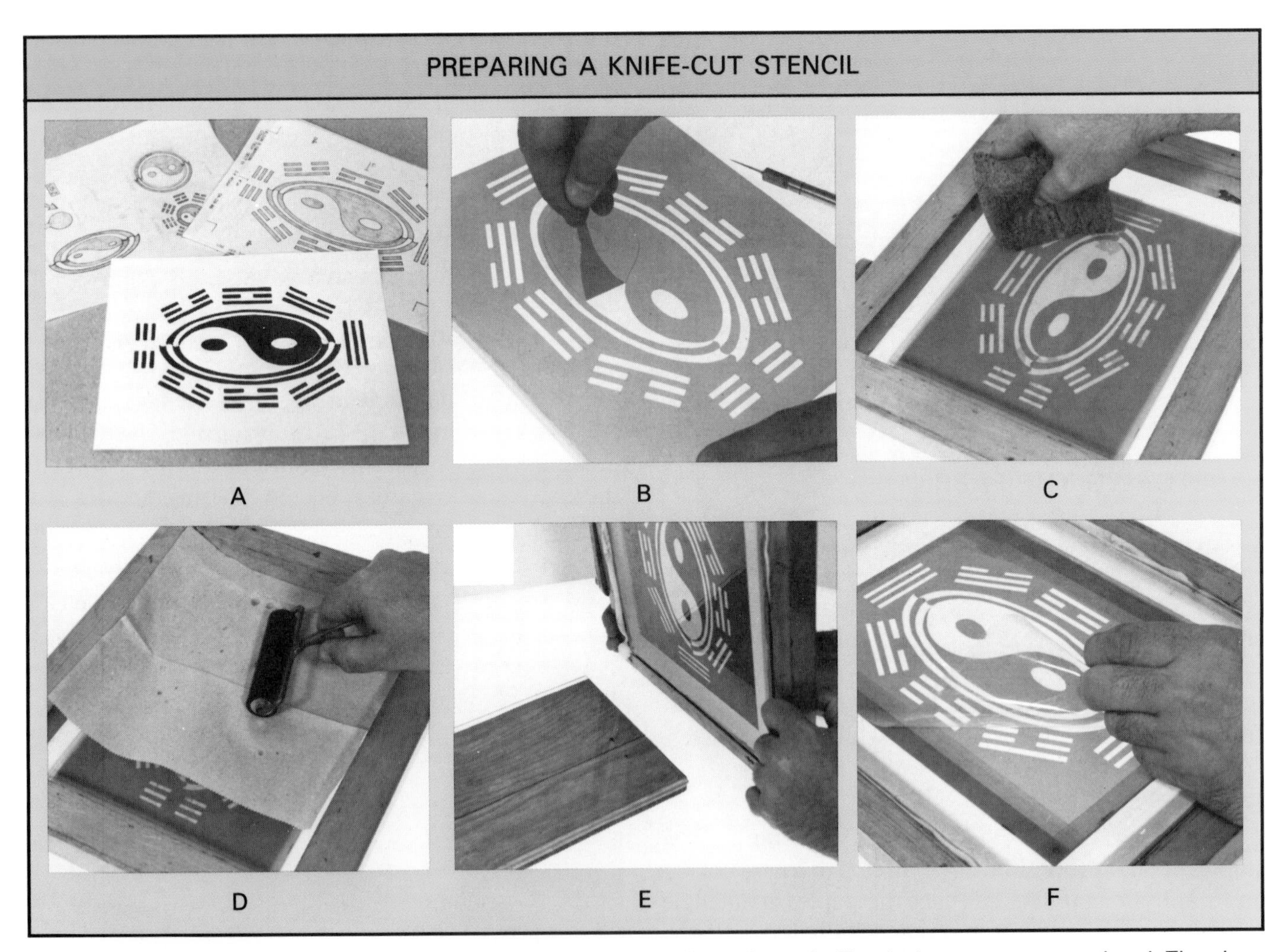

Fig. 24-9. The steps involved in preparing a knife-cut stencil are shown here. A–The design steps are completed. Thumbnail sketches and a rough layout for a poster are shown here. B–After cutting, the emulsion in the image area is carefully removed. C–The damp screen is placed on the stencil and a damp sponge is used to lightly rub over the stencil. D–A paper towel and brayer can be used to remove excess water. E–The screen and stencil are lifted off the buildup and left to air-dry. F–The backing is peeled from the stencil.

design. Using masking tape, fasten the stencil emulsion (dull side) up on the artwork. With a sharp artist knife, trace the edge of the design. Use paper under your hand to keep the stencil clean. Overcut the corners, Fig. 24-10. If too much pressure is used, the knife will cut or groove the backing sheet. This area will collect solvent and melt the cut edge, Fig. 24-11. Start over if you have cut too deeply.

After cutting out the design, use the knife point to help in lifting the film from the image area. Lift the film carefully, recutting small areas that may still be attached.

The stencil is adhered to the bottom of the screen with the correct solvent. Begin by placing the stencil, emulsion up, on a buildup. The buildup is slightly larger than the stencil, but smaller than the inside of the frame. Plywood with glass on top works well. Cardboard can also be used.

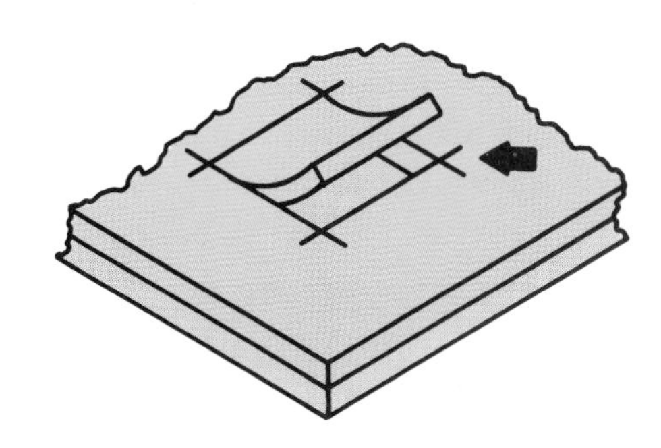

Fig. 24-10. The stencil corners are overcut.

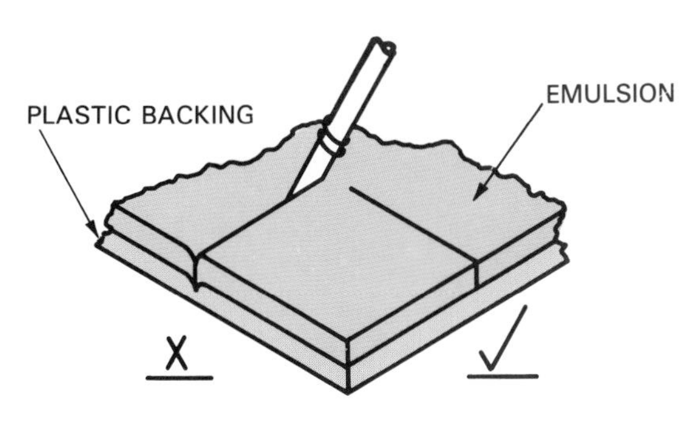

Fig. 24-11. Cutting too deeply may groove the backing sheet.

WATER-SOLUBLE STENCILS

For the water-soluble stencil, place a clean, damp screen on the stencil, which is on the build-up. With a damp sponge, rub over the stencil with long strokes in one direction. Repeat this in an area only if the emulsion is not adhering. Quickly place a paper towel or newsprint over the stencil and remove the excess water. A brayer (small roller) or cloth can be used over the paper. Do this gently so that only excess water is removed. Use more paper as needed. Lift the screen gently off the buildup after two minutes. Place the screen stencil up to air-dry, or place it in front of a fan for faster drying.

LACQUER-SOLUBLE STENCILS

The lacquer-soluble stencil is adhered with a lacquer solvent. Place a clean, dry screen over the stencil. Weights can be used at the frame corners to hold the screen stationary. Saturate a small cotton pad or cloth with solvent. Starting at a corner, apply the solvent in a four-in. square area. Immediately follow with a dry pad to wipe off the excess solvent. Continue until the stencil is adhered. An area that is not adhered will be lighter in color. Go over this area again. After two minutes, remove the screen from the buildup and let it dry for about 10 minutes.

After the stencil is dry, carefully peel off the backing. Start at the corner and pull flat against the stencil. If the stencil pulls away from the screen, stop and place the frame back on the buildup. Adhere and dry the stencil again.

PHOTOGRAPHIC STENCILS

Photographic stencils are light-sensitive. They are exposed through a film positive. This hardens the non-image areas. The image area is washed away after the stencil is developed.

There are three major kinds of photo stencils. These include:

- Direct.
- Indirect.
- Direct-indirect.

The **indirect stencil** has a light-sensitive emulsion coated on a plastic backing sheet. The **direct stencil** uses a wet emulsion that is coated "directly" on the screen before exposure, Fig. 24-12. The **direct-indirect stencil** is a combination of the other photo stencils. A plastic backing sheet is placed against one side of the screen. A wet emulsion is coated on the screen from the opposite side. The indirect method will be explained in detail in this chapter.

Indirect stencil

A transparent positive is used for exposing the indirect stencil. This can be prepared manually or photographically, Fig. 24-13. The photographic method was explained in Chapter 23. Manual methods include masking film, transfer images, and inking. Masking film is placed over the artwork and cut with an artist knife, Fig. 24-14. The emulsion is then stripped away

Fig. 24-12. A photographic emulsion is being coated directly on the screen. (Bowling Green State University, Visual Communication Technology Program)

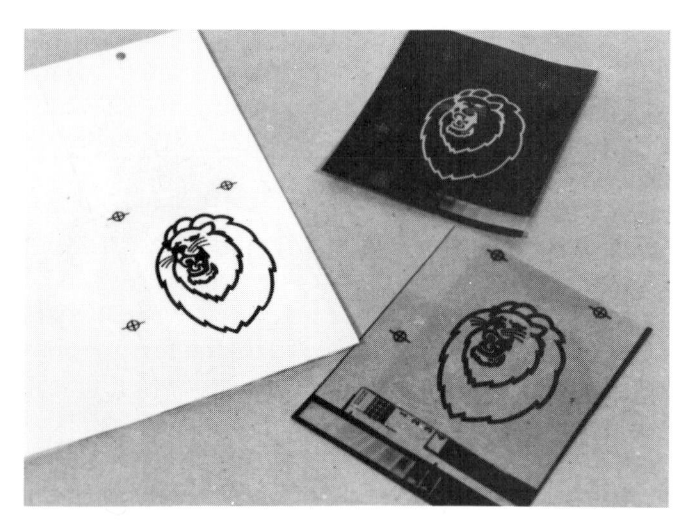

Fig. 24-13. The photographic positive (lower right) was prepared by first making a negative from the pasteup on the left. This negative was then used to make a contact print of the positive.

Fig. 24-14. Masking film is being cut to make a transparent positive.

from the non-image area. Transfer images and inking can be done directly on a plastic sheet. Drafting film works well for inking.

The exposure unit for the stencil should have a light source that is high in ultraviolet light, Fig. 24-15. A #2 photoflood bulb can be used for this purpose. A platemaker is also very good. If an exposure time is not known, use a step test or a commercial step-wedge. Try 1 1/2 minutes for each step for the step test.

In subdued light, cut the indirect film at least 1 in. larger than the image all around. Place the indirect film in the exposure unit, emulsion down. Place the positive in reverse position, over the film. Good contact must be made. Use a sheet of glass or a contact frame. Make the exposure. Place the indirect film, emulsion up, in prepared developer. Agitate the tray gently for the recommended length of time.

After developing the stencil, place it emulsion up, in the sink on a slanted surface. If a slanted surface is not available, the film can be held under the faucet. Using a gentle spray, wash the surface with hot water for the recommended length of time. The image area should be clean, Fig. 24-16. Next, cool the stencil with cold water.

Fig. 24-15. This exposure unit produces light that is high in ultraviolet rays. An exposure is being made. (Bowling Green State University, Visual Communication Technology Program)

Fig. 24-16. Wash out the stencil with a gentle spray of water.

Place the stencil on a buildup and place a clean, damp screen over it. Use paper towels or newsprint to remove the excess water. Gently wipe the paper with a brayer or cloth. Do not use pressure. Repeat with more paper if necessary. Let the frame remain on the buildup for about five minutes and then remove it. After the emulsion is dry, gently remove the backing sheet.

PINHOLES AND BLOCKOUT

After the stencil is made, unwanted open screen areas must be covered. **Pinholes** are small, unwanted image areas on the stencil. **Blockout liquid** or **blockout paper** can be used to cover these areas. With a small paint brush, paint blockout liquid over the pinholes, Fig. 24-17. Apply the liquid to the stencil side of the screen. Make sure it is dry before using the screen.

Fig. 24-17. Pinholes in the stencil can be covered with a small brush. (Bowling Green State University, Visual Communication Technology Program)

Blockout liquid or paper can be used to cover the open areas between the stencil and frame. The liquid blockout is applied to the bottom (stencil side) of the screen with a brush, chipboard paddle, or small emulsion coater, Fig. 24-18. The blockout liquid must be compatible with the type of stencil and ink being used. Dry the blockout with a fan. After the screen is dry, place wide masking tape on the bottom of the frame. Also, tape the inside corners. This will help with cleanup.

A paper blockout or mask can be used when a limited number of copies will be printed. For small frames use 8 1/2 in. x 11 in. paper folded lengthwise. Place one piece at the bottom of the frame, and tape it on the edge nearest the stencil. The tape should be at least 1/2 in. from the image area. Tape two sheets on the sides of the stencil. Next, place a folded sheet at the top. Place masking tape on the paper edges and up each corner.

Fig. 24-18. Blockout liquid is being used to cover the open area between the stencil and frame. (Bowling Green State University, Visual Communication Technology Program)

REGISTRATION

Before printing, the image must be aligned on the substrate. Paper substrates are aligned with at least three guides. Two are placed on a long side, and one on the short side. Commercial guides are available. Chipboard and Z-tabs can also be used. These guides are shown in Fig. 24-19.

For single color prints, place the artwork on the base under the screen. Visually align the artwork with the stencil. Now tape the three guides in position. A more

Fig. 24-19. Three guides are used for the substrate. Paper Z-tabs and chipboard, shown on the right, can also be used.

Fig. 24-20. Print with the squeegee in this manner. (Bowling Green State University, Visual Communication Technology Program)

precise method for multicolor printing will be discussed later.

PRINTING

Before printing, tape two chipboard tabs to the base so they are at the frame corners away from the hinges. This is called **off-contact printing**. When actually printing, the screen only touches the paper where the squeegee is pressed down. Refer again to Fig. 24-5. This technique helps avoid paper sticking to the screen, and slightly stretches the screen for better precision. Mix the ink and place a small bead of ink across the top of the screen. If it is available, spray some adhesive on the base to hold the paper in position. Align the paper inside the three guides and lower the frame.

A square-edged squeegee is used for flat surfaces. Hold it at 45° to horizontal, press down and bring the ink across the image with the squeegee, Fig. 24-20. At the end of the stroke, use the squeegee to scoop up the ink and return it to the top of the frame. Lift the frame and remove the print. If it is printing properly, continue printing. If an unwanted area begins to print, place a small piece of masking tape under the stencil in that area.

Printing with air dry ink should be done rapidly so the ink will not dry on the screen. If the screen does become clogged, place the correct solvent on a rag. Holding a dry rag under the screen, rub the solvent rag on top in the clogged area. The first few prints after this procedure will usually be of poor quality because of the excess solvent on the screen.

As you are printing, the prints should be laid out to dry. Make sure the wet image does not touch another sheet. A drying rack is often used for large quantities of prints, Fig. 24-21.

Fig. 24-21. A drying rack is used to air-dry the prints.

CLEANUP

After printing, remove the excess ink with chipboard. Also remove all paper masking from the frame.

To remove the ink, place the frame on a stack of newspapers. Pour a small amount of ink solvent on the screen and wipe the surface with a rag, Fig. 24-22. Let the solvent soak on the screen briefly. Now wipe over the screen to remove the ink. Continue this procedure until the screen appears open in the image area when held up to the light. Remove the top layers of newspaper as they absorb the ink. Carefully remove the tape from the screen. The screen can now be stored and used again later.

If necessary, the stencil and blockout can be removed after the ink is removed. Lacquer stencils are removed in a similar manner to ink removal. Place the screen on newspaper and pour the correct solvent over the stencil. Let it soak several minutes. Now rub over the surface with a cloth to remove the stencil. Add more solvent as

Fig. 24-22. Ink is being removed from the screen.

necessary. Check the screen by holding it up to the light. Any clogged screen areas must be cleaned again.

Water type stencils use water as the solvent. The screen can be placed on a slanted surface or a sink. Spray hot water on the stencil and let it soak. The stencil will gradually dissolve. Brush and spray the screen until the stencil is removed. A special enzyme powder may also be necessary with some water-soluble stencils.

Once the stencil is removed, the last step is cleaning the screen. Clean it as you did before printing. This removes any ink and stencil residue that might not allow the next stencil to adhere.

PRINTING MULTIPLE COLORS

When printing is to be done in multiple colors, the artwork is prepared differently. Three register marks are used to align the colors when printing. These can be drawn in black ink as small crosses, or commercial register marks can be used.

When colors do not touch, a single pasteup or artwork can be used. Label the colors on a cover sheet. For knife-cut films, prepare a stencil for each color. Be sure to cut and remove the register marks as well. For photographic stencils, make a positive for each color with register marks. If the design is simple, masking film can be used to make each positive manually instead of photographically.

If colors touch or overlap on the artwork, an overlay method is used. The steps involved in producing a two-color screen printing job, with the colors overlapping are shown in Fig. 24-23. Start with the base or key sheet of the pasteup. This normally has the most detail and is a darker color. Add the register marks. Tape a clear plastic overlay by one edge to the key sheet, or use a pin register system. Place the second color image on the

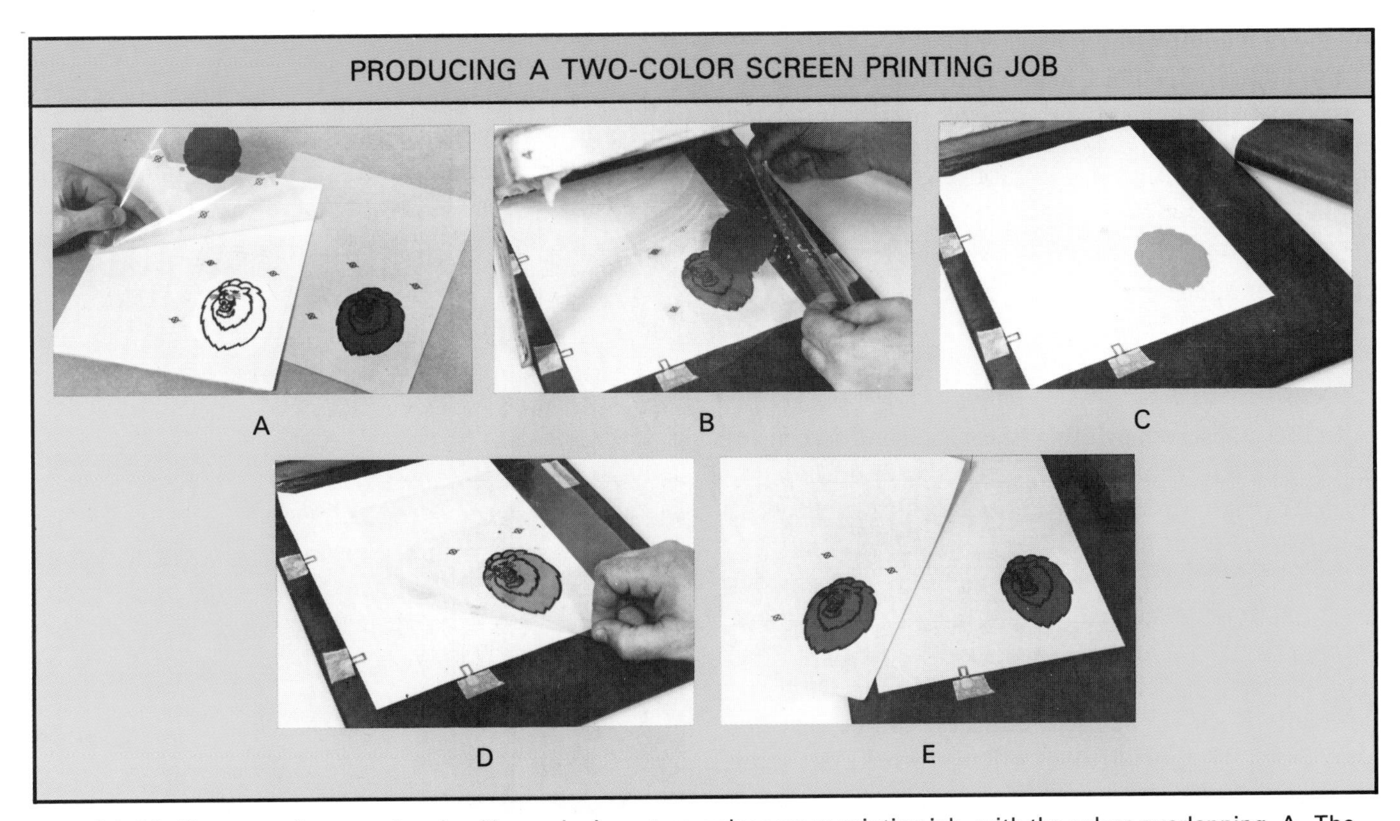

Fig. 24-23. These are the steps involved in producing a two-color screen printing job, with the colors overlapping. A–The pasteup and overlay are shown along with a finished print with register marks. B–The first color is printed on plastic, and the pasteup is registered underneath. C–The first color is printed. D–The second color is registered with the first color. E–The finished print is shown at the right without the register marks.

overlay. Additional colors can be prepared in a similar manner. Add register marks to each overlay in alignment with the key sheet.

For knife-cut films, cut a stencil for each color. Remember to cut out the register marks.

For photographic stencils, first prepare a photographic positive of the base sheet. If the overlays are transparent positives, they can be used without conversion. These are visually prepared with masking film, ink, or transfer images. If necessary, make a photographic positive of the overlays as well. This can be done by sliding white paper between the base sheet and overlay before it is photographed. If no register marks are on the overlay, these can be added by leaving the marks uncovered on the base sheet. The resulting negative or positive will have register marks.

Printing for multiple colors is different than for a single color. Remember to print an overlapping color last. Prepare the printing unit with the stencil and ink. Tape a plastic sheet on the printing base by one edge. Make a print on the plastic. Slide the pasteup under the plastic and align the image using the register marks. Tape the three alignment guides in position as before. Remove the plastic, and print five copies with the register marks. Use adhesive on the printing base so the sheets will not slip while printing. Cover the register marks with masking tape on the bottom. Finish printing the first color without register marks.

For additional colors, make a print on plastic as before. Now slide a print with register marks under the plastic for alignment. Tape the guides in position. Print the next color on the five copies with register marks. Adjust the guides if needed. Now cover the register marks on the screen and finish printing. Continue this process for additional colors.

THERMAL SCREEN PRINTING

In **thermal screen printing**, the stencil and screen are one unit. Heat from a thermal copier is used for exposure. Thermal screens provide a very fast method of doing screen printing.

As with other screen printing methods, the artwork is first designed, keeping in mind the maximum image area for the frame used. A margin of at least 1/2 in. should be left between the artwork and frame. Only line illustrations should be used, but these can be very detailed, Fig. 24-24. Halftones will also work, but some of the dot structure may be lost.

When several images will be used, a pasteup can be made. Tape the image on white paper with removable clear tape.

Fig. 24-24. All of these products were produced by thermal screen printing. (Welsh Products, Inc.)

The thermal process requires that images either have carbon in them, or be converted to a carbon image using a photocopier. Usually, the photocopy method works very well. The copy should be as dark as possible without any carbon in the non-image area.

Cut a thermal screen to size for the screen frame that will be used. Place the illustration faceup between the backing sheet and the screen, Fig. 24-25. The smooth (plastic) side of the film must touch the carbon side of the illustration.

Before exposing the screen, set the thermal copier to the transparency setting. This is normally between the middle and hottest setting. Turn on the copier. With the screen on top, feed the artwork and screen into the thermal copier, Fig. 24-26.

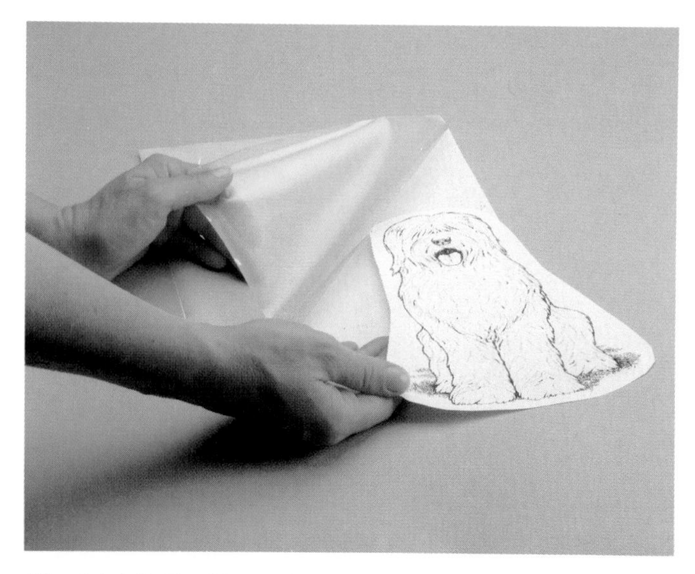

Fig. 24-25. The illustration is placed between the backing sheet and screen. (Welsh Products, Inc.)

When using thermal screens for the first time, the heat setting may have to be adjusted for the best stencil. Check the quality of an imaged screen by checking it for cloudiness in the image area. This is an indication that some of the plastic did not melt away in the image area and more heat will be needed. Check the screen by lifting one corner only and then holding it up to the light. The thermal screen can be sent through the copier a second time if more heat is needed. If the screen tears when separated from the copy, this is an indication that too much heat was used. The heat setting should be reduced for future screens.

When processed, check the screen for unwanted pinholes on a light table. These can be covered with small pieces of transparent tape, or fingernail polish.

A plastic frame is normally used with thermal screens. Place two-sided tape around the inside edge of the frame and then attach the screen to the frame, Fig. 24-27. For off-contact printing, tape the screen to the frame with the plastic side of the film down, just as it was when making your exposure. Keep the screen in this position (plastic side down) for printing.

After the screen is attached to the frame, printing proceeds as with other screen printing methods. If available, a hinged base can be used with the screen to aid in registration. A water-based ink is used with this type of stencil. A water-based textile ink can be used for clothing, Fig. 24-28.

Wash the screen with warm water when done. The screen can be reused. If you are finished with the stencil, it can be removed from the frame. The frame can then be reused.

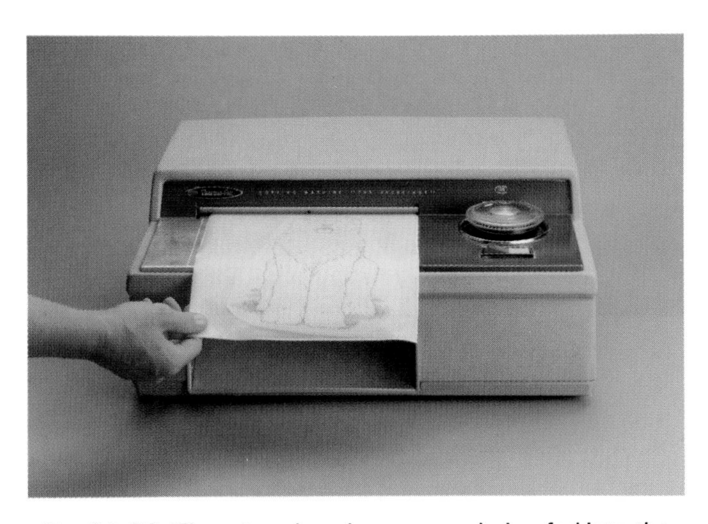

Fig. 24-26. The artwork and screen are being fed into the thermal copier. (Welsh Products, Inc.)

TEXTILE PRINTING

Textiles, such as T-shirts, can be printed directly or by using heat transfers. Both of these methods will be explained.

PRINTING DIRECTLY ON TEXTILES

For printing directly on textiles use a textile ink and compatible stencil. Many textile inks will air-dry, but they need to be heat-set if the print is to be washed, Fig. 24-29. An iron or heat transfer press can be used for heat-setting. Follow the time and temperature directions for the specific ink.

If the cloth to be printed is a single layer, it can be visually aligned on the base and printed. Shirts require a stiff backing, Fig. 24-30. Cut a cardboard or hardboard form that will fit inside the shirt. Adjust the screen hinges for this new thickness. Place the form in the shirt and visually align the shirt on the press. Make the print with a round-edged squeegee. This type of

Fig. 24-27. The screen is being attached to the plastic frame. (Welsh Products, Inc.)

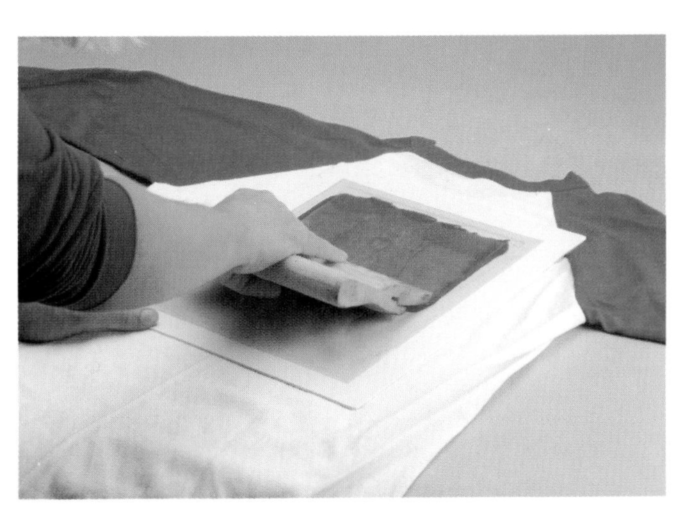

Fig. 24-28. Printing is being done on clothing. (Welsh Products, Inc.)

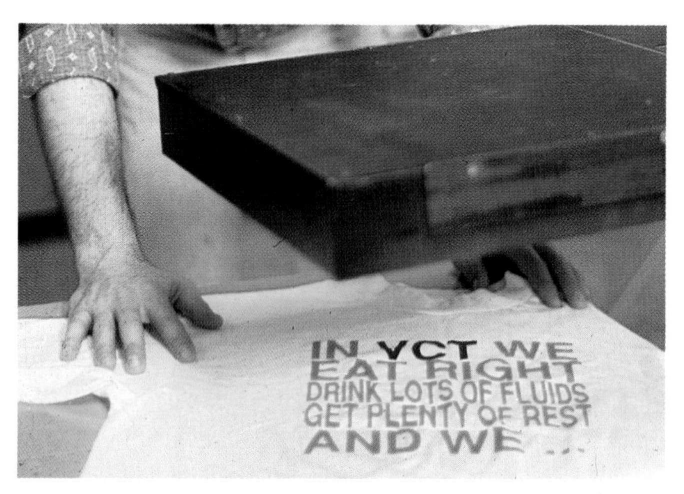

Fig. 24-29. A special dryer is being used to heat-set the ink. (Bowling Green State University, Visual Communication Technology Program)

Fig. 24-31. This golf course flag is being printed on a multicolor printer.

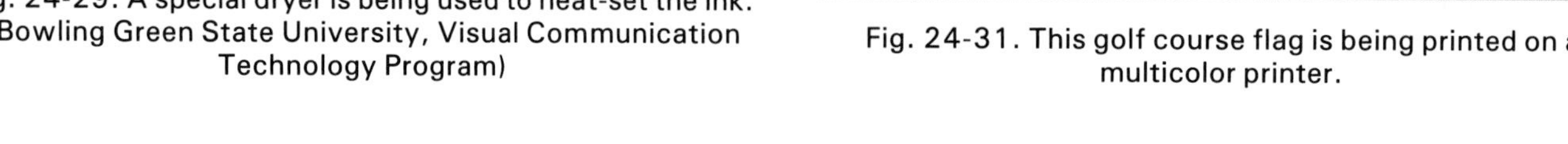

edge gives a heavy ink coverage, which is needed for textiles.

If additional colors are to be printed, these can be visually aligned by looking down through the stencil. For a quantity of shirts or other textiles, it is best to use a special multicolor printer, such as the one shown in Fig. 24-31, or heat transfer. This will provide greater precision and less waste.

HEAT TRANSFERS

Heat transfers are printed on a special paper. When heat and pressure are applied, the print can be transferred to another surface, Fig. 24-32. A special transfer paper or parchment paper is used for this process. The best ink to use is a **plastisol**, which is a type of vinyl ink that will not air-dry. A fairly coarse screen is used, such as an 8XX. Make sure the ink used is compatible with the textile. Plastisol normally works well on cotton and cotton-blend fabrics.

Prepare the stencils so they will print in reverse. For photo stencils, this can be done by placing the positive "right reading" over the stencil. Adhere the stencil and set up the printing unit.

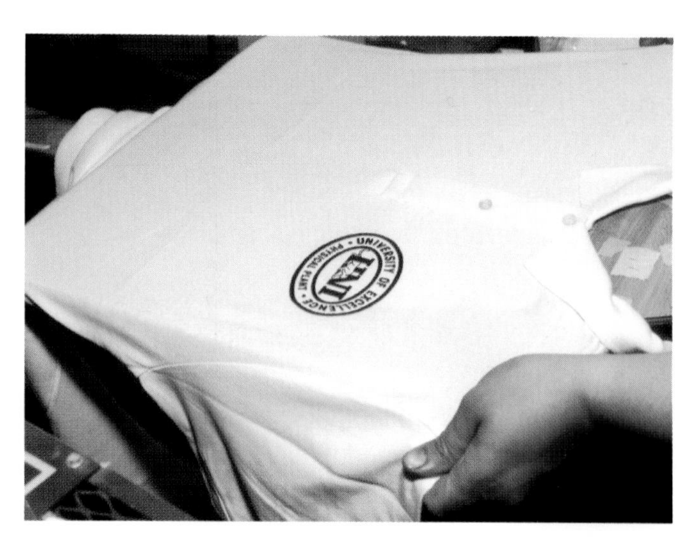

Fig. 24-30. This shirt was screen printed by first placing it on a special base.

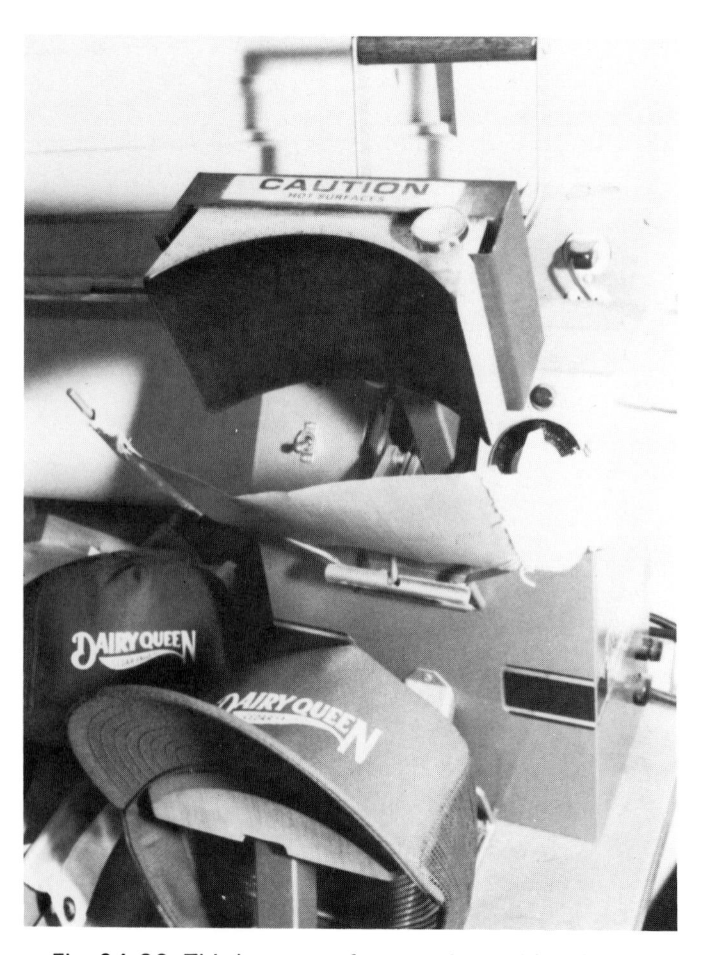

Fig. 24-32. This heat transfer press is used for placing transfers on hats.

Fig. 24-33. A conveyor-type dryer is shown here. (American Screen Printing Equipment Co.)

Print the colors in reverse order. The first color printed on the paper will be the top color on the cloth. Print the first color and partially cure it. Undercuring usually requires around 20 seconds at 300°. A conveyor type dryer is normally used for this, Fig. 24-33. The ink should not be wet after this procedure, but should release from the paper and feel rubbery. If not, change the time or temperature. Print additional colors in this same way.

To transfer the image, use an iron or heat transfer press. Position the transfer on the garment and press for the recommended time and temperature. Let the paper and garment cool completely. Carefully remove the paper.

SUMMARY

Using the techniques described in this chapter can result in good printing with rather inexpensive equipment. Whether you are using a small printing frame or a large industrial screen press, the process is still very similar.

Since a great deal of information has been covered in this chapter, it may be helpful to summarize the steps in screen printing:

1. Prepare the artwork or pasteup.
2. Make the stencil (image carrier).
3. Prepare the printing frame and clean the screen.
4. Adhere the stencil and mask the open area around the stencil with paper or liquid blockout.
5. Apply ink and print with the squeegee.
6. Clean ink from the screen.
7. Remove the stencil, if needed.
8. Degrease the screen with special cleanser.
9. Print additional colors, if needed.

WORDS TO KNOW

All of the following words have been used in this chapter. Do you know their meanings?

blockout liquid
blockout paper
cord and groove method
direct stencil
direct-indirect stencil
heat transfers
indirect stencil
knife-cut stencils
lacquer-soluble knife-cut stencil
mesh count
off-contact printing
photographic stencil
pinholes
plastisol
screen fabric
screen numbers
screen printing
staple method
stencil
thermal screen printing
water-soluble knife-cut stencil

REVIEWING YOUR KNOWLEDGE

Please do not write in this text. Write your answers on a separate sheet.

1. True or false? In screen printing, ink passes through the image carrier.
2. True or false? A screen number of 3 is finer (has more threads per inch) than a screen number of 5.
3. Name two common methods of attaching fabric to a screen frame.
4. List the steps in preparing a knife-cut stencil.
5. Which of the following stencils is light-sensitive?
   a. Photographic stencil.
   b. Lacquer-soluble stencil.
   c. Water-soluble stencil.
   d. None of the above.
6. Name the three types of photographic stencils.
7. True or false? When using blockout liquid to cover pinholes on a stencil, it is applied to the stencil side of the screen.
8. When screen printing, what should you do if an unwanted area begins to print?
9. True or false? With off-contact printing, the screen only touches the pages where the squeegee is pressed down.
10. When producing a two-color screen printing job, with the colors overlapping, the second color is _________ with the first color.
11. True or false? In thermal screen printing, the stencil and screen are two separate units.
12. When making a heat transfer using plastisol ink, you will get the best results when printing on which type of fabrics?
    a. Silk fabrics.
    b. Cotton and cotton-blend fabrics.
    c. Polyester fabrics.
    d. Wool fabrics.

APPLYING YOUR KNOWLEDGE

1. Prepare a display of products printed using the screen printing method.
2. Visit a screen printing company. Prepare a list of some questions prior to the visit. Write a report about your experience.
3. Research screen printing fabric and printing applications. Write a report and present it to the class.
4. Mass-produce screen printing frames for a class.
5. Collect current literature from major screen printing equipment suppliers. Share any new trends with the class.

25

CHAPTER

LITHOGRAPHY

After studying this chapter, you will be able to:
- *Describe the three types of lithographic image carriers.*
- *Compare producing a flat with the emulsion up to producing one with the emulsion down.*
- *Explain how plates are exposed and processed.*
- *Outline the steps in lithographic image transfer, including clean-up.*
- *Identify image problems and possible solutions.*

Printing done from a smooth surface is called **lithography**. It is also known as planography. The process of lithography is possible because oil or grease and water do not mix.

Alois Senefelder discovered lithography around 1795. Using a greasy substance, he drew an image on flat limestone. He then dampened the stone with a roller. Ink was rolled over the surface and adhered to the greasy image area only. Paper was pressed to this surface to make a print. The stone was dampened and inked again for more prints.

Lithography stones are still used today by some artists. However, lithography in industry is done with a thin metal plate. This plate is wrapped on a cylinder for printing. Lithographic presses print thousands of copies per hour using these plates, Fig. 25-1.

LITHOGRAPHIC PLATES (IMAGE CARRIERS)

There are three major categories of lithographic plates. They are:
- Direct image.
- Photo direct.
- Photo indirect (presensitized).

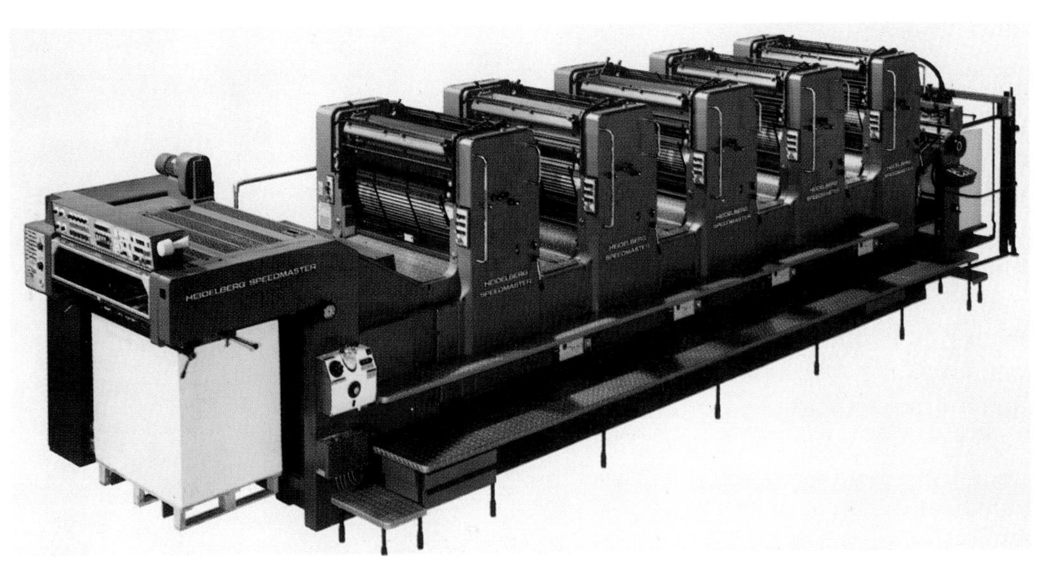

Fig 25-1. This lithographic press is capable of printing five colors. (Heidelberg Eastern, Inc.)

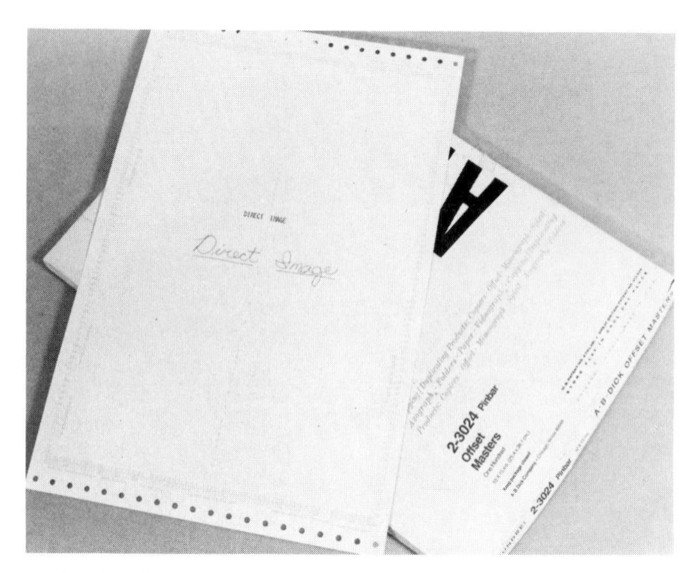

Fig. 25-2. A direct image plate can be imaged in several ways, as shown.

DIRECT IMAGE

The **direct image plate** is most like the lithography stone. The image is manually placed on the surface, Fig. 25-2. This type of plate is for short runs (a small number of copies). These short run plates are also known as **masters**.

Common methods of imaging a direct image plate are grease pencils, pens, and typing. Use a sheet of paper under your hand when imaging this type of plate. If layout lines are needed for image location, a special non-reproducing pencil can be used. Corrections can be made with a soft eraser. Since you can rub through the top surface, rub gently.

When this plate is placed on the press, a special etch, or starter fluid is placed on the surface. This makes the background areas receptive to water.

PHOTO DIRECT

Photo direct plates are made directly from a pasteup. Those types that are used for short runs are sometimes called masters.

Prepare the pasteup on a board or paper that is the same size as the paper to be printed. In addition, make sure the image margins are correct for the press being used.

The electrostatic type of photo direct plate is imaged similarly to paper in a plain paper copier, Fig. 25-3. The plate is given an electrical charge. An exposure to the pasteup releases the charge in the non-image area. An oppositely charged toner is adhered and then fused to the image area. A starter fluid or etch is placed on the plate surface prior to printing. Electrostatic imaging is explored further in Chapter 28.

Fig. 25-3. This is an electrostatic platemaker. (A.B. Dick Co.)

The silver plate works similarly to silver halide film. Exposed areas, after processing, are converted to black metallic silver. After processing, the plate is ready for printing, Fig. 25-4. These plates are sometimes called **silver masters**.

The **diffusion transfer plate** is made by first exposing a negative (donor) sheet to the pasteup. This negative is then sandwiched with a receiver, which is the plate. These are sent through a processor. When they are peeled apart, the image will be transferred to the plate. A fixer is placed on the plate to make the image on the plate more permanent.

Most photo direct plates are produced in special exposure units. The plate material is often dispensed from a large roll. After exposure, needed processing is done automatically.

PRESENSITIZED (PHOTO INDIRECT)

Presensitized (photo indirect) plates require an extra step after the pasteup is made. A flat is produced that

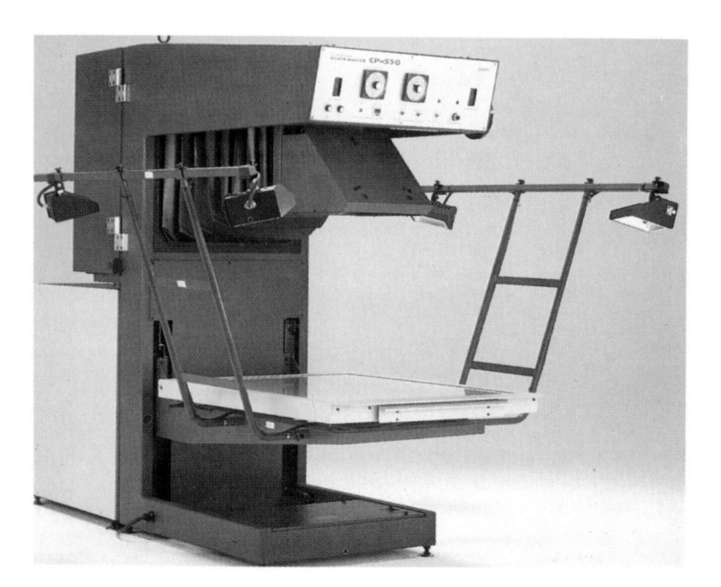

Fig. 25-4. This camera and processor can be used for making silver masters. (Visual Graphics Corp.)

contains a negative of the pasteup. This flat is sandwiched with a light-sensitive plate. Then an exposure is made. The plate is processed with chemicals to create a visible image. Since this technique is the most detailed, it will be covered in a separate section.

PRODUCING A FLAT

A **flat** is a special masking sheet with negatives attached to it. Placing the negatives on the masking sheet is known as **stripping**, Fig. 25-5.

Before a flat can be made, a negative of the pasteup must be produced. Full page pasteups should have inked corner marks that will help later in positioning the negative on the masking sheet. The procedure for making the negative is covered in Chapter 23 of this text.

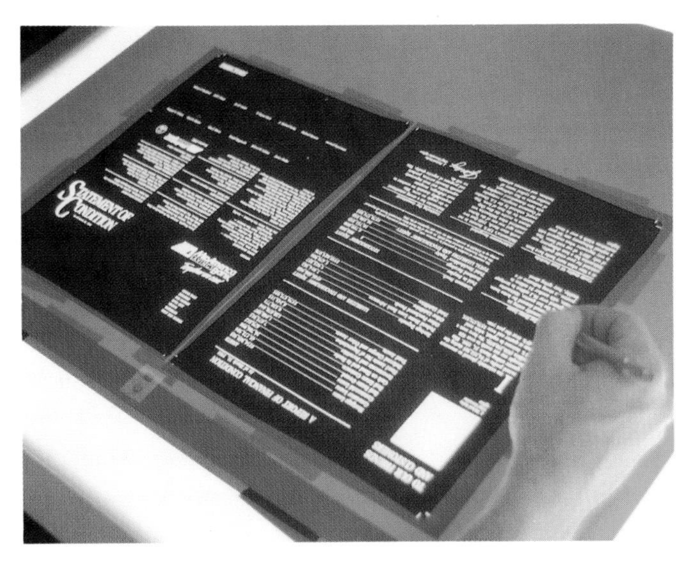

Fig. 25-5. Attaching negatives to a masking sheet is known as stripping.

MASKING SHEETS

Masking sheets are usually sheets of goldenrod paper or plastic used to block light that would expose the non-image areas of the plate. Masking sheets come in different sizes for different presses. They are available as preprinted sheets or plain sheets.

Preprinted masking sheets

Attach preprinted masking sheets horizontally to the light table, Fig. 25-6. Using a T-square, align the sheet with any printed line. This is more accurate than the sheet edge. Tape the corners with masking tape. Using a T-square and triangle, draw a rectangle for the press sheet size. Start at the line marked for the top of your paper. Center the press sheet size on the masking sheet. The negative can now be attached to the masking sheet.

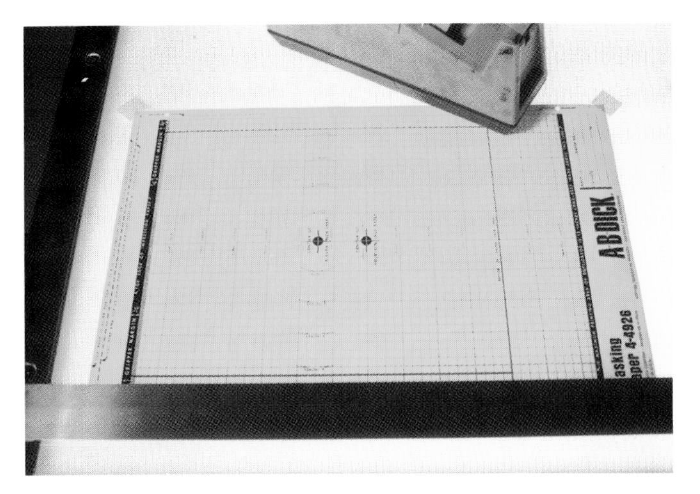

Fig. 25-6. The masking sheet is taped down, and a rectangle has been drawn for the press sheet size.

Plain masking sheets

Align plain masking sheets to the sheet on the light table using the edge of the sheet, Fig. 25-7. Tape the corners. The lines to be drawn are shown in Fig. 25-8. Begin by drawing the lead edge of the paper. The area above the lead edge is known as the **plate bend margin**. This is where the plate will be attached to the plate cylinder. Draw a **gripper margin** next to the lead edge. No image can be placed in this area. This is where the press grippers pull the paper into the press for printing. From the lead edge, measure down and draw the **trailing** or bottom paper edge. Draw a center line down the width of the masking sheet. Measuring out from this line, draw the lines for the sides of the press sheet. The sheet is now ready.

ATTACHING NEGATIVES

The easiest method for attaching negatives is right-reading or emulsion down. Untape the bottom edges of

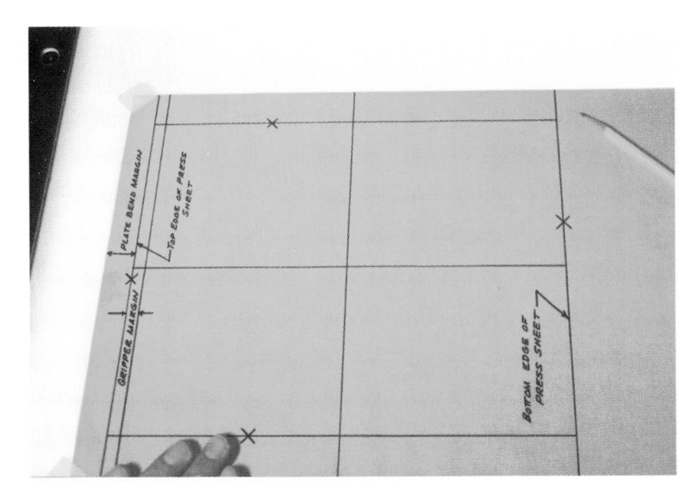

Fig. 25-7. A plain masking sheet is shown with lines completed.

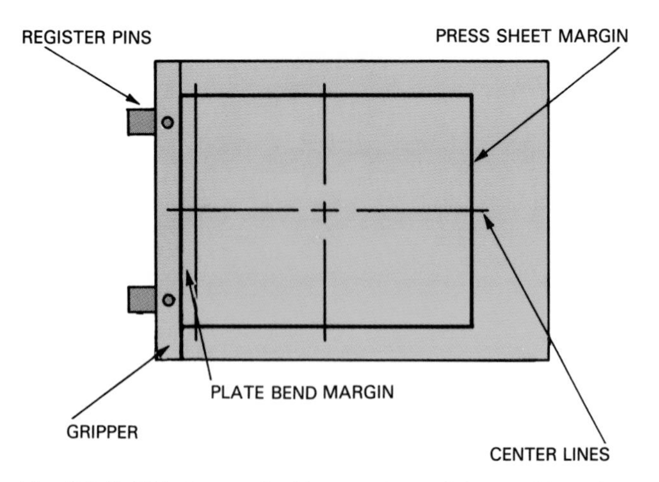

Fig. 25-8. This is a typical layout for a plain masking sheet.

the masking sheet. Place the negative, right-reading, under the sheet. Align the corner marks on the negative with the press sheet edge on the mask. When smaller negatives are combined for a page, corner marks are not available. Draw location lines on the mask for each negative.

Cut two small triangles in the mask, over the non-image area of the negative. Place red lithographic tape over these to attach the negative temporarily. Flip the mask and tape around the negative with transparent tape.

Flip the masking sheet back over and cut windows about 1/8 in. from the image area. Remove the windows. On the light table, opaque any unwanted images (pinholes) on the window side of the flat, Fig. 25-9. A cotton swab and water can be used to remove opaque that accidentally covers an image.

Halftones

An easy method of adding halftones to a negative is to make a clear window on the negative, Fig. 25-10. This window is produced by attaching a black or red rectangle in position on the pasteup. The halftone should be slightly larger than the window. It is taped in position on the emulsion side of the line negative using transparent tape. Be careful not to trap dust between the film layers. Also, make sure that other images are far enough from the window to allow for taping.

STRIPPING WITH EMULSION UP

In industry, flats are often stripped emulsion-up. The masking sheet is lined in the usual manner. The negative is then placed in position *over* the sheet, with the emulsion up or wrong-reading, Fig. 25-11. The negative is taped in place. This technique saves some steps and is fast. However, it is more difficult to align negatives with this method because guidelines are covered by

Fig. 25-9. The negative is opaqued after the windows are removed.

Fig. 25-10. A halftone is taped in position over a clear window on the negative, which is produced by the black paper on the paste-up at left.

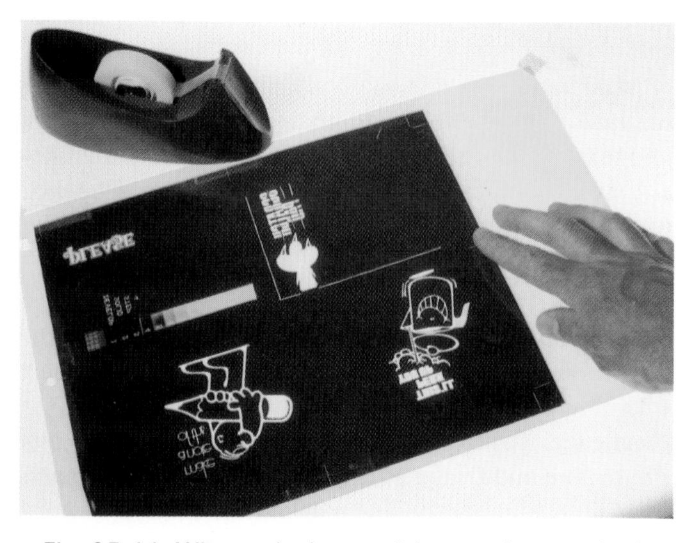
Fig. 25-11. When stripping emulsion up, the negative is taped in position upside down.

the negative. Full page negatives with corner marks are easy to align with either method.

FLATS FOR MULTIPLE COLORS

When printing is done in multiple colors, the colors must be in register, or in exact position. First, three register marks are added to the pasteup and overlays, Fig. 25-12. Second, a pin register system is normally used to ensure exact positioning of the flats and plates.

For pin register, set the punch for the width of the masking sheet. Punch all the masking sheets needed for the job. Place register pins through the holes. Then place the sheet on a light table. Using a T-square and triangle, align the masking sheet using the register pins, Fig. 25-13. Tape the pins and the edges of the masking sheet. Continue producing the flat in the normal manner. Trim the negative if it covers the register pins.

Masking technique

This technique is used when colors do not touch. Prepare a single flat, using the pin register system. While it is still attached to the light table, place a second masking sheet on the pins. Outline images for a single color and the register marks. Remove the masking sheet. Then cut out the windows and label the color. Masks for additional colors are made in the same way. The flat and one of the masks are used to produce a plate for printing in one color.

Fig. 25-12. Three register marks are added to the paste-up when printing more than one color.

Multiple flats

When colors touch or overlap, multiple flats must be made. A pasteup with overlays is used for this technique. Refer again to Fig. 25-12. Make the negatives. Prepare the master flat from the base pasteup. This should have the most detail. Place another masking sheet on the pins. Align a second negative with this masking sheet using the register marks. Tape the negative in place and label the color. Continue this process for additional colors. When finished, cut the windows and complete the flats as usual. Each flat will be used for a separate printing plate, Fig. 25-14.

PLATEMAKING (PRESENSITIZED PLATES)

Presensitized plates are additive or subtractive. **Additive plates** require chemicals that build up the image area. **Subtractive plates** already have this coating. Processing removes this coating from the non-image areas.

Plates are exposed with an ultraviolet light source. A special platemaking unit, such as the one in Fig. 25-15, or a #2 photoflood bulb is used. If the exposure time is not known, a step test can be made. Another option is to strip a film sensitivity guide on the flat, Fig. 25-16. Find the correct exposure according to the directions with the guide.

In subdued light, place the plate emulsion side up on the exposure unit. For pin register, punch the plate edge and attach register pins. Place the flat, right-reading

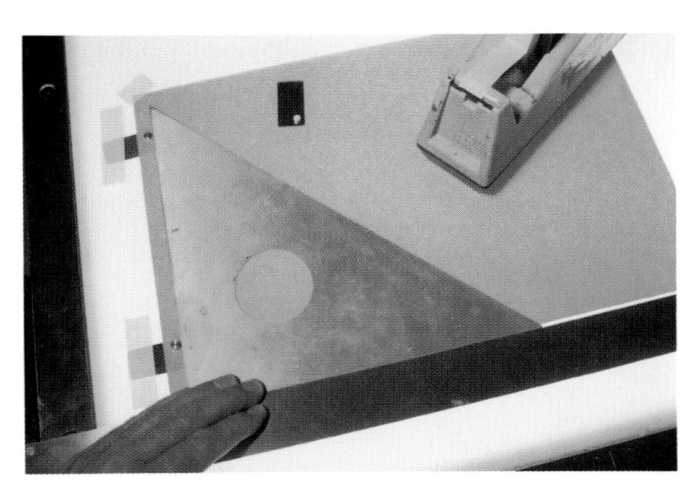

Fig. 25-13. Align the register pins with the triangle. Then tape the pins and sheet in place.

Fig. 25-14. Since the colors on this product overlap, two flats are made.

Fig. 25-15. After placing materials in this exposure unit, the top is flipped for exposure. (nuArc Co. Inc.)

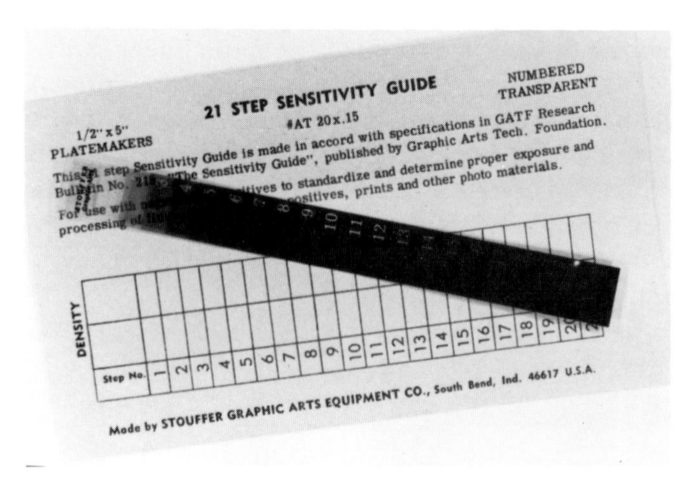

Fig. 25-16. A sensitivity guide for determining correct plate exposure is shown here.

over the plate, Fig. 25-17. Align and center the top edge of the flat and plate. Attach them with masking tape or pins at the top edge. Clean the cover of the exposure unit, if needed, and close it. If available, turn on the vacuum. Set the exposure time and make the exposure. *Never* look at the exposure light.

After the exposure, place the plate in a developing sink or on a newspaper. Process the plate according to directions with the necessary chemicals. Automatic plate processors can be used that apply the chemicals automatically, Fig. 25-18.

Fig. 25-17. The flat is placed over the plate on the exposure unit.

PROCESSING ADDITIVE PLATES

Using a clean, damp sponge, add desensitizing liquid and rub this on the plate surface in a circular motion. This removes the light-sensitive coating. Add a small amount of developer outside the image area and rub this on with another clean, damp sponge. Rub until a dark, uniform image is obtained. The developer makes a more durable plate image. Rinse the plate and place it on edge to dry. If it is not to be used right away, apply a gum arabic solution to the surface with a cotton pad. Buff it almost dry with the other side of the pad, Fig. 25-19. This protects the metal surface of the plate. Place the plate in a large envelope for storage.

PLATE DELETIONS

Small deletions can be made with a soft pencil eraser that is wet with dampening solution used on the press. Rub gently. Larger areas, such as register marks, require the use of a deletion fluid on the eraser. After

Fig. 25-18. The plate shown has just been processed with this automatic plate processor.

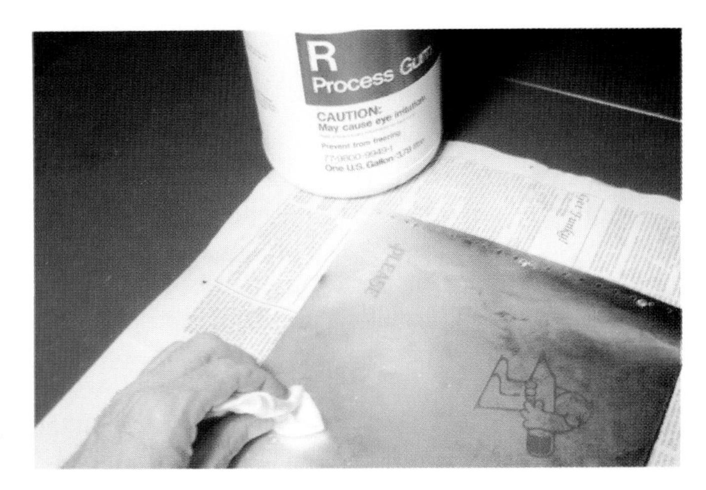

Fig. 25-19. A plate preservative is being applied to this plate.

deleting the image, the fluid is removed from the plate with a cotton pad and dampening solution. When there are small scratches over the entire plate, scratch remover is used. This will not harm images if used according to directions.

IMAGE TRANSFER

For better understanding, let's divide the lithographic press into four systems, Fig. 25-20. (A registration system will also be explained later.) These systems include:

- Feeding.
- Cylinder.
- Ink and dampening.
- Delivery.

The **feeding system** is where the paper is placed for entering the press. There are two types of feeding systems. These are web-fed and sheet-fed. A feeding system that uses a roll of paper is called **web-fed**. With **sheet-fed** presses, sheets are removed from the feed tray for printing with a friction wheel or vacuum type sucker feet. Sheet-fed presses are normally used in schools.

The cylinder system contains a plate, blanket, and impression cylinder, Fig. 25-21. On a three-cylinder press, the **plate cylinder** holds the plate. The **blanket cylinder** contacts the plate and reverses the image. The **impression cylinder** has grippers which pull the paper between it and the blanket. The paper receives the image from the blanket. The terms **offset** and **offset lithography** are sometimes used for this type of printing. This is because the image is "offset" to the blanket before printing occurs.

The **ink and dampening system** applies ink and dampening fluid to the plate. On combined or integrated systems, the ink and fluid are applied to the plate from the same form roller. Conventional or separate systems apply the ink and fluid on separate form rollers. In either system, there is a separate fountain and **fountain roller** for ink and dampening fluid. A **ductor roller** moves back and forth to the fountain roller. This furnishes ink and fluid to the distribution rollers. **Form rollers** deliver the ink and dampening fluid to the plate when printing. The dampening fluid is mixed to a specific pH for the plate. On a pH scale of O (very acid) to 14 (very alkaline), a mild acid 4.5 to 5.5 is normally used.

The **delivery system** stacks the paper after printing. In **tray delivery systems**, the paper falls in a tray. Chain delivery systems have grippers on a chain that carry the paper and drop it on the paper stack. This technique helps avoid **ink set-off**. Ink set-off results when ink from one sheet is transferred to the back of the next sheet.

Smaller lithographic presses are sometimes called **duplicators**, Fig. 25-22. Duplicators have a small plate size, fewer ink and dampening rollers, and are sheet-fed.

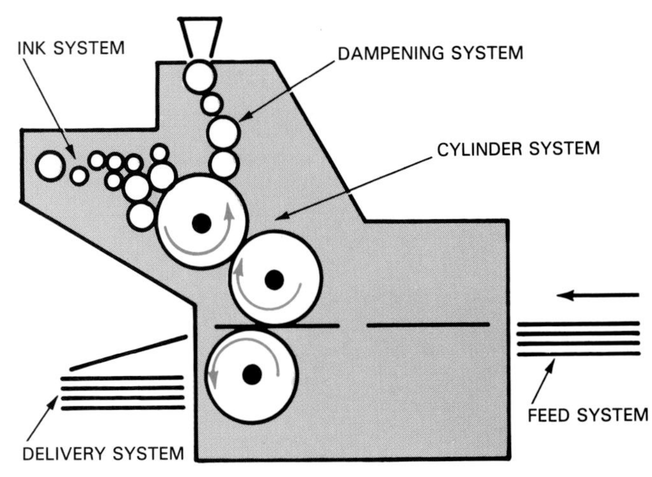

Fig. 25-20. Systems of a lithographic press are shown here.

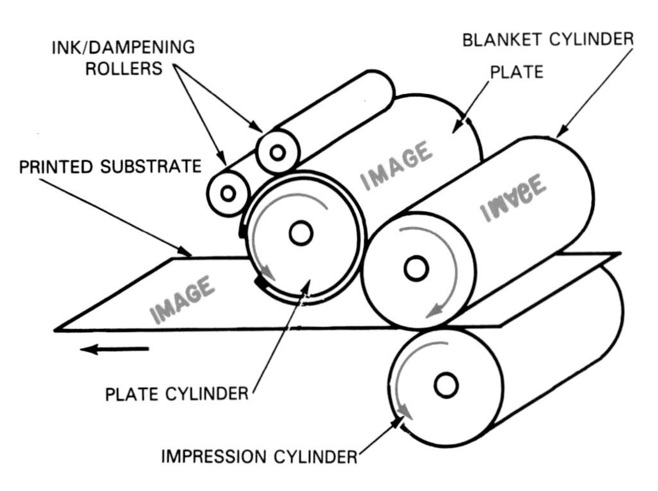

Fig. 25-21. This is a typical cylinder system on a lithographic press.

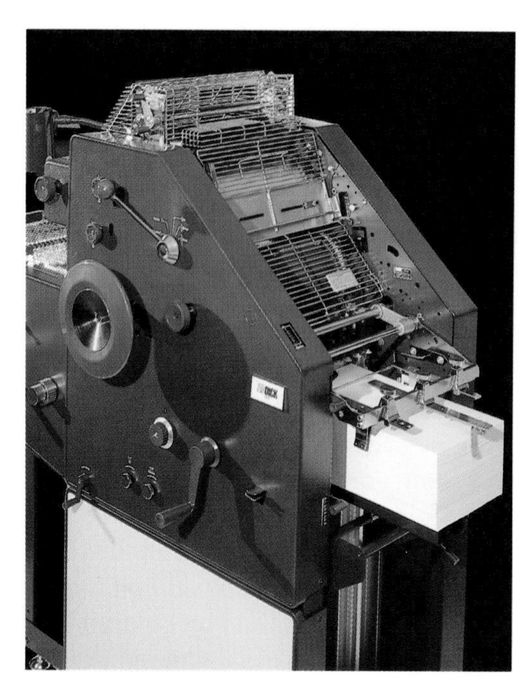

Fig. 25-22. A small lithographic press is sometimes called a duplicator. (A.B. Dick Co.)

PRESS SETUP

Twelve major steps are involved in setting up a lithographic press. These steps are outlined here:

1. Add ink to the fountain. Turn the handwheel counterclockwise to bring the fountain and ductor roller in contact. Turn the fountain roller counterclockwise. Starting in the center, adjust the fountain screws for an even, "orange peel" effect on the roller. See Fig. 25-23. Run the press to distribute the ink. Turn the ink off until you are ready to print.
2. Add the dampening solution. Adjust the dampening control to a medium position or the recommended setting. Cloth-covered (molleton) rollers must be pre-moistened. Use the handwheel to bring the fountain and ductor roller in contact. Turn the fountain roller until the ductor is damp, Fig. 25-24. With the control in neutral, turn the press on and let the moisture distribute to other cloth-covered rollers. Turn the press off and check dampness with the back of the fingers. The rollers should be damp but not dripping wet.
3. Lower the feed tray and insert jogged paper. See Fig. 25-25 and Fig. 25-26. Setup or make-ready paper is usually placed on top of the stack. Adjust paper guides so the paper is centered. Turn on the press and let the stack rise to the top position.
4. On presses without a register board, feed a sheet through the press with the handwheel. Adjust the delivery tray sides within 1/4 in. of the paper. Also, adjust the end stop for paper length, Fig. 25-27. Adjust the jogging guides on chain delivery systems.
5. Attach the plate by the lead edge to the plate cylinder. Holding the center of the trailing edge, roll the

Fig. 25-24. The molleton rollers are being pre-moistened.

Fig. 25-23. The ink fountain screws are adjusted so that an even layer of ink coats the fountain roller.

Fig. 25-25. Paper should be jogged before it is placed in the feed tray. This is the process of getting air between sheets so that the stack of paper will be straight.

Fig. 25-26. The paper is placed in the feed tray.

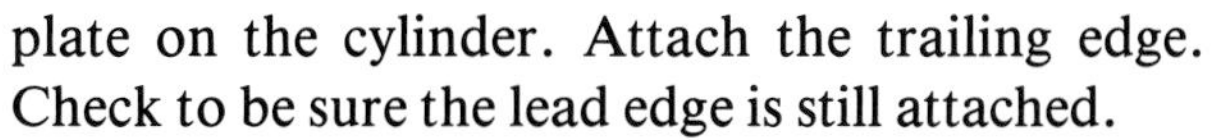

plate on the cylinder. Attach the trailing edge. Check to be sure the lead edge is still attached.

6. Etch the plate with dampening solution or the recommended starter solution. Apply the fluid with a cotton pad in a back and forth motion as the cylinder is turned counterclockwise, Fig. 25-28. The plate should be re-etched with dampening fluid when the machine is turned off long enough for the plate to dry.
7. Turn on the press. Apply dampening solution and ink to the plate, return to neutral, and turn off the press. Inspect the plate. The non-image areas should be damp, but not dripping. Ink should only be adhering to the image areas. If necessary, make adjustments until the ink and dampening system are roughly in balance.
8. Print a few copies. Check for image position first, not quality. The image on the press sheet can be adjusted horizontally, vertically, and diagonally, Fig. 25-29. Most adjustments should be made with the press off. Continue printing and checking

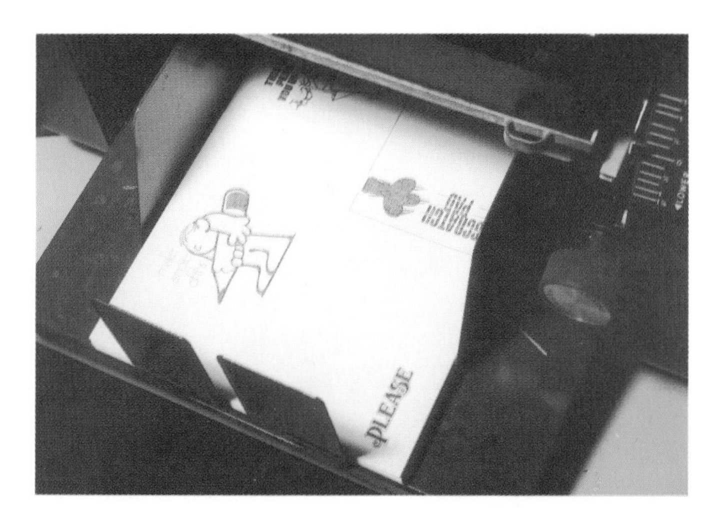

Fig. 25-27. This delivery tray is adjusted properly.

Fig. 25-28. The plate is being etched with dampening solution.

Fig. 25-29. If the image is uphill or downhill, a diagonal adjustment should be made.

sheets until the image is aligned on the sheet. Some adjustments will require that the blanket be cleaned.

9. Turn the ink to the recommended setting. Print some sheets to check for balance. The goal is to achieve a dense, sharp ink image with a clear background. Keep ink at a minimum and try to adjust the dampening solution to achieve balance. After an adjustment, sheets will have to be printed until the change goes through the system. The sheets that are used in achieving balance are called the **spoilage allowance**. Figure about a 10 percent spoilage allowance when deciding how many sheets to purchase.
10. Set the counter and print the desired number of copies. Watch the feed tray. Stop the paper feed if poor image quality occurs. Make any corrections and continue printing.
11. Remove the printed sheets carefully and place them in a drying rack. Pressing the sheets can cause ink set-off.
12. Remove and preserve the plate if it is to be reused. Clean the blanket with a rag and ink solvent, Fig. 25-30. Dry it with a clean rag.

Fig. 25-30. Clean the blanket with the appropriate ink solvent.

Fig. 25-32. These vacuum tubes lift the paper into the register system.

SETTING UP A REGISTER SYSTEM

Some presses have register boards that must be set up along with the feeding system. Although register systems may vary, the following general procedure should be followed:

1. Place the paper in the feed tray slightly off-center (about 1/8 in.) to the jogger side. Adjust the guides in the feed tray.
2. Set the double sheet detector with a folded press sheet, Fig. 25-31. Adjust it so a single sheet will pass and a double sheet will be deflected into a tray.
3. Move the side guides or the register board to the outer-most edges.
4. Turn off the blower and vacuum and turn on the machine. Turn on the blower until the paper floats up to the sheet separators. Now turn up the vacuum until the sheets feed steadily through the press, Fig. 25-32. Turn off the press with a sheet just entering the cylinder system.

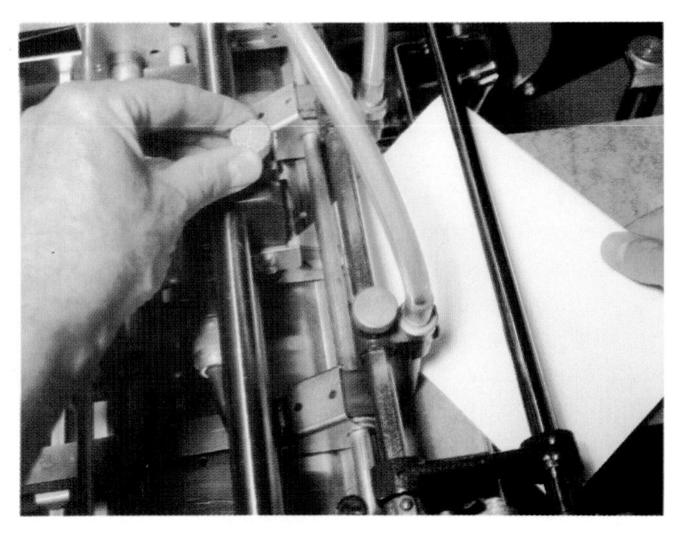

Fig. 25-31. Adjust the double sheet detector as shown.

5. Turn the handwheel until the jogger guide is in to the paper. Adjust it to the sheet edge, Fig. 25-33. Adjust the spring guide with slight (1/64 in.) tension against the other side.
6. Continue with the press setup and print some trial copies. Small adjustments can be made with the jogger. For larger adjustments, the paper stack will need to be moved. Make sure the jogger is always contacting the paper when printing.

TROUBLESHOOTING

Sometimes an image problem occurs on a printed sheet. These include:

- Gray or washed out copy.
- Scumming.
- Hickies.
- Ghosting.

Fig. 25-33. The jogger guide is being adjusted to the sheet edge.

As you are troubleshooting, you may come across some of these problems. There are some general solutions that can be used to remedy them.

GRAY OR WASHED OUT COPY

Gray or washed out copy is often the result of too little ink or too much moisture. Try changing the dampening adjustment first. If cloth-covered rollers are too damp, turn off the ductor roller and let the press run with the dampening form roller on the plate. This will cause the moisture to evaporate. If these adjustments do not help, the problem may be glazed rollers or incorrect pressure between cylinders.

SCUMMING

Scumming is ink that prints in the non-image areas. Check for over-inking first. If the rollers "hiss" and are highly textured with ink, the press is over-inked. Paper can be rolled between the rollers to remove the excess ink. Another problem may be too little dampening solution. Check and readjust the dampening system. Dirty cloth roller covers can also cause scumming. Clean or replace them if necessary.

HICKIES

Hickies are foreign particles on the plate or blanket. Clean the plate and blanket. Also, check to be sure a press sheet is not caught in the ink rollers. If so, it will have to be removed or the press will have to be cleaned.

GHOSTING

Ghosting is a faded image area that normally follows an area that requires a lot of ink. Small presses often have form rollers that cannot supply enough ink for large solids. The ghost occurs when the form roller cannot recover the ink lost on a solid. It is called a ghost because the faded area is usually the same pattern as the solid. It may not be possible to do anything about ghosting. Try to avoid ghosting by not lining up solids on a pasteup. Also, be aware of the press capability concerning solids.

CLEANING THE PRESS

The press should be cleaned daily if possible. If you decide to leave the press uncleaned overnight, be sure the ink is a type that will not dry on the rollers. Also, cover the ink system with plastic to keep the ink clean and moist.

The following general procedure is followed in cleaning up a press:

1. Drain the fountain solution and rinse the fountain tray. Remove the cloth-covered rollers from the system.
2. Remove the ink fountain and remove the ink with an ink knife. Clean it with solvent. Remove excess ink on the fountain roller with a piece of chipboard. With the ductor and fountain roller separated, clean the ink fountain roller with a rag and solvent. Turn off the ink control.
3. If the press has an automatic cleanup feature, turn on the press, apply solvent, and squeegee it off. Continue this until the ink rollers are clean. Clean the squeegee edge with a rag and solvent.
4. Attach a cleanup sheet or mat to the plate cylinder and turn on the press. Apply solvent to the ink rollers, wait five seconds, and move to the inking position. The ink will transfer to the mat. Continue with more mats until the rollers are clean.
5. Wipe the ink rollers with a cloth and solvent. Clean the plate cylinder and blanket.
6. Place the rollers in the night latch position, Fig. 25-34. This separates the rollers so flat spots are not formed.
7. Clean up around the machine and cover it.

SUMMARY

This chapter has provided the basic information necessary to make a printed product using a lithographic press. The steps covered included stripping a flat, making a plate, and printing. As you may recall, this modern, industrial process had its beginning when Senefelder discovered that printing could be done from a flat stone surface.

Fig. 25-34. Place the press rollers in night latch position when they will not be used.

WORDS TO KNOW

All of the following words have been used in this chapter. Do you know their meanings?

additive plates
blanket cylinder
delivery system
diffusion transfer plate
direct image plate
ductor roller
duplicators
feeding system
flat
form rollers
fountain roller
ghosting
gray or washed out copy
gripper margin
hickies
impression cylinder
ink and dampening system
ink set-off
lithography
masking sheets
masters
offset
offset lithography
plate cylinder
photo direct plates
plate bend margin
presensitized (photo indirect) plates
scumming
sheet-fed
silver masters
spoilage allowance
stripping
subtractive plates
trailing
tray delivery system
web-fed

REVIEWING YOUR KNOWLEDGE

Please do not write in this text. Write your answers on a separate sheet.

1. True or false? Lithography is based on the principle that oil and water don't mix.
2. Name the three categories of lithographic plates.
3. The easiest method for attaching right-reading negatives is:
   a. Over the mask, emulsion down.
   b. Under the mask, emulsion down.
   c. Over the mask, emulsion up.
   d. Under the mask, emulsion down.
4. When colors touch or overlap, what is needed?
   a. One flat with masks.
   b. Direct image plates.
   c. Multiple flats.
   d. None of the above.
5. True or false? Subtractive plates require chemicals that build up the image area.
6. On a three-cylinder press, which cylinder contacts the plate and reverses the image?
   a. Plate cylinder.
   b. Blanket cylinder.
   c. Impression cylinder.
   d. None of the above.
7. When deciding how many sheets to purchase, about what percent should be allowed for spoilage allowance?
   a. 5.
   b. 10.
   c. 20.
   d. 40.
8. Gray or washed out copy is often the result of:
   a. Too little ink.
   b. Too much moisture.
   c. Not enough moisture
   d. Either a or b.
9. Foreign particles on the plate or blanket are referred to as:
   a. Scumming.
   b. Ghosting.
   c. Hickies.
   d. None of the above.
10. If possible the press should be cleaned:
    a. Daily.
    b. Weekly.
    c. Monthly.
    d. Yearly.

APPLYING YOUR KNOWLEDGE

1. Using the Dictionary of Occupational Titles, or other source, investigate careers available in lithographic printing.
2. Investigate the history of lithography. Prepare a written report.
3. Visit a printing firm that uses lithographic equipment. Ask questions about the operation of the equipment and prepare a report.
4. Investigate and discuss the future of lithographic technology.
5. Start a paper recycling project at school. Produce notepads from the paper. Using lithography, print conservation messages on the notepads and distribute them.

26

CHAPTER

RELIEF PRINTING

After studying this chapter, you will be able to:
- *Compare various types of relief printing.*
- *List the major uses of relief printing.*
- *Outline the steps in producing an image carrier when using foundry type.*
- *Discuss the steps in producing a rubber stamp.*
- *Explain how hot stamping is done.*

Relief printing is printing from a raised surface, Fig. 26-1. The raised surface is inked and pressed against the substrate to be printed.

The first books were written by hand or printed by the relief method using a carved wooden printing surface. Since handwriting and carving a wooden printing surface took a great deal of time, books were costly and only available to the wealthy. Then, in about 1450, a German printer, Johann Gutenberg, developed an invention that revolutionized the relief printing method. He invented a movable metal type to replace the carved image carriers. This type could be used over and over again. He also invented a high-speed printing press. This faster and lower cost printing method eventually made books accessible to all people.

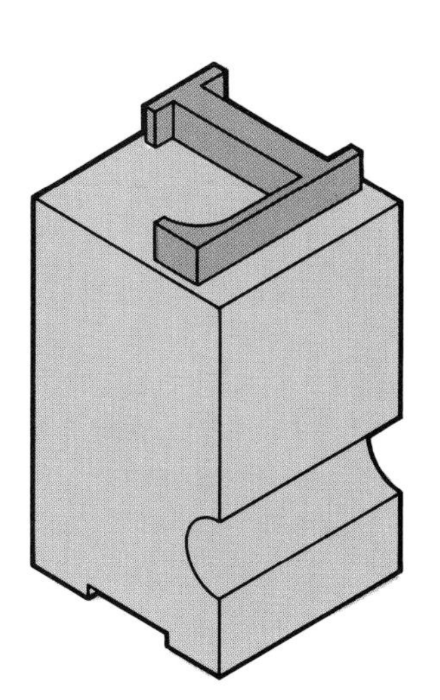

Fig. 26-1. Printing from a raised surface is known as relief printing. (John Walker)

Relief printing was the first industrial printing method used in America. Ben Franklin, called the father of American printing, used this method. Relief printing remained as the major printing method in the United States until about 30 years ago. At that time, the other methods, such as lithography, began to increase in use.

The major uses for relief printing today are package printing, imprinting, and finishing operations. **Imprinting** is done to personalize already printed material. **Finishing operations** on relief equipment include scoring, perforating, die-cutting, and foil stamping. Catalogs and some newspapers are still printed using the relief method.

FLEXOGRAPHY

Flexography is the most popular form of relief printing today, and one of the fastest growing printing methods in general, Fig. 26-2. It is primarily used for printing on packaging materials such as folding cartons, paper bags, and plastic films. However, newspapers and catalogs are increasingly printed using this process.

The flexographic image carrier (plate) is made of flexible rubber or plastic, Fig. 26-3. Rubber plates are produced by first making a mold from an existing relief image such as metal type or a metal sheet with an image etched into it. Using heat and pressure, a rubber sheet is pressed against the mold to form the plate.

Fig. 26-2. This six-color flexography press is being shown at a trade show. The printed web of paper can be seen in the lower left of the photo.

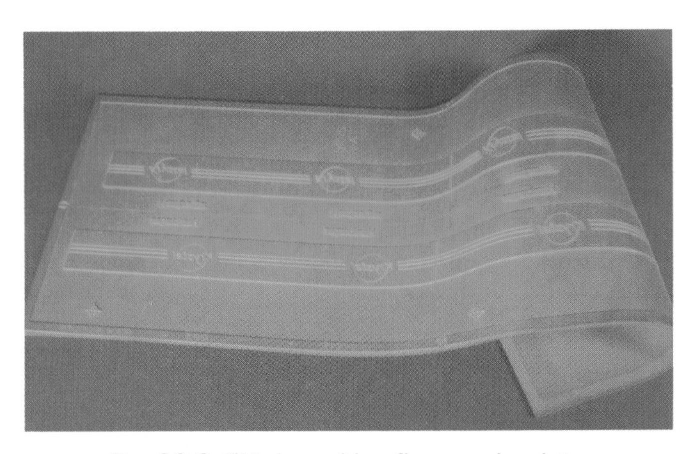
Fig. 26-3. This is a rubber flexography plate.

Photopolymer plates are most often used today for flexography. **Photopolymers** are light-sensitive plastics that harden when exposed to ultraviolet light. A negative is placed in contact with the photopolymer material and an exposure is made. Processing causes the non-image area to be washed away, leaving a relief image carrier.

Flexographic printing is done by putting the plate on a plate cylinder and applying liquid ink with an **anilox roller**. This is a roller with an engraved surface that holds the layer of liquid ink. A web (roll) of paper or other substrate passes between an impression cylinder and the plate cylinder to transfer the image, Fig. 26-4.

Flexography has several advantages that account for its growing popularity. First, the image carrier is very durable and relatively inexpensive. Millions of impressions can be made with one plate. Second, fast-drying liquid inks can be used. These inks are often water-based instead of solvent-based. Water-based inks cause less environmental pollution since solvent fumes are not emitted into the atmosphere as the ink dries.

Flexographic printing also has some disadvantages. The main problem is that the quality of the printing is not as good as that done by lithography and gravure. However, progress is being made to overcome this problem.

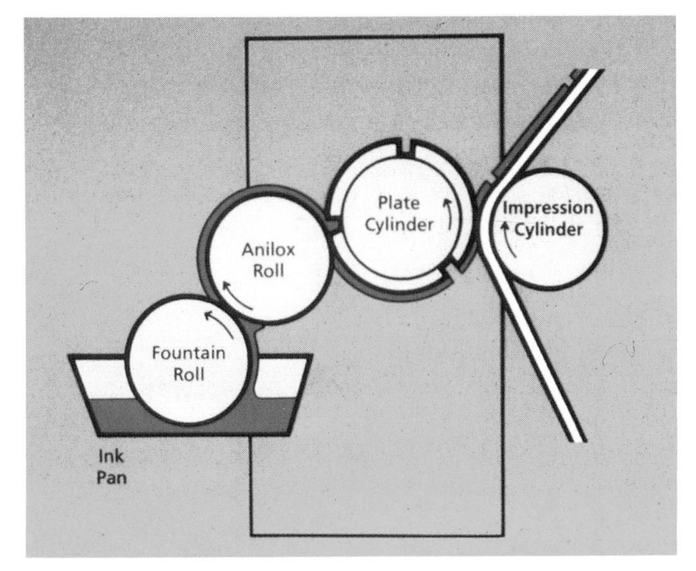

Fig. 26-4. Flexographic printing is done in this manner.

LETTERPRESS PRINTING

Letterpress printing, the oldest of the relief methods, is primarily done with cast metal type or plates as the image carrier. Originally, the type was composed by hand, similarly to the method used by Gutenberg. As technology advanced, other image carriers were developed such as machine cast type and cast relief plates, Fig. 26-5.

Several different printing presses are used for letterpress printing. The **platen press**, named for the area where the item to be printed is placed, uses a flat image carrier for printing. A **chase**, which is a metal frame that holds the image carrier, comes in contact with the platen to make the print, Fig. 26-6. The **flatbed cylinder** press also uses a flat image carrier, but the chase moves under a revolving impression cylinder, Fig. 26-7. The

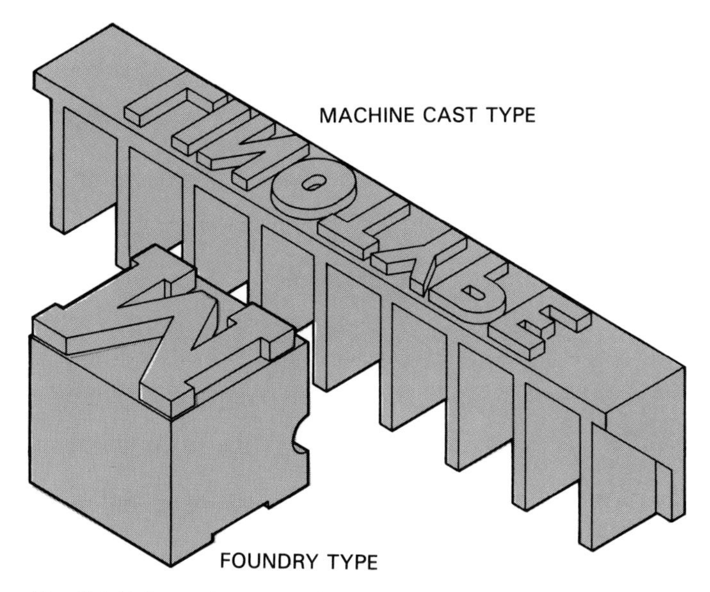

Fig. 26-5. Foundry type, shown at the left, is composed by hand and later reused. Machine cast type, shown at the right, is composed by machine and then cast from liquid metal. (John Walker)

impression cylinder, with substrate attached, moves over the image carrier to make the print.

Rotary presses use a curved image carrier so the plate can revolve, which speeds up the printing operation, Fig. 26-8. The substrate is printed as it passes between the revolving plate and impression cylinder.

Letterpress printing equipment is still being used for some printing operations such as newspapers. However, in modern printing facilities, its greatest use is for finishing operations. Embossing, hot stamping, and die cutting can all be done with letterpress equipment, Fig. 26-9. These topics will be explored later in this chapter and in Chapter 29.

FOUNDRY TYPE

Foundry type consists of individual metal characters that can be assembled to form an image carrier, Fig. 26-10. Although this technique is no longer used as the primary means for making image carriers, it does still

Fig. 26-6. This is a platen press.

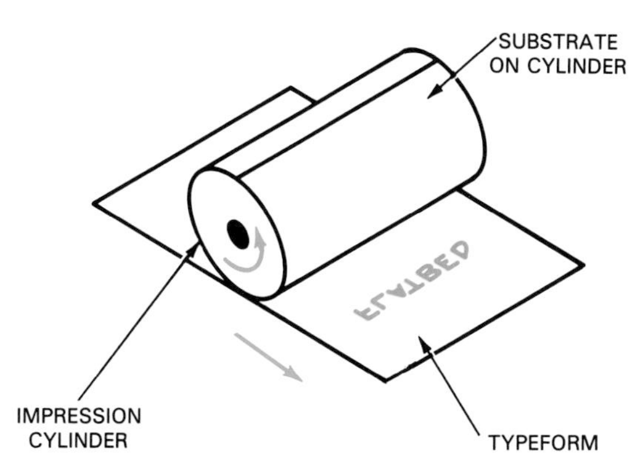

Fig. 26-7. The flatbed cylinder press uses an impression cylinder.

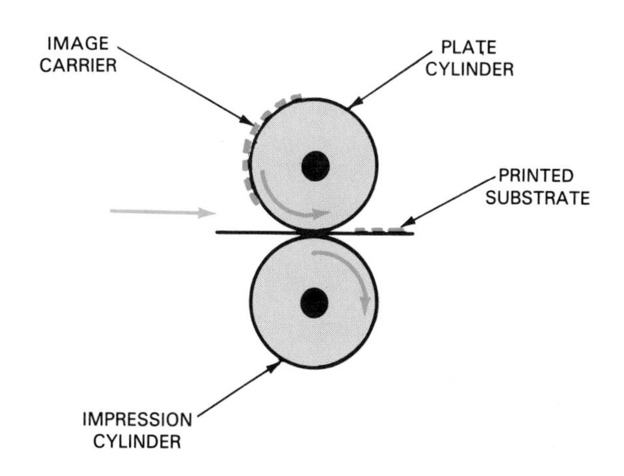

Fig. 26-8. A rotary press uses a curved image carrier to speed up the printing operation.

Fig. 26-9. A platen press is being used to die-cut the signs shown on the table at the right. The die can be seen on the press.

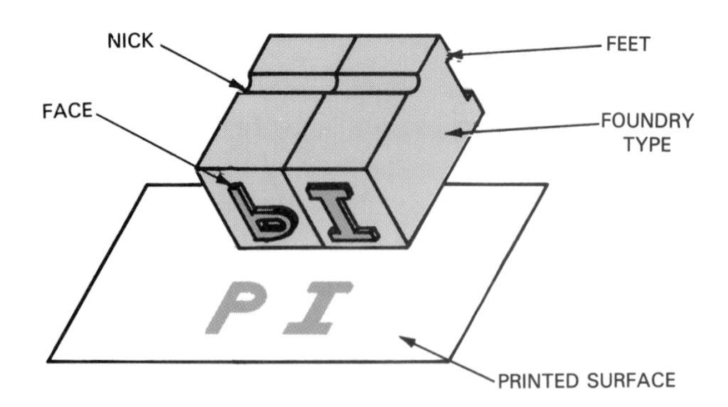

Fig. 26-10. The face on foundry type is reversed so it can print correctly on the substrate.

Fig. 26-11. Demon letters are those that are easily confused.

have applications in the graphic arts industry. Rubber stamps and hot stamping are two examples that will be explored later in this chapter.

Foundry type is handset in a composing stick. Since it is made from metal that had to be cast, this process is sometimes called **hot-type composition**. The type is made from a mixture of lead, tin, and antimony, which is a similar formula to the one used by Gutenberg.

Since type is reversed, some characters can be confused. These are called **demon letters**. Some examples are shown in Fig. 26-11.

Ligatures are special characters that have two or more letters on the same piece of type. These are letters that are often used together such as ff and ffi.

Spacing material is less than type high so it will not print. Spacing within a line of type is based on the **em quad**, Fig. 26-12. This space is the same width as the type size. The 2-em quad is twice as wide as an em quad. A 3-em quad is also available.

Divisions of less than half an em-quad are known as **spaces.** A 3- em space is 1/3 an em-quad. There are also 4-em and 5-em spaces. Smaller spaces are called thin or **hair spaces.**

The em-quad can be used for paragraph indentations. The **en quad** is used for spacing words in all upper case. The 3-em space is used to space words in upper and lower case. A variety of spaces is used for justifying or filling up a line of type in the composing stick.

Line spacing is done with leads and slugs. **Leads** are one or two points thick, and **slugs** are six points thick, Fig. 26-13.

Rules look similar to leads, but they are type high. These are used to print lines in various widths. They can be used for underlining and for borders.

HAND-SETTING TYPE

Before type is set, a rough layout should be made. In addition, a type font should be chosen. Remove the type case carefully using both hands.

Set the clamp on the composing stick to the desired line length. Place a same length slug in the stick. Begin setting type in the lower left corner. The thumb holds the type in place. Letters are placed in the stick with nicks up. If the character location is not marked in the case, refer to a chart showing the layout of the case, Fig. 26-14. Justify the line of type with spacing materiai so it is snug. Place large spacing material at the end of the line.

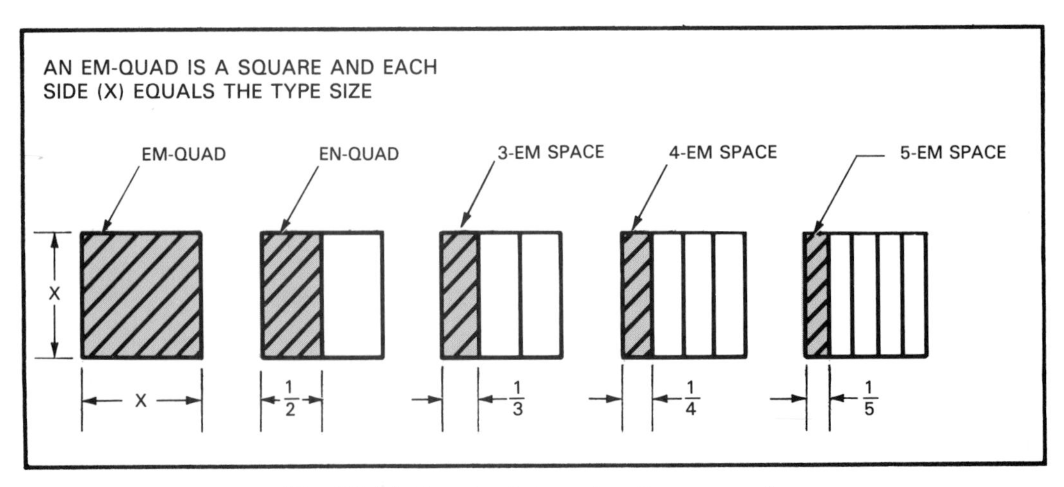

Fig. 26-12. Spacing is based on the em quad.

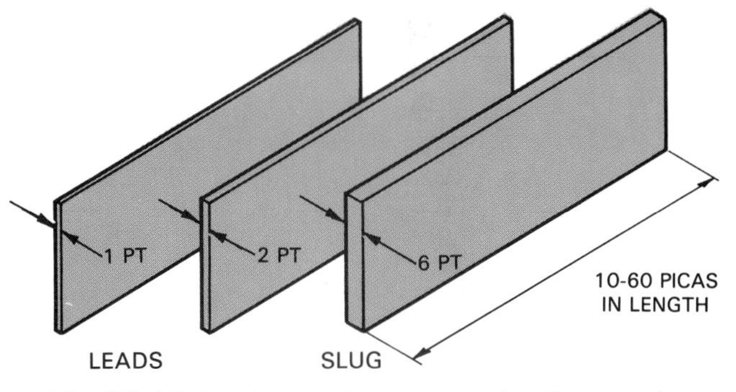

Fig. 26-13. Leads and slugs are used as line spacing materials. (John Walker)

If the type is set solid, no line spacing is used. If spacing is to be used, add a lead or slug.

Continue with additional lines. Place a slug in the stick after the last line.

Several other steps are involved in hand-setting type, as shown in Fig. 26-15. These steps include:

- Dumping and tying the type.
- Proofing.
- Making corrections.
- Distributing the type.
- Locking up the form.

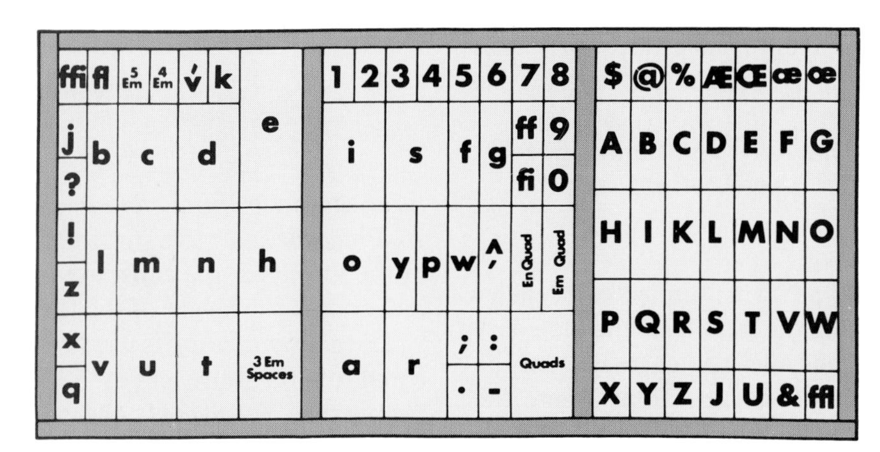

Fig. 26-14. This is the layout of a California Job Case.

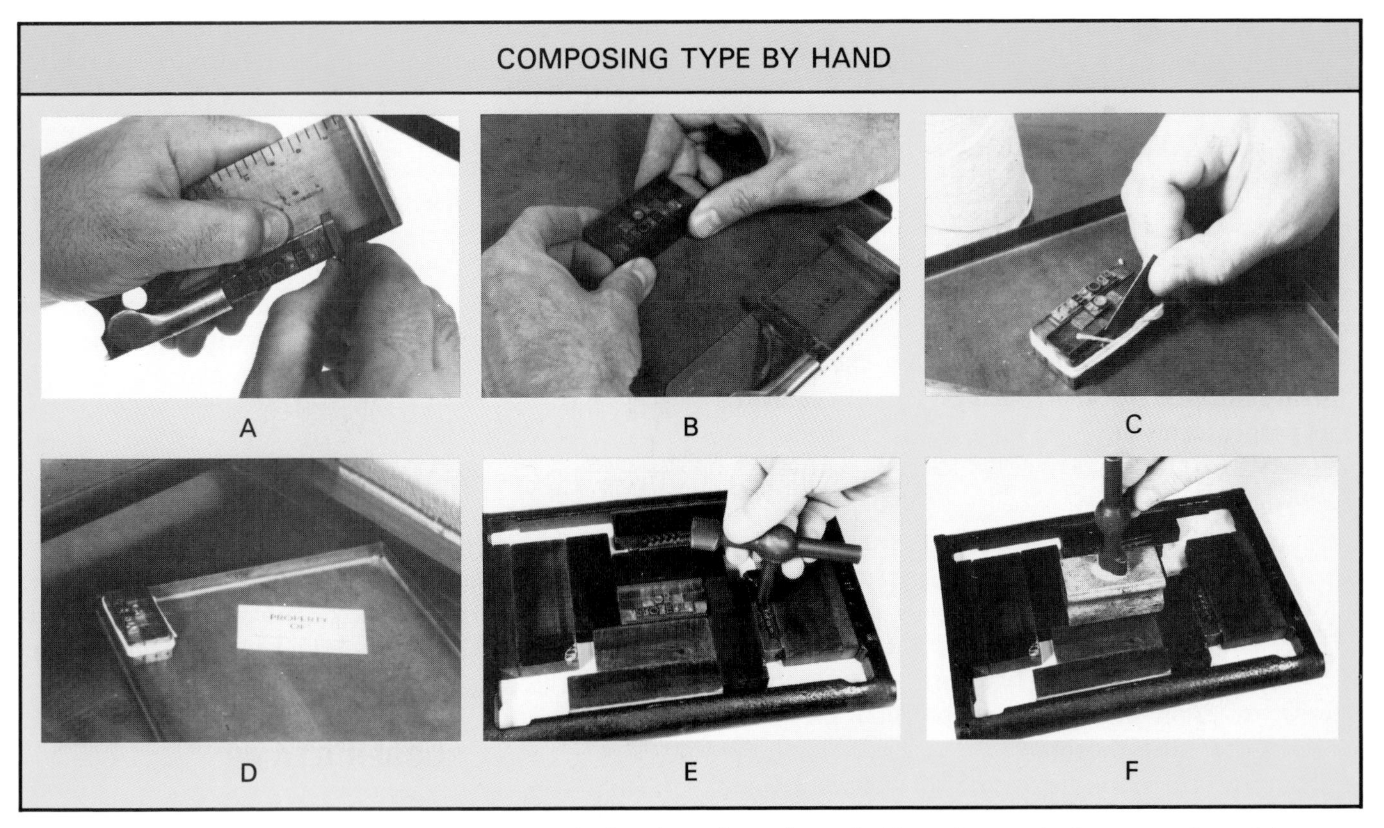

Fig. 26-15. The steps involved in composing type by hand are shown here: A–Place the type in the composing stick, starting at the lower left corner. B–Slide the type form into a galley tray. C–Tie the type form with a printer's knot. D–Make a proof of the type form. E–Lock the type form in the chase with a quoin key. F–Plane the type form.

Dumping and tying the type

Dumping type involves emptying the composing stick into a galley tray. Hold the type form and slide the type to a back corner of the galley. Tie the type form by wrapping string around it, overlapping the first wrap. Use a lead to tuck the end of the string underneath the wraps. This is called a **printer's knot**.

Proofing

Place the tied type form against the back edge and side of the galley. Place the galley on a proof press at a slight angle. Using a brayer, ink the type form and place paper over it. Starting from the open side of the galley, bring the roller over the galley to make the proof. Clean the ink form with a cloth and solvent after making the proof.

Proofread the print carefully. Use standard proofreader marks for marking needed corrections.

Making corrections

For corrections, the type form is untied in the galley tray. When characters are the same size, the old character can be removed and the new character can be placed in position. For other corrections where the spacing may change, the incorrect line should be returned to the composing stick. Separate the line with several slugs on each side. Place the line in the composing stick and make the needed corrections. Rejustify the line. Return the line to the type form. Make another proof to check for errors.

Distributing type

After the type is used for printing, it is placed back in the type case. Be *absolutely sure* that you redistribute the type in the same case where it was removed. The easiest way for a beginner to redistribute type is from the galley or composing stick. Work in reverse order, distributing the last line first. Distribute the type one word at a time. Be sure that spacing material is returned to its proper location.

Locking up the form

For printing, the type form is locked in a metal frame called a chase. First, slide the type form from the galley onto the **imposing table**. This table has a smooth, flat surface. The top of the form should be at the bottom or left. Place the chase over the type form with a long side nearest you. This near side is the bottom of the chase. The opposite side is the top.

Furniture is used to fill areas of space in and around type forms in the chase. Place wooden or metal furniture around the type form. If the line length of the type form corresponds to a furniture length, then the *furniture-within-furniture method* can be used. See Fig. 26-16. Place the same size furniture on the top and bottom of the form. For the ends, use furniture that is slightly longer than the height of the form.

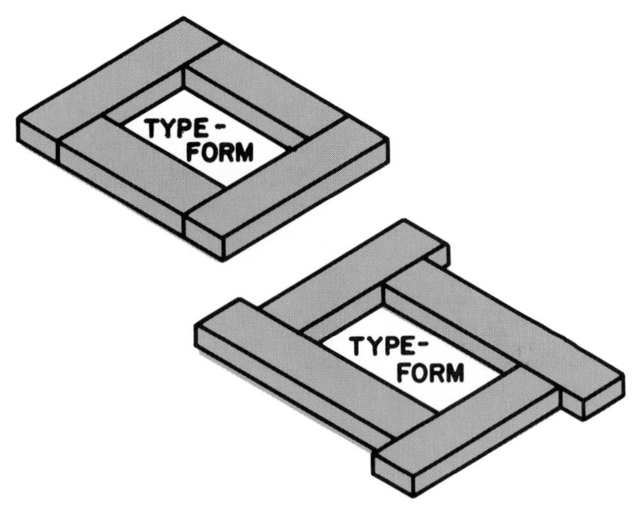

Fig. 26-16. The furniture-within-furniture and chaser method for placing furniture are shown here.

The *chaser method* is used when the type form is not equal to a furniture length. Refer again to Fig. 26-16. In this case, furniture is chosen for each side which is slightly longer than the side. The furniture appears to be "chasing" one another because the ends overlap.

Place **quoins** (locking devices) to the outside of the furniture on the top and right side. Fill the remaining space between the form and chase. Furniture should be larger towards the chase. Use **reglets** (6- and 12-point wood spacing) to fill small gaps.

Using a **quoin key**, tighten the two quoins until a slight resistance is felt. A planer block and mallet are used to plane or level the type form. Tap very lightly or the type may be damaged. Tighten the quoins snugly. Check the form for "lift." Raise one corner of the chase a small amount and press the type with the fingers. If the type moves, loosen the quoins and rejustify the loose lines.

TYPE FORM OPTIONS

Photopolymer plates and cast relief plates can be placed in a chase for printing in place of the hand-set type. The plate is mounted on a block of wood so it is type high. The block is then locked in the chase using the methods described earlier.

RUBBER STAMP

A **rubber stamp** is a good example of relief printing. It is made in the same way that a cast relief plate is made. First, a **matrix** or mold is produced. The rubber is heated and pressed into the mold to form the rubber stamp. Foundry type can be used for making a rubber stamp. Brass type is recommended because of the heat involved. Set the type in a composing stick. Place two slugs on each side of the type form. Place *type-high bearer strips* outside the slugs. These protect the type from being crushed. Lock the type in the chase used with the rubber stamp press. Metal furniture is normally used in the chase.

Three major steps are involved in producing a rubber stamp. These steps are illustrated in Fig. 26-17. The steps include:

- Producing the matrix.
- Producing the stamp.
- Mounting.

PRODUCING THE MATRIX

Preheat the press to 300°F. Place the chase in the press for two minutes. Remove the chase and place the plastic material with the red side against the type. Make sure it is large enough to cover the bearer strips. Place paper over the plastic and place the chase back in the press. Raise the bed until a slight resistance is felt. Leave it in this position for one minute to soften the molding material. Raise the bed to the stops on the chase and leave it for 10 minutes. After 10 minutes, lower the bed and remove the chase. Carefully pry the matrix loose and let it cool. Inspect it. A shallow image or broken letters will require that a new matrix be made.

PRODUCING THE STAMP

Cut a piece of stamp gum the same size as the matrix. A cloth layer will usually need to be removed from the good side of the gum. Dust the matrix and rubber with soapstone powder and tap off the excess. Place the matrix, image side up, on the vulcanizing tray. Place the stamp gum, good side down, on the matrix. Place paper over the top and slide the tray into the press. Raise the

Fig. 26-17. The steps involved in making a rubber stamp are shown here. A–The type form is locked in the chase. Notice that metal furniture is being used. B–The chase is being placed in the rubber stamp press. C–The matrix is carefully pried loose. D–The rubber stamp is being checked. E–The stamp gum is trimmed and attached to the handle. F–The completed rubber stamp is shown.

bed all the way. After six minutes, remove the tray and let the rubber cool. Peel the stamp off and check it.

MOUNTING

Mounting is the next step. Using scissors, trim around the image. Choose a handle that is just wide enough for the stamp. Cut it 1/16 in. longer than the stamp. Attach the stamp with rubber cement. Use the stamp to make an impression. Place the print in the window of the handle.

HOT STAMPING

Hot stamping is considered to be a relief printing process and a finishing operation. Instead of ink, the type form presses a decorative foil on the surface to be printed. This hot stamping technique is used for decorative printing on a variety of surfaces, such as book covers and stationery.

Lead type can be used for hot stamping, but brass is recommended because of the heat involved. Lead type used for the hot stamping process should not be used for other purposes.

Hot stamping can be done with a hand pallet or a hot stamping press. The pallet is heated on a hot plate, while the press has its own heating source.

The steps for hot stamping are shown in Fig. 26-18. Begin by composing the type as usual, making sure the form is within the size limitations of the press. Transfer the type from the composing stick to the type holder on the press, with nicks up. Lock the type in the holder. Lower the holder to print position and lock it in place.

Select a foil that is slightly wider than the type form. For feeding units, the foil is fed, bright side up, under the type form. Otherwise, cut off the length needed.

Heat the press for 10 minutes on medium, or until 250°F is reached. While it is heating, adjust the guides on the table.

Make a trial impression on scrap stock. Press the type form against the surface with medium pressure for one second. Check for image location and quality. More time and pressure can be used, if needed.

Make the final print. When finished, clean the hot type with a cloth.

SUMMARY

Relief printing, in the form of letterpress, was the most important printing method used worldwide for many centuries. It became less important when other printing methods, such as lithography and gravure, were developed and became more popular. Relief printing is still used for a variety of finishing operations.

Flexography is continuing to grow as an important printing method. A rubber stamp provides a simple simulation of this technique.

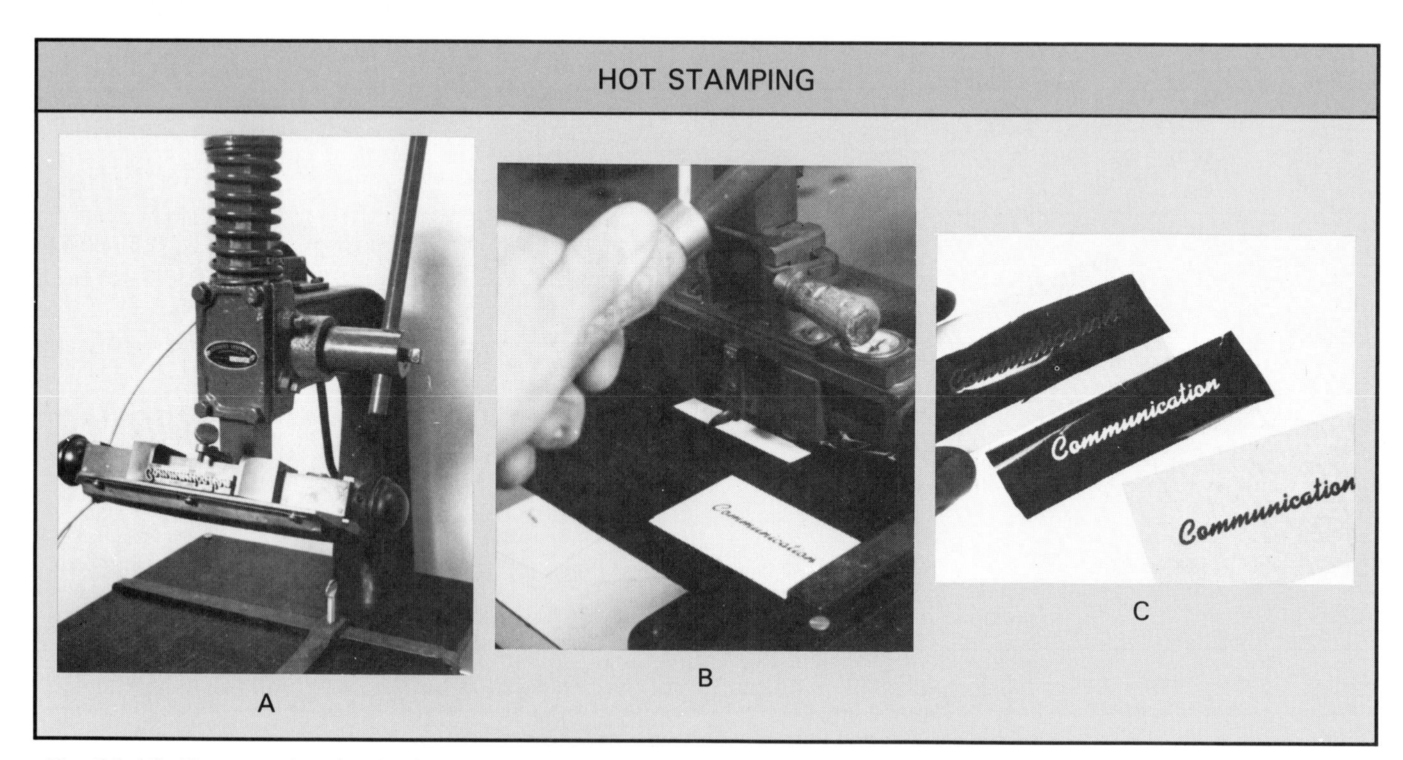

Fig. 26-18. The steps involved in hot stamping are shown here. A–The type form is locked in the holder of the hot stamping press. B–The heated type form is being used to make an impression. C–On the left, the foil is shown still stamped to the card. On the right, the foil and card are then separated.

WORDS TO KNOW

All of the following words have been used in this chapter. Do you know their meanings?

anilox roller
chase
demon letters
dumping type
em quad
en quad
finishing operations
flatbed cylinder
flexography
foundry type
furniture
hair spaces
hot stamping
hot-type composition
imposing table
imprinting
leads
letterpress printing
ligatures
matrix
photopolymers
platen press
printer's knot
quoin key
quoins
reglets
relief printing
rotary presses
rubber stamp
rules
slugs
spaces

REVIEWING YOUR KNOWLEDGE

Please do not write in this text. Write your answers on a separate sheet.

1. True or false? Relief printing was the first industrial printing method used in America.
2. Name three major uses of relief printing today.
3. Which of the following is NOT an advantage of flexography printing?
   a. High-quality printing.
   b. The image carrier is durable and relatively inexpensive.
   c. Fast drying liquid inks can be used.
   d. None of the above.
4. The greatest use of letterpress printing equipment is for:
   a. Newspapers.
   b. Finishing operations.
   c. High-quality textbooks.
   d. Magazines.
5. True or false? Ligatures are characters that can be easily confused with one another.
6. Which of the following is used to fill areas of space in and around type forms in a chase?
   a. Quoins.
   b. Furniture.
   c. Slugs.
   d. Leads.
7. Which type is recommended for making a rubber stamp?
   a. Iron type.
   b. Brass type.
   c. Lead type.
   d. Plastic type.
8. True or false? The hot stamping method uses ink.

APPLYING YOUR KNOWLEDGE

1. Investigate the past, present, and future of relief printing.
2. Visit a museum with exhibits on early relief printing. Note how it differs from printing systems of today.
3. Visit a printing firm that uses relief printing equipment. Observe how the printing is done and write a report about your experience.
4. Collect examples of relief image carriers. Display them in the classroom.
5. Survey (by phone) local printing firms to determine how relief printing is being used today. Report your results to the class.

This electronic scanner is used to produce gravure cylinders by electromechanical engraving. Originals on the scanning cylinder (left) are scanned and a signal is sent to the engravers. As the gravure cylinder (right) rotates, the image is cut into its surface.(Hell Graphics Systems)

27

CHAPTER

GRAVURE PRINTING

After studying this chapter, you will be able to:
- *Describe the gravure printing process.*
- *Name the two major processes used to create gravure image carriers.*
- *Identify the steps involved in producing a gravure cylinder by electronic engraving.*
- *Differentiate between gravure and intaglio.*
- *Discuss the differences between etching and engraving an intaglio image carrier.*

Gravure is an industrial method of printing from a sunken surface, Fig. 27-1. The image carrier is a cylinder. The surface of the cylinder has tiny depressions or cells in the image area that hold the ink for printing. This printing process is also known as **rotogravure**, since the cylinder *rotates* for printing.

The gravure cylinder is very durable and can be used for printing millions of copies. In addition, the image quality is very good. Products printed by this method include fine magazines, catalogs, newspaper supplements, and packaging. Also, specialty products such as wall coverings, floor coverings, and gift wrap are printed using this method.

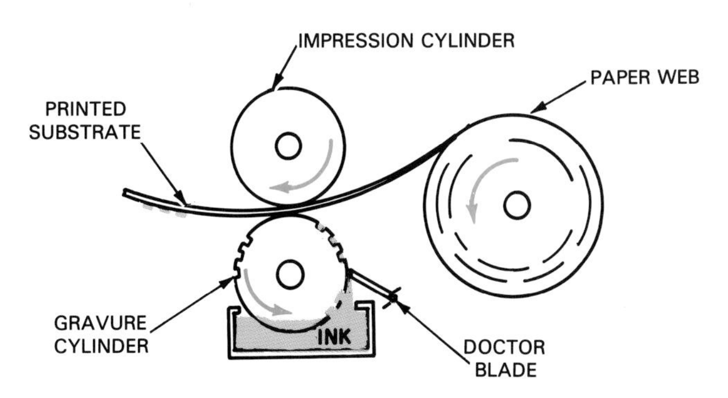

Fig. 27-1. The process of printing using gravure is shown here.

Intaglio is also a method of printing from a sunken surface. In this case, lines instead of cells are placed on the surface for printing. The image carrier is usually handcrafted, using an engraving or etching technique.

GRAVURE IMAGE CARRIERS

The major type of image carrier used in gravure printing is the **gravure cylinder**. This is a steel or aluminum cylinder that is coated with copper. The gravure cylinder is very durable, but is more costly to produce than other types of image carriers. This is the reason that this technique is used when millions of copies need to be printed.

Tiny ink cells are cut or etched into the surface of the cylinder. There are about 22,000 of these cells per square inch on a gravure cylinder. The cells vary in size and are used for printing all line and continuous tone images. This is different from lithography where only the continuous tone images are printed as dots. Also, unlike lithography, two ink dots can vary in intensity by varying the cell depth, Fig. 27-2.

Gravure image carriers are produced through two major processes. These are:
- Chemical etching.
- Electronic engraving.

CHEMICAL ETCHING

With **chemical etching**, the cells are etched into the surface of the cylinder using an acid. Although several chemical methods were used in the past, the method most widely used today is known as direct transfer.

Direct transfer is a process that involves coating the gravure cylinder with a light-sensitive photopolymer

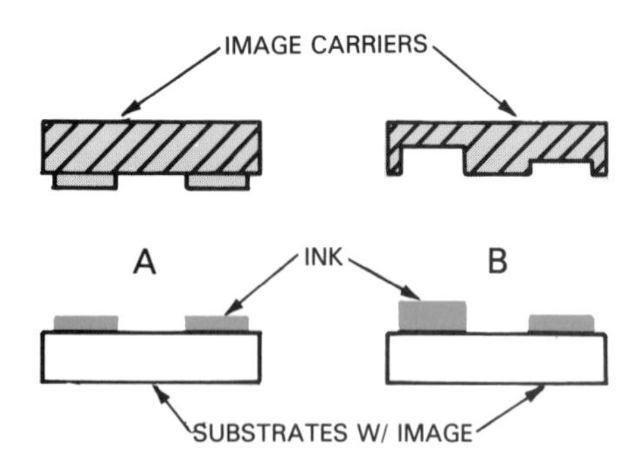

Fig. 27-2. Unlike lithographic plates (A), some gravure image carriers (B) can apply varying amounts of the same ink color.

emulsion. After the emulsion is dry, a film positive is placed in contact with it and an exposure is made. All the images on the positive have been screened or broken up into dots.

An ultraviolet light is used to make the exposure through the positive onto the emulsion. The exposure hardens the emulsion in the non-image areas. After the exposure, a developer and then water rinse is applied to the cylinder. The developing process removes the unexposed emulsion from the image areas, leaving the copper underneath exposed.

Next, an acid solution is applied to the cylinder. The acid etches away the copper in the exposed areas making the ink cells. The cells on the cylinder will vary in size, but have a uniform depth, Fig. 27-3.

The emulsion mask is then washed off, and the cylinder is chrome plated to increase the life of the cylinder. The cylinder is now ready for printing.

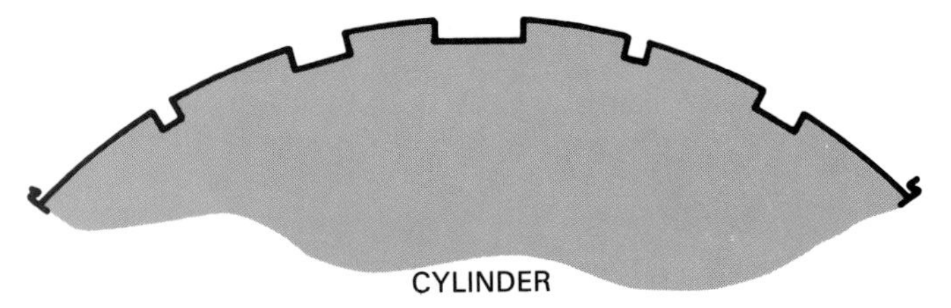

Fig. 27-3. Cylinder cells produced by direct transfer vary in width, but have uniform depth.

ELECTRONIC ENGRAVING

With **electronic engraving**, the copy is placed on an electronic scanner. A light beam, reflected off the copy, is used to change the image into an electronic (digital) signal. This signal is sent to a computer that is used to control a cutting mechanism held against the gravure cylinder.

Electromechanical engraving, a type of electronic engraving is the most common process used for preparing gravure image carriers, Fig. 27-4. This is the method used for almost all publications printed by gravure and a large percentage of packaging as well.

As the gravure cylinder rotates on the engraving machine, the ink cells are cut into the surface. A diamond cutting tool, called a **stylus**, is used for cutting the cells into a copper-coated cylinder. These cells vary in both width and depth, Fig. 27-5.

After the cylinder is made, it is inked, and a proof copy is made. Corrections are then made, if necessary. The cylinder is then chrome plated. This provides more durability.

Fig. 27-4. The scanned images on the left are being engraved onto the gravure cylinder on the right. (Linotype-Hell)

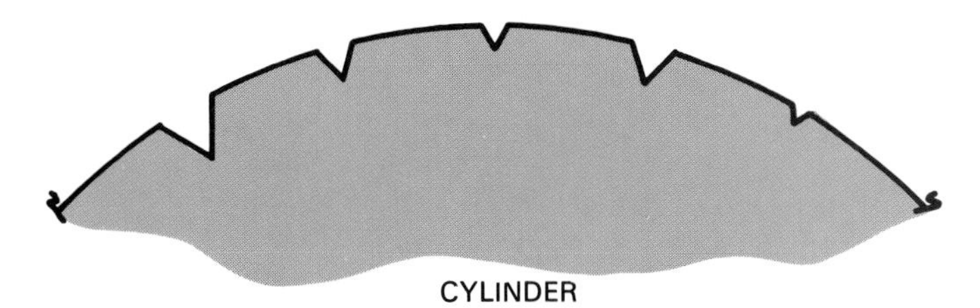

Fig. 27-5. Ink cells produced on a cylinder by electromechanical engraving vary in both width and depth.

Other types of electronic engraving include the laser and electron beam. With **laser beam engraving**, the cylinder is coated with plastic and the laser is used as the cutting tool. The laser melts the plastic to form the cells. **Electron beam engraving** utilizes a very hot electron beam that melts the copper on the gravure cylinder to make the cells.

IMAGE TRANSFER

The gravure cylinder is placed on the press for printing. As was shown in Fig. 27-1, ink is applied to the cylinder and a doctor blade removes the excess ink. The ink is very liquid and dries quickly.

The substrate to be printed passes between the gravure cylinder and impression cylinder to make the print. Gravure printing is often done on webs (rolls) of paper. In multicolor printing, a web passes through a different cylinder unit for each color, Fig. 27-6.

Gravure cylinders can be used over again after printing. This is done by placing a new layer of copper on the cylinder by electroplating, Fig. 27-7. The cylinder is then ready for a new image, using one of the methods discussed.

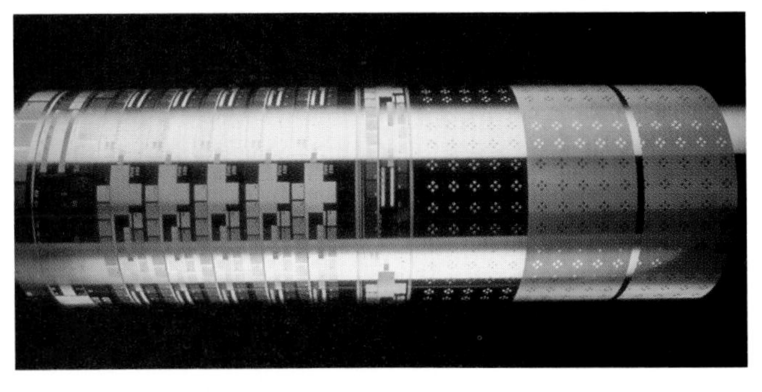

Fig. 27-7. Gravure cylinders can be reused after use by electroplating new copper on the surface. (Ohio Electronic Engravers, Inc.)

INTAGLIO

Intaglio, like gravure, is printing from a sunken image area. However, with intaglio printing, lines are used instead of ink cells. These are cut or etched into the image carrier. This technique is used as an art technique, but it is also used in industry for high-volume printing.

Because of the hand labor and skill involved, this method is only used for printed products of very high quality. Paper money, postage stamps, stock certificates, and fancy stationery are examples of products that are printed using this method.

The two most common methods that can be used to produce intaglio plates are etching and engraving.

Fig. 27-6. This multicolor gravure press prints from a web, shown on the right. (Bobst Group, Inc.)

ETCHING

In the etching process, an acid resist is coated onto a flat metal plate, Fig. 27-8. The image is drawn through the resist, which exposes bare metal. When the plate is placed in an acid solution, the exposed metal in the image area is eaten away. This leaves a sunken image area. The resist is removed, and the plate can be used for printing.

A **photo-resist method** can also be used for etching, Fig. 27-9. The plate is coated with a light-sensitive emulsion. A transparent positive is placed in contact with the plate and an exposure is made. The plate is processed and placed in an acid solution. The non-image area is protected by a hardened emulsion layer. The acid removes the image area where the emulsion is not hardened.

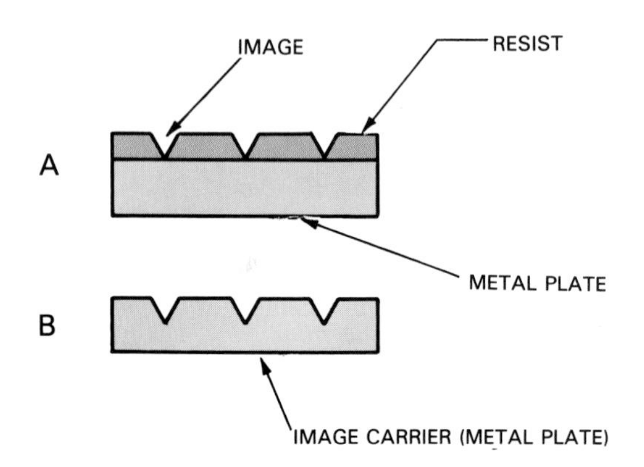

Fig. 27-8. An intaglio plate is etched by first drawing the image in the resist (A) and then placing the plate in an acid bath that removes the image area (B).

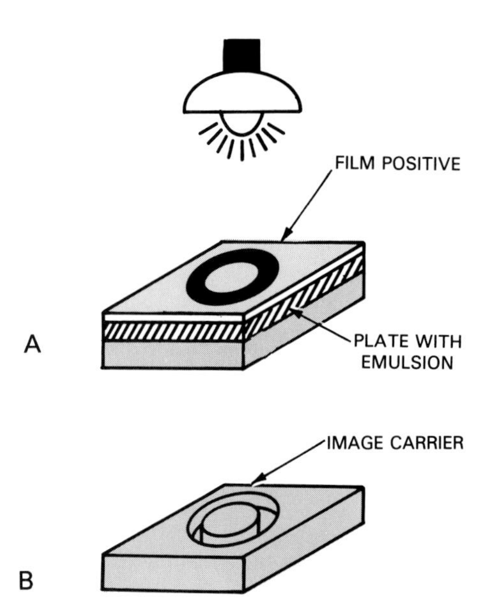

Fig. 27-9. The photo-resist method of etching is done by exposing a light-sensitive emulsion through a film positive (A) and then using an acid bath to remove the image area (B).

ENGRAVING

In the engraving process, a metal plate is engraved by hand, using engraving tools. The scratched surface creates the depressed image areas. This plate can then be used directly for printing.

If multiple images are needed of an engraving, then a master die is engraved, Fig. 27-10. On a special transfer press, a relief (raised image) of the die is made by pressing the die into a plastic or steel sheet. This intermediate is then used to stamp multiple recessed images

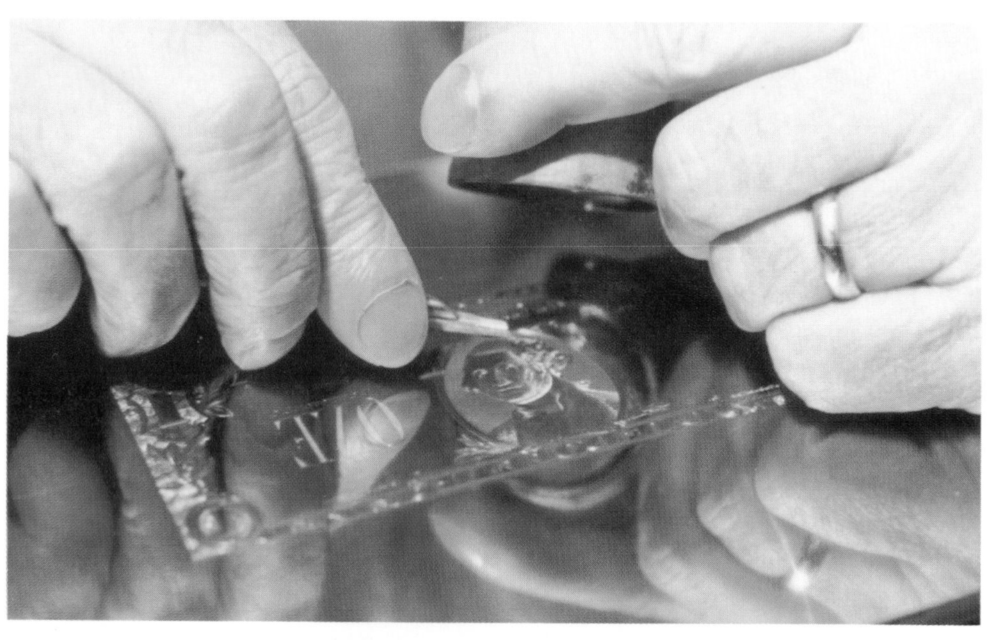

Fig. 27-10. A master die for a one-dollar bill is being engraved. (Bureau of Engraving and Printing)

Fig. 27-11. An intaglio plate is being inspected prior to printing. (Bureau of Engraving and Printing)

on an intaglio plate. See Fig. 27-11. Each image produced is an exact duplicate of the master die. This process of producing multiple images from a master die is known as **siderography**.

Intaglio plates produced in this manner are usually placed on a plate cylinder and printed on a rotary press, Fig. 27-12. This process is, essentially, the same as gravure printing.

Fig. 27-12. This intaglio plate is used to print postage stamps. (Bureau of Engraving and Printing)

SUMMARY

Gravure is a printing process used for high-quality, large- volume printing. Gravure image carriers are produced using two major processes. These are chemical etching and electronic engraving. Gravure printing will continue to play a large role in the printing industry in the future.

Intaglio printing, like gravure, is printing from a sunken image area. However, with intaglio printing, lines are used instead of ink cells.

WORDS TO KNOW

All of the following words have been used in this chapter. Do you know their meanings?

chemical etching
direct transfer
electromechanical engraving
electron beam engraving
electronic engraving
gravure
gravure cylinder
intaglio
laser beam engraving
photo-resist method
rotogravure
siderography
stylus

REVIEWING YOUR KNOWLEDGE

Please do not write in this text. Write your answers on a separate sheet.

1. Gravure is an industrial method of printing from a __________ (raised, sunken) surface.
2. True or false? The gravure cylinder is very delicate and is used mainly for small press runs.
3. Name two major processes used to produce gravure image carriers.
4. True or false? With direct transfer, the cells are etched into the surface of the cylinder using an acid.
5. True or false? Gravure cylinders can be used over again after printing by electroplating a new layer of copper on the cylinder.
6. How does intaglio printing differ from gravure printing?
7. When producing intaglio plates by etching:
   a. An acid resist is coated onto a flat metal plate.
   b. A metal plate is engraved by hand.
   c. The metal plate is coated with a light-sensitive emulsion.
   d. None of the above.
8. The process of producing multiple images from a master die is known as __________.

APPLYING YOUR KNOWLEDGE

1. Collect samples of gravure printing. Prepare a display featuring the samples.
2. Investigate the history and present techniques used in intaglio printing.
3. Research current literature on gravure production. Prepare a written report.
4. Schedule a field trip to a printing company that has a gravure press.

28

CHAPTER

SPECIALTY PRINTING AND REPROGRAPHY

After studying this chapter, you will be able to:

- *Discuss specialty printing and reprography.*
- *Define ink jet printing.*
- *Explain how mimeograph and spirit duplication are done.*
- *Explain how electrostatic copying, digital laser scanning/printing, and direct thermal copying are done.*
- *Describe monochrome copying techniques and color copying techniques.*

A large volume of printing is done using methods other than the four major printing processes. Some examples include:

- Ink jet printing.
- Mimeograph duplication.
- Spirit duplication.
- Electrostatic copying.
- Digital laser scanning/printing.
- Direct thermal copying
- Color copying.

Processes used for printing a small number of copies from a master are called **duplicating processes.** Mimeograph and spirit duplication are examples. Processes for printing copies directly from the original are often called **copying processes**, quick copy, or office copying processes. Electrostatic and thermal are examples of copying processes. The entire category of duplicating and copying processes is called **reprography**.

INK JET PRINTING

In **ink jet printing**, drops of ink are sprayed onto the surface to be printed. Personalized advertising and addressing are often done with this method, Fig. 28-1. In

Fig. 28-1. This ink jet system is used for addressing envelopes. (Domino Amjet, Inc.)

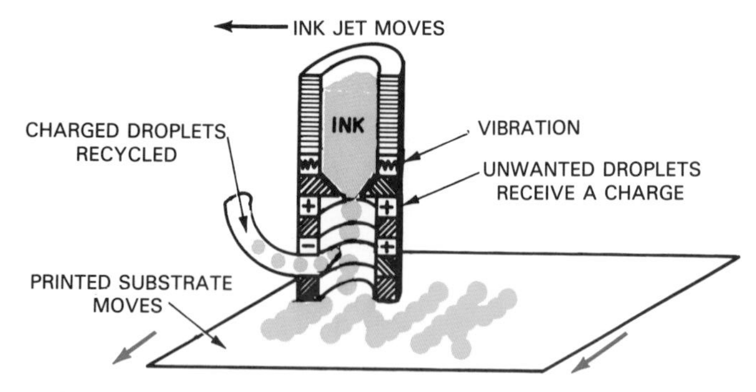

Fig. 28-2. This ink jet moves back and forth for printing.

addition, products with irregular surfaces can be printed because only the ink touches the surface.

Ink jet printing is controlled by a computer. Ink, passing through a nozzle, is broken into tiny drops. Ultrasonic vibration is used to ensure uniform drop size. These drops of ink are sprayed onto the substrate to produce the image.

There are two major ink jet types. In one system, the droplets used for printing pass straight through the ink jet, Fig. 28-2. Droplets that are not used for printing receive a charge and are deflected to be reused. If a single ink jet is used, it travels back and forth to print one line at a time. This same system can be used with multiple, stationary jets. In this case each jet covers a single point on the line.

In a second system, each drop is given a negative electrical charge that varies based on where the ink is to be printed, Fig. 28-3. Next, the drop passes through deflection plates that have a positive electrical charge. This causes the negatively charged ink drops to deflect from the straight line path. The amount of deflection varies depending on the amount of negative charge given the drop. These ink drops are deflected to print characters made from dots of ink.

Ink jet printers for use with home and office computers have rapidly improved in quality and declined in price during the 1990s. For most users, they are the most affordable means of doing color printing of documents.

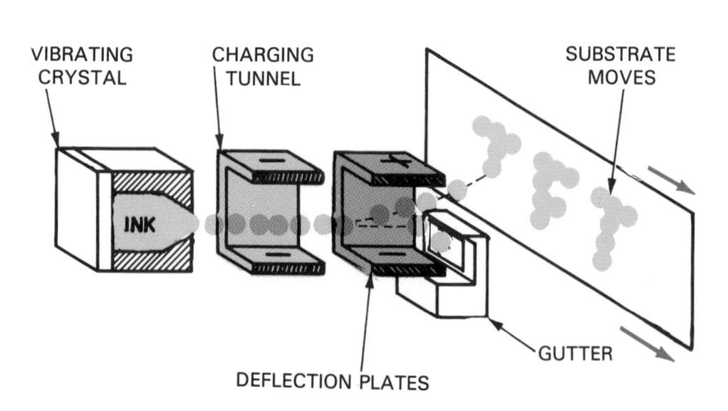

Fig. 28-3. This stationary ink jet prints on a moving substrate.

MIMEOGRAPH DUPLICATION

Mimeograph duplication is a stencil process, like screen printing. In this process, the stencil is a prepared master that has open image areas. The stencil is placed on a stencil cylinder that has an ink pad on the surface. Ink passes through the stencil onto the paper when printing, Fig. 28-4.

Mimeograph can be used to print up to about 3,000 copies. The image quality is very good.

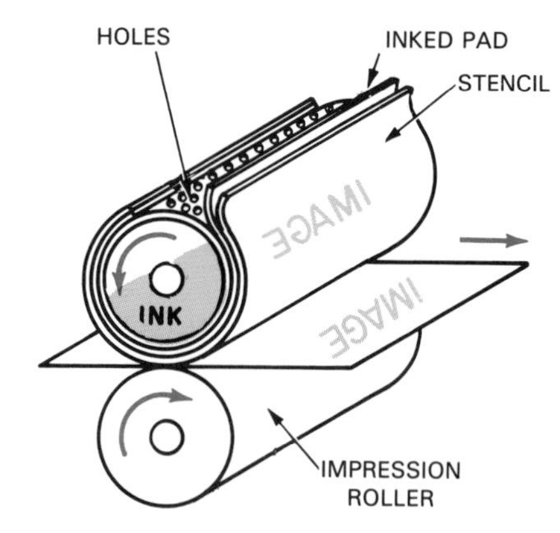

Fig. 28-4. Mimeograph duplication is a stencil process.

MIMEOGRAPH STENCIL PREPARATION

The **mimeograph stencil** is made from a tough, porous tissue that is wax-coated. When pressure is applied, the wax is pushed aside, exposing the tissue. Ink will pass through the tissue. The stencil can be imaged by drawing, typing, or with other impact methods such as dot matrix.

Stencils have a typing cushion and backing sheet attached, Fig. 28-5. The typing cushion can be removed when drawing directly on the stencil. Place artwork

Fig. 28-5. A mimeograph stencil is shown here.

under the stencil and trace over it on a light table. Use a special stylus with a rounded point. A lined sheet may also be placed under the stencil for handwriting. Make sure that you stay within the image margins that are marked on the stencil surface.

Most stencils are prepared by an impact method such as typing or dot matrix printing. The typing cushion is left on the stencil for this procedure. Some stencils also have a plastic top sheet. This prevents the typewriter from removing the centers of letters such as "O."

Type the stencil without a ribbon. This can be done by removing the ribbon or moving the selector to "white" or "stencil" position. Make sure you stay within the paper size to be printed. Use a stencil correction fluid to fill mistakes. This fluid can be used for corrections when drawing stencils as well.

Stencils can also be prepared by electronic scanning, Fig. 28-6. A single drum holds the original on one end and the stencil on the other end. As the drum rotates, a light reflected from the copy changes the image into an electrical signal. This signal operates a stylus that burns the image into the stencil. Originals with a large amount of detail can be used for stencils when this technique is used.

DUPLICATION PROCESS FOR MIMEOGRAPH

Once the stencil is prepared, it is placed on a mimeograph duplicator, Fig. 28-7. These duplicators are hand or motor operated. These are the procedures for mimeograph duplication:

1. Remove the ink pad cover and, if needed, place ink in the stencil cylinder.
2. Attach the stencil to the head clamps. Turn the handwheel while holding the trailing edge of the stencil. Clamp the trailing edge.
3. Lower the feed table and place the paper in position. If necessary, adjust the side guides. Evenly space the feed rollers inside the side guides above the paper.
4. Set up the delivery tray for the paper size to be printed.

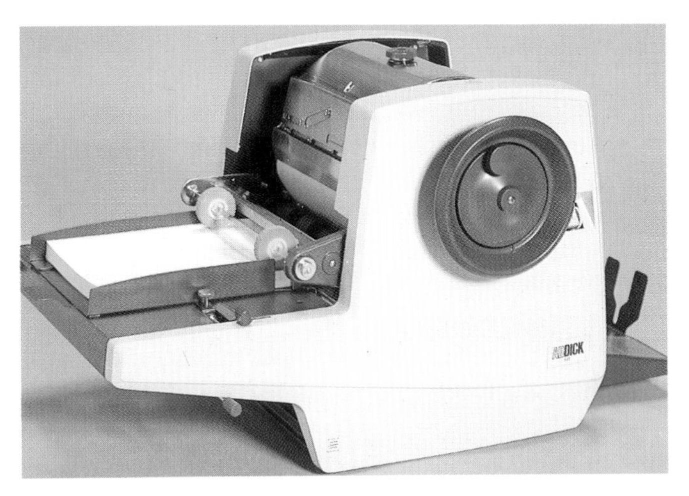

Fig. 28-7. A mimeograph duplicator is shown with paper in the feed tray on the left. (A.B. Dick Co.)

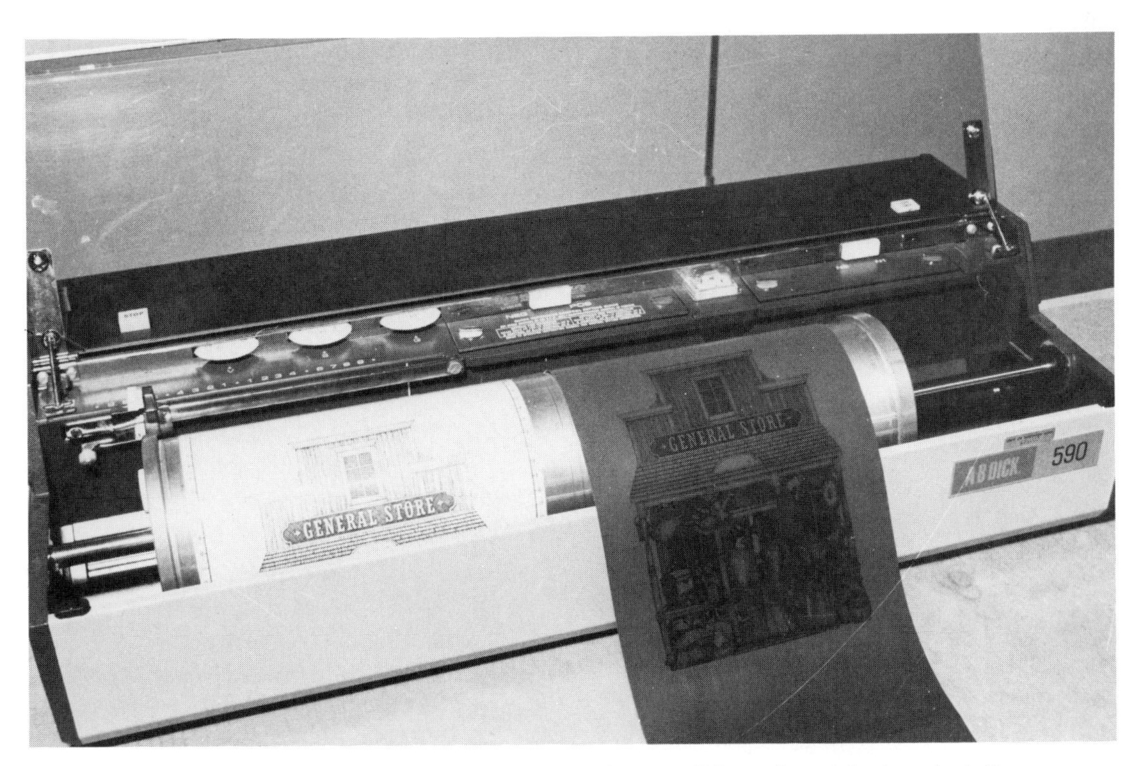

Fig. 28-6. This electronic scanner produced a stencil from the original on the left.

5. With the feed lever off, turn on the duplicator. Feed several sheets through and stop. Check for copy position. Make machine adjustments as needed. Continue this procedure until the image is in position.
6. Set the counter on the duplicator. Print the desired number of copies.
7. Remove the stencil. If it is to be reused, place it in a stencil folder.
8. Place the ink pad cover back on the duplicator stencil cylinder.

SPIRIT DUPLICATION

In **spirit duplication**, no ink is involved. The spirit master contains an aniline dye that is used for printing. When paper passes through the duplicator, it is slightly dampened with alcohol. Upon contact with the spirit master, the alcohol causes the dye on the master to be transferred to the paper, Fig. 28-8. The dye is quickly used up, so this process is only useful for printing 200 to 300 copies. The image is of low quality compared with the major printing processes.

SPIRIT MASTER PREPARATION

The spirit master is purchased as a three-part unit, Fig. 28-9. The master sheet is separated from the aniline dye sheet by a protective tissue paper. When the tissue is removed, pressure on the master causes the dye to transfer to the back of the master. This reversed image is used for printing. The master can be prepared by drawing, typing, or with other impact methods.

When hand drawing masters, first lightly sketch the drawing with pencil. For a tracing, unfold the master sheet and place it over the artwork on a light table. Make sure the tissue protects the dye sheet. An option is to prepare the sketch directly on the three-part master. Be sure the protective sheet is in place. When the sketch is finished, remove the tissue and image the master with a stylus, rounded pencil point, or ball-point pen. Place the tissue back in place until you are ready to use the master.

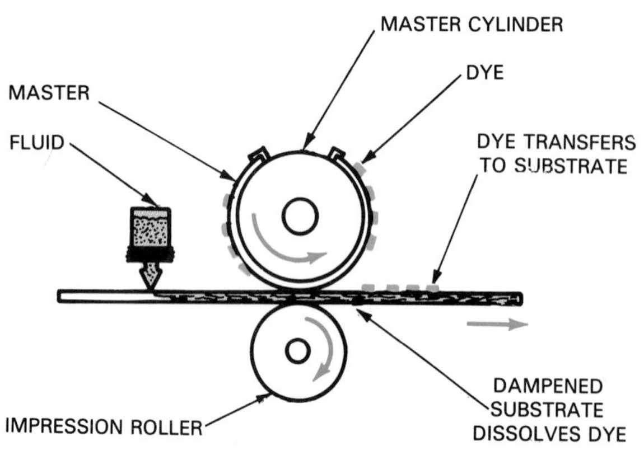

Fig. 28-8. The spirit duplication process uses a dye for printing.

Fig. 28-9. This is a spirit master.

When an impact unit is used to create the master, the tissue is removed and the master is placed in the imaging unit with the master sheet up.

Corrections are made by scraping the dye from the back of the master. Use an artist knife or razor blade. Place an unused piece of dye sheet in position and make the correction.

Multiple colors can be printed using a single master. Simply change the dye sheets for each color desired. All the colors will print at the same time.

DUPLICATION PROCESS FOR SPIRIT MASTER

Once the spirit master is prepared, it is placed on a spirit duplicator, Fig. 28-10. These are hand- or motor-operated. The following procedure is used for printing:

1. Add duplicator fluid, if needed, and dampen the wick.
2. Load the feed tray with paper and adjust the side guides. Evenly space the feed rollers inside the side guides.
3. Using the clamp lever, open the master clamp and place the master on it, dye paper up. Smooth the master and clamp it in position.
4. Adjust the receiving tray.
5. Adjust the fluid control lever for the paper size.
6. Adjust the pressure control to a medium setting. This adjusts the pressure between the master and paper.

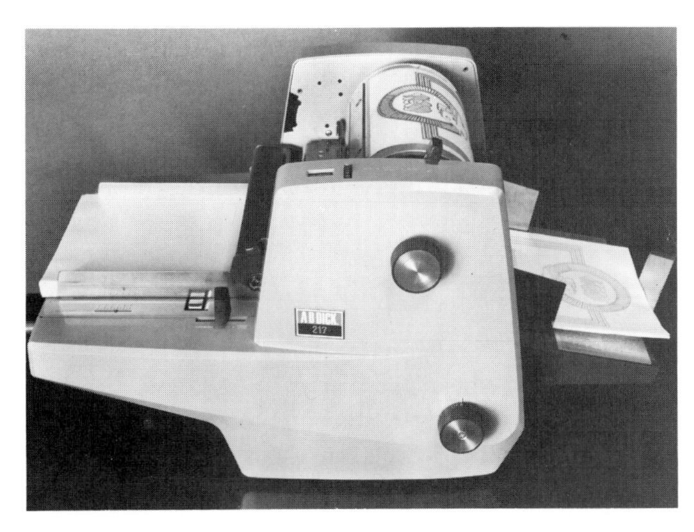

Fig. 28-10. This spirit duplicator has a master attached and printed sheets on the right. (A.B. Dick Co.)

7. Turn on the duplicator and feed a few sheets to check image position. Make any adjustments, if necessary.
8. Set the counter and duplicate the copies needed.
9. Remove the master. It can be saved by placing the tissue against the back and placing both sheets in a folder.
10. Turn off both the pressure control and fluid control levers.

ELECTROSTATIC COPYING

Electrostatic copying is probably the most well-known copying method. Often referred to as **xerography**, electrostatic copying involves photocopying using an electrostatic charge and heat processing. Electrostatic copiers are common in schools, offices, and libraries. These machines are often called plain paper copiers.

Electrostatic copying can produce a high quality image. Some machines are capable of printing multiple colors.

There are three major kinds of electrostatic printing. These are:

- Transfer process.
- Direct process.
- Electrofax.

TRANSFER PROCESS

The **transfer process** uses plain paper for copies, Fig. 28-11. First, the original is placed image down on a glass surface. In darkness, a selenium-coated drum in the machine is given a positive electrical charge. The original is exposed to the drum. The non-image areas reflect light to the drum and cause the charge to be released. The image area still has the electrical charge. A toner powder, which is negatively charged, is attracted to the image on the drum. Plain paper is given a positive charge and placed on the drum. The toner is attracted to the paper. The paper then passes through a heating unit that fuses the powder to the paper.

DIRECT PROCESS

The **direct process** uses a paper that has a metal oxide coating, Fig. 28-12. Some lithographic printing plates are made using this technique.

First, the original is placed in the machine. At this point, the paper is given an electrical charge. When exposed to the original, the charge is released in the non-image area. The oppositely charged toner is then placed on the image area. The toner is fused to the paper with heat.

ELECTROFAX

Electrofax is a type of electrostatic printing used for irregular surfaces, such as boxes and bottles. A screen, which has a charge in the image area, attracts an oppositely charged toner. A charged plate is placed below the

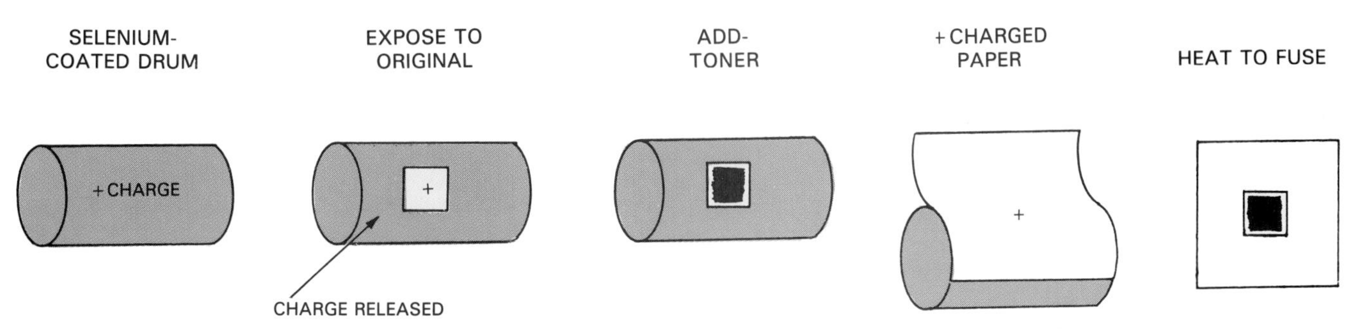

Fig. 28-11. The transfer electrostatic process is shown here.

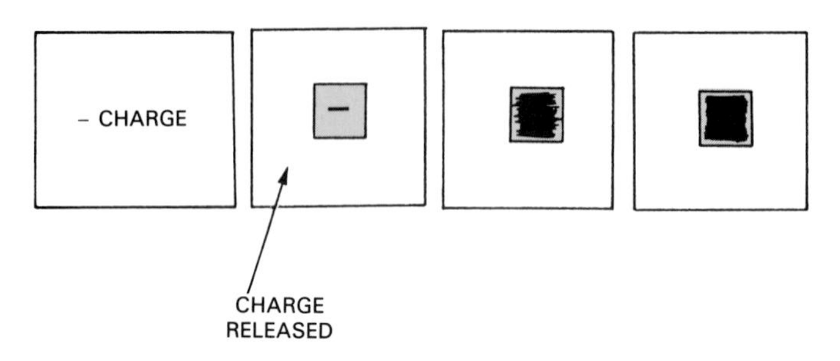

Fig. 28-12. This is the direct electrostatic process.

object to be printed. When the object passes under the screen, the toner, attracted to the plate, adheres to the object. Heat fuses the toner.

DIGITAL LASER SCANNER/PRINTERS

A **digital laser scanner/printer** uses a scanning unit to convert images into digital information and prints the image with a laser printer. The scanner may look like the platen of a traditional photocopier, but it functions differently. The laser printer works in the same way as xerography except a laser is used for placing the electrostatic charge on the light-sensitive drum. A scanner printer is shown in Fig. 28-13.

There are several advantages to a scanner printer. First, since the image is in digital format, it can be manipulated. For example, a picture can be moved on the page or scaled while keeping the rest of the page the same size. Another advantage is the capability of being connected to other computers, so that it can be used as an input and output device. A third advantage is that the image is stored in digital format, on disk, for example, and thus, can be used later without the original.

Fig. 28-13. This sophisticated scanner/printer can scan images (left), manipulate them (center), make copies with a laser printer, and even produce finished booklets. (Xerox Corp.)

DIRECT THERMAL COPYING

In **direct thermal copying**, heat is used to create an image on the paper. The original must have a carbon image, such as pencil or a carbon typing ribbon. If not, an electrostatic copy can be made of the original. Most electrostatic toners have carbon in them.

A heat-sensitive paper is used for copying. It is placed on top of the original and fed into the thermal copier. Infrared radiation causes the carbon image to become hot. This heat is transferred to the heat-sensitive paper. An image is then formed, Fig. 28-14.

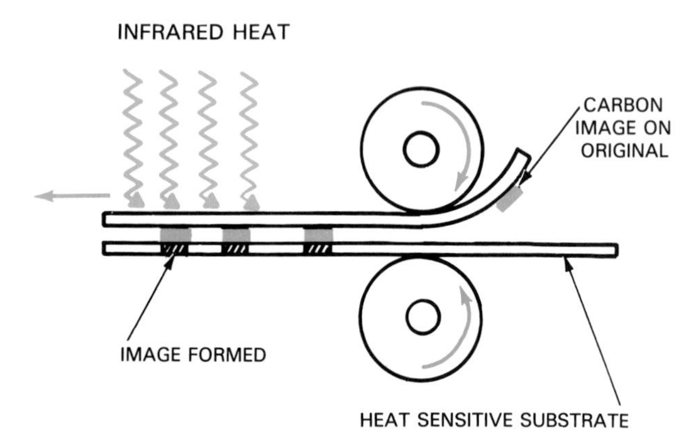

Fig. 28-14. The direct thermal process requires the use of a heat-sensitive substrate.

COLOR TECHNOLOGIES

A variety of technologies is being used for color copying. In most cases, the subtractive primaries, consisting of magenta, cyan, yellow, and black, are used to create full color on the substrate. In combination, these colors can create almost any other color in a color picture that you could imagine.

In several cases, the same basic technology is used for color copying as for monochrome. Thermal and electrostatic are two examples.

The **color electrostatic process** is most often used in color copiers, Fig. 28-15. It is similar to xerography except that a writing head is used to charge the paper. Cyan, magenta, yellow, and black toners are applied to the paper and fused with heat.

The **thermal transfer process** uses a ribbon consisting of plastic film with alternating colors of cyan, magenta,

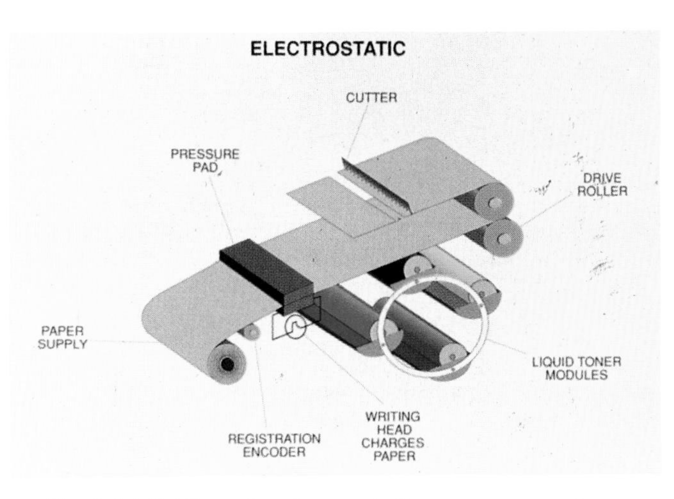

Fig. 28-15. The color electrostatic process uses four toners. (Ariel Communications)

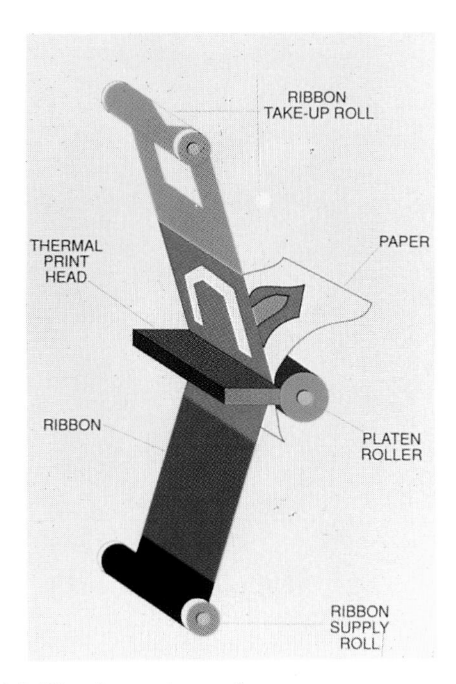

Fig. 28-16. The thermal transfer process uses a multicolor ribbon. (Ariel Communications)

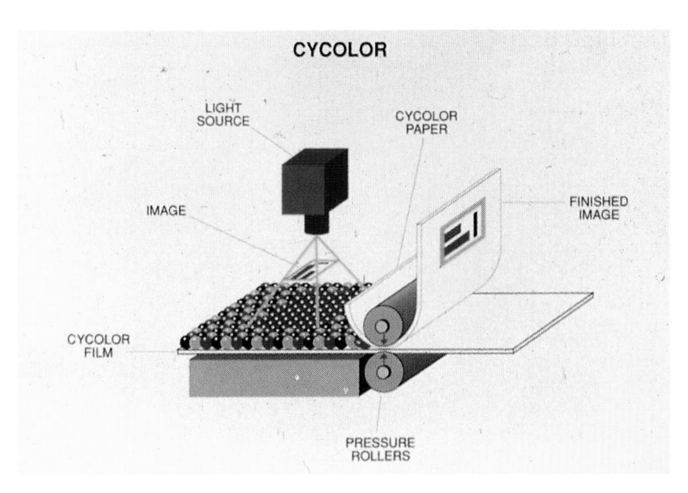

Fig. 28-17. This is the cycolor process. (Ariel Communications)

yellow, and black on the surface, Fig. 28-16. A thermal print head heats the wax so that it transfers to the substrate to be printed.

The **cycolor process** uses special film with dye capsules embedded in it, Fig. 28-17. Exposure causes these capsules to harden. After exposure, the film is pressed against a chemically treated paper and the dye is transferred to the paper wherever the capsules have not been exposed to light.

SUMMARY

A variety of copying and printing techniques have been explored in this chapter. As new technology develops, some of these techniques may become more important than the four major printing processes.

The topics explored in this chapter included ink jet printing, mimeograph duplication, spirit duplication, electrostatic copying, digital laser scanning/printing, direct thermal copying, and color copying.

WORDS TO KNOW

All of the following words have been used in this chapter. Do you know their meanings?

color electrostatic process
copying processes
cycolor process
digital laser scanner/printer
direct process
direct thermal copying
duplicating processes
electrofax
electrostatic copying
ink jet printing
mimeograph duplication
mimeograph stencil
reprography
spirit duplication
thermal transfer process
transfer process
xerography

REVIEWING YOUR KNOWLEDGE

Please do not write in this text. Write your answers on a separate sheet.

1. True or false? Processes used for printing a small number of copies from a master are called copying processes.
2. The entire category of duplicating and copying processes is called __________.

3. Which of the following is NOT true about ink jet printing?
   a. Drops of ink are sprayed onto the surface to be printed.
   b. It is controlled by computer.
   c. It cannot be used to print on irregular surfaces.
   d. The printing quality is low.
4. Mimeograph duplication is a(n):
   a. Ink spraying process.
   b. Stencil process.
   c. Inkless process.
   d. Electrostatic process.
5. When paper passes through a spirit duplicator, it is dampened with:
   a. Alcohol.
   b. Ink.
   c. Water.
   d. Acid.
6. Name the three major kinds of electrostatic printing.
7. List three advantages of the digital laser scanner/printer.
8. Which of the following is most often used in color copiers?
   a. Thermal transfer process.
   b. Cycolor process
   c. Color electrostatic process.
   d. None of the above.

APPLYING YOUR KNOWLEDGE

1. Investigate current technology in reprography. Write a report to be presented in class.
2. Collect printed samples of the various copying and duplicating processes. Create a display, labeling each one.
3. Write a report comparing quick copying to more conventional printing methods.
4. Investigate ways in which copying methods are used in telecommunications. Report your findings to class.

29

CHAPTER

FINISHING OPERATIONS

After studying this chapter, you will be able to:
- *Identify various finishing operations.*
- *Explain paper cutting and trimming operations and safety rules.*
- *Describe folding, perforating, slitting, die-cutting, embossing, thermography, and laminating methods.*
- *Explain how punching and drilling are done.*
- *Discuss assembling methods.*
- *Identify binding options.*

After printing, most products require additional operations before an item is delivered, Fig. 29-1. These are known as **finishing operations**. Finishing operations make the final product more attractive and useful. Some examples include:
- Cutting and trimming.
- Folding.
- Perforating and slitting.
- Die-cutting.
- Embossing.
- Thermography.
- Laminating.

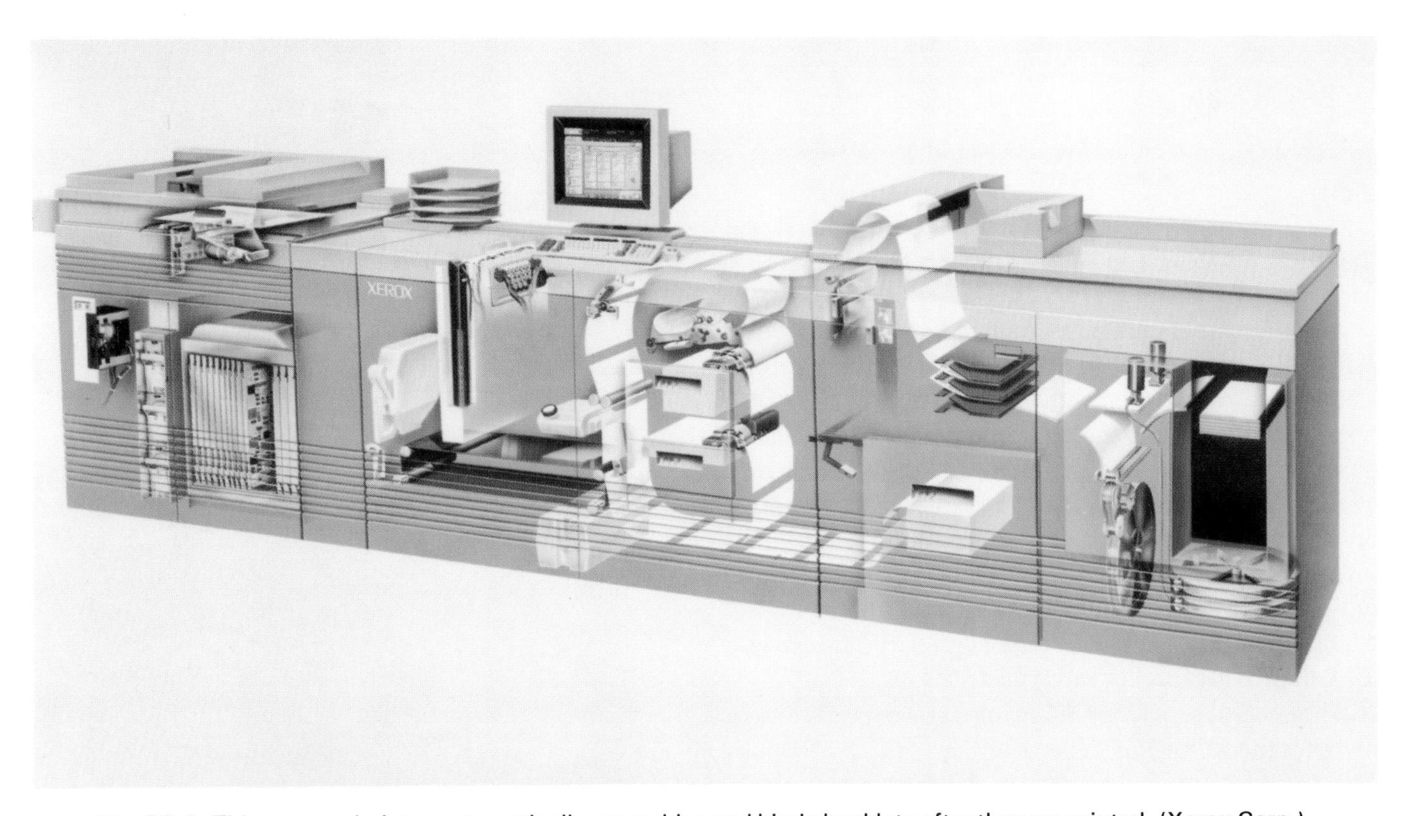

Fig. 29-1. This scanner/printer automatically assembles and binds booklets after they are printed. (Xerox Corp.)

- Punching and drilling.
- Assembling.
- Binding.

CUTTING AND TRIMMING

Paper cutting is done before and after printing. Before printing, large sheets of paper are cut into press sheets for printing. After printing, the press sheet is often cut into finished page size. After sheets are bound together, the edges are trimmed to make them even.

Paper cutters come in a variety of types and sizes. *Paper trimmers* are used to cut a small number of sheets at a time, Fig. 29-2. The *hydraulic paper cutter* is used for cutting a large quantity of sheets in one cut. This is also known as a *guillotine cutter* because of the cutting method.

Fig. 29-2. Paper trimmers are used when only a few sheets need to be cut.

POWER PAPER CUTTER OPERATION

A power paper cutter is shown in Fig. 29-3. As when using all tools, safety rules should be learned before using the power paper cutter. For instance, only one operator should use the equipment at a time. Those observing should stand safely away (about three feet away). Keep hands away from the table when clamping or cutting paper. Also, the paper cutter should be turned off when adding or removing paper. These are only a few of the safety rules that must be followed when working with a power paper cutter. Your instructor will provide further safety instructions.

The general procedure for using a power paper cutter involves the following steps:

1. Jog the paper to even the edges. Measure and mark the top sheet for the cuts.

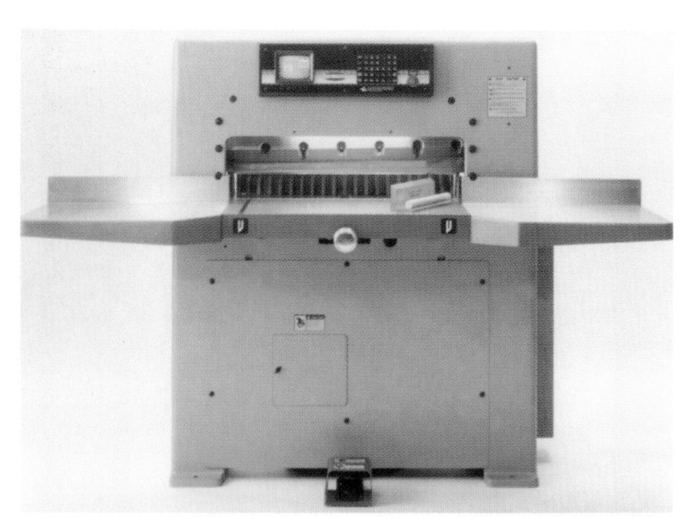

Fig. 29-3. A power paper cutter should always be used in a safe manner. (Challenge Machinery Co.)

2. Set the back fence to the desired measurement with the hand wheel, Fig. 29-4. Lock the hand wheel.
3. Place the paper against the left side of the back fence (back gage). If necessary, use chipboard on top of the paper to avoid clamp marks, Fig. 29-5.
4. Turn on the cutter and clamp the paper. Cut the paper by depressing the two buttons at the same time, Fig. 29-6.
5. Unclamp the paper. Turn off the cutter. Remove the paper.

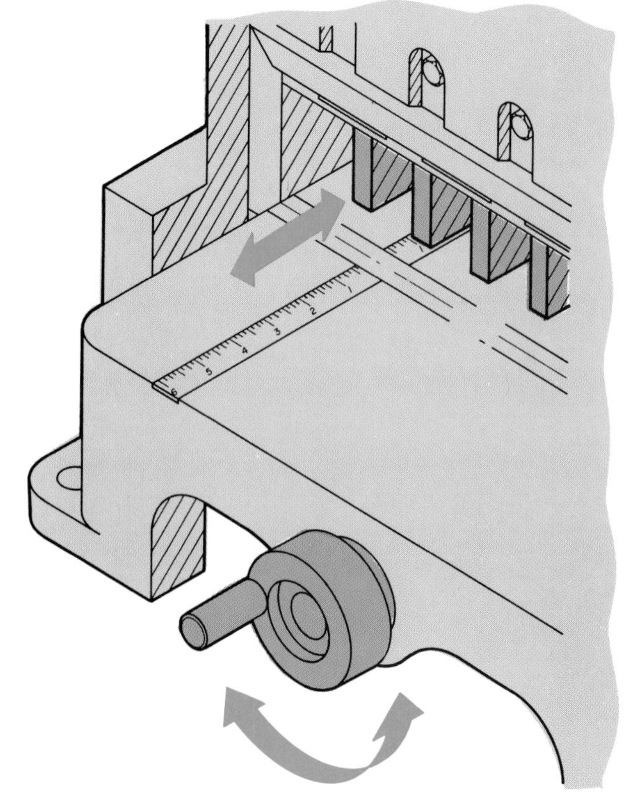

Fig. 29-4. The back fence is moved by turning the hand wheel. (John Walker)

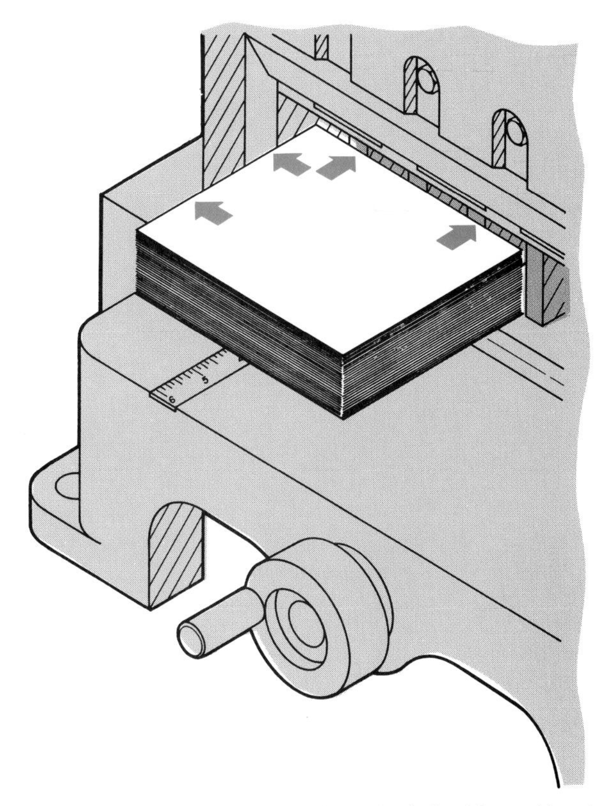
Fig. 29-5. Place the paper against the left side and back fence. (John Walker)

Fig. 29-7. The operator of the hand-lever paper cutter uses a lever (right) to cut the paper. (Challenge Machinery Co.)

Fig. 29-6. Cut the paper by depressing both buttons at the same time.

6. Turn on the cutter and lower the clamp. Turn off the power.

HAND-LEVER PAPER CUTTER OPERATION

The hand-lever paper cutter works in a similar manner to the power paper cutter, Fig. 29-7. However, additional safety precautions are necessary. You must be very cautious when cutting paper with the lever. Make sure that you understand the rules of operation before using this machine.

The general procedure for using the hand-lever paper cutter is as follows:

1. Open the paper clamp and place the paper against the left side and back fence.
2. Move the back fence hand wheel for the depth of cut. When adjusted, lock the hand wheel with the thumbscrew.
3. Clamp the paper. Make sure that observers are outside the machine safety zone.
4. Release the lever safety and pull the lever down with both hands to cut the paper. Return the lever to the top position.
5. Unclamp the paper and remove the cut stack.

FOLDING PAPER

Folding can be done manually or with a power folder. For manual folding, a **bone folder** is used, Fig. 29-8. Fold the paper and align the edges. Then crease the folded paper, starting at the middle and creasing to each side.

The two major kinds of power folders are the knife folder and the buckle folder, Fig. 29-9. The **knife folder** uses a steel knife that pushes the paper between two rollers for folding. The **buckle folder** uses rollers to push the paper against a stop. At this point, the paper is buckled and is caught between two more rollers. These rollers crease the fold.

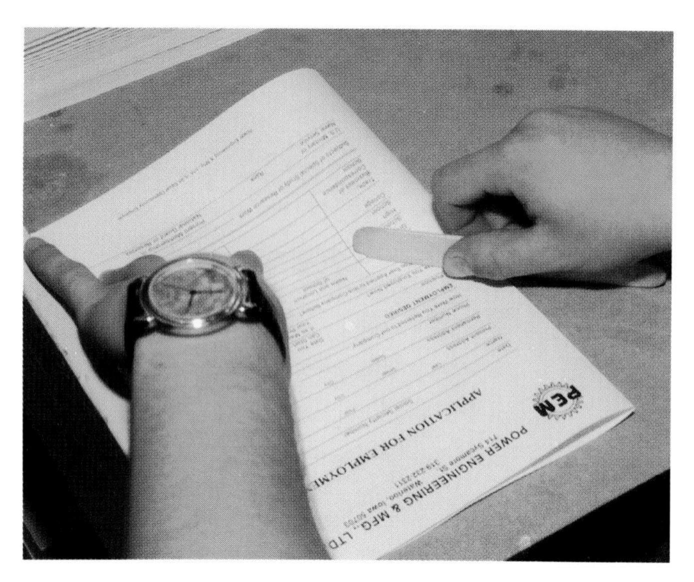

Fig. 29-8. A bone folder is used in this manner.

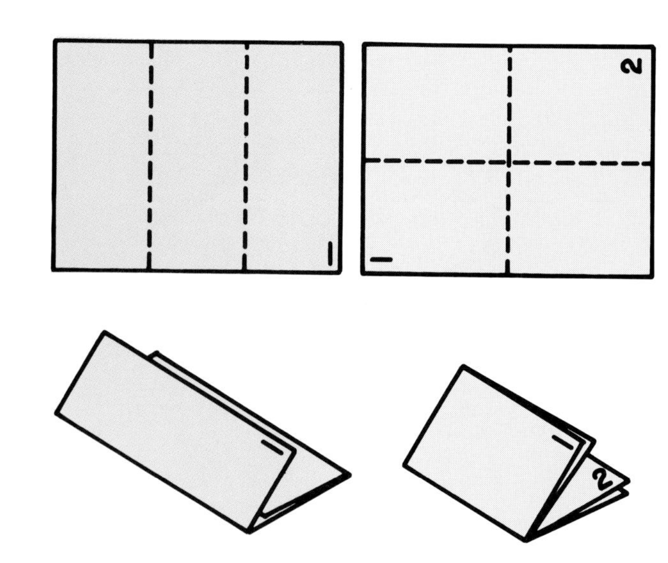

Fig. 29-10. A letter fold is shown on the left and a French fold is shown on the right.

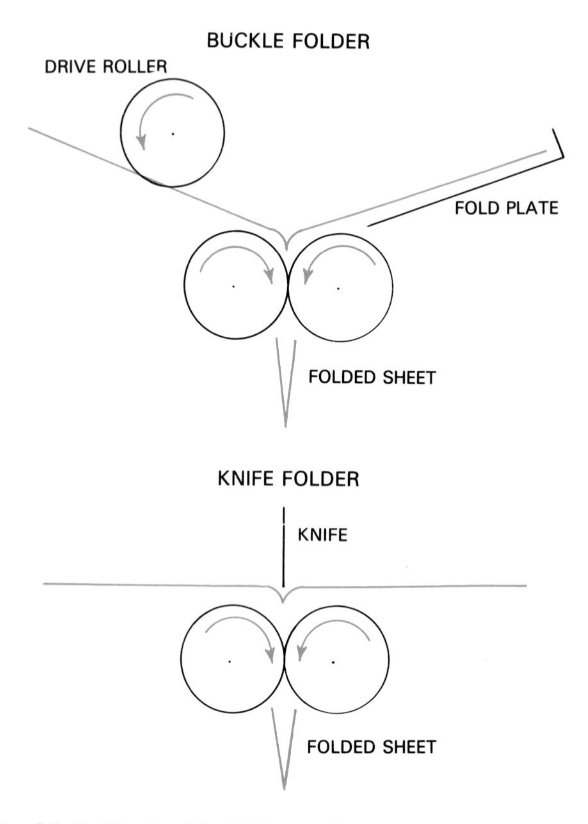

Fig. 29-9. The buckle folder and knife folder are shown here. (A.B. Dick Co.)

The two basic folds are the parallel fold and the right angle fold. See Fig. 29-10.

A **parallel fold** is made by aligning the folds parallel with one another. A **letter fold** is an example of this type of fold. Whenever possible, make these folds parallel with the paper grain for the best crease.

A **right angle fold** has at least one fold at 90° to the other folds. A **French fold** is an example. This fold is often used for greeting cards.

Table-top buckle folders are often used in schools, Fig. 29-11. The general procedure for using this type of folder is described below and is shown in Fig. 29-12.

1. Make the desired folds on a sheet of the paper to be folded. Measure the folds.
2. Set the fold plates for the folds to be made.
3. Set up the feeding system. Adjust the side guides so the paper has a small amount of clearance. The feed wheel should be in the center of the paper.
4. With a single sheet, adjust the feed wheel tension. Feed a sheet into the feed wheel with the hand wheel. Adjust the tension screw while pulling back on the sheet. Adjust for slight resistance.
5. Feed a sheet through by hand and set up the delivery system.
6. Turn on the folder and feed a few sheets. Check the folds and adjust the fold plates if needed.

Fig. 29-11. This is a table-top buckle folder. (A. B. Dick Co.)

SETTING UP A BUCKLE FOLDER

A

B

C

D

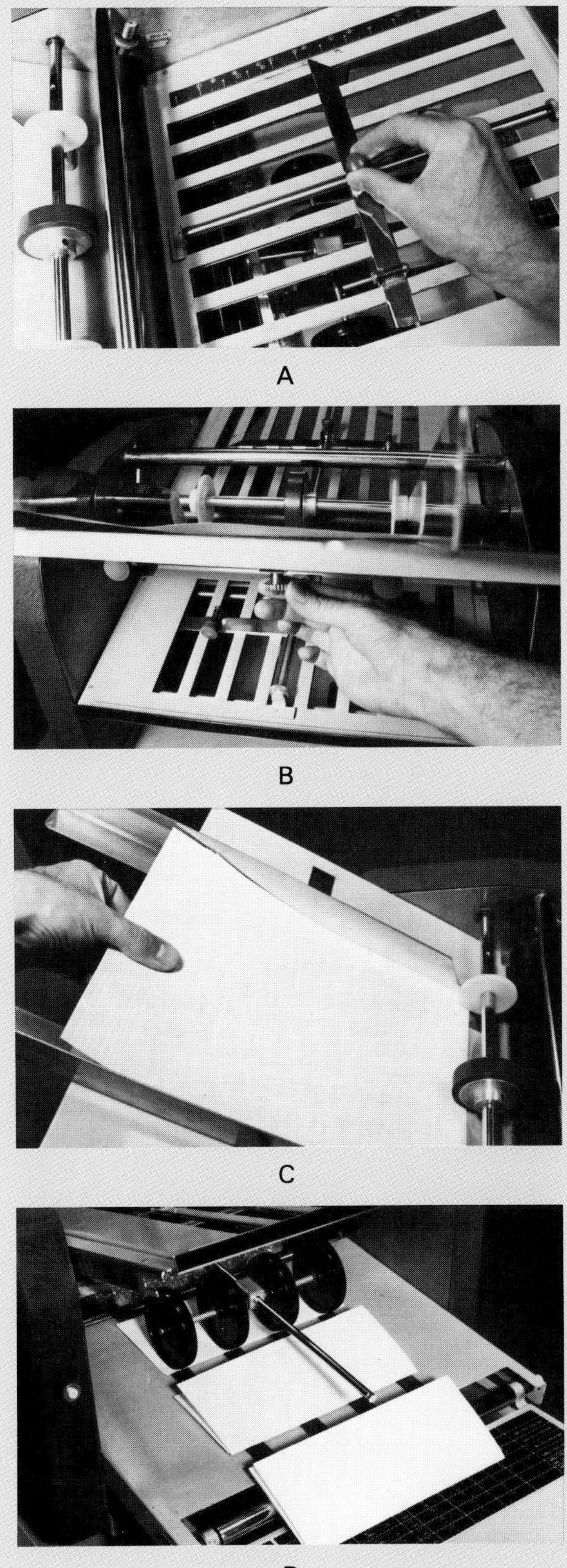

Fig. 29-12. The steps in setting up a buckle folder are shown here. A–Set up the fold plates. B–Adjust the feed wheel tension. C–Feed the paper. D–Inspect the folded sheets and then finish folding.

7. Fan the sheets so they are 1/8 in. apart at the edge. Place the paper in the feed tray so the paper edges are visible in the back. Hold the paper with the index finger and thumb.
8. Turn on the machine at slow speed. Place the paper against the feed roller and let the paper feed from under your thumb. Continue until all the sheets are folded.

SCORING

Thick papers must be scored before they are folded. Otherwise, the fold may be uneven. **Scoring** involves crushing paper along the fold line. This is especially important when folding across the grain. Fold the paper away from the indented surface, Fig. 29-13.

Scoring can be done with a scoring rule or scoring wheels, Fig. 29-14. A **scoring rule** is rounded on the top.

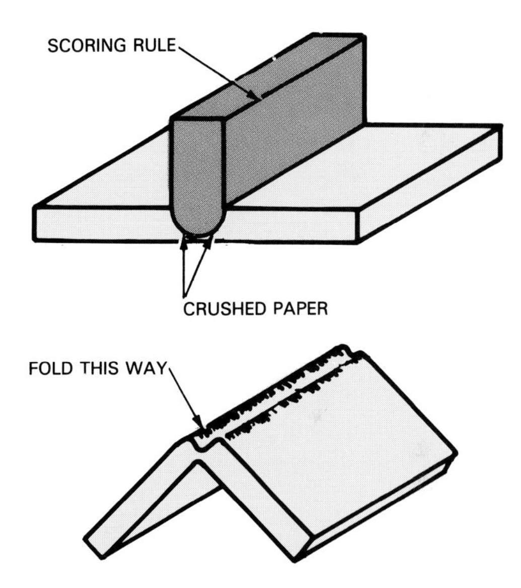

Fig. 29-13. Fold paper away from the scored surface.

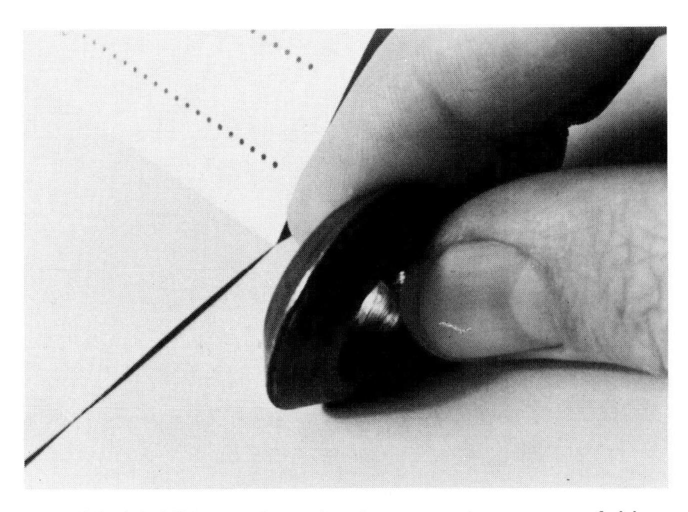

Fig. 29-14. This scoring wheel was used on a paper folder to make the fold on the left.

It is locked in a chase and placed on a platen press. When the platen comes in contact with the rule, the paper is scored.

Scoring may also be done with scoring wheels. These wheels are mounted on folders or on the delivery side of presses.

PERFORATING AND SLITTING

Perforating involves cutting small slits into paper so it can be separated. Tickets are often perforated. See Fig. 29-15.

Perforating can be done in the same way scoring is done. A perforating rule can be used on a platen press. Also, perforating wheels can be used on folders and some presses.

Slitting involves cutting paper. This can be done on a folder or press with slitting wheels, Fig. 29-16.

DIE-CUTTING

Die-cutting is done by pressing a sharp steel rule into the paper surface, Fig. 29-17. This can be done on a platen press or special die-cutting equipment.

Whenever paper is cut into unusual shapes, die-cutting is necessary. Paper boxes are die-cut and then folded.

EMBOSSING

Embossing is a method of raising the paper surface. Embossing is done with a convex (raised surface) die and a concave (sunken surface) die. When these are pressed together, the paper is molded or embossed, Fig. 29-18. Embossing is done on a platen press.

Fig. 29-15. This ticket was perforated using a perforating wheel.

Fig. 29-16. A slitting wheel cuts the paper.

Fig. 29-17. The die shown on the press is being used to cut the arrow shape from the paper.

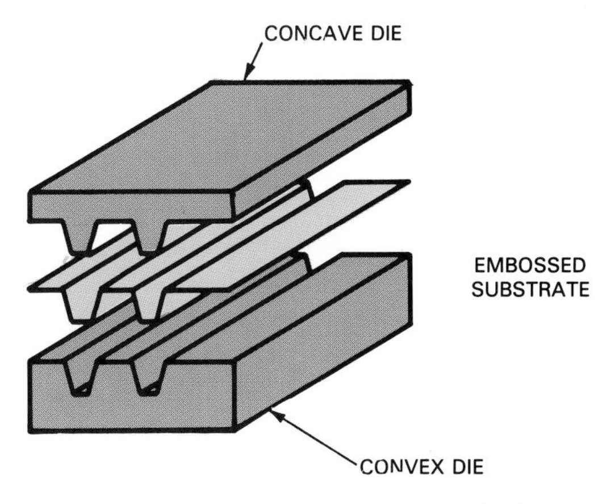

Fig. 29-18. Embossing is done with dies.

Regular embossing has a raised printed image. **Blind embossing** has a raised surface. However, it involves no printing.

THERMOGRAPHY

Thermography, like embossing, is a method for raising a printed surface. In this process, a resin powder is placed on an image with wet ink. When heated, the powder expands and fuses with the ink.

Resin powder is available in a variety of regular colors as well as metallic colors. Translucent powder can be used that allows the ink color to show. Fine image detail requires a finely ground powder. Resin powder is available in various grinds, from coarse to fine.

Thermography is very popular for business cards and greeting cards. The procedure for using this technique on a small card is as follows:

1. Print the product.
2. Place the resin powder in a pan. Lift the powder on the card surface and shake it over the surface, Fig. 29-19. Tap the excess off into the pan.
3. Heat the powder until it fuses evenly on the surface, Fig. 29-20. This can be done on a conveyor-type heating unit, over a hot plate, or in an oven. Follow all safety precautions with heating units.
4. Lay the cards out individually and stack them when they are cool.

LAMINATING

Laminating is the process of attaching plastic to one or both sides of a printed sheet, Fig. 29-21. Heat and pressure are used to apply the plastic, which gives the surface a protective coating.

PUNCHING AND DRILLING

Punching involves pressing a die through the paper to form a hole, Fig. 29-22. Any hole shape can be made with this technique.

Fig. 29-19. Thermography involves placing resin powder on the wet ink.

Fig. 29-20. A close-up of a finished card shows how the resin fuses to the card and produces a raised surface.

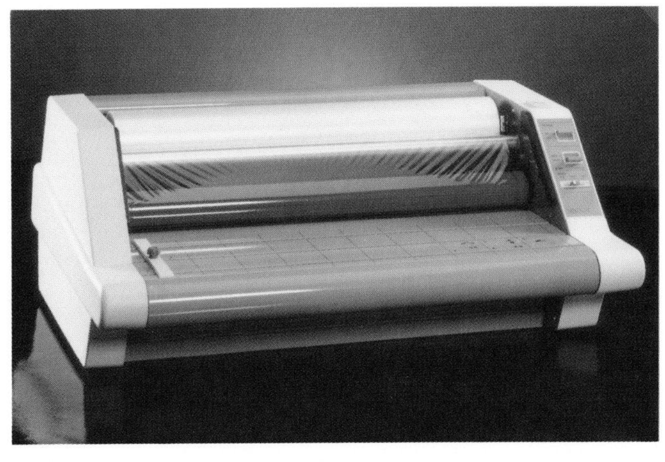
Fig. 29-21. This laminator places a thin layer of plastic on the surface of printed products. (General Binding Corp.)

Fig. 29-22. A three-hole punch is commonly used to punch paper for a ring binder.

Fig. 29-24. A drill is being placed in the chuck.

Drilling is used for round holes, Fig. 29-23. This method is used when a large number of sheets need holes in them. A paper drill is used as follows:

1. Select a drill and press it into the chuck, Fig. 29-24. Be sure to keep your fingers away from the cutting edge.
2. Adjust and clamp the back fence. Also set the stops for multiple holes.
3. Place the paper stack in position.
4. Turn on the drill and wax the cutter tip.
5. Using the lever, drill through the paper. Immediately bring the drill back up with the lever.
6. Turn off the machine and remove the paper.

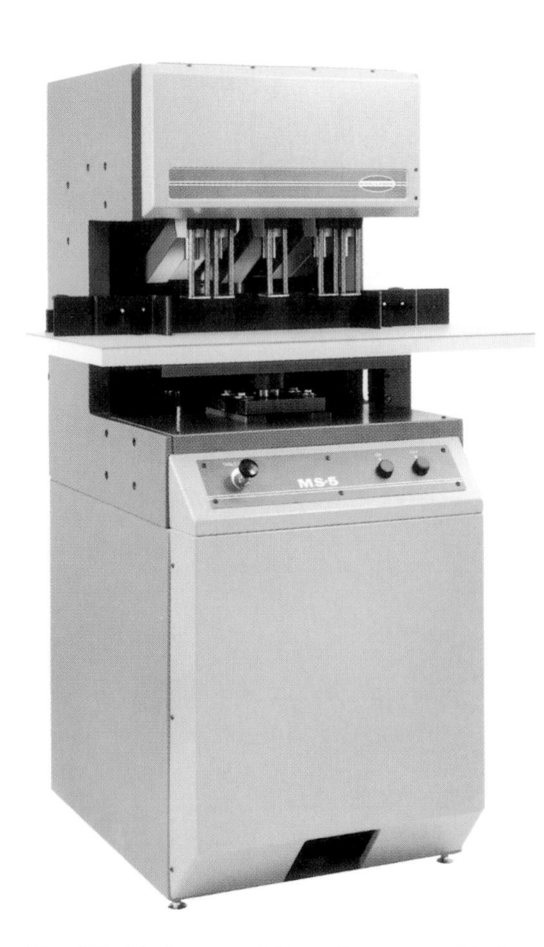

Fig. 29-23. A paper drill produces round holes. (Challenge Machinery Co.)

ASSEMBLING

Assembling takes place when sheets are placed together for binding. Assembly can be done by hand or by machine. The three methods of assembly are:

- Collating.
- Inserting.
- Gathering.

COLLATING

Collating is the process of placing individual sheets in order for binding. A semi-automatic collator is shown in Fig. 29-25. When the sheets exit the bins, the operator removes them and places them in the tray below. The collated sheets are cross-stacked.

INSERTING AND GATHERING

Inserting and gathering refer to assembling signatures for binding. A **signature** is a large printed sheet that is folded into pages.

Inserting takes place when signatures are placed inside one another for binding. Thin booklets and magazines are usually assembled in this manner, Fig. 29-26.

Gathering refers to signatures that must be stacked for binding. Thick products, such as books, are usually assembled by gathering.

Fig. 29-25. This is a semi-automatic collator. (Challenge Machinery Co.)

BINDING

Binding is the process of fastening pages together. It is done after assembly. There are a variety of binding techniques. These include:

- Adhesive binding.
- Mechanical binding.
- Loose-leaf bindings.
- Wire staple binding.
- Sewn binding.
- Welding.

ADHESIVE BINDING

Adhesive binding is used for paperback books as well as note pads. Another name for this type of binding is **perfect binding**.

Using adhesive binding to make note pads involves the following steps:

1. Use a sheet counter to make evenly sized pads, Fig. 29-27. Place a piece of chipboard on the back of each pad.
2. Stack the pads, and jog them to the binding edge.
3. Use a weight or clamp to hold the sheets together, Fig. 29-28. A brick will work well for this. A padding press can be used for larger quantities of pads, Fig. 29-29.
4. Apply the adhesive with a brush. Brush out from the center. Apply two thin coats, 10 minutes apart.
5. After the pads are dry, separate them with a padding knife, Fig. 29-30.

MECHANICAL BINDING

Mechanical binding methods include *spiral wire* and *plastic cylinder*, Fig. 29-31. In both cases, holes are drilled or punched in the binding edge, Fig. 29-32. The binding is then placed through these holes. Products bound in this way will lie flat when opened. School notebooks and cookbooks are two common examples.

Fig. 29-26. These magazines were assembled by inserting folded signatures inside one another.

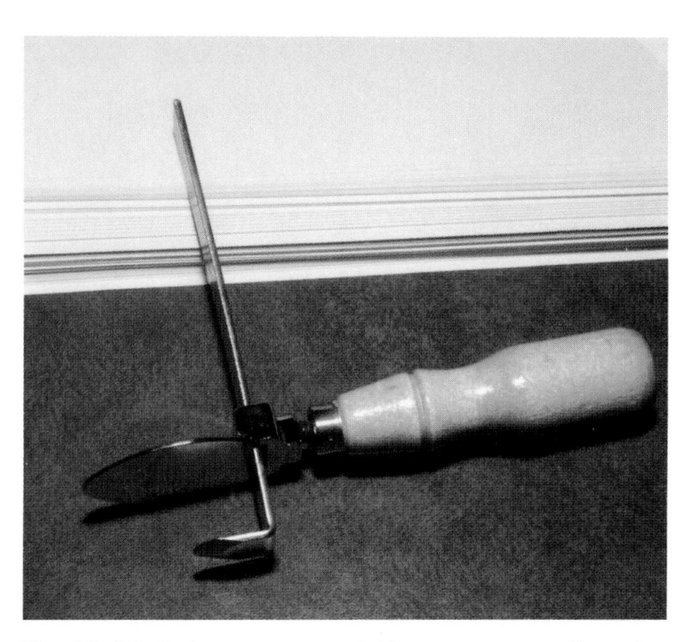

Fig. 29-27. A sheet counter quickly measures small stacks of paper.

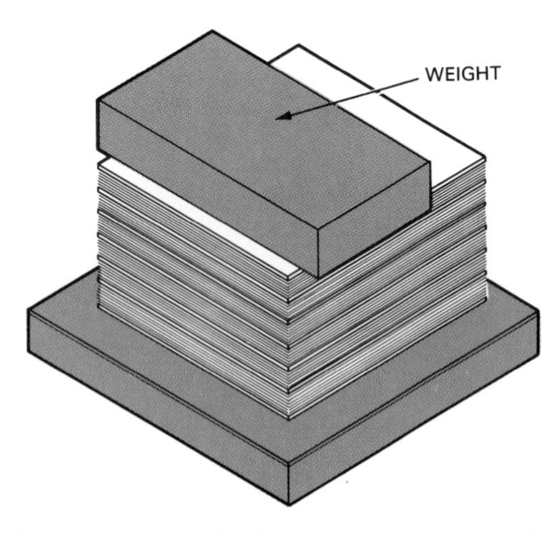

Fig. 29-28. A weight is placed on the pads to hold them together while the adhesive is applied. (John Walker)

Fig. 29-30. Pads are separated with a padding knife.

The steps involved in plastic cylinder binding are as follows:

1. Set the binding punch so holes are evenly spaced on the paper. Punch a few sheets at a time, until all sheets are punched.
2. Choose a plastic cylinder that is at least 1/8 in. larger than the stack to be bound. Insert the cylinder onto the binding machine and open the rings with the lever.
3. Place the sheets on the rings, Fig. 29-33.
4. Close the rings with the lever and remove the bound booklet.

Fig. 29-31. These are examples of spiral wire and plastic cylinder binding. (John Walker)

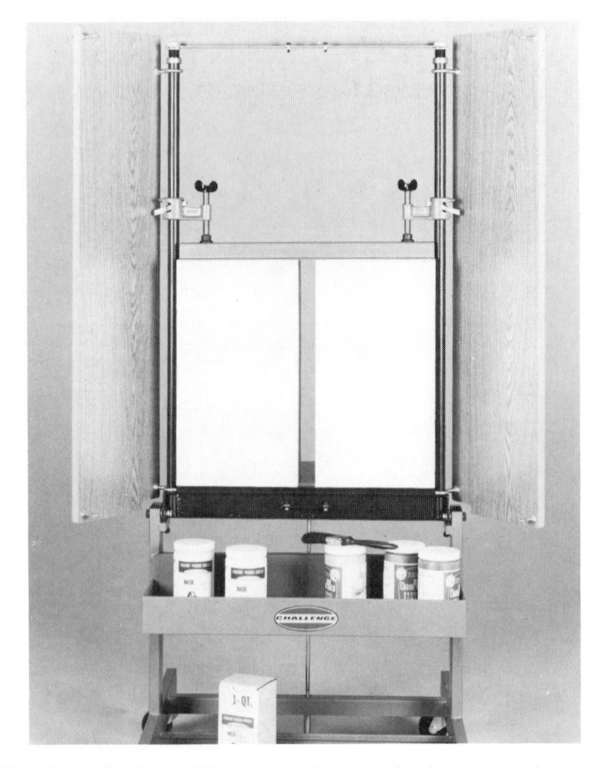

Fig. 29-29. A padding press is used for large numbers of pads. (Challenge Machinery Co.)

Fig. 29-32. This plastic cylinder binding unit is used for both punching holes and placing the plastic cylinder on the edge. (General Binding Corp.)

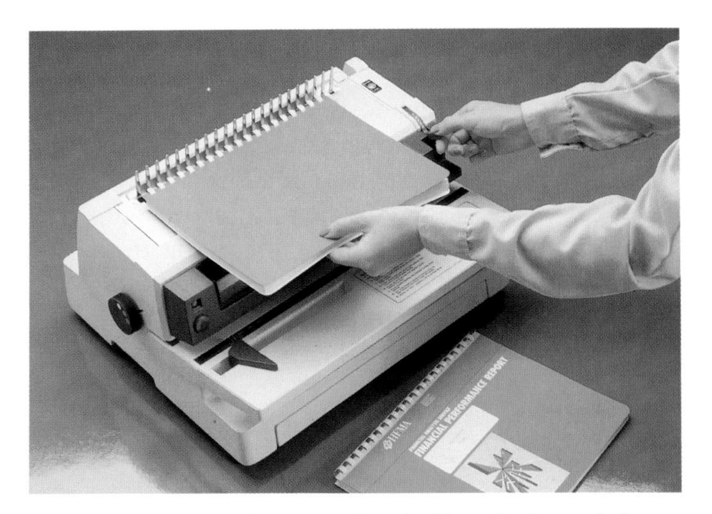

Fig. 29-33. Place sheets on the plastic cylinder and close the rings. (General Binding Corp.)

Fig. 29-34. Loose-leaf binding methods include ring binding (left) and prong fasteners (right).

LOOSE-LEAF BINDINGS

Loose-leaf bindings allow the easy addition and removal of sheets. Loose leaf bindings include:

- Three-ring binders.
- Post binding.
- Prong fasteners.
- Friction binding

Three-ring binders

Three-ring binders hold paper with rings that easily open. The three-ring binder, commonly used in schools, is the most common ring binder, Fig. 29-34.

Post binding

Post binding is done with a post and cap. The post is inserted through a hole in the paper. The cap is screwed on to the post to hold the paper together.

Prong fasteners

Prong fasteners have a base with prongs attached. These prongs are inserted through holes in the paper and a compressor is placed on the prongs. The prongs are bent down against the compressor and locking slides hold them in place.

Friction binding

Friction binding uses a plastic U that slips over the binding edge. This method is often used with plastic covers to bind reports.

WIRE STAPLES

Wire staples can be used for saddle-wire binding or side-wire binding. **Saddle-wire binding**, also called saddle-stitching, is often used to bind signatures that are inserted into one another. Magazines are often saddle-stitched. See Fig. 29-35.

Fig. 29-35. This saddle-wire binding machinery is used to produce stitched booklets, magazines, and catalogs. (McCain Manufacturing Corp.)

Side-wire binding is often used for gathered signatures. Staples are placed along the binding edge. See Fig. 29-36.

When the staple is made from a roll of wire, the machine is called a **stitcher**, Fig. 29-37.

Fig. 29-36. Side-wire binding is being used to staple this printed product.

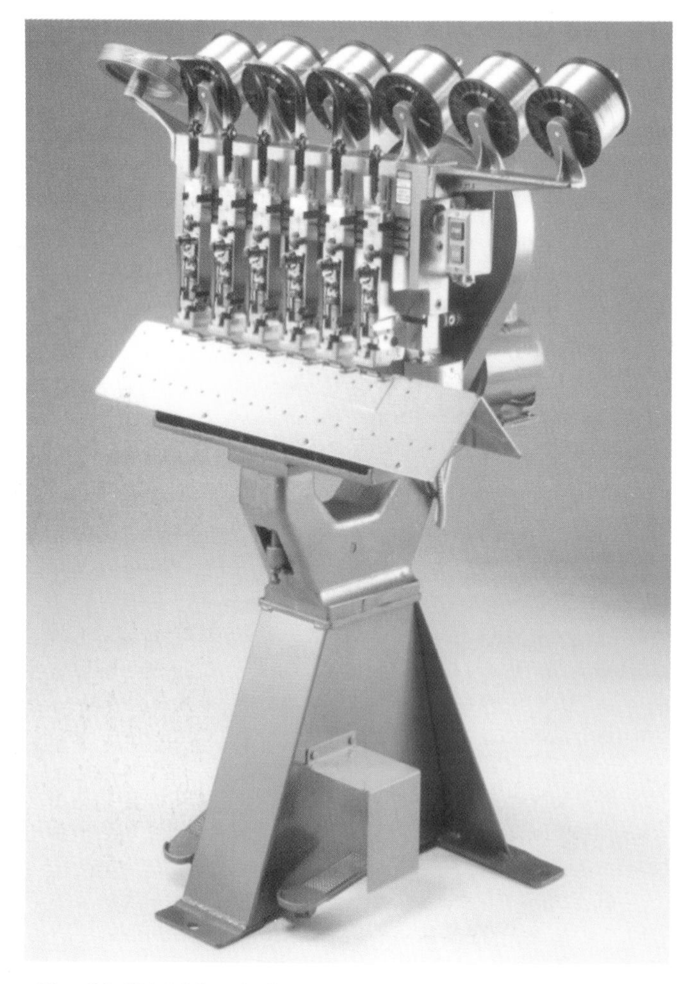

Fig. 29-37. This stitcher can place staples in six places at one time. (Interlake Packaging Corp.)

SEWN BINDING

Sewn binding is the most durable form of binding. Thread is used for binding the sheets together. Like staple binding, pages can be saddle-sewn or side sewn. For thick books, the signature to signature method is used. Signatures are gathered and sewn together.

Hard covers are often used with sewn binding. The cover is placed on the book after binding. This is known as **case binding** and the book is called a *case-bound book*, Fig. 29-38.

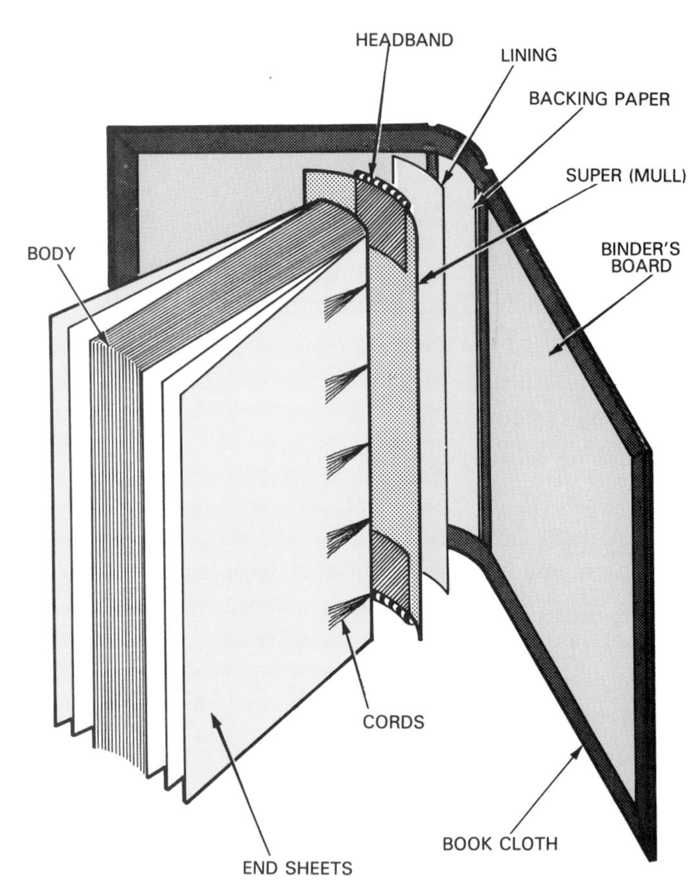

Fig. 29-38. This is an example of a case-bound book. (John Walker)

WELDING

Welding involves exposing the binding edge to ultrasonic waves. This fuses the sheets together.

SUMMARY

People often think that image transfer is the last step in printing. However, many operations are performed after printing. These finishing operations have been described in this chapter. Finishing operations discussed in this chapter included cutting and trimming, folding, perforating and slitting. Also included were die-cutting, embossing, thermography, laminating, punching and drilling, assembling, and binding.

WORDS TO KNOW

All of the following words have been used in this chapter. Do you know their meanings?

adhesive binding
assembling
binding
blind embossing
bone folder
buckle folder
case binding
collating
die cutting
drilling
embossing
finishing operations
French fold
friction binding
gathering
inserting
knife folder
laminating
letter fold
loose-leaf binding
mechanical binding methods
parallel fold
perfect binding
perforating
post binding
prong fasteners
punching
regular embossing
right angle fold
saddle-wire binding
scoring
scoring rule
sewn binding
side-wire binding
signature
slitting
stitcher
thermography
three-ring binders
welding
wire staples

REVIEWING YOUR KNOWLEDGE

Please do not write in this text. Write your answers on a separate sheet.

1. Name ten finishing operations described in this chapter.
2. True or false? Paper cutting is done before and after printing.
3. Which of the following folders uses rollers to push the paper against a stop?
   a. Bone folder.
   b. Knife folder.
   c. Buckle folder.
   d. None of the above.
4. _________ involves cutting small slits into paper so it can be separated.
5. True or false? Regular embossing has a raised surface, but involves no printing.
6. _________ involves placing a resin powder on an image and fusing it with the ink to produce a raised printed surface.
7. _________ is the process of attaching plastic to one or both sides of a printed sheet to protect the surface.
8. Which of the following involves pressing a die through paper to create a variety of hole shapes?
   a. Punching.
   b. Perforating.
   c. Drilling.
   d. None of the above.
9. Which of the following assembly methods involves placing individual sheets in order for binding?
   a. Gathering.
   b. Inserting.
   c. Collating.
   d. None of the above.
10. Name six binding techniques.

APPLYING YOUR KNOWLEDGE

1. Analyze periodicals or books and determine the steps involved in finishing.
2. Produce a graphic arts product that includes finishing operations such as perforating.
3. Investigate industrial methods of finishing. Write a report.
4. Produce greeting cards and note pads using various finishing techniques described in this chapter.

Section ACTIVITIES 5 GRAPHIC COMMUNICATION-GRAPHIC ARTS AND DESKTOP PUBLISHING

BASIC ACTIVITY:
Thermal Screen Printing

Introduction:

Thermal screen printing is a quick and easy way to try a printing method. See Section Fig. 5-1. Although it is not used in industry, this technique closely simulates photographic screen printing which is used in industry.

Section Fig. 5-1. A variety of products can be printed with thermal screens. (Welsh Products, Inc.)

Guidelines:
- Work individually or in small groups.
- Use one color of ink.
- Your instructor will specify the maximum image size.

Materials and equipment:
- Thermal screen, screen frame, double-sided tape, and in some cases, removable clear tape.
- Artwork.
- Thermal copier.
- Substrate.
- Water-based screen printing ink.

Procedure:
1. Design a one-color screen printing job, using the maximum dimensions provided by your instructor.
2. Use the computer or appropriate manual techniques to complete the final design. If the image is not carbon-based, make a photocopy on a copy machine.
3. For a small illustration, place the illustration face-up between the backing sheet and screen. In the case where the sheet containing the artwork is larger than the thermal screen, remove the backing sheet and place the film rough (screen) side up over the artwork. Center the screen and tape it on the front edge with removable clear tape.
4. Expose the screen on the transparency setting. Readjust the setting if needed. Remove the backing sheet from the stencil.
5. Attach the screen to the frame with the double-sided tape. Check for pinholes and cover them if needed.
6. Proceed with printing on the substrate.
7. Wash the screen with warm water when finished. Remove the stencil from the frame so the frame can be reused.

Evaluation:

Considerations for evaluation are:
- Correct design procedure followed.
- Screen printing work habits.
- Quality of finished print.

ADVANCED ACTIVITY:
Survival Guide to Campus

Introduction:

Students who have been on a school campus for awhile learn strategies for survival. Information such as a campus map, descriptions of various clubs and activities, and even the best places to eat can be invaluable to new students. This activity will give you a chance to help new students make a smooth adjustment to your school.

Guidelines:
- Group activity.
- Guide will be produced on 8 1/2 x 11 paper, folded into a four- page signature.
- Finished size before trimming will be 5 1/2 x 8 1/2.
- Pages should include both text and graphics.

Materials and equipment:
- Appropriate composition equipment.
- Printing equipment if the guide will be printed.

Procedure:
1. Identify the audience for the guide and the topics that should be included.
2. Make decisions on margins and page elements that will be repeated throughout the publication. For

example, pages will need a page number that is in the same location on every page. Also, make decisions on type style and size, and leading.
3. Divide responsibilities and begin designing the publication. Follow standard design procedures.
4. Compose the finished pages and make a copy on a copy machine to be used as a proof copy. Put all the proof pages together in booklet form and check for errors. Make the needed corrections.
5. If the booklet is to be printed, proceed with the necessary steps. If not, make a photocopy of the guide for each member of the group.
6. For printed booklets, fold the pages into signatures and insert the pages in order for each booklet. Saddle stitch the booklets. Finish the booklets by trimming the outside edges.

Evaluation:

Considerations for evaluation are:

- Design quality.
- Quality of written information, including grammar and spelling.
- Overall quality of the publication.
- Individual and group effort.

ABOVE AND BEYOND ACTIVITY: Purchasing a Color Prepress System

Introduction:

Pretend that you are going into business for yourself, specializing in the design and layout of printed products. Realizing that many publications are using color, you want to investigate that option as part of your equipment and software package.

Section Fig. 5-2. The computer will be a central element in the color prepress systems you choose. (Agfa Compugraphic)

Guidelines:

- Small group activity.
- Computer, peripherals, and software will be required. This can be a package deal from a single vendor or products from various vendors.
- Outside services can be investigated as an alternative to some part of the system.
- Needs include input of full color slides or photos, color proofing, and output of color separations.
- You have a total of $30,000 to $50,000 to spend on the system, depending on how well you convince the bank of the necessity of the system.
- System should prepare publications up to the printing stage.

Materials and equipment:

- Periodicals that provide information on color prepress systems.
- Literature on the products.

Procedure:

1. Review recent literature on color prepress systems. Brainstorm and list the features that your system should contain.
2. Divide up responsibilities in your group and begin the investigation. For example, one person might look at input devices while another looks at computers. See Section Fig. 5-2. Begin by looking at all possible options and then narrow the field to those that are affordable.
3. Select the two best options for each piece of equipment and software, with one system being in the $30,000 range and the other being in about the $50,000 range. Also, if time allows, investigate outside services that might allow you to subcontract parts of a job (such as slide scanning).
4. Write a report comparing the two options, based on the major features. If you think you will need outside services, mention those as well. Include in the report how much business you will have to generate to pay for the system, figuring 50% profit on each printing job. The bank loan will be for 5 years at 12% interest.
5. Since you need a loan for the system, you will need to make a presentation to your financial institution explaining what each system will allow you to do. The rest of the class can act as the financial institution. At the end of the presentation have the other class members decide for which system, if any, they would provide the loan.

Evaluation:

Considerations for evaluation are:

- Quality of the investigation.
- Report quality.
- Convincing and well-prepared presentation.

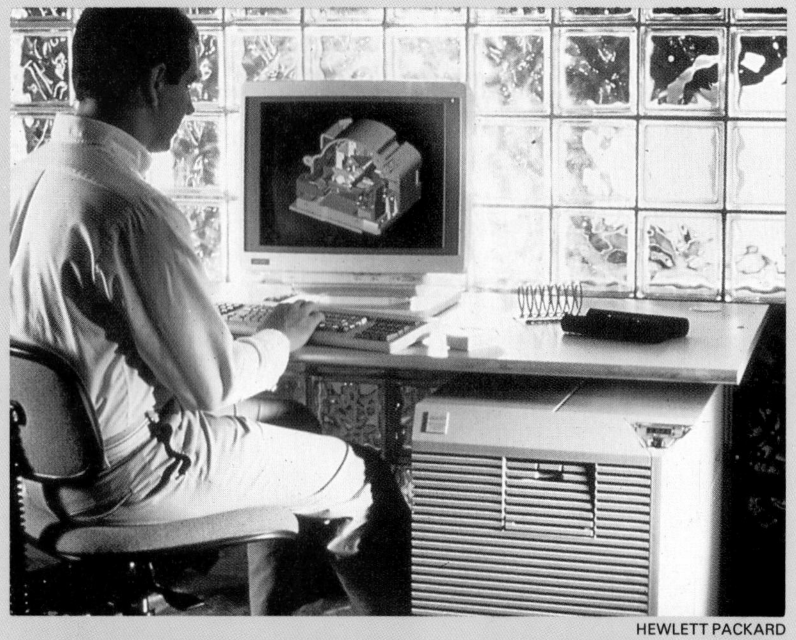

HEWLETT PACKARD

MOTOROLA, INC.

BOWLING GREEN STATE UNIVERSITY, VISUAL COMMUNICATION TECHNOLOGY PROGRAM

TRW INC.

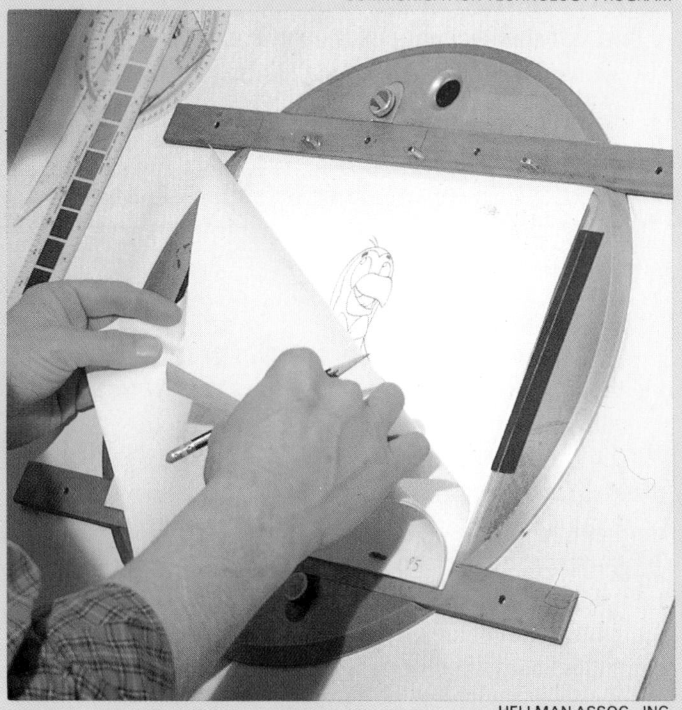

HELLMAN ASSOC., INC.

section 6

Electronic Communication

The advances in electronic communication are constantly occurring. These engineers are working to develop interactive, multicolor onscreen menus that provide instant reference for picture, sound, signal source, picture-in-picture, and other functions. (Zenith Electronics Corp.)

INTRODUCTION TO ELECTRONIC COMMUNICATION

After studying this chapter, you will be able to:
- *Define electronic communication.*
- *Explain the concepts of current, voltage, and resistance.*
- *Identify major landmarks in the development of electronic communication systems.*
- *Describe the electronic communication system process.*

Many devices that you use every day, from an alarm clock to a microwave oven to a computer, have electronic components. Some of these items may or may not be communication devices. In order to be classified as **electronic communication**, a primary purpose of the system should be sending messages by electrical signal, Fig. 30-1. For example, a facsimile machine, which can transmit pictures over telephone lines, would be part of an electronic communication system. However, a washing machine, which may have electronic components, would not be classified as an electronic communication system.

In this chapter, some of the basics of electricity and electronics will be covered. This will help you to better understand how they are used in communication. Also, components of an electronic communication system will be discussed. This information will provide a background for the other chapters in this section.

ATOMS AND ELECTRICITY

To understand electronic communication, it is helpful to know some basics about electricity. What *is* electricity? One definition states that **electricity** is the flow of electrons in a conductor. These electrons perform some function or job. For example, in a flashlight, electrons are used to light the bulb. However, you may have an even more basic question at this point, such as, "What are electrons?" This is why we will first explore atoms.

ATOMS

Atoms make up everything around us. Although atoms cannot be seen, scientists have provided us an understanding of them. Thanks to Niels Bohr, we have a rough idea of what the atom looks like. In 1913, Bohr used the solar system as a way to describe the atom. This analogy is still used today, Fig. 30-2.

Fig. 30-1. A phone is an obvious electronic communication device. (Radio Shack)

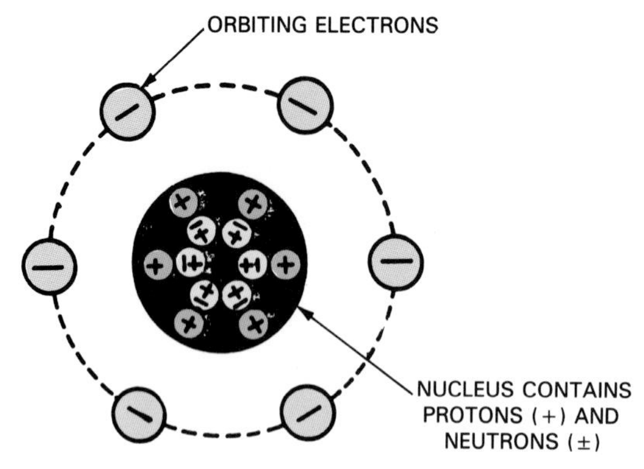

Fig. 30-2. The atom can be described much like a solar system, with electrons orbiting around the nucleus.

The atom has a *nucleus* in the center that contains particles known as neutrons and protons. The *protons* have a positive electrical charge, while *neutrons* have no charge. *Electrons* orbit around the nucleus and have a negative charge. The opposite charges of the protons and electrons attract one another. Normally, there are an equal number of protons and electrons that keep the atom in balance.

Some types of atoms have electrons that are not tightly bound to the atoms. In these cases, the atoms can give up or accept extra electrons. If an atom has extra electrons, it is negatively charged. If it has too few electrons, it is positively charged. Materials that allow the free movement of electrons in their atoms are called **conductors**. Materials that have atoms with tightly bound electrons are known as **insulators**, Fig. 30-3. For instance, copper is a good conductor, and plastic is a good insulator.

ELECTRIC CURRENT

An **electric current** consists of a flow of electrons through a conductor, Fig. 30-4. A power source, such as a battery, sends out a stream of electrons. These electrons are passed along the atoms of the conductor. Actually, the process is more like a "domino effect." This is because the electron that enters may not be the one that exits the conductor. As an electron enters the orbit of the atom, another electron is bumped or pushed out of orbit and to the next atom.

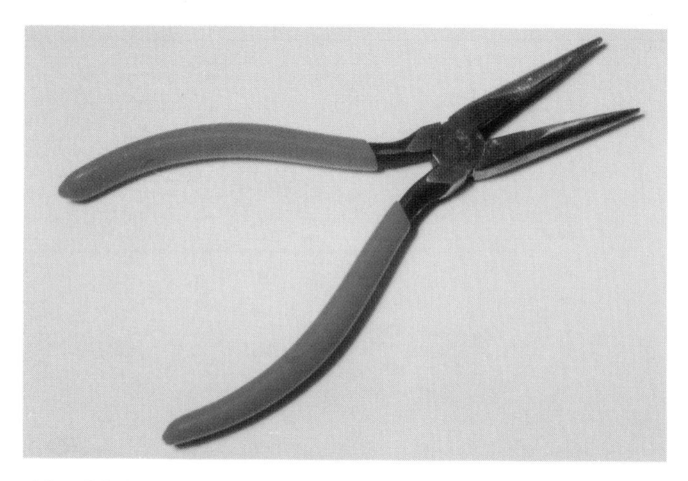

Fig. 30-3. An insulating material is used on the handles of this tool.

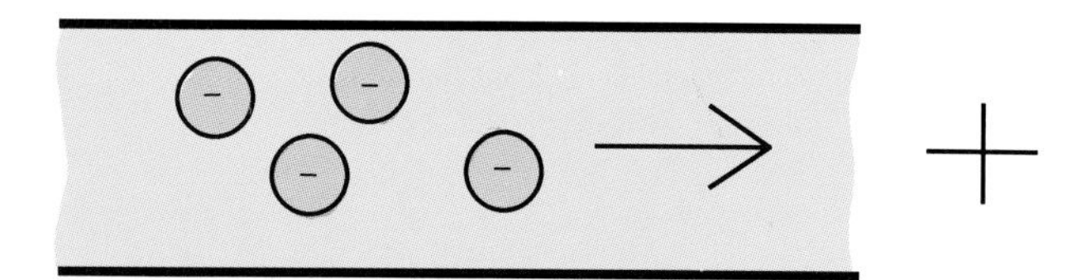

Fig. 30-4. Current consists of the flow of electrons through a conductor.

Electrons that travel through the conductors are attracted to the positive side of the power source, which has too few electrons. This electron stream, from negative to positive, is electric current.

Electric current is measured in **amperes**, which is often shortened to amps. One ampere is equal to over 6 quintillion (6 billion billion) electrons flowing past a point in an electrical circuit each second.

In order to achieve a flow of electrons, a force is applied. This is known as **electromotive force (emf)** or **voltage (volts)**, Fig. 30-5. Most outlets in homes have 120 volts available. A "D" size battery is 1.5 volts, while most car batteries are 12 volts.

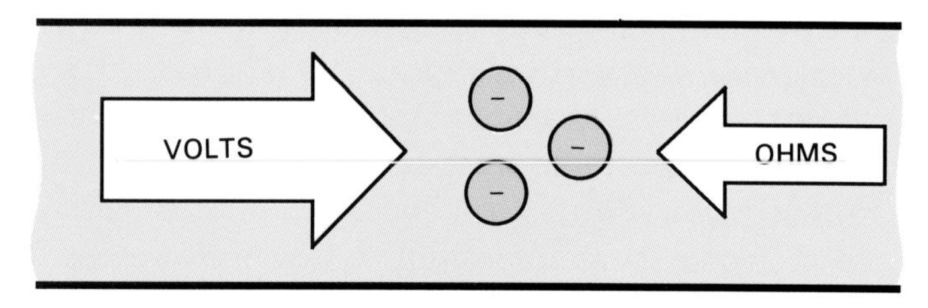

Fig. 30-5. Voltage is the force that is applied to the electrons. Ohms are the resistance to the flow.

You might think that if the voltage is high, more current will flow. However, there is another consideration called resistance. **Resistance** is the opposition to the flow of electric current. All materials through which current flows have some resistance. However, those materials that have very low resistance are known as conductors, as already stated. The unit of measure for resistance is the **ohm**.

Thus, there are two considerations when measuring current. They are voltage and resistance. This relationship between current, voltage, and resistance is known as **Ohm's Law**, Fig. 30-6. In the early 1800s, Georg

OHM'S LAW

$$\text{Current (Electron Flow)} = \frac{\text{Volts (Force)}}{\text{Ohms (Resistance)}}$$

Fig. 30-6. Ohm's Law describes the relationship between current, voltage, and resistance.

Ohm discovered that the greater the voltage, the greater the current, as long as resistance stayed the same. However, if resistance is increased, while voltage stayed the same, the current would be decreased. In other words, current is directly proportional to voltage, but inversely proportional to resistance.

Another useful term regarding current is watts, named after James Watt, who invented the steam engine. A **watt** is a measure of electrical power. **Power** is the rate of doing work or using energy. Watts are an indication of how much power is required by an electrical device each second, Fig. 30-7.

FORMULA FOR WATTS

W (Watts) = (Volts) × I (Current)

Fig. 30-7. This is the formula for watts.

For example, a 100-watt light bulb will require a greater amount of power to operate each second than a 60-watt light bulb. However, more work is performed by the 100-watt bulb, as can be seen by the fact that the 100-watt bulb is brighter.

As well as indicating how much power is required by an electrical device, watts are also used to indicate output power. For example, a guitar amplifier may require 200 watts for operation, but the watts of output that drives the speakers may be 65 watts.

Electricity is purchased from the power company in watts. The measure for the amount of watts used over time is **kilowatt hours**. For instance, if you use a 1500-watt toaster oven for one hour, you have used 1.5 kilowatt hours. If a kilowatt hour costs $.06, then the cost for operating the appliance is $.09.

Direct current and alternating current

Electric current can flow in one direction, **direct current**, or in alternate directions, **alternating current**. A battery produces direct current. The circuits in homes utilize alternating currents, Fig. 30-8.

Fig. 30-8. Devices plugged into this outlet use alternating current.

Alternating current is used when sending the current over long distances. The reason for this is that the voltage level for ac can be easily changed. High voltages are used for transmitting the ac, which results in less energy being lost in the form of heat in the wire. A transformer is then used to reduce the voltage before the current is used, Fig. 30-9.

Alternating current is produced by rapidly changing the direction of the voltage. The frequency of this alternation is stated in **hertz** (cycles per second). Most circuits in homes use 60 hertz ac.

Circuits in many devices, such as radios, require direct current. If ac is used to power the radio, this must be converted to dc. A circuit that changes ac to dc is a rectifier circuit.

Fig. 30-9. This transformer reduces the voltage coming into homes.

ELECTRIC CIRCUITS

An **electric circuit** is a complete path for current, Fig. 30-10. Normally, the circuit consists of a voltage source, a conductor, a load, and some kind of control device.

The voltage source provides the electric energy to the circuit. It is the power source. A battery is an example. In the case of household current, the power company generates the electricity by various means and sells it to us, the customers.

The conductor is the path for the electric energy. Copper is an example of a good conductor.

The **load** in the circuit is what utilizes the current to perform some useful purpose. In a flashlight or a lamp, the bulb is the load. Current is converted to light in the bulb.

Often, there is a control device in the circuit. For example, a switch is used for turning lights on and off. This is one of the simplest kinds of control devices.

Circuits can be series or parallel, Fig. 30-11. A **series circuit** has devices that require current in a line, so that the current passes through the first device, then the second, and so on. A **parallel circuit** has branches, with each branch leading to a device needing current. In this case, each device is directly connected to the power source. The branches converge to a single conductor before connecting to the power source.

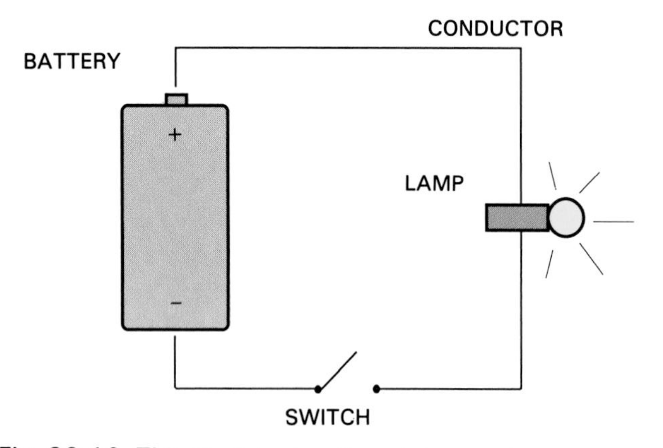

Fig. 30-10. This electric circuit contains a voltage source, a conductor, a load, and a control device.

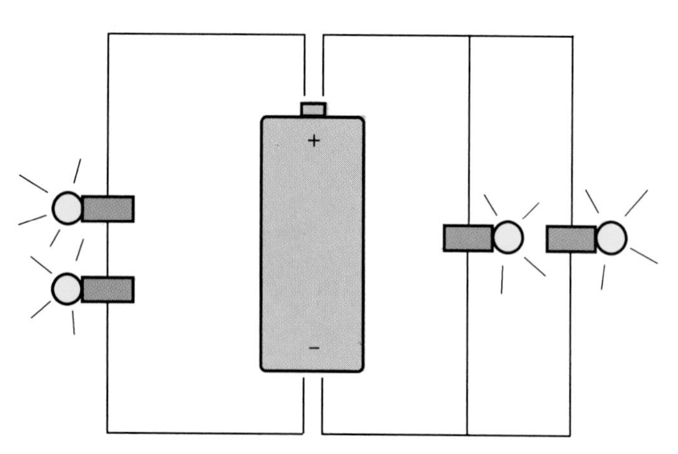

Fig. 30-11. A series circuit is shown on the left and a parallel circuit is on the right.

In electronic devices, such as radios and televisions, there are usually a large number of individual circuits, each performing a different function. **Amplifier circuits** take weak signals and increase, or amplify them. For example, a radio signal is received by the radio and amplified so that you can hear it. An oscillator is another example. This circuit provides a signal on which information can be placed. Transmitters for communication devices always have oscillators.

ELECTRONIC MILESTONES

Numerous discoveries have led to devices such as radios, televisions, and computers. In fact, they are so numerous it would take many books to explain all of them. However, there are a few of these discoveries with which you should be familiar.

A device that had a major impact on the field of electronics was the **vacuum tube**, Fig. 30-12. The vacuum tube was the result of Thomas Edison's experiments with making a light bulb in 1883. He found that the heated filament of the light bulb actually gave up electrons in the vacuum of the bulb. These could be collected by a positive plate at the top of the bulb. This became known as the *Edison effect*. About 20 years later, two researchers independently developed two types of vacuum tubes based on this discovery. John A. Fleming developed a tube capable of detecting radio waves. Lee DeForest developed a tube that acted as a control device.

After its invention, the vacuum tube was used in a variety of ways in electronic devices. For instance, it could be used to amplify a signal or to generate radio

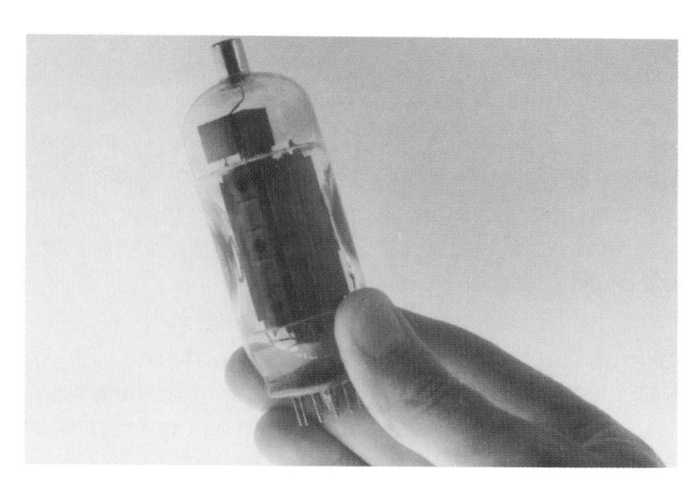

Fig. 30-12. This is a typical vacuum tube.

waves. In order to see how important tubes were, you only have to look inside an old electronic device. A radio built in 1950 might have 15 or 20 tubes in it. The most common device that uses a vacuum tube today is the television. The picture tube is actually a large vacuum tube.

Vacuum tubes were important in the development of another major electronic development, the computer. Switches are important in computers since they are counting machines. Early computers used vacuum tubes as these switches. The first electronic digital computer, which used these tubes as switches, was the Atanasoff Berry Computer (the ABC), Fig. 30-13. It was invented by John Atanasoff at Iowa State College (now known as Iowa State University) and built by a graduate assistant, Clifford Berry, in the early 1940s. In 1946, the ENIAC was developed by John Mauchly and J. Presper Eckert at the University of Pennsylvania. It used 18,000 tubes. The term, **debugging**, is derived from these early days of computers. Moths were attracted to the warm vacuum tubes and had to be removed, which was called debugging. The term now means finding problems in computer software.

A device that replaced the vacuum tube was the **transistor**. In 1947, the scientists at Bell Laboratory invented the transistor, Fig. 30-14. William Shockley was the leader of the team and had help from John Bardeen and Walter Brittain. The transistor is capable of amplifying a signal or acting as a switch, like a vacuum tube. In addition, the transistor performs these functions without high temperatures. It is also much smaller than a vacuum tube.

The transistor was the beginning of solid state electronics. **Solid-state devices**, such as transistors, are made from a semiconductor. **Semiconductors** are made up of crystal material, such as silicon, that is normally an insulator. However, by doping (adding impurities), the material can be made to conduct electricity under certain conditions. Typically, thin layers of semiconducting material are sandwiched together to make up electronic devices such as transistors.

Fig. 30-13. The ABC was the first electronic digital computer. Vacuum tubes can be seen in the lower right corner. (Iowa State University)

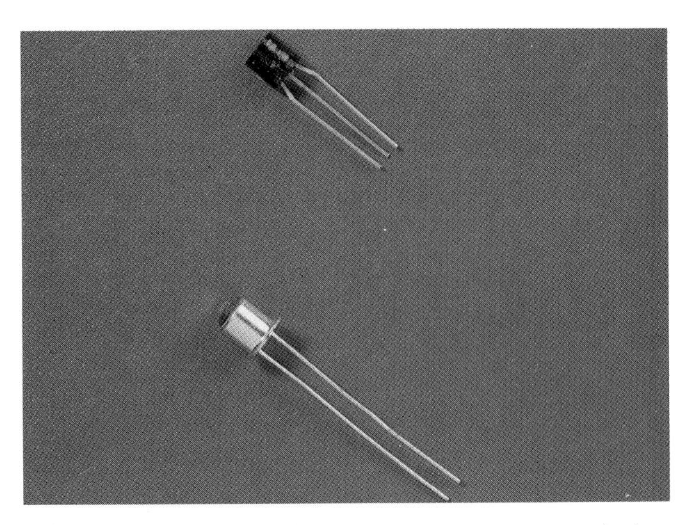

Fig. 30-14. The transistors replaced the vacuum tube in many devices. (Radio Shack)

When semiconductors began to be used as electronic devices, it was possible to greatly reduce the size of electronic circuits. In 1958, Jack Kilby, at Texas Instruments, invented the first integrated circuit that contained two transistors on one semiconductor. At this time in history, there was a need to miniaturize circuits for space and defense purposes, so the number of components on integrated circuits increased rapidly. In fact, today, hundreds of thousands of electronic components can be placed on an integrated circuit.

Integrated circuits are often called **microchips**, since they are very small and made of chips of silicon and other materials. An important chip used for computing functions is the **microprocessor**. Chips are often placed on a board with printed copper paths, known as a **printed circuit board**, Fig. 30-15.

ELECTRONIC COMMUNICATION SYSTEMS

Now that some of the basics of electricity and electronics have been discussed, let's turn our attention to electronic communication systems. Again, keep in mind that a system simply consists of inputs, processes, and outputs. In the case of a communication system, the purpose of the system is to send and receive messages. In electronic communication systems, the messages are converted to electrical form for transmitting and receiving. These messages might be sound, pictures, or data.

Fig. 30-15. Printed circuit boards contain microchips with printed copper paths. (Univision Technologies, Inc.)

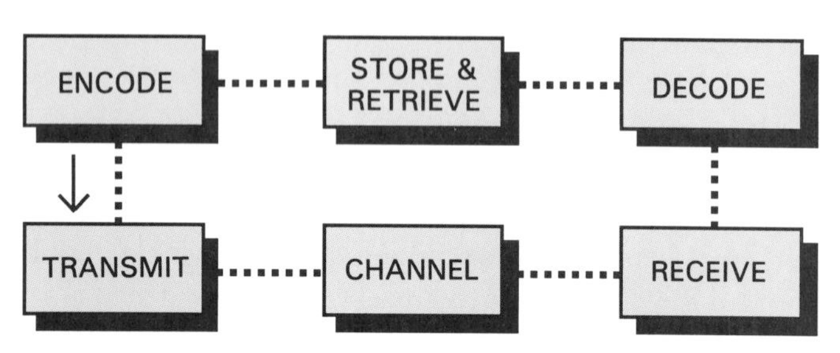

Fig. 30-16. These are the steps involved in sending a message in an electronic communication system.

Regardless of the message that is being sent, the process is basically the same, Fig. 30-16. The message is encoded, transmitted on a channel, and then decoded. For example, the telephone is used to encode sound into electrical signals. These signals are transmitted over a channel, such as copper wire, to a receiver. The signal is then decoded back into sound. Since these basic concepts in electronic communication systems are so important, we will look at each one individually.

ENCODING

In electronic communication systems, encoding is the conversion of some form of energy to electrical energy. The two most common signals that are converted to electrical signals are sound and light. The general term for devices that perform this energy conversion is **transducers**.

In the case of sound, sound waves are converted to electrical signals. A microphone is an example of a device that performs this function.

When pictures are converted to an electrical signal, light is the original form of energy. Light rays, reflecting from a picture, are what give us the capability of seeing a picture. Thus, a picture can be thought of as a light signal. A typical device that converts light to electrical energy is a video camera, Fig. 30-17.

In communication between machines, such as computers, the signal is already in electrical form. However, in order to be transmitted over long distances, the signal must be converted further. In this case, an electrical signal is converted to a different type of electrical signal.

TRANSMISSION

Transmission involves placing the electric signal on a channel. The electric signals will normally be in analog or digital form. Since these types of signals are very

Fig. 30-17. This video camera converts light into electrical energy. (Bowling Green State University, Visual Communication Technology Program)

important in electronic communications, they require further explanation.

Analog and digital signals were being used long before electricity had been discovered. An **analog signal** is any signal that continuously varies. A voice is a good example. Your voice varies in loudness as you speak. In addition, the frequency varies. This is the variation in high and low notes used in speaking.

What do analog signals, in electrical form, look like? As a hint, an analog is something that looks like something else. So, if we represent sound as a wave that constantly varies, an electrical signal could be represented by the same wave. For example, as a voice grows louder, the electrical signal would be stronger. Telephones usually transmit analog signals.

A **digital signal** consists of an *on* or *off* state, Fig. 30-18. An early form of digital signal was the drum beat. Heard from a distance, the message was conveyed by the frequency of beats. Morse code is another type of digital signal that was used for the telegraph. Long and short electrical signals, called dashes and dots, were transmitted. Computers are also devices that generate digital signals. These are in **binary code**, consisting of on and off states of electricity.

Digital signals, in electrical form, can be thought of as pulses of electricity. The length of the pulses or frequency of pulses can be varied to send the signals.

In some cases, analog and digital signals can be transmitted directly without change. However, most electrical signals, in digital or analog form, are placed on a carrier wave for transmission. This process of placing information on a carrier wave for transmission is known as **modulation**.

Carrier waves are generated by a circuit in transmitters known as an **oscillator**. The oscillator generates waves at a certain frequency. By itself, the carrier wave has no information.

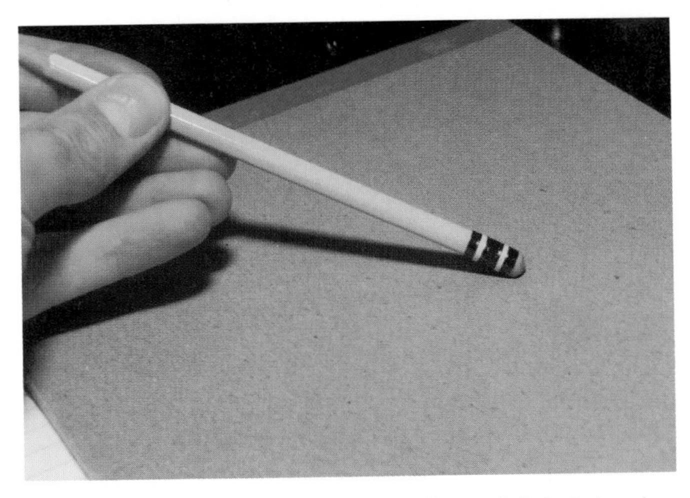
Fig. 30-18. A pencil tap is a simple form of digital signal.

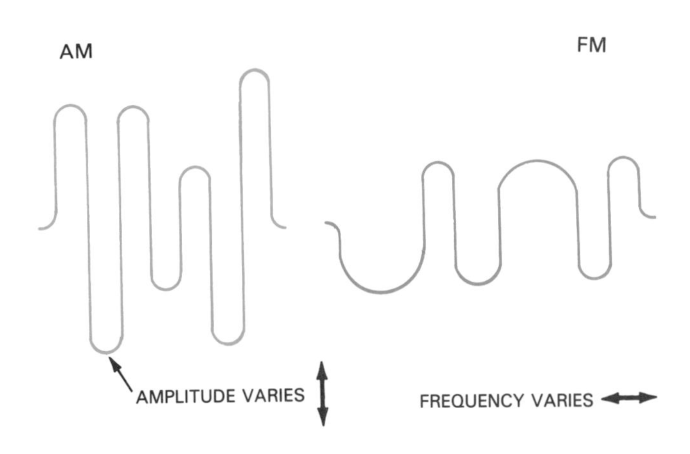

Fig. 30-19. Amplitude or frequency can be varied to carry the signal.

Information is placed on the carrier wave in one of two ways, Fig. 30-19. In one case, the amplitude (height) of the wave is modulated (varied) to contain the signal, while the frequency of the wave stays the same. This is known as **amplitude modulation**. The second option for placing information on the wave is to vary the frequency of the wave while keeping the amplitude (height) constant.

Analog signals, such as voice, can be sent by varying the amplitude or the frequency of the carrier wave. Remember, an analog signal is one that constantly varies. This variation can be in height of the wave or frequency. A typical example of this is AM and FM radio. Both carry analog signals. They simply modulate the information in a different way.

Digital signals can be placed on a carrier wave in a similar manner to analog signals. Both amplitude modulation and frequency modulation can be used to carry these signals. For example, binary code is represented by 0's, and 1's. In order to modulate these 0's and 1's by amplitude, a low amplitude could be assigned for a 0 and a high amplitude for a 1. An analogy would be a light bulb that is first dim and then bright to represent "0's" and "1's." The carrier wave can also be modulated by frequency, by having a 0 be a different frequency than a 1.

A typical method of transmitting digital signals is by **pulse code modulation (PCM)**. This is the technique most often used to transmit computer data. In PCM, a pulse of current could be a "1" in binary and the absence of the pulse could be a "0." If the pulses are placed on a wave for transmission, the frequency or amplitude could be varied as a code for "0's" and "1's."

It is becoming very common for analog signals to be converted to digital signals before transmission. In this case, an **analog to digital (A/D) converter** is needed in the transmitter. Also D/A converters may be used. For

Fig. 30-20. Computer modem change digital signals into analog signals for transmission over phone lines. Modems can be external (left) or internal (right).

example, computer modems change digital signals to analog signals for transmission over the phone lines. See Fig. 30-20.

Another consideration in transmission is **multiplexing**. This is where more than one signal is transmitted through a single channel, such as a wire. In one type of multiplexing known as **frequency division multiplexing (FDM)**, each signal has a different frequency. For example, cable television uses this type of multiplexing. Each station is assigned a different frequency to which you tune in by changing the channels.

COMMUNICATION CHANNELS

A channel is the pathway for the signal. These channels can be a physical link between the transmitter and receiver. Air can also be used as a channel.

Some channels physically link the transmitter and receiver. In the case of the telephone, a pair of copper wires are used to connect homes with the central office. Cable television is brought into the home with **coaxial cable.** This is a cable that is specially shielded to avoid interference.

Electromagnetic waves can be transmitted through fiber optic cables and wave guides, Fig. 30-21. **Fiber optic cables** are used for transmitting signals of visible light. **Wave guides** are used for transmitting electromagnetic waves that are of a lower frequency than visible light. The wave guide is basically a hollow metal tube.

Electromagnetic waves are often transmitted through the air. Radio and television signals are transmitted by antennas in this manner. Signals are sent to communication satellites for transmission using air as the channel.

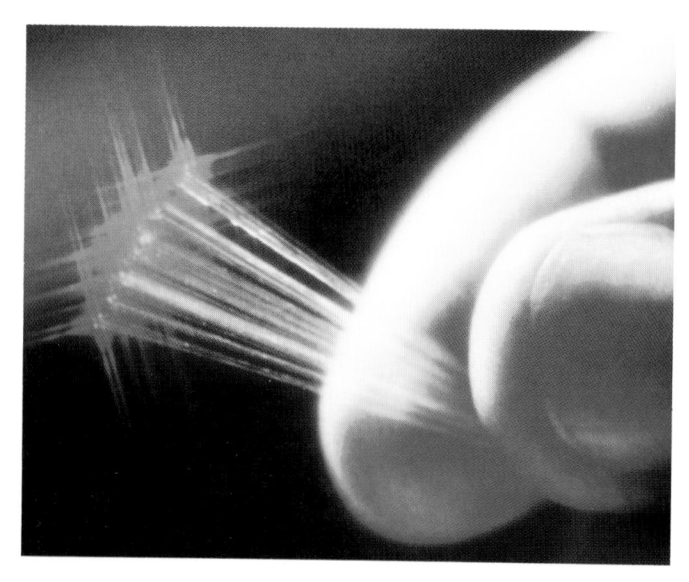

Fig. 30-21. Fiber optic cables are one form of channel.

RECEIVING

After transmission of a signal through a channel, the signal must be received, Fig. 30-22. In the case of an electromagnetic wave, an antenna is used for detecting the wave. This signal is then sent to an electronic receiver that amplifies and demodulates the signal. **Demodulating** the signal involves separating the carrier wave from the information that is being carried.

Other functions may also be performed by the receiver. For example, an analog signal that was converted to digital for transmission will need to be converted back to analog.

Fig. 30-22. The antenna on this radio is used for signal reception.

DECODING

Decoding involves changing the electrical signal into a form that humans understand. As you will recall, encoding devices change light or sound to electrical signals. Decoding devices change electrical signals back into light or sound.

A common decoding device is the speaker, Fig. 30-23. In this case, an electrical signal causes the speaker cone to vibrate. These vibrations are sound waves. A television takes electrical signals and converts them into a picture. This is an example of conversion of electrical energy to light.

Fig. 30-23. This speaker acts as a decoding device.

STORING AND RETRIEVING

Storing and retrieving are important concepts in an electronic communication system. Storage and retrieval can take place before transmission or after transmission. A show produced for radio or television can be recorded at an earlier time and then be broadcast one or several times. In the case of television, the show is normally recorded on videotape. In the case of radio, audio tape is used.

When data is received by a computer, it is often stored on magnetic disks. It is then retrieved by the computer user. Retrieving involves bringing the information up from storage.

SUMMARY

This chapter has provided you with an introduction to electronic communication. First, some basics of electricity and electronics were explored. Then the basic elements of an electronic communication system were discussed. These concepts will be further developed in the following chapters.

WORDS TO KNOW

All of the following words have been used in this chapter. Do you know their meanings?

alternating current
amperes
amplifier circuits
amplitude modulation
analog signal
analog to digital (A/D) converter
atoms
binary code
coaxial cable
conductors
debugging
demodulating
digital signal
direct current
electric circuit
electric current
electricity
electromotive force (emf)
electronic communication
fiber optic cables
frequency division multiplexing (FDM)
hertz
insulators
integrated circuits
kilowatt hours
load
microchips
microprocessor
modulation
multiplexing
ohm
Ohm's Law
oscillator
parallel circuit
power
printed circuit board
pulse code modulation (PCM)
resistance
semiconductors
series circuit
solid-state devices
transducers
transistor
vacuum tube
voltage (volts)
watt
wave guides

REVIEWING YOUR KNOWLEDGE

Please do not write in this text. Write your answers on a separate sheet.

1. True or false? In order to be classified as electronic communication, a primary purpose of a system should be sending messages by electrical signal.
2. Electricity is the flow of which of the following through a conductor?
   a. Protons.
   b. Neutrons.
   c. Electrons.
   d. None of the above.
3. The unit of measure for resistance is the:
   a. Ohm.
   b. Volt.
   c. Amp.
   d. None of the above.
4. True or false? Electric current that flows in one direction is alternating current.
5. Name and describe an electronic milestone that led to devices such as radios, televisions, and computers.
6. Describe the electronic communication system process.
7. Computers generate digital signals in ________ code.
8. ________ takes place when more than one signal is transmitted through a single channel such as a wire.
9. Which of the following is NOT an example of a channel?
   a. Coaxial cable.
   b. Air.
   c. Copper wire.
   d. An antenna.
10. Which of the following involve demodulating a signal?
    a. Decoding.
    b. Receiving.
    c. Storing.
    d. Retrieving.

APPLYING YOUR KNOWLEDGE

1. Using the system elements provided in the chapter, analyze an electronic communication system and record your results.
2. Investigate several electronic communication devices to determine their needs in regard to current, voltage, and watts. Use formulas, if needed, to obtain the information.
3. Write a report on an inventor who has made a significant contribution in the development of electronic communication. Examples would be Atanasoff, Edison, Morse, and Bell.
4. Interview an electrician about the basic principles of electricity. Prepare a list of questions in advance. Share this information with the class.

31

CHAPTER

COMPUTERS

After studying this chapter, you will be able to:
- *Define the term "computer."*
- *Identify the major parts of a computer and their functions.*
- *Describe common input and output devices used with computers.*
- *Explain the importance of the use of microchips in computers.*

What is a computer? A **computer** is an electronic device that can help you perform work with greater speed and accuracy than you would normally be able to do. It does this by processing information, in the form of data, very rapidly, Fig. 31-1. This data can be letters, numbers, or even pictures. A **computer program** is needed to provide directions on what to do with the data. For example, a program might direct the computer to add (process) a group of numbers (data) together. Thus, a computer can be described as an electronic device that processes information under the control of a program.

Fig. 31-1. A computer processes information.

The **human brain** is another example of an information processor. When you see a picture of something, an apple for instance, the image is sent to the brain from your eyes. The picture, a type of data, is only meaningful to you if your brain can interpret it. From past experience, you have learned that an apple is good to eat, so you decide to eat it. Your past experience, in this case, is the program that allows you to process the data into information. A computer, functioning in a similar manner, inputs data, and under the direction of a program, processes it.

USES OF COMPUTERS

Can you think of some uses of computers in your life? Much of the food you eat is sold in packages with bar codes. These codes are "read" by a computer, making checkout easier. Cars are designed with the help of computers and made with assistance from computer-controlled robots. When you dial the telephone, a computer helps in placing the call. Music on the radio is produced with the help of computers. Of course, some more obvious examples are video games and home computers, Fig. 31-2.

Computers are being used all around us and new uses are being found constantly. Why? Because certain jobs can be performed faster and more accurately with computer assistance.

DIGITAL COMPUTERS

Most computers with which you are probably familiar are most likely digital. **Digital computers** process

Fig. 31-2. A personal computer can be used in many work situations. (Hewlett-Packard Co.)

Fig. 31-3. The minicomputer on the left is connected to the terminals shown here. (IBM Corp.)

data known as **bits** (binary digits). A bit is stored in computer memory as on (1) or off (0), related to whether the electrical circuit is on or off. Bits, in combination, make **bytes** (eight bits). Using a binary code, these bytes can be used to symbolize numbers and letters. Why are bits necessary for symbolizing information? Computers are electronic devices, and on/off states of electricity, like a light switch, are a practical method of creating codes for this type of equipment.

Digital computers are classified according to their size and processing capabilities. The three main types of computers are:

- Mainframe computers.
- Mini computers.
- Personal computers.

MAINFRAME COMPUTERS

Mainframe computers, which were produced in large numbers starting in the 1960s, are very large and powerful computers. These computers are used by organizations that need assistance with large quantities of data such as financial records. Organizations that might use mainframes are insurance companies, banks, and schools. Within the organization, the computer is wired to a number of workstations (terminals) so that many people can work with the computer. Mainframes require an air conditioned environment. They are placed on a raised floor so all the cables can be placed underneath them.

MINICOMPUTERS

In the 1970s, as parts for computers became smaller, minicomputers were developed, Fig. 31-3. **Minicomputers** are just as powerful as the mainframe in many cases, but require less space. Minicomputers are also used in organizations and are wired to a number of terminals (workstations) so that many people can use them.

PERSONAL COMPUTERS

The computers with which you are probably most familiar are **personal computers.** They are small computers that are normally used by one person at a time. At home, they are used for activities such as games, typing (word processing), and budgeting. At school, the personal computer can be used to assist with instruction and teach programming. Businesses use them for tasks such as word processing, inventory, and CAD.

OPERATIONS AND PARTS OF A COMPUTER

A computer consists of parts that work together as a system to process information. This system consists of operations that must be performed to change or process the data into useful information. These operations are:

- Input.
- Process.
- Output.
- Storage.

You, yourself, are an example of a system. With sight and your other senses, you send information to your central processor, your brain. You also have a memory for storing information, and speech and writing capabilities for "outputting" information. In fact, many of the computer terms, such as memory, can also be related to people.

A computer system has parts, or hardware, that perform the operation in the information processing cycle, Fig. 31-4. Input is the way in which data enters the system, such as a keyboard. Processing takes place in the central processing unit. Output devices, which display the processed information, include monitors and printers. Storage of information takes place in devices, such as disk drives, so information can be saved.

INPUT

An **input device** is simply a method of getting information into the computer, Fig. 31-5. The most common input device for a computer is the keyboard and display screen. The keyboard is used for typing letters and numbers as input to the computer.

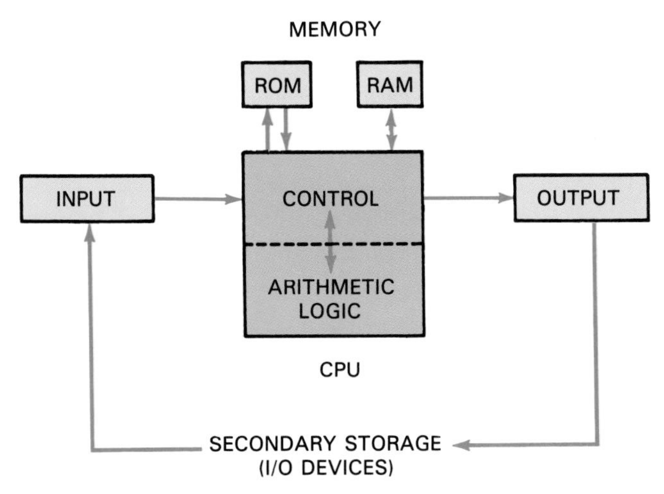

Fig. 31-4. These are the main parts of a computer system.

Information that is input for the computer is often in coded form. For example, magnetic tape and disks are read by devices that can pick up the magnetic signals. A disk drive is commonly used for this purpose. **Optical readers** are used for a variety of purposes such as scanning bar codes and optical disks. Another input device is the **modem** that allows the computer to send and receive information through the phone line.

The **mouse** is one of the most common input devices, Fig. 31-6. When the mouse moves across a flat surface, there is a corresponding movement of a cursor on the display. Buttons on the top of the mouse are used for special functions, such as making selections from a menu. The **mechanical mouse** has a ball or wheel underneath that rolls against the work surface. An **optical mouse** uses light and a special reflective surface to determine location.

Drawing devices

A variety of input devices can be used to create drawings by moving a cursor on the screen. The mouse, previously mentioned, is used for this purpose. The joystick and track ball are similar devices in that they remain stationary on a table and a handle or ball on top is used to control movement of the cursor. The **track ball** is sometimes described as an upside down mouse.

A **graphic tablet** has a flat surface with a wire grid underneath that sends electrical signals to the computer. When a stylus or puck moves over the surface of the tablet, the cursor moves on the display. A device that is similar to the graphic tablet is the digitizer. The **digitizer** has a large surface on which a drawing can be placed and traced for input to the computer. This is known as digitizing the drawing.

Fig. 31-5. This touch-sensitive monitor allows you to use your finger as an input device. (IBM Corp.)

Fig. 31-6. The keyboard and mouse are two of the most common input devices. (Apple Computer, Inc.)

A **light pen** is an input device that is used by touching the display or monitor. A sensor in the pen is used for determining the location on the screen and sending this information to the computer.

A variety of input devices can be used for automatically inputting graphics or text into the computer. Scanners, such as the laser scanner used in grocery stores to read bar codes, are examples. The **optical character recognition (OCR)** device is a type of scanner that automatically inputs pages of text matter. Still another type of scanner is used for inputting graphics, Fig. 31-7.

Other types of input include video and audio. Video input is accomplished by way of a video camera that is adapted for use with a computer. Voice input can provide instant translation of voice into text.

PROCESSING

The main part of the computer, where work is performed on the data, is called the **central processing unit (CPU)**, Fig. 31-8. The work that is performed is called "processing." The CPU is actually made of small silicon chips known as microprocessors or microchips. Electric current, representing data, flows through the chips in coded form known as **binary code**. This code has only two symbols, 0 and 1, that represent *on* (current passing) or *off* (current stopped), similar to a light switch. However, by combining these on-off combinations, this simple code can be used to represent letters, numbers, etc. Another code consisting of two symbols is the Morse Code that was used for sending information by telegraph before the telephone was invented. Information was sent as dots and dashes.

Fig. 31-8. The picture on the monitor is possible because of the CPU, which is inside the case on the left. (Agfa Corp.)

The CPU actually consists of two separate areas. These are control and arithmetic-logic. (Refer back to Fig. 31-4.)

Fig. 31-7. This is a flatbed color scanner. (Scitex America Corp.)

The **control unit** coordinates all the activities between input and output and can be considered the "brain" of the computer. In other words, this section ensures the various parts of the computer work together. For example, when using a program, the control section runs the program in an orderly manner and makes sure that data is processed as needed. It does this by repeating four steps:

1. Retrieving program instructions from memory.
2. Decoding the instructions by converting them into machine language.
3. Executing the instructions in sequence, including sending data to the arithmetic-logic unit.
4. Storing the processed information in memory.

The **arithmetic-logic section** performs mathematical functions as needed. Most mathematics are processed using addition and subtraction. For example, 4 x 4 can be processed as 4 + 4 + 4 + 4 = 16. Logic operations involve comparing data and determining if one is less than, equal to, or greater than the other.

Memory

Memory is where information is stored for use by the computer. All data that enters the computer is placed in **random access memory (RAM)**. This is also known as read-write memory because you can obtain information from it (read it) and put information in (write to it). Random access means that information in memory can be removed as needed, rather than by the order in which it was stored. RAM is a temporary memory that is erased when you turn off the computer.

One of the first items stored in RAM is a program or software, which is a list of operations for the computer to perform. This software might be on a separate magnetic disk that is read into RAM or on the computer's hard drive. As the program runs, additional data is placed in RAM and processed according to program directions. Since all information in RAM will be lost when the computer is turned off, the final results must be sent to permanent storage, such as a disk.

Read-only memory (ROM) is installed when the computer is assembled. This memory permanently stores instructions necessary to perform tasks. For example, the first instructions the computer receives when turned on are stored in ROM. Like RAM, this memory can be read by the computer, but it cannot be changed or erased when the computer is turned off. Another term for ROM is "firmware."

RAM and ROM are examples of main memory which is also known as **semiconductor memory**. This memory is stored directly on the microchips. However, data can also be permanently stored outside the computer and used as necessary. This is known as mass, secondary, or auxiliary storage. Magnetic disks and recordable CD-ROMs are examples of secondary storage.

OUTPUT

After information has been processed by the computer, it is sent to an output device that displays the results or stores results for later use.

One of the most common devices for viewing output is the video display or monitor. However, results that are displayed are in RAM and will be erased if the computer is turned off. To permanently save processed information, another type of output device is necessary.

Some output devices store information in visual form. For example, a printer can be used to produce typed information. The ink jet and laser printers are most common. In graphic arts, typesetters and imagesetters are used for high-quality images. For computer drafting, special pen plotters and photo plotters are normally used to produce detailed drawings.

Another category of devices is one that can be used with information stored in coded form outside the computer. These are generally called **input/output (I/O) devices** because the information can be both input and output by them. Examples are disk drives and modems. Information input or output by these devices is in binary code.

Usually, more than one output device is used in the same system. For example, a letter can be viewed on the display screen, printed for mailing, and stored on a magnetic disk for later use or changes.

Monitors

The **monitor**, or screen of the computer, displays information that is in RAM. The **cathode ray tube (CRT)** display is most common, and works in a similar fashion to most televisions. An image is scanned onto the screen from top to bottom by means of an electron beam in the tube. These electrons strike phosphors on the screen, which then causes it to glow. Since the phosphors will stop glowing without electrons, the electron beam must continue to scan the image.

Screens are either monochrome (single color) or color. Color monitors utilize red, green, and blue phosphors to produce all other colors.

Computer monitors can be divided into raster and vector displays. This refers to the two ways images are placed on the screen.

The most common type of monitor is the **raster** or bit-mapped display monitor. The image is created by a series of **pixels** (dots) on the screen. Each pixel on the screen can be turned on or off to create the image.

Higher resolution (clarity) is achieved by having more pixels on the screen.

One problem with raster displays is that images can suffer from the "jaggies." For example, an angled line might have a stair-step appearance, since it is made up of dots.

A **vector display** avoids this problem. Lines are drawn from point to point, like a person actually draws a line. Rather than scan the entire screen, the electrons only follow what is drawn. Since the phosphor coating is not in dot form, lines are smooth.

Some displays are flat, Fig. 31-9. These are often found on lap top computers. Although the technology will vary in flat displays, the concept is basically the same. The screen is divided into a grid of pixels. Pixels that are given an electrical charge, create the image by either glowing or allowing light to shine through.

One such display is the **liquid crystal display (LCD)**, which is similar to the display on many calculators and digital watches. In this case, each pixel consists of a crystal that changes position when given an electrical charge. This new position allows light to pass through the crystal, thus forming the image.

The plasma displays have pixels made of clear electrodes in a gas. An electrical charge to the electrode causes the gas to light up in that location. These screens have a red color.

STORAGE

Since computers process large amounts of data, it is necessary to have a way of storing this information. The devices used are called secondary or auxiliary storage devices. Initially, paper tape and cards punched in coded form were used. However, magnetic and optical storage methods are primarily used today. Almost all of these are input/output devices in that they can also be used for inputting information to the computer.

Magnetic storage methods include tapes and disks that store information as magnetic pulses, similar to the way music is stored on an audio cassette. The diskette, often called a *floppy* because of the thin plastic medium that holds the information, is the most common storage device. While 5 1/4″ disks were once common, they are now seldom used. The 3 1/2″ disk, with a hard plastic case protecting the actual floppy data disk, is almost universal. It is available in two capacities, 720 kilobytes (720,000 bytes of storage) or 1.44 megabytes (1,440,000 bytes). The 1.44Mb capacity is most widely used.

Larger-capacity secondary storage disk systems have been introduced in recent years because of the need to store and transport graphics files too large for the 1.44Mb diskette. They require special disk drives that can be either built into a computer or plugged in (external drives). The most widely used of these drives, Fig. 31-10, accepts disks with a capacity of 100Mb (more than 50 times the capacity of a 1.44Mb diskette).The disk used in this drive is the same 3 1/2″ size as a floppy, but is thicker. Even larger capacity drives and disks (1Gb or more) are available from various manufacturers.

Fig. 31-9. This lap top computer has a flat monitor.

A

B

Fig. 31-10. Secondary storage. A—The 1.44Mb floppy disk (at right) is used with built-in drives in most computers. The slightly thicker disk at left is a large capacity (100Mb) disk that must be used in a special drive. B—An external drive used to read and write from the 100Mb disk shown above. Some computer manufacturers are supplying units with these drives built in. Similar drives are available for even larger capacity disks.

A disk drive is used to write data to or read data from the diskette. A write-protect notch or tab on the diskette can be closed to prevent making changes to the data.

Precautions should be taken when using diskettes. Some of these precautions are as follows:

- Disks should not be dropped or bent.
- Never place disks near a magnetic field or heat.
- When writing on the label, use a felt-tip pen.
- The plastic disk, which can be seen through the access hole, should never be touched.
- Never force a diskette into a drive.
- After use, place the disk in a storage box.

The primary storage device for computers is a built-in hard drive that is capable of holding very large amounts of information. Most computer systems sold today for home, office, or school use are equipped with hard drives whose capacity is measured in gigabytes (billions of bytes).

A hard drive consists of several metal platters coated with a magnetic medium to allow storage in the form of magnetic pulses. The disks spin at high speed inside a sealed case, between pairs of read/write heads that permit rapid access to information.

The read/write head float only a few thousandths of an inch above the disk surface. A surface contaminant on the disk (such as a particle of dust), or a physical shock can make the head bounce and make damaging contact with the disk surface. This is called a **crash**.

To prevent loss of important information in the event of a crash, regularly *backing up* the drive's contents is recommended. This involves high-speed copying of the information to diskettes or tapes.

A widely used information storage device is the CD or compact disc, which resembles the disc used for music. Many computers are equipped with **CD-ROM (Compact Disc-Read Only Memory)** drives that allow the computer to use (read) information, such as a program or a digital photograph. See Fig. 31-11. New software is often packaged today on CD-ROM, rather than diskettes. Some new computers include **DVD (Digital Video Disc)** drives.

To provide the ability to both read and write information, special types of CD drives are available. The CD-R recordable disc can be written to once, but read back many times. A more versatile format is the CD-RW (compact disc-rewritable) that can be used much like a diskette. CD-RWs can be used for approximately 1000 read/write cycles.

Hardcopy output

Some output devices store information in visual form and are called **hardcopy devices**. The most common of these printers, which can be classified as impact or non-impact.

Fig. 31-11. This optical compact disc is used for storing photographic images. (Eastman Kodak Co.)

Impact printers create an image by pressing a carbon ribbon against a substrate, **Dot matrix printers** make the image with a series of dots. The print head contains tubes with pins inside that strike the ribbon and paper. Dot matrix printers can be used for graphics as well as text, but the resolution is not as good as with other methods. The **daisy wheel printer** is an impact printer that is like a typewriter. It is seldom used today.

Non-impact printers include thermal printers, laser printers, and ink jet printers. The **thermal printer** uses heat to form the image, while the **direct thermal printer** uses a chemically coated paper. Heat causes an image to form on the paper. The **thermal transfer printer** is another type of thermal printer. It can be used for color output. A ribbon, when heated and placed against the substrate, causes the ink to transfer.

The ink jet printer has become very popular as a low-cost, high quality output device. It forms an image by spraying drops of ink onto a page through small nozzles or jets. Some printers use only a single color of ink; others provide full-color output by using ink cartridges in black, red, blue, and yellow.

The **laser printer**, another type of non-impact printer, makes an image similar to a copying machine, Fig. 31-12. A light-sensitive drum or belt is given a positive electrical charge. Then the image is scanned on the drum using the laser light beam. This makes this area negatively charged. A positively charged toner powder is then placed on the drum and adheres to the image.

Fig. 31-12. This laser printer is set up to print on envelopes or paper. A printed sheet is shown on the top. (Hewlett-Packard, Co.)

The drum is then rolled against a substrate that transfers the toner. Heat then is used for fusing.

Pen plotters

A **pen plotter** is used for outputting technical drawings, and other drawings where high-quality line work is needed, Fig. 31-13. A pen is used for drawing.

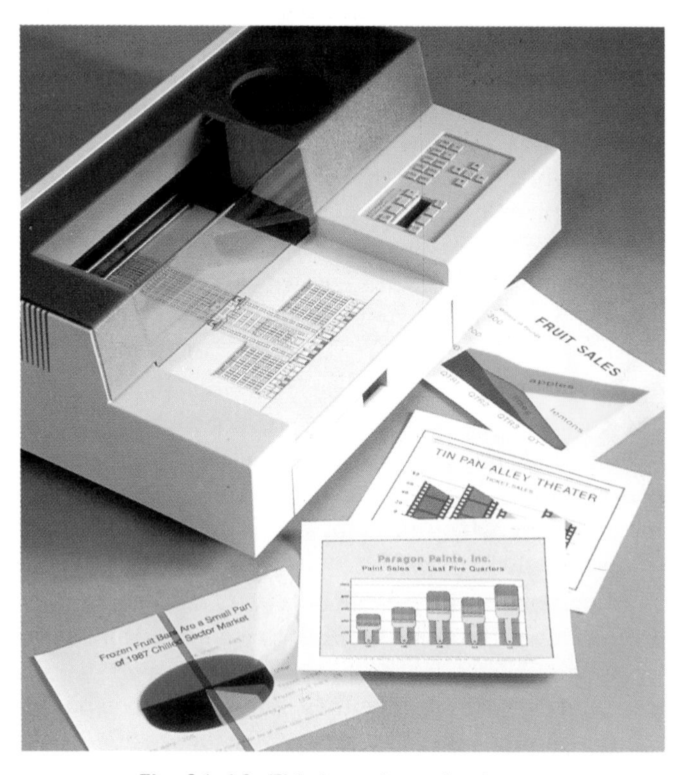

Fig. 31-13. This is a microgrip plotter. (Hewlett-Packard Co.)

A **flatbed plotter** is designed so that the substrate remains stationary while the pen moves over the surface horizontally (x axis) and vertically (y axis). A **drum plotter** uses a cylinder on which the paper moves forward and back. The pen then only needs to move left to right to draw.

Another type of plotter, which is similar to the drum plotter, is the grit-wheel or **microgrip plotter**. Instead of a drum, small wheels grip the paper at each side. These wheels turn to move the paper.

COM devices

Some hardcopy units can produce microfilm as output, which is known as **computer output microfilm (COM)**. Since less space is required for microfilm, this technique is often used where large numbers of documents must be saved, such as in insurance companies or drafting firms.

PERIPHERAL CONNECTIONS

Connections must be made between the computer and peripherals. These connections are different, based on whether they are used for serial or parallel processing, Fig. 31-14. In **serial processing**, data is sent in single file. In **parallel processing**, data is sent in parallel streams. For example, one byte (eight bits) would be sent by serial processing one bit at a time. In parallel processing, eight wires might carry all eight bits at the same time.

As you might guess, parallel processing is faster than serial processing. However, it is not as reliable over long distances.

Keyboards and modems are usually serial devices. Printers can be serial or parallel, depending upon the

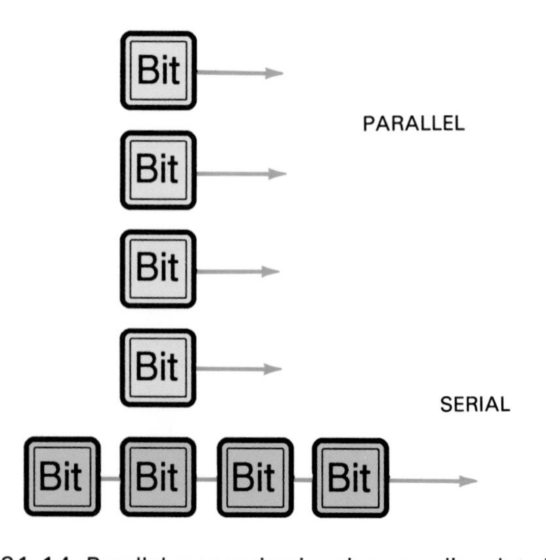

Fig. 31-14. Parallel processing involves sending data in parallel streams (top), while serial processing involves sending data in single file.

type of printer and computer. It is important to have the right cable for the device used.

MICROCHIPS

A **microchip** is a very small integrated circuit (IC) that is a major component in computers and other electronic devices. Each circuit board in a computer contains many microchips.

The microchip is made from very pure silicon. Sand on the beach is also silicon. However, silicon crystals for chips are grown in a laboratory and sliced for use. Integrated circuits, containing thousands of electronic components, are placed on the chips using photographic and chemical techniques. In order to design these circuits, they are first drawn over 200 times oversize using computer systems and then reduced photographically, Fig. 31-15. For example, a circuit design that covers a drawing board will later be used to produce a chip that can fit through the eye of a needle. After chips are made, they are mounted in plastic for easier handling with pins for electrical connections.

A variety of chips are used in a computer. The chips are connected on printed circuit boards that have copper paths leading between the chips. If you look in a computer, you will see these circuit boards with chips on them, Fig. 31-16.

A **microprocessor** is a microchip that contains the circuits necessary to process data or compute. The central processing unit of a computer consists of a microprocessor. Microprocessors are normally classified by the number of bits that can be processed simultaneously, which is known as the word size. A 16-bit microprocessor can process 16 bits of data all at once, and has a 16-bit word size.

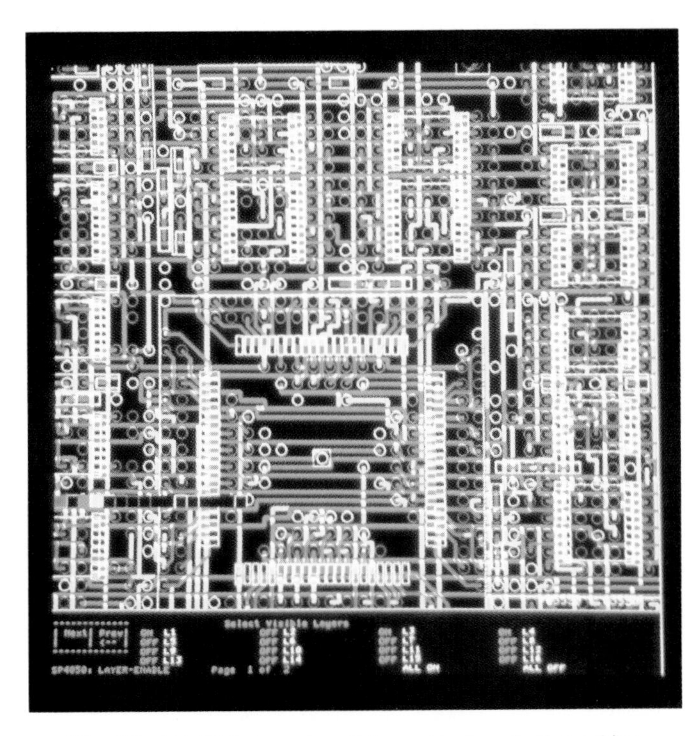

Fig. 31-15. This is a circuit board being designed by computer. (IBM Corp.)

Fig. 31-16. Microchips and other electronic components can be seen on this circuit board. (IBM Corp.)

Clock speed is a measure of how fast the computer processes data. The control unit in the computer has a system clock that sends out electrical pulses at a certain rate of speed, measured in megahertz (1 million cycles per second). Since word size and other factors can affect total processing capability, clock speed alone cannot be used to determine the ability of the computer to process data. However, when comparing two computers with the same word size, one being 8 Mhz and the other being 12 Mhz, the faster unit will have greater processing capabilities.

An accurate measure of processor capability is in **million instructions per second (MIPS)**. Personal computers are typically capable of 1 to 4 MIPS.

SUMMARY

This chapter has provided some general background information about computers. Other chapters in the book will explore specific uses for computers in business and industry.

When you work with a computer, it is sometimes easy to believe that it "thinks" as you do. However, remember that computer "thinking" is a program, created by people. Until scientists completely understand the brain, it will be difficult to create a machine that thinks as you do. Research in this area is known as artificial intelligence.

WORDS TO KNOW

All of the following words have been used in this chapter. Do your know their meanings?

arithmetic-logic section
binary code
bits
bytes
cathode ray tube (CRT)
central processing unit (CPU)
clock speed
CD-ROM (Compact Disc-Read Only Memory)
computer
computer output microfilm (COM)
computer program
control unit
crash
daisy wheel printer
digital computer
digitizer
direct thermal printer
dot matrix printer
drum plotter
DVD (Digital Video Disc)
flatbed plotter
floppy diskette
gigabyte
graphic tablet
hard drive
hardcopy devices
impact printers
ink jet printer
input device
input/output (I/O) devices
kilobyte
laser printer
light pen
liquid crystal display (LCD)
mainframe computer
mechanical mouse
megabyte
memory
microchip
microcomputer
microgrip plotter
microprocessor
milling instructions per second (MIPS)
minicomputer
mini floppy diskette
monitor
mouse
nonimpact printer
optical character recognition (OCR)
optical mouse
optical readers
parallel processing
pen plotter
phone modem
pixels
random access memory (RAM)
raster
read-only memory (ROM)
sectors
semiconductor memory
serial processing
thermal printer
thermal transfer printer
track ball
tracks
vector display

REVIEWING YOUR KNOWLEDGE

Please do not write in this text. Write your answers on a separate sheet.

1. Computer data can be in the form of:
   a. Letters.
   b. Numbers.
   c. Pictures.
   d. All of the above.
2. True or false? Certain jobs can be performed faster and more accurately with computer assistance.
3. Name the three main types of digital computers.
4. Which of the following is an example of an input device?
   a. Printer.
   b. Keyboard.
   c. Monitor.
   d. Disk drive.
5. All data that enters the computer is placed in:
   a. Random access memory (RAM).
   b. Read-only memory (ROM).
   c. Both a and b.
   d. None of the above.
6. True or false? One problem with vector displays is that images can suffer from the "jaggies."
7. List six precautions you should take when using diskettes.
8. Which of the following is an impact printer?
   a. Thermal printer.
   b. Dot matrix printer.
   c. Ink jet printer.
   d. Laser printer.

9. True or false? Parallel processing is faster than serial processing.
10. A __________ is a very small integrated circuit that is a major component in computers and other electronic devices.

APPLYING YOUR KNOWLEDGE

1. Have a group brainstorming session. List all the ways computers affect your daily life.
2. Analyze a computer advertisement, and in your own words, describe the computer's capabilities. For example, what size is the drive or drives? What is the clock speed? Would you recommend this computer to your friends?
3. Write an article describing the future of computer technology. Predict how you think computers will be used 25 years from now.
4. Prepare a display of common computer input and output devices. Label them.

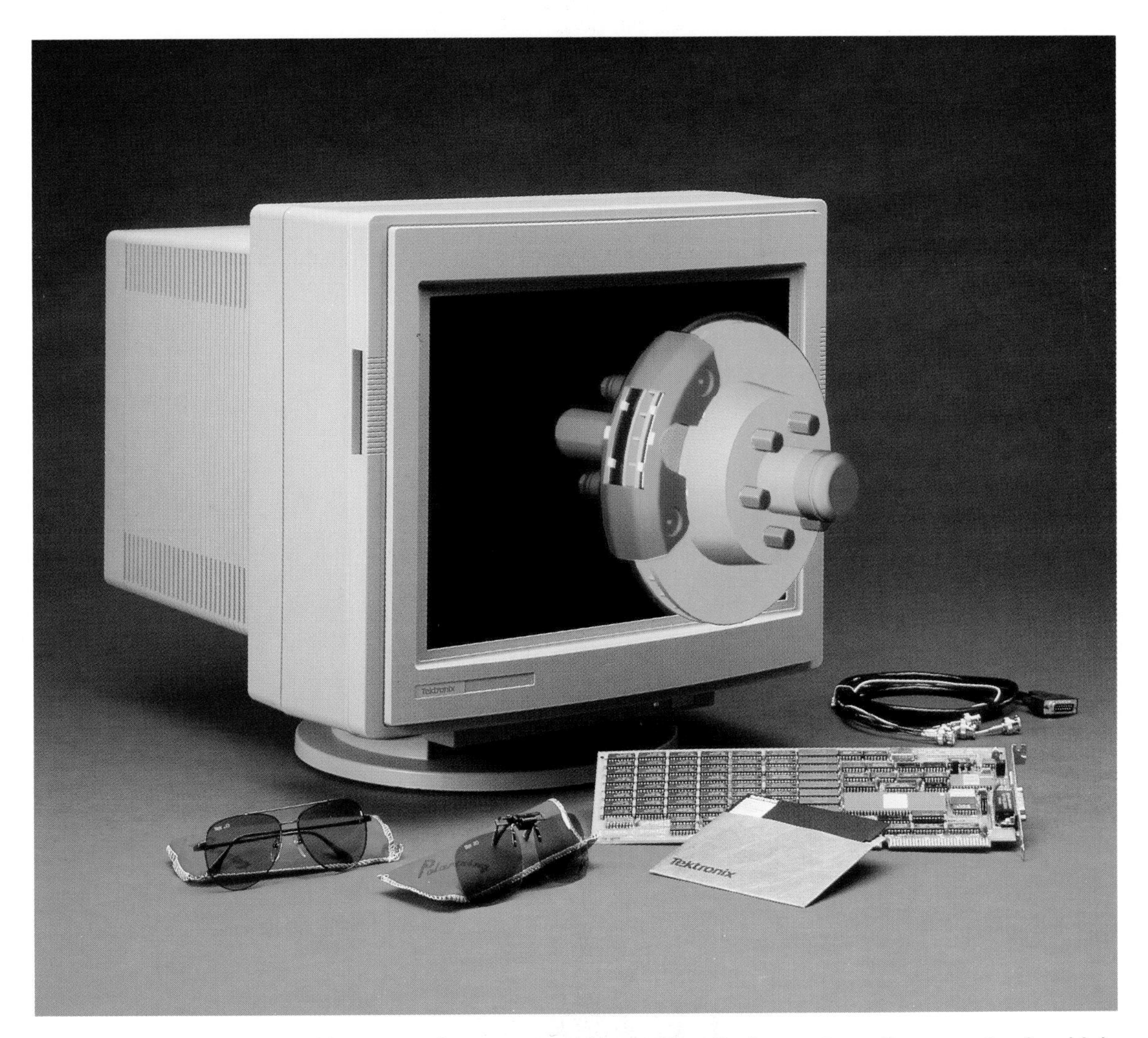

Combined with a compatable personal computer, this graphics display system allows you to view high quality stereoscopic three-dimensional images. Special polarizing glasses worn by the user decode the stereogram, providing a true stereoscopic image. This system has applications in areas such as CAD, education, visualization of complex data, and medical imaging. (Tektronix, Inc.)

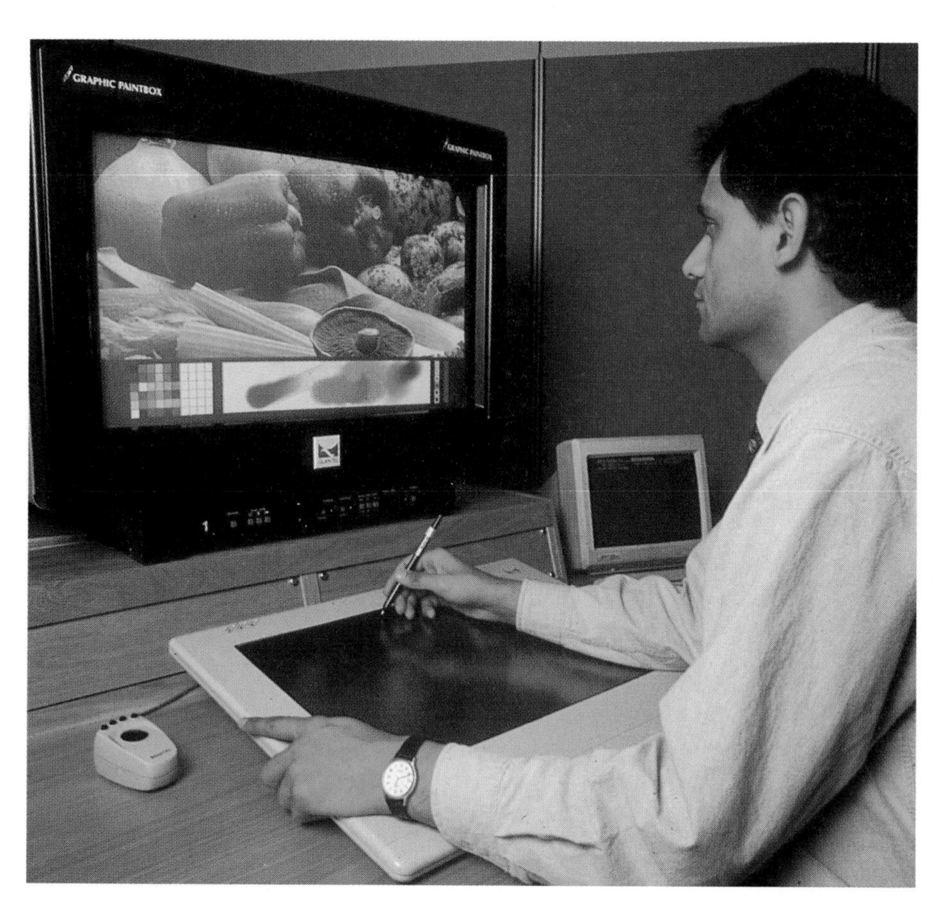

Computer software, such as Graphic Paintbox, can extend the creativity of the user. (Quantel)

32

CHAPTER

COMPUTER SOFTWARE

After studying this chapter, you will be able to:
- *Differentiate between machine language and high level language.*
- *Describe how system software is utilized.*
- *Identify the major types of application software that are available.*
- *Define the two categories of graphics software.*
- *Identify common user interface features.*

Computer hardware is of no use without the necessary software or computer programs, Fig. 32-1. **Computer software** is basically the specific instructions for directing the operation of the computer.

Fig. 32-1. Different software packages are used to perform work with the computer.
(Bowling Green State University, Visual Communication Technology Program)

This chapter will give you an overview of software. Since all software is written in special codes, known as computer language, we will begin at that point. In addition, system software and applications software will be explored.

COMPUTER LANGUAGE

Computer language consists of the various codes that are used for providing instructions for the computer. Development of these instructions is called **programming**. At the most basic level, computers only work with bits (binary digits). A program consisting of binary codes is called **machine language**. Since machine language code is challenging, higher level languages have been developed that make the job easier. **High level language** more closely represents the way we normally communicate in writing. However, all high level language must be converted to low level language (machine language) before it can be processed.

MACHINE LANGUAGE

In order to process information, machine language is used. This is in binary code that can be handled by the chips as on and off electrical current.

You are probably accustomed to working with a decimal number system, Fig. 32-2. In this system, each digit to the left is ten times greater. For example, 1538 = 8 *ones*, 3 *tens*, 5 *hundreds*, and 1 *thousand*. However, this system is inadequate for electronic circuitry. A base 2 system, the binary system, is used with computers because its numbering corresponds to on (1) and off (0) states.

In the binary system, each digit to the left in a number is two times greater. For example, 1110 = 0 *ones*, 1 *two*, 1 *four*, and 1 *eight*. To convert this to a decimal

BASE 10 SYSTEM (DECIMAL)

$$785 = (5 \times 10^0) + (8 \times 10^1) + (7 \times 10^2)$$

BASE 2 SYSTEM (BINARY)

$$1010 = (0 \times 2^0) + (1 \times 2^1) + (0 \times 2^2) + (1 \times 2^3) = 10$$

Fig. 32-2. The decimal number system is a base 10 system, while the binary number system is a base 2 system. Base refers to the number of symbols used in the system.

number, these are added together to equal 14. Thus, 1110 could be used as binary code for 14. Binary numbers, which are all 0 and 1 to correspond to on and off, can be used in combination as a code for any symbol. Common symbols are the alphabet and numbers.

Each binary digit is known as a bit. The CPU has a maximum capacity in bits that can be processed together as a "byte" (8 bits) or larger "words." For example, there are 8 bit, 16 bit, and larger microprocessors. More bits mean more combinations of 0 and 1, so more work can be done by the computer.

Several systems have been developed to standardize the binary code for various characters and numbers. One such system is the **American Standard Code for Information Interchange (ASCII)**. The basic 128-character (letters, numbers, and figures) set can be represented with 7 bits, Fig. 32-3.

Since it is easy to make mistakes when working with large numbers of bits, 00111000111000, for example, other number systems are sometimes used that represent binary numbers in a more compact form. Hexadecimal (base 16) is one of the most common of these codes. In hexadecimal, 16 single numbers and letters are used as code for 16 combinations of 4 bits each. For example, the binary code for 10 is 1010. In hexadecimal this is represented as the letter A. This system is used to reduce human error when writing out binary numbers. The computer itself works with binary, not hexadecimal numbers.

HIGH LEVEL LANGUAGE

High level languages are in English rather than binary code and make communication with the computer easier. When these languages are used by the computer, they are translated to machine language before processing takes place.

BASIC is an example of a popular high level language. This is an acronym that stands for Beginner's

SYMBOL	ASCII CODE
1	0110001
2	0110010
3	0110011
4	0110100
5	0110101
A	1000001
B	1000010
C	1000011
D	1000100
E	1000101
?	0111111

Fig. 32-3. Several letters and numbers are shown in ASCII code.

All-Purpose Symbolic Instruction Code. Other examples are FORTRAN and Pascal.

When you use the computer, you must either program the machine to do what you want or use a prepared program. A good programmer writes a program in as few steps as possible and in a logical order. Programmers use algorithms and flow charts to help in programming. An **algorithm** is a written, step-by-step procedure for solving a problem, Fig. 32-4. A flow chart uses special symbols to show these steps, Fig. 32-5.

Let's write a computer program in BASIC language. Before beginning you should know what "system commands" and "instructional statements" mean.

System commands tell the computer what to do with the program. For example, *RUN* will start a program. *NEW* erases RAM memory so a new program can begin. Other examples are *SAVE*, *END*, and *LIST*.

Instructional statements are words in each line of the software program that ask the computer or person at

ALGORITHM

1. Computer asks your name.
2. Input your name.
3. Computer displays "Hello, Name".
4. Computer asks for 2 numbers to add.
5. Input 2 numbers.
6. Computer displays sum.
7. Program ends.

Fig. 32-4. An algorithm is used to show the steps needed in a program.

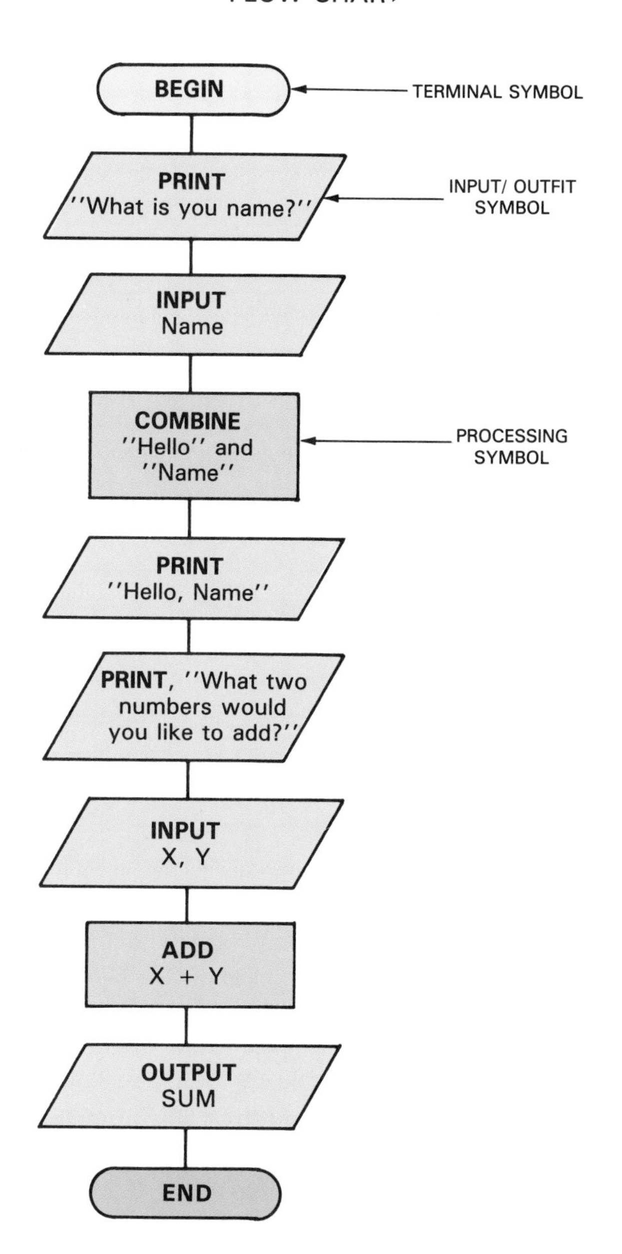

Fig. 32-5. A flow chart uses special symbols to map out the program.

the keyboard to do something. *PRINT* tells the computer to display what follows the statement. For example *PRINT "JOHN"* will display *JOHN* on the screen. *INPUT* tells the computer you should enter information. Each statement is numbered, and the program is completed by the computer starting with the lowest number. Statements are usually numbered by tens (10, 20, 30, etc.) so that other statements can be added if necessary.

Now let's look at a simple program, Fig. 32-6. Your name will be Chris. In order to actually use this program, you would have to modify it for your computer system.

SYSTEM SOFTWARE

System software, or the operating system, consists of a collection of programs that allow the parts of the computer to work together as a single unit. Another name for this software is **disk operating system (DOS)**, although it does much more than just control disk drives. Specifically, this software manages the computer, peripheral devices, and software so they all work together correctly, Fig. 32-7.

The operating system is the first set of programs that are loaded into the computer when it boots up. In some cases, this software is on the same disk as the application program. In other cases, it can be on a separate disk. When a hard drive is used, the system software is normally placed on it.

Many of the files in DOS are not visible to the user. For example, the input/output system files (programs) ensure that a program, such as word processing, operates correctly with all the hardware. In other words, these programs make the hardware and software work together.

The programs in the operating system that are most visible to the user are the **utilities**. These programs carry out useful functions such as formatting disks, copying disks, deleting files, and viewing directories, just to name a few.

Presently, there are different operating systems for Macintosh and IBM-compatible computers, the two most common platforms. Some work has been done on trying to develop a common system. The Windows operating system is a graphical user interface (GUI) that allows users to launch applications (programs) with a mouse click.

As with any software, operating systems will have version numbers that refer to revisions that have been made. Major revisions are a whole number, for example, 3.0. Small corrections in the program result in smaller increments, such as 3.1 and 3.2.

As operating systems are updated, application programs also change to take advantage of the upgrades. Therefore, newer software will often not work with older versions of operating systems.

APPLICATION SOFTWARE

Application software performs specific tasks, such as word processing. Although simple application programs may be written by the user, most sophisticated software is purchased. This is often called **commercial application software**. The most common application programs include:

- Word processing.

A SAMPLE PROGRAM

COMMANDS AND STATEMENTS	WHAT HAPPENS
RUN	Starts the program.
10. PRINT ''What is your name''	Display shows the statement in parentheses.
20. INPUT NAME	Computer waits for you to type in your name and return.
30. PRINT ''HELLO''; NAME	Display shows HELLO CHRIS
40. PRINT ''What two numbers would you like to add?''	Display shows statement.
50. INPUT, X, Y	Computer waits for two numbers which will be called X and Y.
60. LET SUM = X + Y	Instruction for the computer to add the numbers.
70. PRINT ''The sum is''; SUM	Sum is displayed.
80. END	Computer displays the program.

Fig. 32-6. This is an example of a program written in BASIC. The right column explains what will happen as the program is executed.

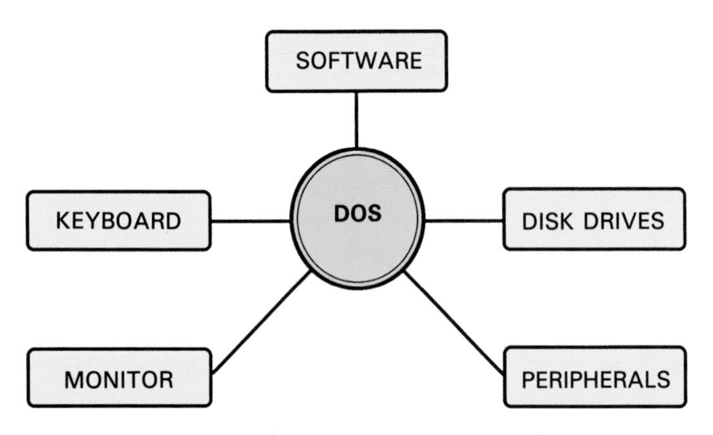

Fig. 32-7. An operating system is necessary for the hardware components to work together. It is also used to enable the hardware to work with the software.

- Data base management.
- Electronic spreadsheets.
- Integrated software.
- Graphics software.

WORD PROCESSING

A **word processor** is a program that simplifies writing and rewriting, Fig. 32-8. When you write a paper by hand or on a typewriter, later revisions often require a rewrite of the entire paper. Since a word processor can store the writing on disk, changes can be added with very little effort. Paper formatting tasks, such as centering headlines and justifying columns, are done easily on the computer. Word processors also normally have spelling checkers to check for errors.

DATA BASE MANAGEMENT

Data base software allows you to file important information in computer memory. This is why it is sometimes called an electronic file. An example would be a

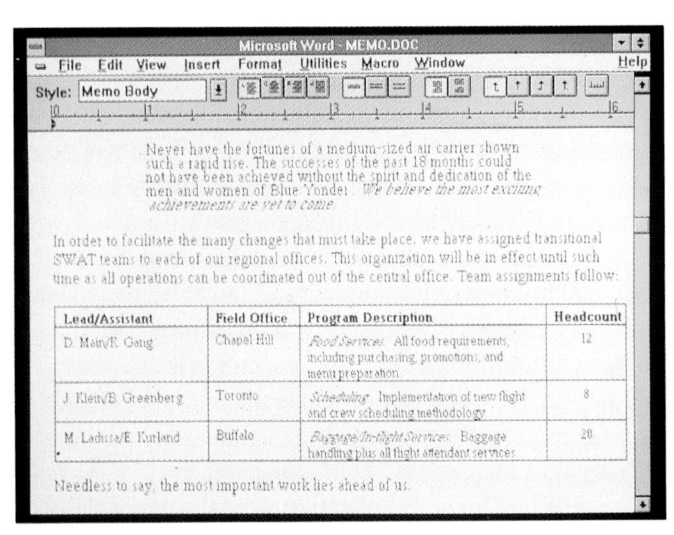

Fig. 32-8. A memo is being created with a word processing program. (Microsoft Corp.)

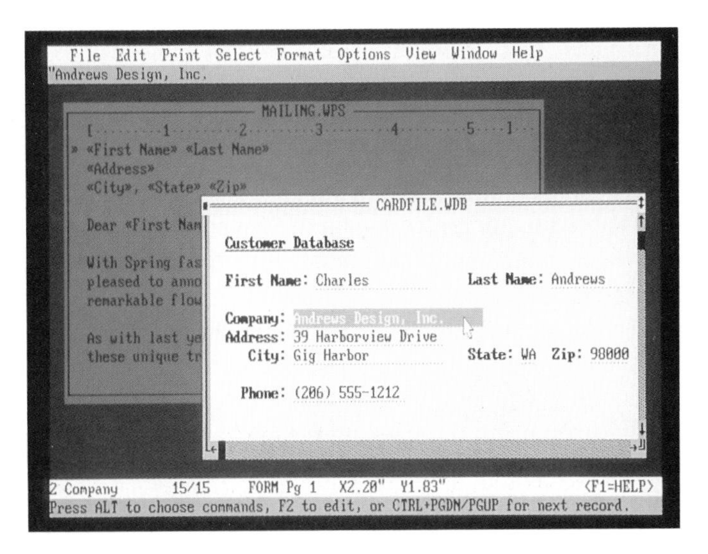

Fig. 32-9. One record for a database is shown. (Microsoft Corp.)

list of potential customers for a large business, Fig. 32-9. When the business is ready to advertise by mail, the labels can be printed out and applied to envelopes. Changes and updates are very simple.

When a file is created in a data base, fields must be determined for the records that will be placed in the file. For example, fields might be last name, first name, company, and address. One record in the file will be one listing of all these fields. Once established, the records can be sorted by any of the fields. For example, all the records could be sorted by city.

ELECTRONIC SPREADSHEETS

Electronic spreadsheets are typically used for financial record keeping, Fig. 32-10. This tool handles these records electronically, just as paper spreadsheets in accounting ledgers are used for the same purpose. Spreadsheets allow companies to manage important records, such as billing, profit calculations, etc.

A spreadsheet is organized in rows (horizontally) and columns (vertically). A **cell** is where a row and column meet.

INTEGRATED SOFTWARE

Many computer programs now can be purchased as **integrated software**. This is software that is combined. For example, word processing, data base management, and electronic spreadsheets can be purchased as a single package.

GRAPHICS SOFTWARE

Graphics software is another common application program. **Graphics software programs** are used for creating pictures. These programs might include charts in reports, or technical drawings.

A wide variety of graphic programs is available. For example, presentation graphics programs are used to create exciting graphics that can be placed in reports or used to present to large groups, Fig. 32-11. Common graphics include bar charts, and pie charts. CAD software, another type, is used for creating precise technical drawings. Drawings of machines and buildings are made in this manner. Another software package, paint software, is typically used for generating freehand images, as an artist might do, Fig. 32-12.

Most graphics software can be divided into two categories, based on how the images are created. These are vector graphics and bit-mapped graphics.

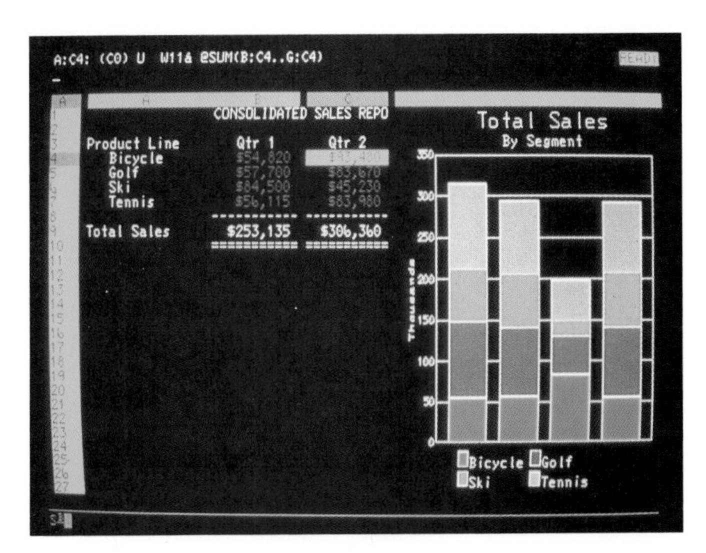

Fig. 32-10. The spreadsheet on the left is being used to generate the chart on the right. (IBM Corp.)

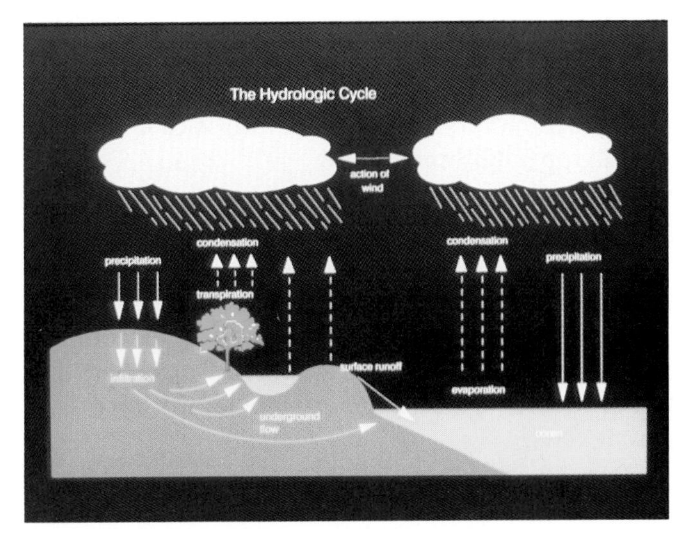

Fig. 32-11. A presentation graphic is shown here. (Polaroid Corp.)

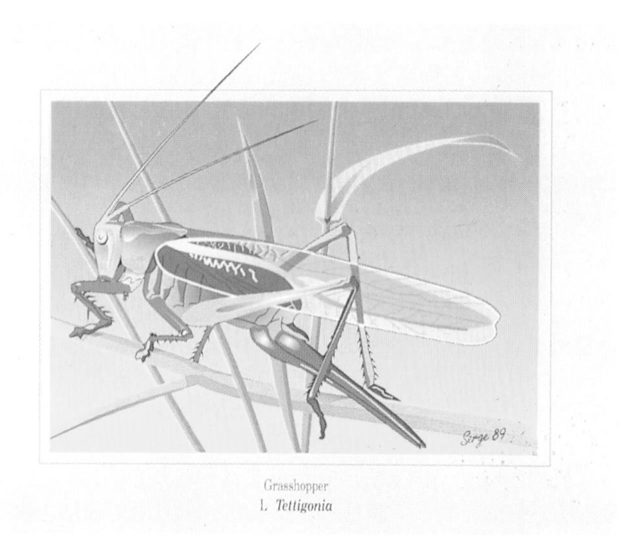

Fig. 32-12. Paint software was used to create this picture. (Corel Systems Corp.)

Vector graphics programs

Vector graphics programs create drawings that you might normally do with technical drawing tools, such as a T-square and compass. These are also called object-oriented or draw-type graphics, since drawings are composed of separate objects, such as lines, circles, and arcs. An example of this type of program is a CAD program. Vector graphics are usually of very high quality when printed.

Bit-mapped graphics programs

Bit-mapped graphics programs create images consisting of a pattern of dots or pixels. Paint-type programs generate bit-mapped graphics. This type of program is often used for generating freehand art, much like an artist might create a picture. The image quality of bit-mapped graphics is usually lower than that of vector graphics.

OTHER APPLICATION PROGRAMS

Other programs that are frequently used include:

- Desktop publishing.
- Data communication.
- Multimedia.

Desktop publishing

Desktop publishing software is used to compose pages consisting of text and graphics. Documents that are produced in this manner include brochures and newsletters.

Data communication

Data communication software is used along with a modem in order to communicate over distances via computer. Data is normally sent over the phone lines to computers.

Multimedia

Multimedia software products allow the computer to be used to mix different media such as sound and video. Moving images can be created with animation techniques or captured with a video camera, Fig. 32-13. Likewise, sound can be recorded or digitally created. Either the sound or moving images can then be combined with text, graphics, and still images for a multimedia presentation, using the computer.

Multimedia software normally requires the use of additional computer peripherals. A video camera and audio speakers are two examples.

USER INTERFACE FEATURES

A **user interface** is the system that is used to allow the computer user to communicate with the computer system. Hardware, as well as software, can be involved in user interfaces. For example, some user interfaces utilize the keyboard, while others require a mouse.

A common feature of many user interfaces is on-screen menus. A menu displays options from which the user can choose. A pull-down menu provides a submenu of options that can be pulled down using the mouse. Software with menus is sometimes called **menu-driven software**.

Software can also be command-driven. With **command-driven software**, the user inputs a command from the keyboard instead of choosing from screen menus. Experienced operators sometimes prefer this system because it can be faster than looking at menus. Some types of software provide either option.

Fig. 32-13. This illustration is part of an animated presentation. (Autodesk, Inc.)

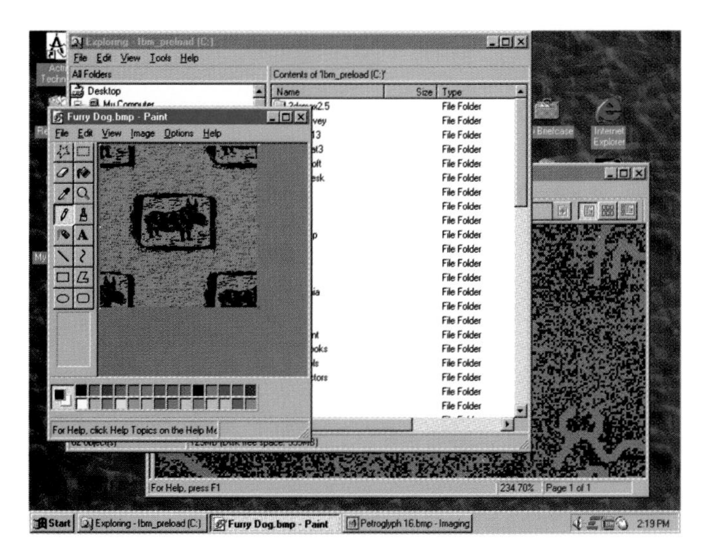

Fig. 32-14. Multiple windows are open on this screen.

Another common feature of user interfaces is a **windowing system**, Fig. 32-14. Windows are a rectangular portion of the screen that display information while the program screen is still underneath. Like the name implies, a window is opened to show something different on the screen. Usually, multiple windows can be placed on the screen, like sheets of paper on a table.

Icons are often used in software, Fig. 32-15. These are graphic symbols that represent choices that can be made. Icons can speed up operation since it is easier to recognize a single symbol than to read the text for the same command.

User interfaces can contain other features that have not been described. In fact, they continue to change at a rapid pace, and each new generation of interfaces is normally easier to use than the previous one.

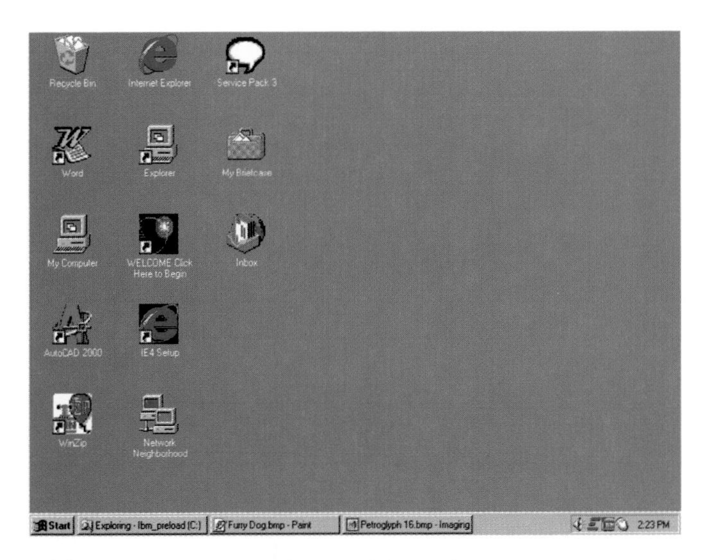

Fig. 32-15. Icons are the graphic symbols that represent choices that can be made.

SUMMARY

This chapter introduced you to computer language and software. Software can be divided into system software and application software, both of which are needed to operate the computer. User interface systems have greatly simplified the act of communicating with the computer.

WORDS TO KNOW

All of the following words have been used in this chapter. Do you know their meanings?

algorithm
American Standard Code for Information Interchange (ASCII)
BASIC
bit-mapped graphic programs
cell
command-driven software
commercial application software
computer language
computer software
data base software
data communication software
desktop publishing software
disk operating system (DOS)
electronic spreadsheets
graphics software programs
high level language
icons
instructional statements
integrated software
machine language
menu-driven software
multimedia software products
programming
system commands
system software
user interface
utilities
vector graphic programs
windowing system
word processor

REVIEWING YOUR KNOWLEDGE

Please do not write in this text. Write your answers on a separate sheet.

1. True or false? All high level language must be converted to low level language (machine language) before it can be processed.
2. What is the purpose of the American Standard Code for Information Interchange (ASCII)?

3. Which of the following tell a computer what to do with a program?
   a. Algorithms.
   b. System commands.
   c. Instructional statements.
   d. System software.
4. The disk operating system (DOS) manages:
   a. The computer.
   b. Peripheral devices.
   c. Software
   d. All of the above.
5. Name five common types of commercial application software.
6. True or false? The image quality of bit-mapped graphics is usually higher than that of vector graphics.
7. ________ software products allow the computer to be used to mix different media such as sound and video.
8. Why do experienced computer operators sometimes prefer command-driven software over menu-driven software?

APPLYING YOUR KNOWLEDGE

1. Develop and test a simple program in BASIC.
2. Investigate how to use an operating system for a computer you will be using. Find out how to use important utilities such as "copy" and "format."
3. Using any program, identify the user interface features and describe them.
4. Write a letter using a word processing program.

DATA COMMUNICATION

After studying this chapter, you will be able to:
- *Define data communication.*
- *Explain how modems are used in data communication with computers.*
- *Explain how mode of transmission, speed of transmission, and direction of transmission are used in data communication.*
- *Contrast the types of computer networks.*
- *Identify some common data communication services.*

Data communication involves sending and receiving information by computer. This is a very common form of communication today. For example, airline personnel check on flight information in this way. Nearly all major companies use some form of data communication.

The typical channel for data communication is the telephone line. However, the telephone line was originally intended for audio transmission, such as voice, which is an analog signal, Fig. 33-1. When transmitted, this signal is an electrical wave that varies to duplicate the sound. In contrast, computers use digital signals for communication, which are characterized by electrical pulses. These pulses represent on or off states of electricity. In order to change the digital signal into an analog signal, a modem is used.

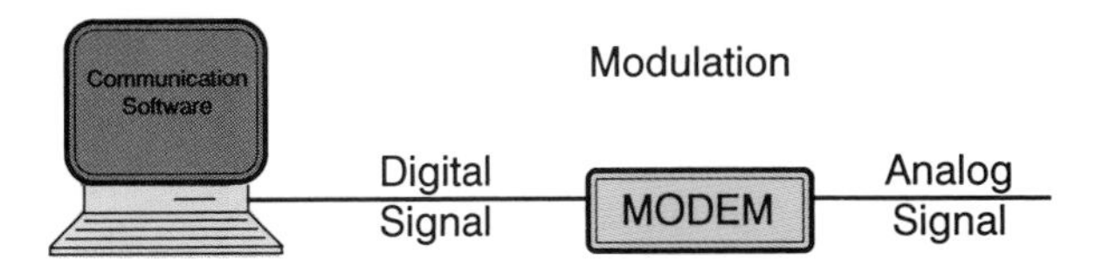

Fig. 33-1. A modem converts digital signals to analog signals (modulation) and converts analog signals back to digital signals (demodulation).

MODEMS

Modems are input/output devices that allow you to send and receive information with computers. Typically, this takes place over telephone lines. Modems perform this function by converting the digital signals from the computer to analog signals. Conversion from digital to analog is a type of **signal modulation**. When the modulated signal reaches a modem at the receiving end, the signal is converted back to digital. This is known as **signal demodulation**. The term *modem* is derived from the two functions of the devices: **mo**dulation - **dem**odulation.

There are two major types of modems. These are the acoustic coupler and the direct connect modem.

The **acoustic coupler** is the older technology. It requires you to place the telephone handset directly on the modem for sending and receiving information.

The **direct connect modem** attaches directly to the phone line for sending information. Direct connect modems can be external or internal, Fig. 33-2. **External modems** sit next to the computer and plug into it, while the **internal modems** are placed inside the computer, Fig. 33-3.

In order to use a modem, special software is needed. This software performs a variety of functions, such as ensuring that the sending and receiving modems communicate correctly.

DATA COMMUNICATION BASICS

In order to understand data communication, there are several concepts that need to be understood. These include:
- Speed of transmission.

Fig. 33-2. This is an external modem. (Apple Computer, Inc.)

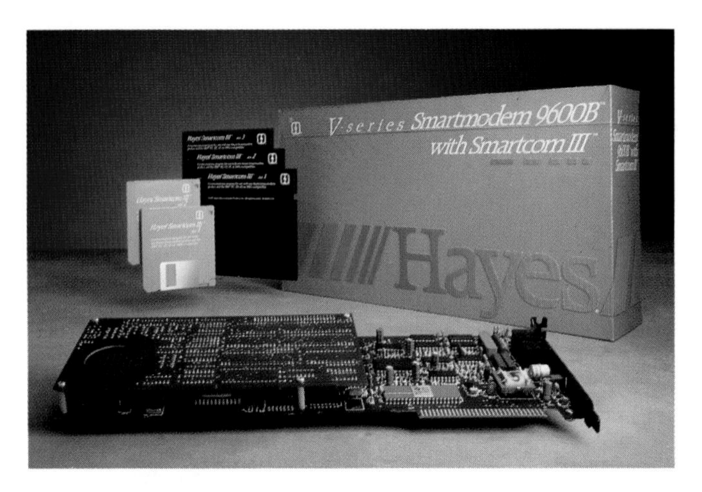

Fig. 33-3. This is an internal modem and the necessary communication software. (Hayes Microcomputer Products, Inc.)

- Mode of transmission.
- Direction of transmission.

SPEED OF TRANSMISSION

The speed at which modems send and receive data is known as the **baud rate**. This normally refers to the number of bits per second that can be transmitted. Typical baud rates for personal computers are 28,000 and 56,000. Larger business computers use modems that transmit at even higher baud rates.

MODE OF TRANSMISSION

Signal transmission by modems can be synchronous or asynchronous. This refers to how two modems function together to send and receive data at the same rate.

In **synchronous transmission,** a transmitter clock sends a timing signal to the receiving modem to ensure that both are synchronized. This system is used when large amounts of data must be sent, such as in a large company. Synchronous transmission is a very fast and accurate system, but also expensive. It is normally used with more powerful computers.

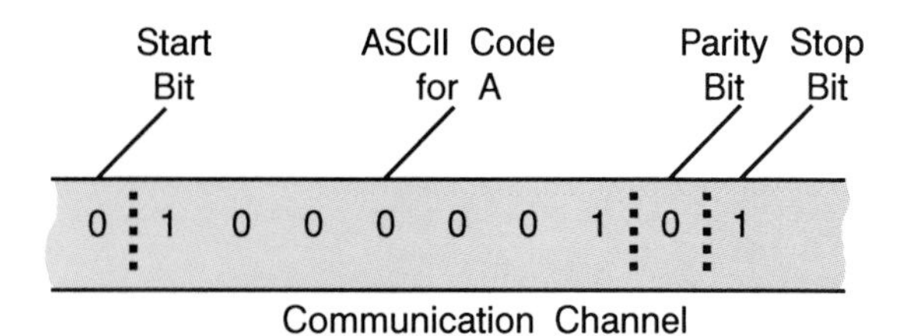

Fig. 33-4. In asynchronous transmission, extra bits are used to transmit each character.

Most personal computers use an **asynchronous transmission** mode, Fig. 33-4. Instead of a timing signal that synchronizes the modems, extra bits known as **synch characters** are used. For example, the letter A, which is represented in ASCII code with 7 bits, would be transmitted with 10 bits. A stop and start bit signal the modem when the character begins and ends. In addition, one extra bit, called a **parity bit**, is often used at the end of a character code to check for errors. Thus, the number of characters that can be sent per second is about 10% of the baud rate (bits per second). A 9600 baud modem can transmit about 960 characters per second. Since asynchronous modems use extra bits, this is a slower system than the synchronous modem. However, it is a much less expensive system. Typical baud rates for asynchronous systems are 9600, 14.4, and 28.8 baud.

DIRECTION OF TRANSMISSION

Data can be transmitted one way or both ways. In **simplex transmission**, data flows in one direction only. For example, a traffic sensor can send data to a computer that controls the traffic signal. A signal is sent from the sensor location but not back.

In **half duplex transmission**, signals can be sent both ways, but in only one direction at a time. A CB radio is a type of half-duplex system.

In **full duplex transmission** systems, sending and receiving are allowed to take place at the same time. A telephone is a good example. Most data communication between computers is full duplex.

COMMUNICATION NETWORKS

A **communication network** consists of a number of computers or terminals that can be linked together. In this way, data can be shared, as well as software and peripherals, such as printers.

Computer networks consist of nodes and links. A **node** is an end point, such as a personal computer. A **link** is the channel between two nodes. If four personal computers were linked together, the network would have four nodes.

A **local area network (LAN)** is a network that resides in a small area, such as several buildings. This system is often used to link personal computers, as well as allowing users to share peripherals. For example, a laser printer in the network could be used by each user. Another use of LANS is to share software, Fig. 33-5. A special network license is normally needed for this, which can be purchased from the software vendor. The computer that is used to store the software is known as the **file server**.

A **wide area network (WAN)** covers a large geographic area and usually involves sending data over long distances. For example, airlines use a WAN to keep track of flight bookings and schedules. Long distance telephone companies also consist of wide area networks.

NETWORK TOPOLOGY

Topology refers to the configuration of nodes and links in the networks. The **star network** is normally used when a central, or host computer is needed, Fig. 33-6. Links go out to each node from the computer, much like a star.

A **bus network** is often used for linking personal computers, Fig. 33-7. A single cable is used, and nodes can be hooked on the cable as needed. This is a simple and reliable system. Data normally travels either direction in the bus.

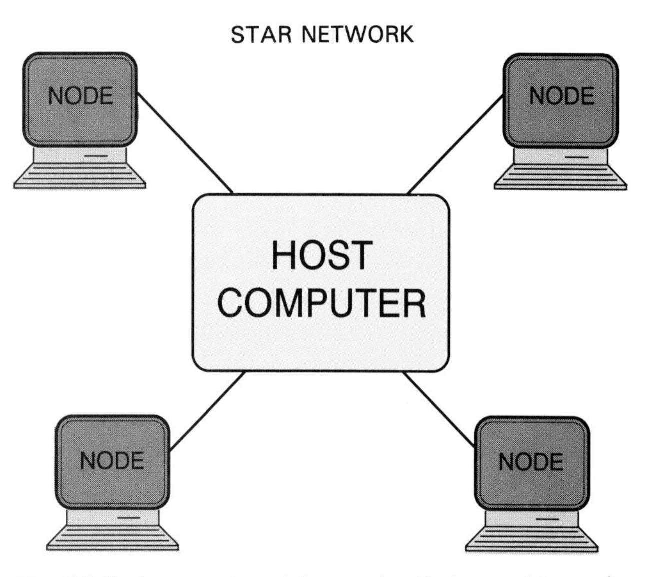

Fig. 33-6. A star network has nodes that go out to each node from the computer.

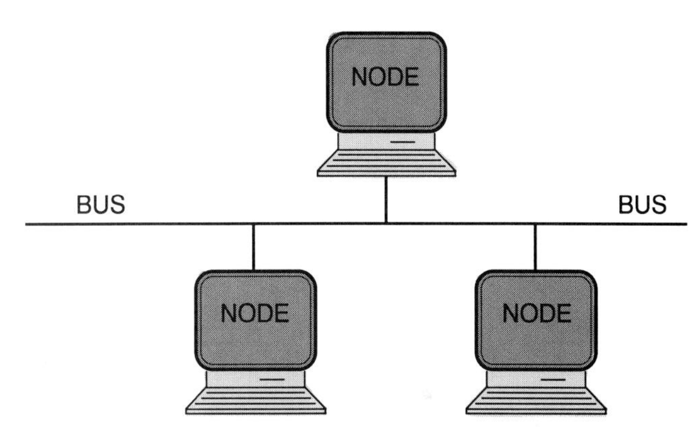

Fig. 33-7. A bus network uses a single cable to link nodes together.

Fig. 33-5. A local area network is being used to link these computers. The computer under the desk acts as the file server. (IBM Corp.)

A **ring network** consists of a number of nodes linked together in circular fashion, Fig. 33-8. Data travels in one direction in the ring network.

DATA COMMUNICATION SERVICES

Although you may not always think about it, data communication services surround you. When an automated bank teller is used, data about your account is transmitted to the bank's central computer. Data bases are available at libraries and for home computers. These provide information on almost any subject, depending upon the data base.

You may be familiar with some popular data communication services. An **electronic bulletin board** or computer bulletin board is a service that allows the user to post messages electronically via computer. Messages might include hints about using a computer program.

Electronic mail is similar to the bulletin board, but more private. Individuals, using computers and special software, can send messages to each other. **Voice mail** is a type of electronic mail that uses voices. Voices are digitized for communication.

Facsimile machines, often called fax machines, are another way to transmit data, Fig. 33-9. **Fax machines** can transmit or receive pictures or text. The sending unit scans the image, converts it to digital form, and transmits it. The receiving fax machine converts the image back into visual form and a hard copy is made.

Some communication companies offer a combination of services. Electronic mail, electronic bulletin boards, electronic shopping, and other services are grouped together and provided for a monthly fee, Fig. 33-10.

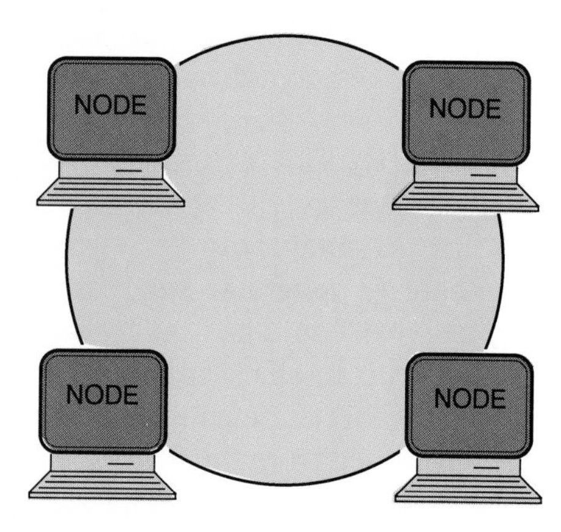

Fig. 33-8. A ring network links nodes together in a circular manner.

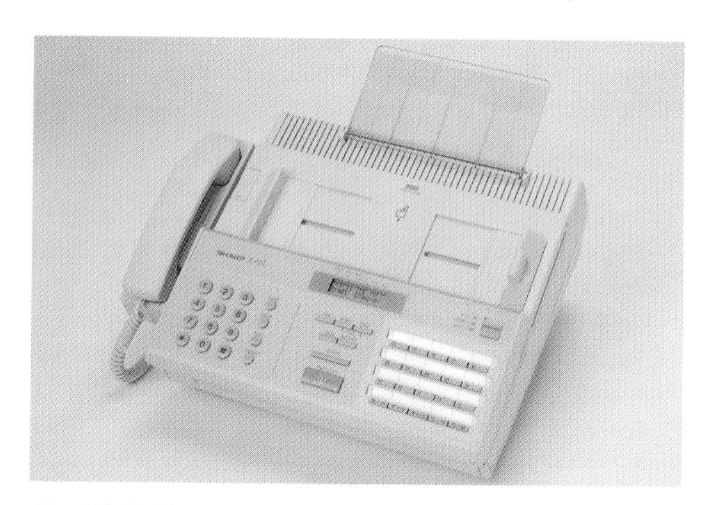

Fig. 33-9. Many businesses rely on fax machines for rapid communication. (Sharp Electronics)

Fig. 33-10. This person is learning to use a communication service. (IBM Corp.)

SUMMARY

Data communication is the transmission of information in digital form via computer. As computer uses have increased, data communication has increased, both at work and at home. This trend is predicted to continue into the future.

WORDS TO KNOW

All of the following words have been used in this chapter. Do you know their meanings?

acoustic coupler
asynchronous transmission
baud rate
bus network
communication network
data communication
direct connect modem
electronic bulletin board
electronic mail
external modems
facsimile machines
fax machines
file server
full duplex transmission
half duplex transmission
internal modems
link
local area network (LAN)
modems
node
parity bit
ring network
signal demodulation
signal modulation
simplex transmission
star network
synch characters
synchronous transmission
topology
voice mail
wide area network (WAN)

REVIEWING YOUR KNOWLEDGE

Please do not write in this text. Write your answers on a separate sheet.

1. A modem performs which of the following major functions?
   a. Signal modulation.
   b. Signal demodulation.
   c. Both a and b.
   d. None of the above.
2. True or false? When using a direct connect modem, you must place the telephone handset directly on the modem for sending and receiving the information.
3. The speed at which modems send and receive data is known as the ________ rate.
4. An asynchronous transmission mode uses:
   a. A timing signal.
   b. Synch characters.
   c. A parity bit.
   d. Both b and c.
5. In which of the following are sending and receiving of signals allowed to take place at the same time?
   a. Simplex transmission.
   b. Half duplex transmission.
   c. Full duplex transmission.
   d. Both a and b.
6. True or false? Long distance companies primarily consist of local area networks.
7. Which of the following is used when a central or host computer is needed?
   a. Star network.
   b. Bus network.
   c. Ring network.
   d. None of the above.
8. ________ machines can transmit or receive pictures or text.

APPLYING YOUR KNOWLEDGE

1. Learn how to use a modem for sending and receiving data.
2. Investigate how a business or organization uses communication networks. Report your findings to the class.
3. Use a data base at your local library to find out more about some aspect of data communication. Prepare a written report.
4. Prepare a list of local data communication services available to you. Describe their services.

A two-way radio extends human abilities to produce and hear sound. (Texaco)

34

CHAPTER

COMMUNICATING WITH SOUND

After studying this chapter, you will be able to:
- *Define sound.*
- *Describe how hearing takes place.*
- *Discuss the major breakthroughs in the development of audio technology.*
- *Give examples of ways in which sound is stored and transmitted.*
- *Explain the process of synchronizing sound with slides.*

Music, voices, traffic, wind–these are sounds you probably hear every day. Take a moment to listen to all the sounds that surround you. Sound is a key part of many forms of communication. This is why many industries make products that can extend our ability to produce and hear sound. Sound and its reproduction will be explored in this chapter.

SOUND

What is sound? **Sound** can be defined as a rapid disturbance in a substance or medium. A medium, such as a solid, liquid, or air is needed to transmit the sound, Fig. 34-1. The sound source must be rapid enough to vibrate this medium or a sound will not be produced.

For example, move your hand through the air. The air is disturbed slightly, but no sound is produced. Now clap your hands. This rapid disturbance causes vibrations in the form of sound waves. Humans are able to hear from 25 to 20,000 of these sound waves per second.

HOW PEOPLE HEAR

The part of the ear you can see is used to direct sound into the ear, Fig. 34-2. The *eardrum*, which is a thin membrane like a drum, vibrates at the same rate as the sound. This vibration passes through three bones (hammer, anvil, and stirrup) to a coiled tube filled with fluid called the *cochlea*. There are over 15,000 nerve fibers in the cochlea. Each hair-like fiber vibrates to a certain sound frequency. This information is sent to your brain by the auditory nerve.

The human ear is sensitive to 30 to 20,000 vibrations per second. This frequency is measured in **hertz (Hz)**. So, the frequency range of the human ear is 30 Hz to 20,000 Hz. Higher frequencies produce a higher pitched sound.

Loudness of sound is measured in **decibels**. Sounds over 150 decibels can damage your ability to hear. In industry, workers wear special earplugs if the decibel

Fig. 34-1. A can telephone transmits sound.

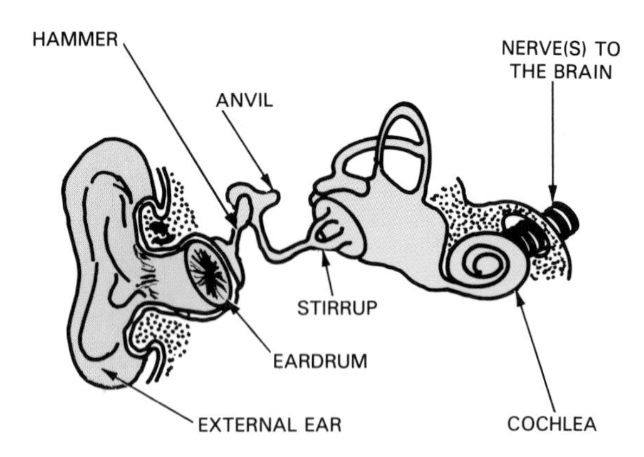

Fig. 34-2. These are some of the parts of the human ear.

level is too high or limit the amount of exposure to loud noises.

Many electronic entertainment devices have the ability to damage hearing if played too loudly. Unfortunately this damage is irreversible. To simulate what you might hear with damaged hearing, place your hands over your ears while listening to the radio. The sound would be muffled and not clear and crisp.

ULTRASONICS

Ultrasound is sound above the frequency range humans can hear. The bat uses ultrasound to "see" with its ears. It sends out ultrasonic vibrations and listens for the echo. The bat can judge the distance to objects by the time it takes for the echo to return. An **echo** is sound waves that bounce back from an object.

Sonar (*so*und *n*avigation *a*nd *r*anging) works in a similar way. A ship can determine water depth by sending out ultrasonic waves. Distance to the bottom can be found by the time it takes for the echo to return.

Ultrasonics are used in businesses for a variety of purposes, Fig. 34-3. For example, ultrasonic cleaners use vibration in a liquid to clean parts. Some dentists use ultrasonic methods to clean teeth. Doctors can use ultrasound to see inside the body of a patient without harm or pain.

THE AMAZING ELECTROMAGNET

The use of the **electromagnet** is very important in communications. This simple, electrical device is used in motors, telephones, radios, and thousands of other electrical devices. For example, a telephone receiver uses an electromagnet to change an electrical signal into the voice you hear, which is made of sound vibrations.

Hans Christian Oersted discovered electromagnetism. He found that a magnetic field is created around a wire carrying current. This knowledge was used in making the electromagnet. A typical electromagnet is a

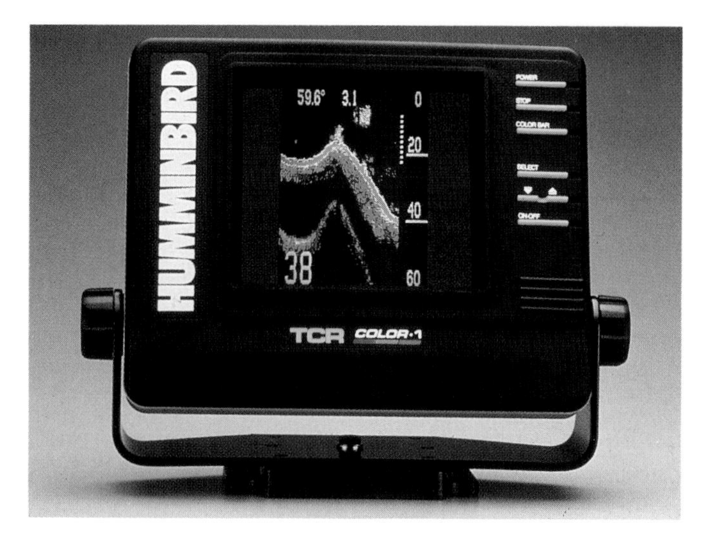

Fig. 34-3. This depth sounder uses ultrasound to find out what is under the water. (Techsonic)

coil of wire, or solenoid, with an iron core. When current flows through the coil, the iron becomes magnetized, Fig. 34-4. As current increases, magnetic intensity increases. This intensity is measured in **oersteds**.

Michael Faraday discovered that a magnet can cause an electric current. This is known as **electromagnetic induction**. In induction, a magnet is moved near a coil of wire. This produces, or induces, the electric current, Fig. 34-5. Either the coil or the magnet may be moved to induce the current.

A compass, which is sensitive to magnetism, can be used to demonstrate electromagnetism. Find true north with the compass. Now turn on a flashlight and hold it near the compass. The current-carrying wires in the light should cause the needle to move.

THE TELEGRAPH

The electromagnetic **telegraph** was invented by Samuel F. Morse in 1835. It was the first rapid method of communicating over long distances.

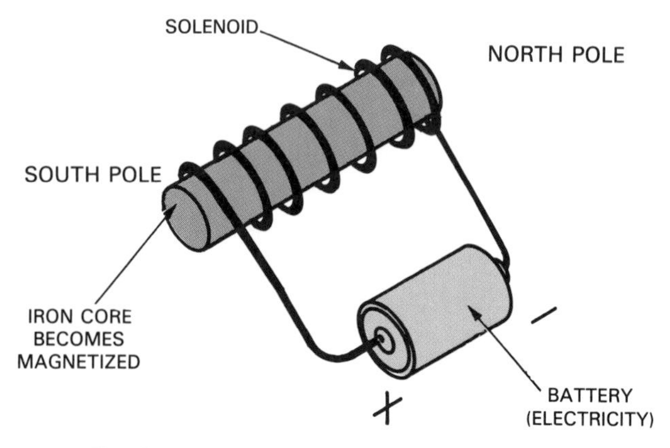

Fig. 34-4. This is a simple electromagnet.

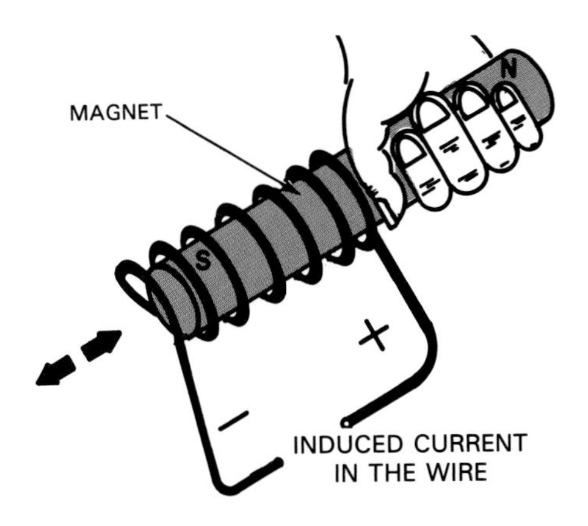

Fig. 34-5. When a magnet is moved near a coil of wire, an electric current is induced.

A telegraph consists of a key and sounder in two locations connected by a wire, Fig. 34-6. When the key, or sending device, is pressed, current is allowed to flow through the wire. This current magnetizes an electromagnet at the receiving station. The metal sounder is attracted to the electromagnet. This makes a clicking sound. Each station can send and receive signals.

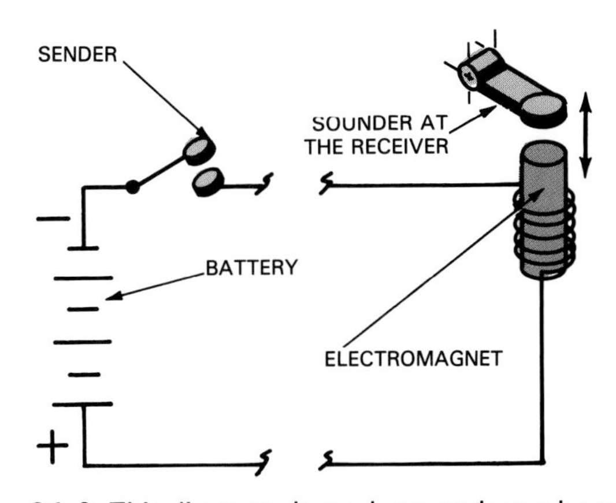

Fig. 34-6. This diagram shows how a telegraph works.

Morse also developed a code for sending messages across the telegraph line. Morse code uses dots and dashes. A dot is a rapid click, and a dash is a click followed by a short pause. Morse assigned a series of dots and dashes for each letter and number. For example, three dots is code for the letter "C."

THE TELEPHONE

Between 1835 and 1876 most long distance communication was done by telegraph. However, in 1876, Alexander Graham Bell invented the telephone, Fig. 34-7. This device allowed voice to be transmitted electrically for the first time.

The **telephone** consists of a transmitter and receiver, Fig. 34-8. The transmitter has a diaphragm, like your

Fig. 34-7. This is the telephone invented by Alexander Graham Bell. (AT&T Bell Laboratories Archives)

eardrum, that vibrates at the same frequency as your voice. As the diaphragm presses in, carbon granules beneath it are pressed together. Current flows through the carbon. As pressure on the carbon increases, more current flows through. So, the sound vibrations are changed (transduced) into modulated electric current. A modulated current varies in flow in relation to the sound vibrations. This is an **electrical signal.**

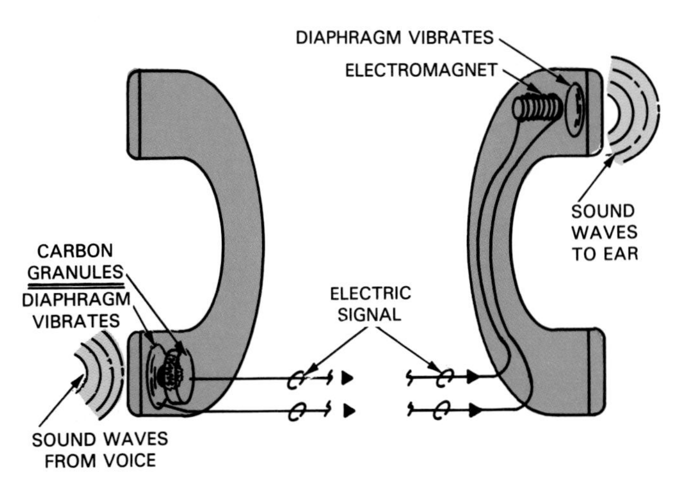

Fig. 34-8. On the left, sound is transduced into an electrical signal. On the right, the electrical signal is transduced back into sound waves.

The receiver of the telephone contains a metal diaphragm and an electromagnet. The changing or modulated current passing through the electromagnet causes the diaphragm to vibrate. These vibrations create sound waves in the same form as the transmitted signal.

New inventions have made the telephone much easier to use. Early models had a sidewinder that signalled the operator. The operator then connected the caller with the other party using a switchboard, Fig. 34-9. Now, switching systems at a central office automatically make a connection between two telephones when a telephone number is punched in or dialed. A computer is used to determine the best path for the call, Fig. 34-10. For example, a call to another state could be routed by microwave or by fiber optic cable, depending on which path has available space for the signal.

There are two methods frequently used to place a call. The older method is known as rotary or pulse dialing. A dial is used that you turn with your finger. When each number is dialed, a series of pulses is sent to the central office. Each number generates a different number of pulses and so can be recognized.

The newer method of placing a call is by audio tone, Fig. 34-11. Instead of using a dial, a button is pressed for each number. When the button is pressed, two tones are generated and sent to the central office. Two tones of specific frequencies are used to eliminate errors. This system is called **dual tone multiple frequency (DTMF)**, or simply **touch-tone**.

Most telephone numbers are seven digits long. The first three digits identify the central office you are call-

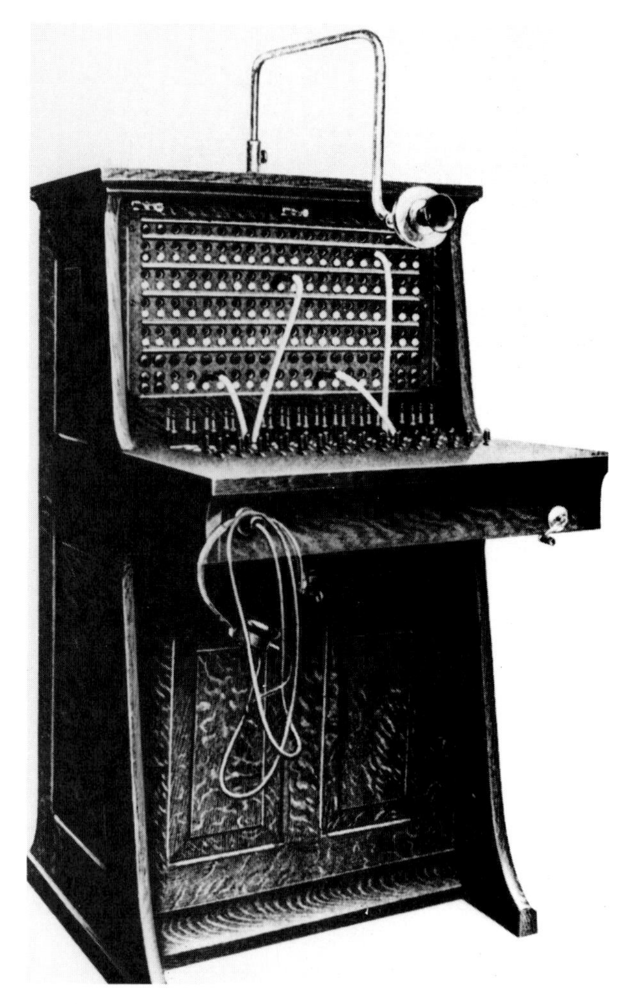

Fig. 34-9. Switchboards contained jacks for each telephone connected to the switchboard. An operator used a patch cord to connect the telephones. (AT&T Bell Laboratories Archives)

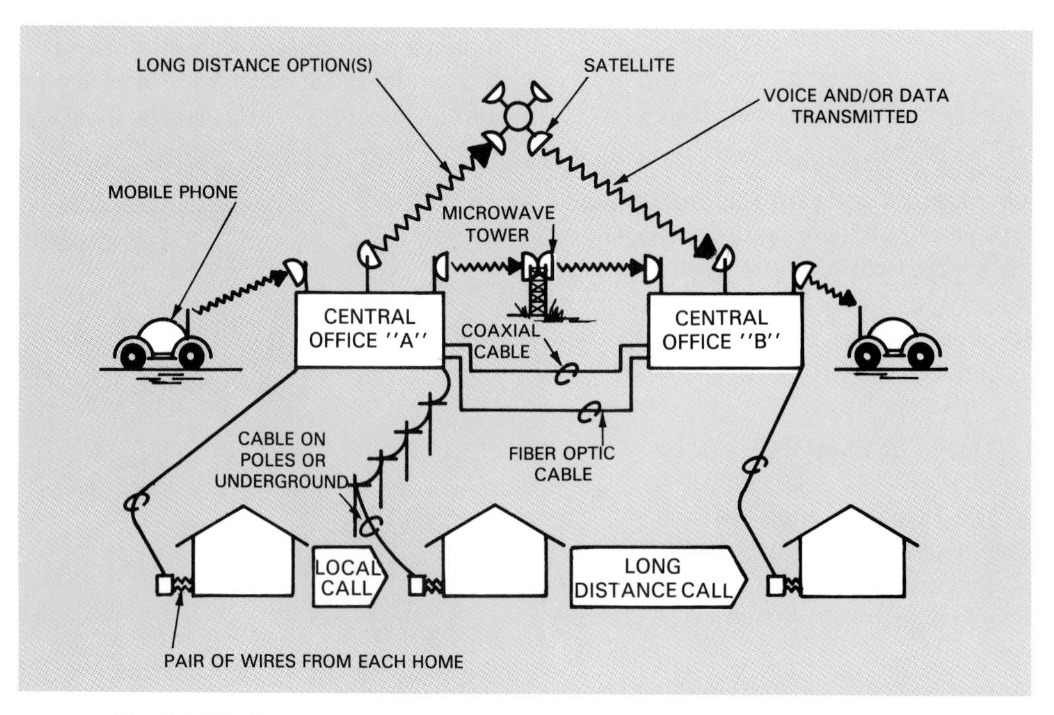

Fig. 34-10. Some of the options for routing telephone calls are shown here.

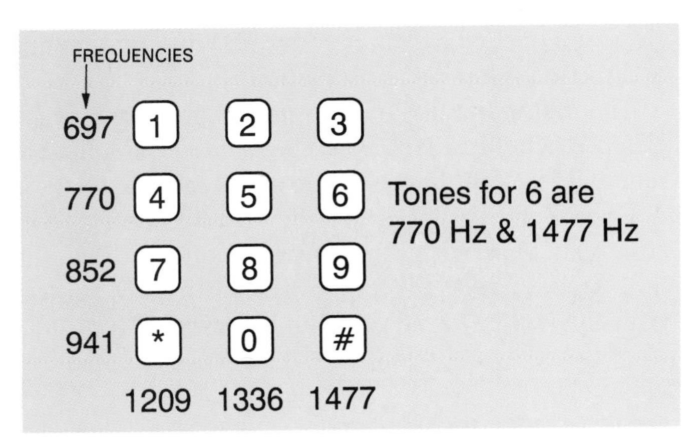

Fig. 34-11. The dual tone multiple frequency system uses two tones for each number.

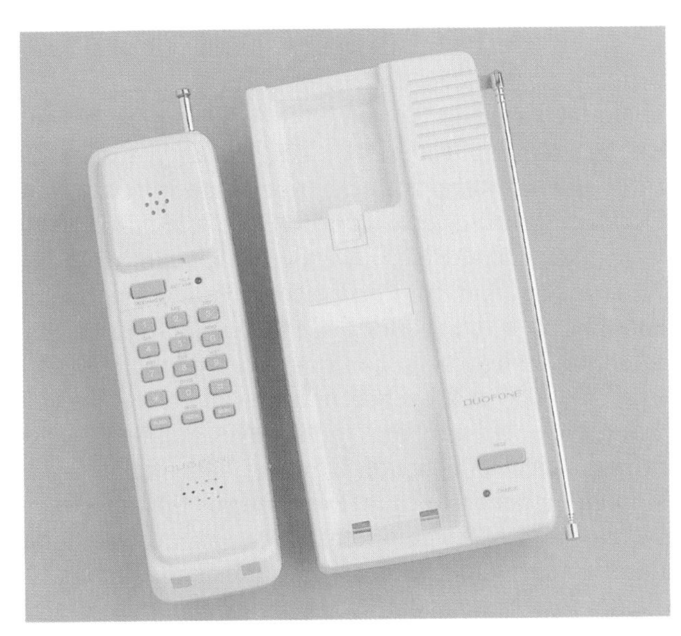

Fig. 34-12. A cordless phone uses radio technology, but is different from a cellular phone. (Radio Shack)

ing. The last four digits identify the specific home or business serviced by the central office.

When calling long distance, an area code is used. This is a three-digit number that allows you to make connections with other parts of the country.

Telephones are used for other communication besides voice communication. Computers can be linked by telephone lines for data communication. Radio and TV networks use the telephone system to send programs to local stations.

CELLULAR TELEPHONES

Cellular telephones use radio technology to provide mobile phone service. They are often used in cars. A cordless phone also uses radio technology, but the cordless phone has a limited range and uses the conventional telephone network, Fig. 34-12. Cellular phones are completely mobile.

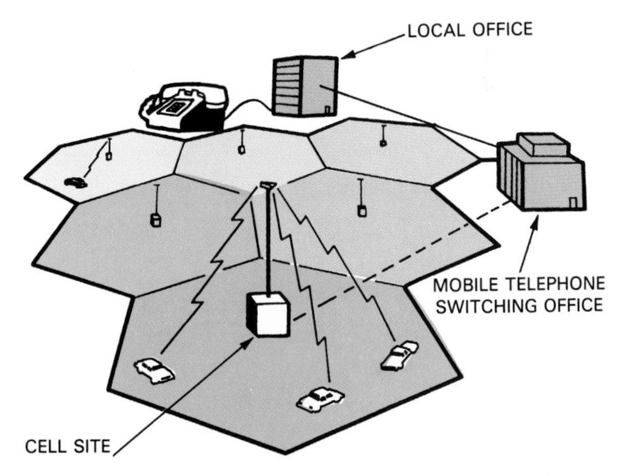

Fig. 34-13. This is how the components of a cellular system fit together as a whole system. (Motorola, Inc.)

Cellular technology involves dividing the country up into cells, with each cell containing a cell site, Fig. 34-13. The **cell site** contains the transmitting and receiving equipment for communicating with cellular phones in that cell. The cell site is also connected to a **mobile telephone switching office (MTSO)**, which, in turn, is connected to the conventional telephone network.

When a call is placed with a cellular phone, the antenna transmits a signal to the cell site. This signal is then sent to the MTSO and on to the central office. If the cellular phone moves to a second cell during a conversation, that cell site is used to send and receive the signal.

Several types of cellular phones are available. The typical car phone, or mobile cellular phone, uses a **transceiver** (transmitter and receiver) unit that is mounted in the trunk and uses the car battery as the power source. The portable cellular phone can be carried with you and contains its own transceiver and batteries, Fig. 34-14.

Fig. 34-14. This is a portable cellular phone. (Motorola, Inc.)

RADIO

Radio is a general term used for communicating with electromagnetic waves. The signal that you pick up through your radio and television is made up of these electromagnetic or radio waves, Fig. 34-15.

In 1860, James Clerk Maxwell used mathematical formulas to figure out how electricity and magnetism could be used to produce these waves. He figured that these waves travel at the same speed as the speed of light, 186,000 miles per second. Because of Maxwell's work, we now know that radio waves and light waves are both electromagnetic. However, light waves are at a much higher frequency. **Frequency** is the number of waves that pass a given point in one second. Also, unlike sound waves, these waves require no medium, such as air, for transmission. This is why these waves can travel through outer space.

In 1886, Heinrich Rudolph Hertz actually transmitted electromagnetic waves. He discovered that when current direction is rapidly changed in a coil of wire, electromagnetic waves are produced. This rapid change in direction is known as **oscillation**.

Fig. 34-15. This radio receives electromagnetic waves through the antenna. (Radio Shack)

The first individual that invented a practical use for electromagnetic waves was Guglielmo Marconi. He sent and received the first radio signal in 1895.

One of the first uses for this new technology was communication with ships in Morse code. In 1906, the S.S. Arapahoe was the first ship to use the S.O.S. (save our souls) signal to call for help. Voice was successfully transmitted by radio in the early 1900's.

HOW A RADIO WORKS

The first step in transmitting a message with the radio is to change or transduce the sound waves to electrical signals, Fig. 34-16. A microphone, similar to the telephone transmitter, can be used for this purpose. The signal then goes to a transmitter that converts the signal to RF (radio frequency) by oscillating the signal. This oscillated signal is sent to an antenna that sends out the radio or electromagnetic waves.

To receive the signal, an antenna is again used. When the receiver is tuned to the same frequency as the signal transmitted, it is received. For example, you turn the dial on a car radio to tune in various frequencies. The weak signal received is amplified and sent to a **transducer** that converts the signal back into sound waves. A loudspeaker is often used for this purpose.

The radio with which you are probably most familiar is a receiver. However, many types of radios are both transmitters and receivers. This includes radios used for marine, aviation, and police purposes, Fig. 34-17. The CB (Citizen's Band) radio for vehicles is also this type.

Commercial radio includes both AM and FM stations. These letters stand for amplitude modulation and frequency modulation. (Refer back to Chapter 30.) Modulation is the way in which the message is placed

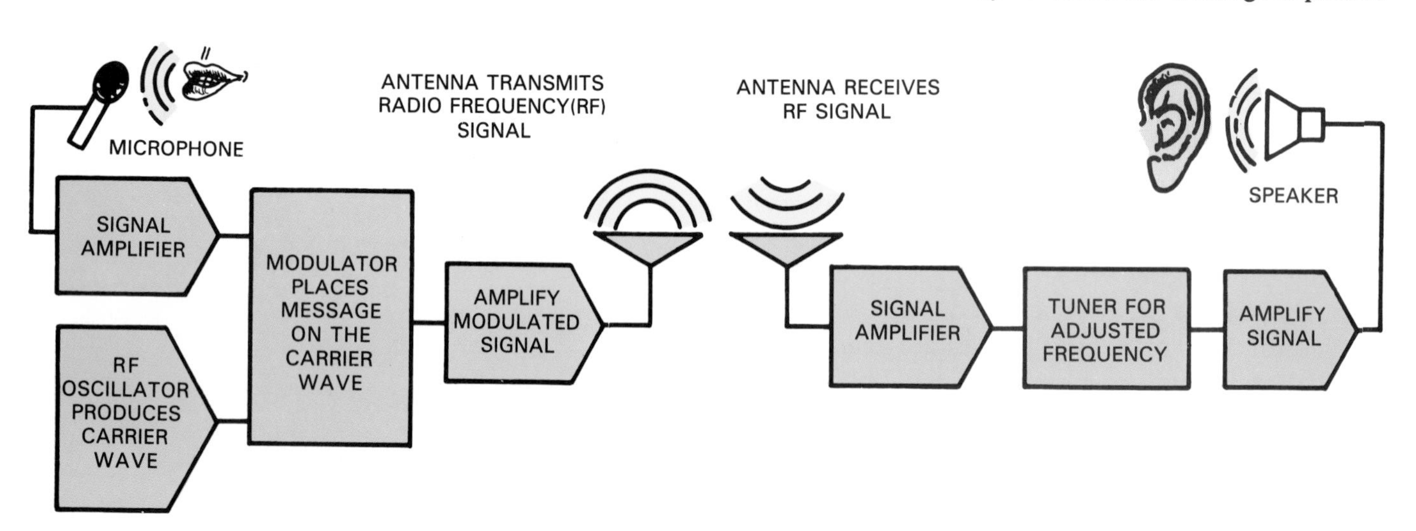

Fig. 34-16. This diagram shows the steps in transmitting and receiving a message via radio.

Fig. 34-17. This radio is for marine use. (Techsonic Industries, Inc.)

on the broadcast waves. For AM, the height of the wave is made to vary in order to carry the signal. For FM, the frequency of the waves is varied while the amplitude (height) of the wave remains the same. The advantage of AM is a longer signal range. The advantage of FM is less noise. This is because most noise is in the form of wave amplitude changes.

SOUND RECORDING

There are three major ways in which sound is recorded or stored. These are:

- Mechanical sound recording.
- Optical sound recording.
- Magnetic sound recording.

MECHANICAL SOUND RECORDING

A phonograph record is an example of storing sound mechanically, Fig. 34-18. **Mechanical sound recording** involves using a stylus to cut grooves in a record. This stylus vibrates exactly as the sound waves vibrate. The record groove contains these vibration patterns.

The phonograph is used to retrieve the sound on the record. The needle on the phonograph vibrates as it rides in the record groove. In early phonographs, the needle was connected to a diaphragm. The sound waves created by the diaphragm were magnified by a large horn. Cupping your hands around your mouth when calling someone has the same effect. Today, the needle vibrations are transduced into an electrical signal. This signal is amplified and then reproduced with a loudspeaker.

The earliest phonograph was invented by Thomas A. Edison in 1877. The disk was made of wax. Ten years later, Emile Berliner developed the gramophone with a shellac disk. This was much like today's phonograph record.

Fig. 34-18. A phonograph record contains mechanically stored information.

OPTICAL RECORDING

In **optical sound recording**, sound waves are recorded as variations in light intensity. This is the method used for sound on movie film. The light is recorded on the side of the film. These light variations are reproduced as sound waves at the movie theater.

Another optical storage method is the **audio disk**, Fig. 34-19. These disks are played by using a laser beam instead of a needle. Digital information recorded on the surface of the disk as bits is read by the laser. The compact disk (CD) is a popular optical disk used for music.

MAGNETIC SOUND RECORDING

Magnetic sound recording is one of the most popular methods of recording sound. Storage devices include audio tapes, video tapes, computer disks, and video disks.

Fig. 34-19. Compact disks store information optically.

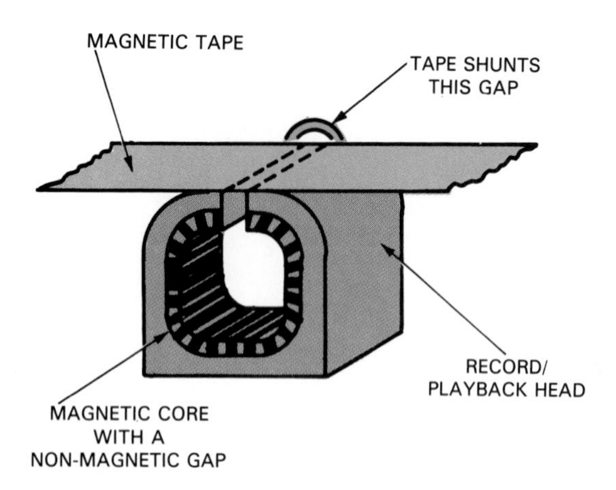

Fig. 34-20. This is an electromagnetic recording and playback head.

Recording sound magnetically is very similar to the other recording methods. Sound waves are first transduced into an electrical signal. A microphone is used for voice recording. This signal passes through an electromagnetic recording head, Fig. 34-20. At this point, the signal is converted into variations in magnetism. This magnetic pattern is transferred to the tape that has a metallic coating on it. This tape is either stored in open reels (reel to reel) or in cassettes.

To play the tape, a playback head is used. The magnetic signals on the tape are converted back to electrical signals through electromagnetic induction. This signal is then amplified and sent on to a speaker.

To protect recorded tapes, remove the plastic tabs with a screwdriver. Now the tape cannot be recorded again. If you change your mind and want to record on the tape again, simply place tape over the tab opening.

DIGITAL AUDIO

When sound is encoded into an electrical signal, it can be represented by a continuously varying wave made up of varying voltage levels. In order to convert this wave to digital, **binary words** are used to represent different voltage levels. A binary word is a group of bits that represent some value such as a voltage level.

Since it is almost impossible to convert every voltage level in an analog wave to digital format, the wave is sampled. Greater accuracy is achieved by increasing the number of samples per wave, Fig. 34-21. For example, using a two-bit word size, four different voltage levels can be designated. A larger word size allows more sampling locations. Sixteen-bit word size is most often used today.

An **A/D converter** is used to change the analog signal to a digital signal. Sound recording devices use this technology. The **D/A converter** changes the digital signal back to analog. A compact disk player utilizes a D/A converter to change the digital signal back to an analog signal.

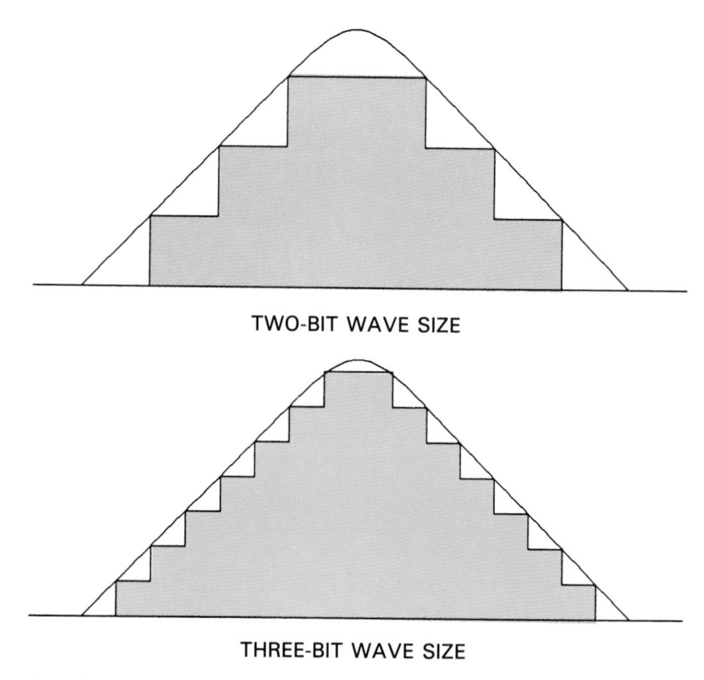

Fig. 34-21. Greater accuracy is achieved by increasing the number of samples per wave. A two-bit wave size is shown on the top and a three-bit wave size is on the bottom.

STEREO SOUND

All early recordings were **monophonic**. In this method, one signal is sent to the speakers.

Stereophonic recording is an attempt to provide more realistic sound. In this method, many signals are recorded by placing microphones in different locations. These "tracks" are then mixed in a studio into two "tracks" that are played back over separate speakers in a stereo system. This gives the impression of actually being in the audience. Many FM stations send out stereo signals.

MICROPHONES

Thomas A. Edison is given credit for inventing the first microphone. His carbon microphone was used by Alexander Graham Bell in the telephone.

A **microphone** changes or transduces sound waves into an electrical signal, Fig. 34-22. In most microphones, a diaphragm is used that vibrates at the same frequency as the incoming sound waves.

The **dynamic microphone** uses electromagnetism to produce the audio signal, Fig. 34-23. The diaphragm is attached to a coil of wire. This coil is between the poles of a magnet. As the coil vibrates, an electric signal is induced. The velocity microphone works in the same way, but the coil is replaced by a flat, metal ribbon.

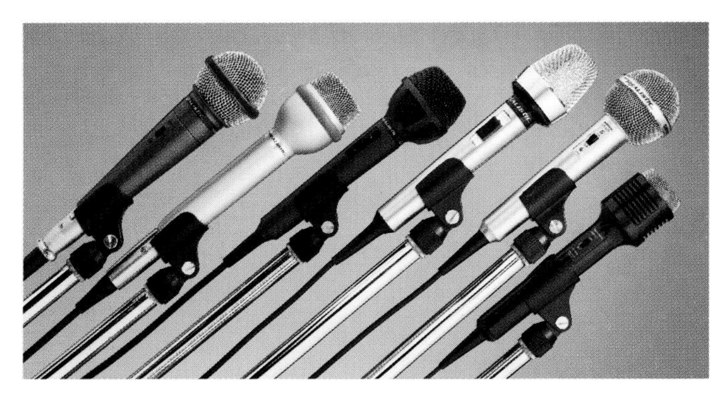

Fig. 34-22. Various kinds of microphones are available. (Radio Shack)

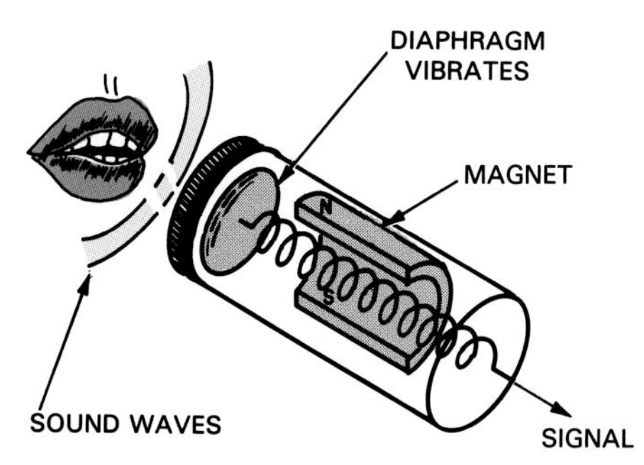

Fig. 34-23. A dynamic microphone is illustrated here.

In the **crystal microphone,** a crystal vibrates. This vibration causes the crystal to produce an electrical signal. This property of some crystals, such as quartz, is known as the **piezoelectric effect**.

The **condenser microphone** requires its own power. Batteries supply power to a metal plate attached to the diaphragm. As the charged metal plate vibrates, the charge discharges to another metal plate. This discharge pattern is the electrical signal. Condenser microphones require more careful handling than other types of microphones.

Microphones have different pick-up patterns, Fig. 34-24. A **multidirectional microphone** picks up sound from all around the microphone. Most built-in microphones are this type.

The **cardioid microphone** picks up sound in a heart-shaped pattern in front of the microphone.

A **directional microphone** picks up sound directly in front of the microphone.

The easiest way to record is to use the built-in microphone. However, for more sophisticated recording you will want an external microphone. A dynamic microphone with a cardioid pattern is a good general purpose microphone. The dynamic microphone is very sensitive to sound and will withstand some rough handling. The cardioid pick-up pattern allows you to pick up sound while leaving out distracting background noises.

To test a microphone, place it in the desired location and talk in a normal voice. It usually helps to talk slightly above the microphone. Never tap or blow on the microphone to test it. A wind screen can be used to eliminate wind noise and other distracting background noise.

SYNCHRONIZING SOUND WITH SLIDES

Audio tape recordings are often used to provide sound with slides and filmstrips. The recording is usually done on audio cassette tapes, Fig. 34-25.

To produce the sound track, you should first develop the script. Mark the script wherever it is necessary to change slides. When finished, read the script into a tape recorder. Use an audible cue, or signal, for changing slides. A small bell ring, or pencil tap will work well.

Another technique uses an inaudible cue to automatically change slides. This requires an automatic cuing device that places the cue on the tape when recording. The cue is placed on a separate track on the tape so only one side of the tape is used. This technique should only be used if special playback equipment is available.

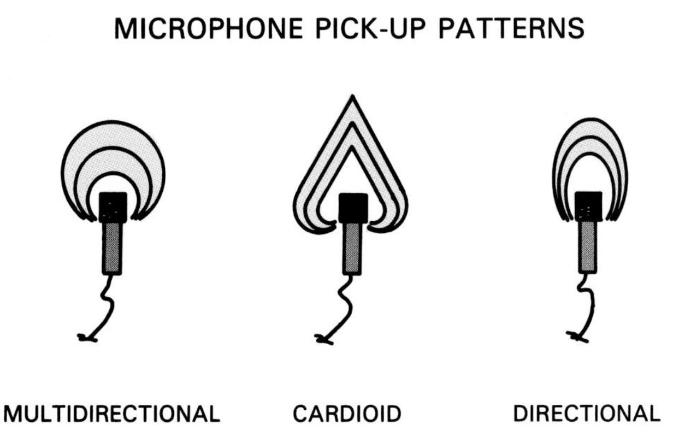

Fig. 34-24. There are various types of microphone pick-up patterns. Shown here are multidirectional (left), cardioid (center), and directional (right).

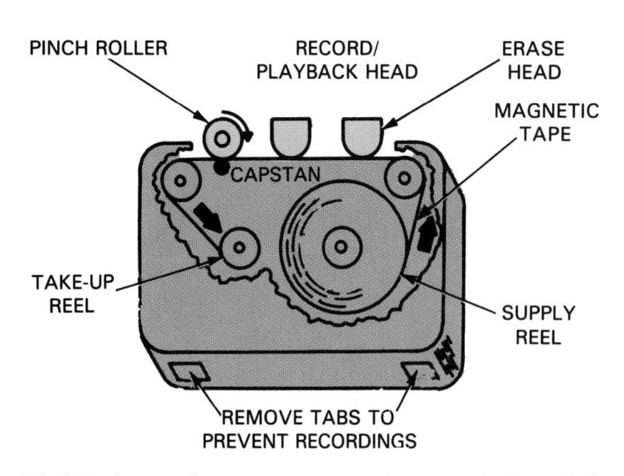

Fig. 34-25. An audio cassette tape is recorded and played back as shown here. When the tabs are removed, the tape cannot be recorded.

SUMMARY

This chapter dealt with sound and how humans hear it. How sound can be transduced (changed) into an electrical signal was also explored. This chapter described how signals can be used to extend our range of communication as well as to store messages. This information should make many of the devices around you, such as the telephone and radio, more understandable. In addition, you can use this knowledge to improve your own audio recordings.

WORDS TO KNOW

All of the following words have been used in this chapter. Do you know their meanings?

A/D converter
audio disk
binary words
cardioid microphone
cell site
condenser microphone
crystal microphone
D/A converter
decibels
directional microphone
dual tone multiple frequency (DTMF)
dynamic microphone
echo
electrical signal
electromagnet
electromagnetic induction
frequency
hertz (Hz)
magnetic sound recording
mechanical sound recording
microphone
mobile telephone switching office (MTSO)
monophonic
multidirectional microphone
oersteds
optical sound recording
oscillation
piezoelectric effect
sonar (*so*und *n*avagation *a*nd *r*anging)
sound
stereophonic
telegraph
telephone
touch-tone
transceiver
transducer
ultrasound

REVIEWING YOUR KNOWLEDGE

Please do not write in this text. Write your answers on a separate sheet.

1. Loudness of sound is measured in:
   a. Hertz.
   b. Decibels.
   c. Oersteds.
   d. None of the above.
2. True or false? The newest method used to place a telephone call is known as rotary or pulse dialing.
3. ________ telephones use radio technology to provide mobile phone service.
4. The number of electromagnetic waves that pass a given point in one second is known as:
   a. Frequency.
   b. Oscillation.
   c. Sound waves.
   d. None of the above.
5. True or false? The advantage of AM is longer signal range, while the advantage of FM is less noise.
6. Name the three major ways in which sound is recorded or stored.
7. In digital audio, which of the following word sizes is used most often?
   a. One-bit.
   b. Two-bit.
   c. Ten-bit.
   d. Sixteen-bit.
8. What type of recording is an attempt to provide more realistic sound?
9. A(n) ________ changes or transduces sound waves into an electrical signal.
10. Which of the following types of microphones picks up sound in a heart-shaped pattern in front of the microphone?
    a. Directional microphone.
    b. Cardioid microphone.
    c. Multidirectional microphone.
    d. None of the above.

APPLYING YOUR KNOWLEDGE

1. Invite an audiologist to class to discuss how humans hear. Prepare a list of questions in advance.
2. Visit a business whose product is sound, such as a telephone company, a recording studio, or a radio station. Prepare a report about your experience.
3. Research the history of sound technology. Prepare a written report.

35
CHAPTER

THE MOVING IMAGE

After studying this chapter, you will be able to:

- *Describe persistence of vision and how it relates to moving images.*
- *Compare film and videotape.*
- *List the different methods that can be used for creating moving images.*
- *Define broadcast and nonbroadcast television.*
- *Identify the parts of a video system.*

Most likely, you see moving images created by technology every day, but you may not be aware of how these images are created. This chapter will provide you with some background about this exciting and rapidly changing technology. Many industries are involved with moving images and sound. For example, most of the motion pictures seen at theaters or on television are produced in large studios. Television stations use video technology to produce programs such as the local news. You might even enjoy creating your own videos at home, Fig. 35-1.

Fig. 35-1. Using video equipment is one way to capture a moving image.

PERSISTENCE OF VISION

Technologies such as motion picture film and television are possible because of a principle known as **persistence of vision**. This was first described by Peter Roget in 1824. The principle stated that a person retains an image for a split second after the image leaves the field of view. Because of this, still images can be shown at a high rate of speed and appear as motion rather than as individual images.

You can use flip pictures to illustrate this principle, Fig. 35-2. Draw a picture of a stick figure on the edge of an index card. Now draw the figure on another card

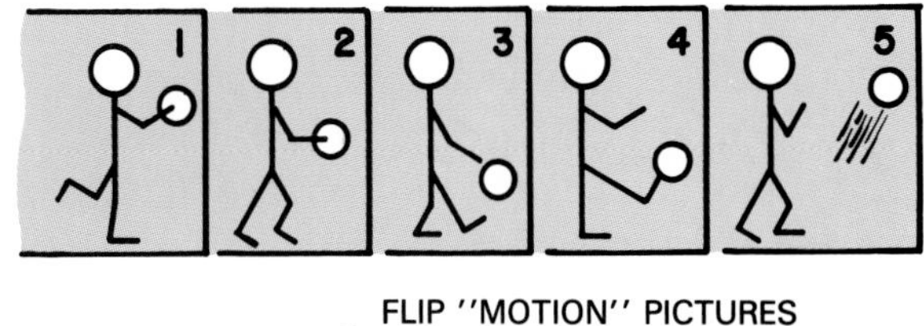

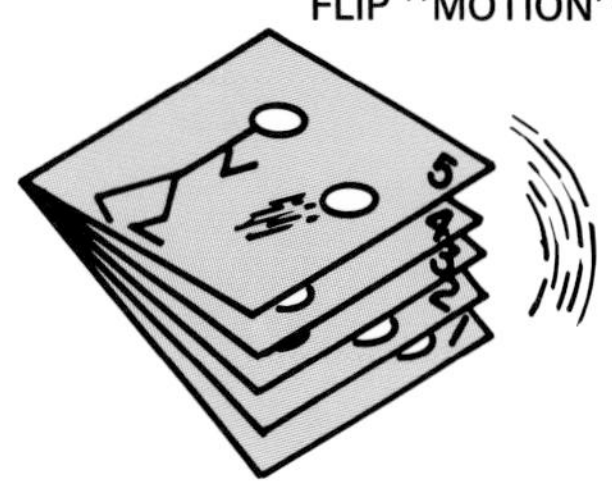

Fig. 35-2. Flip pictures are a simple way to demonstrate persistence of vision.

with the arms slightly moved . Continue this for 20 to 25 cards. When you flip back through the cards the arms appear to move because of persistence of vision.

RECORDING MOVING IMAGES AND SOUND

The use of film and videotape are the two major ways to record and store moving images and sound, Fig. 35-3. Although they both accomplish a similar goal, there are differences between them.

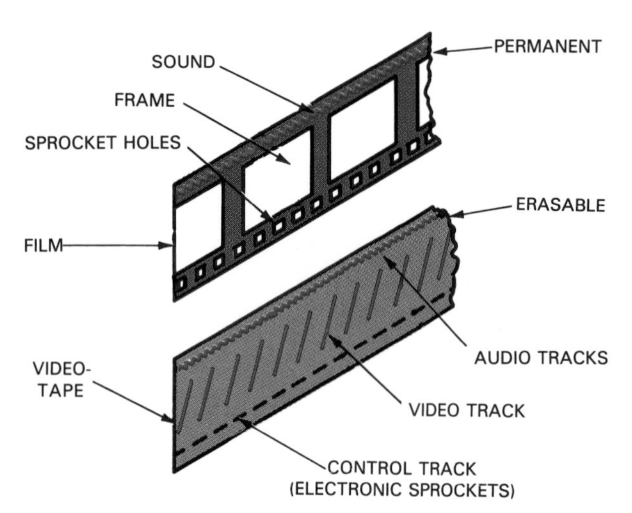

Fig. 35-3. Note the differences between film and videotape.

MOTION PICTURE FILM

A motion picture camera exposes still photos on motion picture film at a rate of 18 to 24 frames per second. After the film is developed, it is viewed by projecting light through it onto a large screen. This screen could be in a movie theater, classroom, or salesroom at a business. Motion pictures are often used for entertainment, education, and advertising.

Motion picture film has very high picture quality or resolution. In fact, a film image has about three times more picture detail than a videotape image, such as television. A disadvantage of motion picture film is that it is expensive and nonreusable. Once exposed, it is used or destroyed.

VIDEOTAPE

Videotape stores pictures and sound magnetically. This is similar to a computer disk or audio tape. Like motion picture film, videotape records and stores a series of still pictures that appear as a moving picture when played back on a television screen or monitor. Videotape records 30 pictures per second.

An advantage of videotape over film is that it can be erased and used over again many times. In fact, this was one of the reasons for the development of videotape. In the early days of television, a method was needed to record a program so it could be shown at different times. Videotape did this quickly and economically.

CINEMATOGRAPHY

Cinematography involves using photographic techniques to make motion pictures. Cinematography is the process used by film studios to produce the movies you see in theaters.

The first motion picture camera was invented by Thomas A. Edison and his assistant, William K. L. Dickson in the late 1800s. Dickson's decision to use another new invention, celluloid film, as the recording medium, made the new technology practical. The first motion picture camera that used film was known as the **kinetograph**.

The motion picture camera was perfected over the years, greatly improving picture quality. A major breakthrough was the addition of sound to film in the late 1920s. These first sound films were known as *"talkies."* Motion picture technology has continued to improve over the years.

ANIMATION

Animation is the addition of motion to drawn images or three-dimensional figures such as puppets. Perhaps the most familiar form of animation is *cartoons.* In cartoons, drawn figures appear to come to life by moving and talking.

Animation works in a similar way to the flip pictures described earlier. Still images are created and photographed onto a frame of film. Each successive image is slightly different than the last. When the processed film is played back at 24 frames per second, the illusion of movement is created.

Originally, animated pictures were created by drawing each individual scene. However, in about 1910, **cel (celluloid) animation** was developed. A background was drawn separately and transparent celluloid was placed on top of the background. Characters drawn on the celluloid could then move across the background. This technique made the creation of animated features much easier, since the background did not have to be constantly redrawn. Walt Disney used cel animation to produce cartoon characters such as Mickey Mouse and animated films such as *"Snow White and the Seven Dwarfs."* Some high-quality animation is still done in this way, Fig. 35-4.

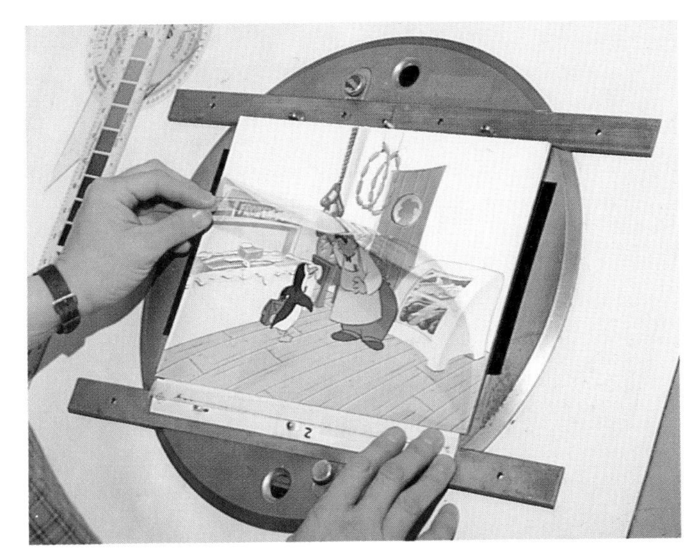

Fig. 35-4. This is an example of one cel that will be used for cel animation. (Hellman Associates, Inc.)

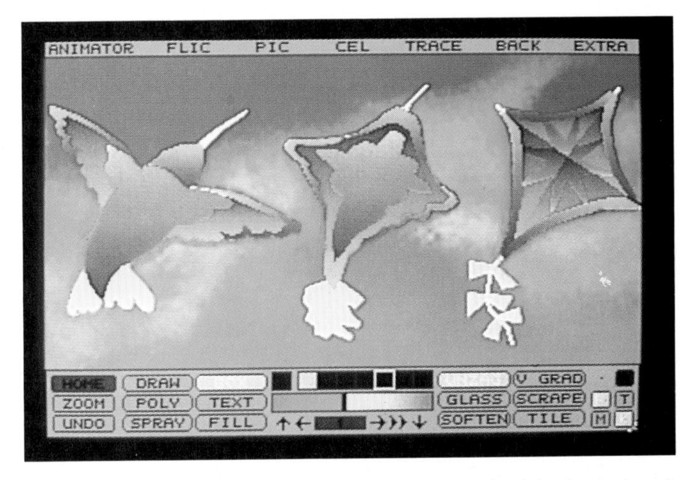

Fig. 35-5. An animation is being generated with the help of a computer and special software. An example of tweening is shown, where one object (the bird) is transformed into another object (the kite) with the help of the computer. (Autodesk, Inc.)

The computer is now being used to assist with animation, Fig. 35-5. Video games include animations created with the help of computers. Many new television cartoons are created with the help of the computer.

Although techniques may vary, computer animation is often done in a similar way to cel animation. A solid model of objects or characters is first drawn on the computer. A background is then created on which the images can be superimposed. The computer can then assist in moving the model along the background. The completed animation is usually stored on videotape.

TELEVISION

Television is the electronic transmission of moving images and sound. This technology first became practical because of earlier inventions, including radio and motion pictures. Two inventors, working independently, demonstrated broadcast television in about 1925. Charles Francis Jenkins demonstrated this technology in the U.S. while John Logie Baird demonstrated a similar system in Britain. Baird went on to also develop the first color television system.

After World War II, black and white television developed rapidly in the United States. In fact, by 1949, there were nearly one million sets being used. In the late 1950s, color television became available. However, at that time color televisions were too costly for most families to afford, and few television programs were produced in color. As color televisions have become less expensive and more television programs were broadcast in color, more people began to acquire color televisions. Today, most homes have a color television.

TYPES OF TELEVISION

The two major forms of television are broadcast and nonbroadcast. Both use videotape as the standard recording medium.

Broadcast television

Broadcast television is so named because programs are sent by signals from one place to another, such as from a television station to your home, Fig. 35-6. The most common form of broadcast television is network television, such as NBC, ABC, CBS, or PBS. In this system, local television stations across the country are

Fig. 35-6. This camera operator is preparing to record the local news, which will be broadcast to the surrounding community. (WBAL-TV, Channel 11, Baltimore, MD)

affiliated with one of the national network companies. Each local station in the network is required to show at least eight and one-half hours of national programming each day in addition to local programs.

The national networks pay expenses through selling advertising. An exception is PBS (Public Broadcasting System) which often solicits donations from the public.

Another type of broadcast television is cable television. Unlike broadcast television, cable television is funded through subscription fees.

Cable television began as a way to send programming into remote areas where it was difficult to receive a good signal. For example, in a mountainous region, a large antenna would be installed on top of a mountain and coaxial cables would bring the signal to homes in a valley below. However, entrepreneurs soon realized this same method could be used to bring in specialized programs all over the country. For a subscription fee, you can now receive a variety of cable channels on your television. Cable channels are often very specialized, providing movies, sports, or music throughout the day.

Nonbroadcast television

Nonbroadcast television is a category of television that is not sent by signal to distant sites. This is often called "video." Video productions are used in industry and education for purposes such as training, advertising, or surveillance. Videotape is played back on one or more monitors. Playing a videotape on various monitors throughout a building, is known as **closed-circuit television**, Fig. 35-7.

VIDEO EQUIPMENT

A video production requires the use of a video camera, a video recorder, a television monitor, a microphone, lighting, and assorted cables. The camera and microphone send the visual and audio signal to the video cassette recorder (VCR) where it is recorded and sent on to the monitor for viewing. The next few sections will explore how each part of this system works.

VIDEO CAMERA

The video camera can be a larger studio model permanently mounted on a tripod and dolly for rolling. Another option is the smaller, portable unit that can be hand-held or mounted on a tripod as needed, Fig. 35-8.

The main purpose of the **video camera** is to change the scene viewed through the lens into an electronic signal to be transmitted to the VCR. This conversion takes place in the camera tube or in semi-conductor chips in newer cameras.

Fig. 35-7. This is the master panel for a closed circuit television system. (Siemens)

The camera tube has a light-sensitive signal plate at one end. See Fig. 35-9. When a picture is focused on this plate, it releases electrons depending on light intensity. At the other end of the camera tube, an electron beam scans the signal plate at the rate of 525 lines each

Fig. 35-8. This portable camera, which also contains a videocassette recorder, is known as a camcorder. (Radio Shack)

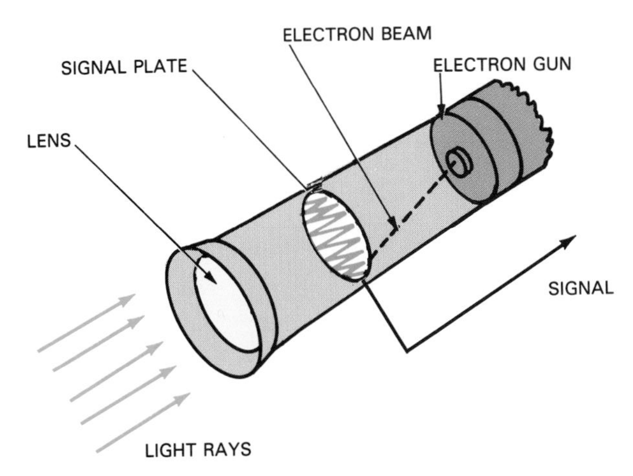

Fig. 35-9. A camera tube is shown here.

Fig. 35-10. The knob on the lower right allows for automatic or manual color adjustment on the camera.

1/30 second. As this beam replaces electrons, a signal is produced that is sent to the VCR. This signal is in the form of an electrical current.

Many video cameras now use a **charge-coupled device (CCD)** in place of a camera tube. The CCD is a light-sensitive microchip with similar characteristics to the camera tube. It changes light in the picture into an electrical signal. The CCD has several advantages, including the fact that it is not easily damaged and is very small in size.

Most video camera lenses have three adjustments. These include the:

- Focus ring.
- Zoom feature.
- Aperture setting.

A **focus ring** is used to create a sharp image. The **zoom feature** allows you to move closer or further from an object while standing still. When using the zoom, you normally focus in the close-up position. As long as the camera-to-subject distance does not change, the focus will be correct while using the zoom. The third adjustment is the **aperture setting**. The aperture or iris allows you to adjust the size of the lens opening for various lighting conditions. As with still cameras, each higher F number cuts the light coming through the lens in half. Many of the newer cameras have auto focus as well as automatic aperture. However, it is often helpful to know how to use the camera manually in certain situations.

The camera has a viewfinder that allows you to see what the lens is seeing. Viewfinders range in sophistication from through-the-lens systems to small monitors mounted on the camera. Major parts of the scene should be near the center of the viewfinder since some televisions will not show as much of the scene as the viewfinder.

Color video cameras record red, green, and blue in the picture. These primary colors of light are then set to the monitor where they recreate various colors in the scene. Some color cameras must be adjusted to different light conditions so the colors appear natural, Fig. 35-10. You will need to consult your camera manual for these adjustments.

From your study of photography, you should be familiar with the following camera shots:

- Long shot (LS).
- Medium shot (MS).
- Close-up (CU).
- Extra or extreme close-up (XCU).

Most shots are medium to close-up so viewers can see the action on the small television screen. Often, a production begins with a long shot to establish location, as shown in Fig. 35-11. Then it goes to a medium shot for more detail. During the production, however, the zoom should be used infrequently. Also, with a zoom lens at telephoto, there is very little depth of field. In other words, it is more difficult to keep the subject in focus. This is why where there is a great deal of movement, be cautious in using the telephoto setting.

Besides camera shots, camera movements are important, Fig. 35-12. Typical movements are:

- The **dolly**- Moving the camera toward or away from the subject.
- The **truck**- Moving the camera left or right.
- The **tilt**- Slanting the camera up or down.
- The **pan**- Pivoting the camera left or right.

These moves can be accomplished with hand-held cameras or those mounted on a tripod and dolly. As with the zoom, these movements should be used sparingly. For professional results, always practice first.

VIDEO RECORDER

The purpose of the **video recorder** is to receive the sound and picture signals from the camera and store them on videotape, Fig. 35-13. Another purpose is to

Fig. 35-11. This is an example of a long shot. (United Telecom)

play back the videotape by sending the signal to a television monitor for viewing.

Many of the controls on a video recorder are similar to those on an audio tape recorder. These include *play*, *stop*, *fast-forward*, *rewind*, *record*, and *eject*. On some units, play and record must be depressed together for recording. The *pause* control allows you to stop the tape for a few minutes without stopping. If the tape is left on pause too long, however, it can be damaged. *Audio dub* is a feature that allows you to tape new sound in place of the existing sound on the tape.

The tape counter is usually returned to zero when beginning a recording. It counts the revolutions of the take-up reel for the tape. Since this will be a different length of tape depending on the circumference of tape on the reel, you cannot translate revolutions into time. You can buy conversion charts for this or make your own by writing down time and counter numbers at one minute intervals. Always check the counter after recording so that you know where to begin for another recording on the same tape.

Video recorders have a setting for changing the tape speed. The fastest speed is best for high quality because fewer signals are "packed" in each area of the tape.

Most video recorders for home use have **RF converters** that change the video and audio signals into a single RF signal. This allows hookup to a television receiver that usually accepts an RF type signal only.

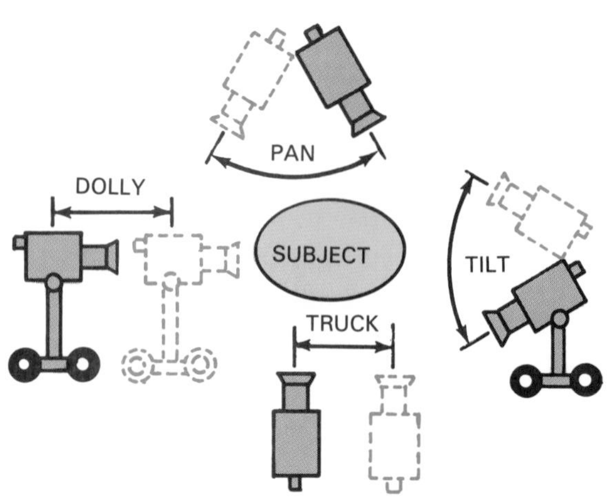

Fig. 35-12. These are typical camera movements.

Fig. 35-13. This is a video cassette recorder and remote control. (Radio Shack)

VIDEOTAPE

Most videotape is packaged in cassette form, much like audio tape. Since there are different cassette types for different machines, it is important to know your equipment before purchasing tapes. In addition, tapes come in various widths, such as 1/2 in. and 3/4 in.

Videotape consists of four layers, Fig. 35-14. Similar to film, a polyester layer is coated on both sides to make the tape. The top side has a metal oxide layer that is magnetic. A topcoat is used to protect the metal oxide layer. An anti-static layer is on the other side of the polyester.

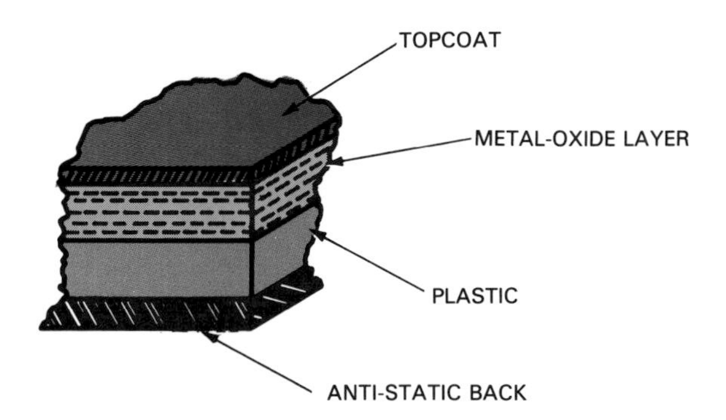

Fig. 35-14. The layers that make up videotape are shown here.

The video recorder is used to record information on the videotape. A *"head"* is used to record the information. A **video head** records the picture in the center of the tape in slanted lines. An **audio head** records sound on an edge of the tape. A **synchronizing head** records a signal that instructs the monitor how to scan. When the tape is played back, all of these signals are sent to the monitor.

Video tapes shouid be stored where it is dry and cool. Also, keep them away from other equipment that use magnetism. Examples of such equipment include motors, monitors, and speakers.

VIDEO MONITOR

A video production can be shown on a television monitor. It can also be shown on a television receiver, or monitor/receiver combination.

A **video monitor** is designed to accept separate video and audio signals. See Fig. 35-15. A **television receiver** is designed to accept signals coming through the air called RF (radio frequency) signals. Many video recorders have an RF converter that allows hookup to television receivers. The RF converter changes the signal into a Channel 3 or Channel 4 signal before sending it to your television.

Fig. 35-15. A video monitor has separate video and audio inputs.

The picture to be in the monitor works like a video camera in reverse. An electron beam scans the front of the picture or cathode ray tube, Fig. 35-16. Phosphors on the front of the screen light up when struck by the electrons. This happens at the same rate of speed as the video camera, 525 lines each 1/30 second.

The color monitor has clusters of blue, red, and green phosphors. One electron beam is used for each color. In combination, these three colors can produce all colors in the visible spectrum. The phosphors are so small and close together that your eyes mix them. You cannot *resolve* the individual dots.

VIDEODISCS

The **videodisc** is another storage medium for video information. In its most popular form, it is an optical

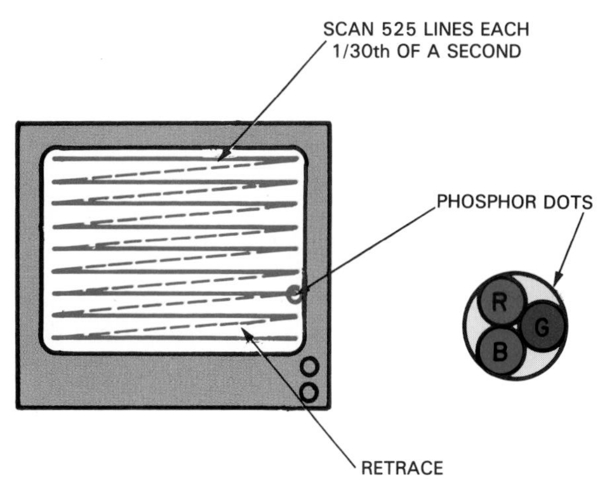

Fig. 35-16. A picture tube is shown here.

Fig. 35-17. This optical disc player can be used for videodiscs as well as compact discs.

storage device. Images are stored digitally on a disc. They are played back with a laser beam, much like a compact disc. Most videodisc units are used only as playback devices for permanently recorded videodiscs. See Fig. 35-17.

Interactive video refers to using the computer with a video player to access information. Videodiscs are often used for this because they provide random access. This means that the computer can quickly access the information on the disk so it can be played back.

Videodisc players are often used as part of an integrated media or multimedia system, Fig. 35-18. A computer and software are used to create and play back different media such as sound and video. The videodisc player is typically used to play back video images because they can be accessed very quickly.

Fig. 35-18. A multimedia system is shown here. A videodisc player is in the background on the right. (IBM Corp.)

SUMMARY

In this chapter you have explored how moving images are produced. The concept of persistence of vision and how it relates to the moving image was described. A comparison between film and videotape was made. Also explored was the different methods that can be used for creating a moving image and the parts of the video system. This background will help you as you learn to produce your own video production in the next chapter.

WORDS TO KNOW

All of the following words have been used in this chapter. Do you know their meanings?

animation
aperture setting
audio head
broadcast television
cable television
cel (celluloid) animation
charge-coupled device (CCD)
cinematography
closed-circuit television
dolly
focus ring
interactive video
kinetograph
nonbroadcast television
pan
persistence of vision
RF converters
synchronizing head
television
television receiver
tilt
truck
video camera
video head
video monitor
video recorder
videodisc
videotape
zoom feature

REVIEWING YOUR KNOWLEDGE

Please do not write in this text. Write your answers on a separate sheet.

1. True or false? Because of the concept of persistence of vision, still images can be shown at a high rate of speed and appear as motion rather than as individual images.

2. Name two ways to record moving images and sound.
3. Which of the following is an advantage video tape has over motion picture film?
   a. It has very high picture quality or resolution.
   b. The image has about three times more picture detail.
   c. It can be erased and used over again many times.
   d. All of the above.
4. True or false? With cel animation, the background must be redrawn for each frame.
5. When a videotape can be played on various monitors throughout a building, this is known as:
   a. Cable television.
   b. Closed-circuit television.
   c. Broadcast television.
   d. Network television.
6. When using a video camera, a production often begins with a _________ (long, medium) shot to establish location and then goes to a _________ (long, medium) shot for more detail.
7. Which of the following means to move the camera toward or away from the subject?
   a. Dolly.
   b. Truck.
   c. Tilt.
   d. Pan.
8. Videotapes should be stored:
   a. In warm, damp areas.
   b. In cool, dry areas.
   c. On top of or near stereo speakers or television monitors.
   d. None of the above.
9. A video production can be shown on a:
   a. Television monitor.
   b. Television receiver.
   c. Monitor/receiver combination.
   d. All of the above.
10. A _________ is an optical storage device on which images are stored digitally on a disk and played back with a laser beam, much like a compact disc.

APPLYING YOUR KNOWLEDGE

1. Make a set of flip pictures to demonstrate persistence of vision.
2. Visit a local television station and describe the stages of television production.
3. Investigate the features on new video cameras, recorders, and television/ monitors.
4. Read the operating manual and then practice operating a video system.
5. Videotape a television commercial in which you demonstrate the features of a piece of video equipment of your choice.

Video productions are often used in the preparation of training programs. (U.S. Navy)

36

CHAPTER

THE VIDEO PRODUCTION

After studying this chapter, you will be able to:

- *List the steps necessary in making a professional-looking video production.*
- *Identify video and audio connections.*
- *Recognize good scene composition.*
- *Compare transition methods.*
- *Describe the use of simple graphics for a video production.*
- *Discuss how to achieve good sound quality in a video production.*
- *Demonstrate various lighting techniques that can be used in producing a video.*

In Chapter 35, you were introduced to video technology. In this chapter, you will learn how to produce your own video. Planning, composition, transitions, graphics, and lighting are all necessary components in producing a video with professional results, Fig. 36-1.

PRODUCTION PLANNING

As with any communication product, video production requires careful planning. Typical steps in planning a video production include:

1. Identifying a topic and developing a rough outline.
2. Producing a storyboard.
3. Writing a script.

Fig. 36-1. A video production can be produced with a minimum of equipment. (Bowling Green State University, Visual Communication Technology Program)

TOPIC AND ROUGH OUTLINE

The first step is to determine the topic or main idea of the production. In industry, this might be decided for you. Once a topic is decided upon, a rough outline for the production is made. The three main parts of most production rough outlines are similar to the parts of a paper you might write. Both begin with an introduction that sets the tone for the content. After the introduction, the content is outlined. A summary then follows. The rough outline is often done rather quickly. This is because further changes can be made later.

STORYBOARD

The **storyboard** helps you to think visually about the type and sequence of scenes in the production, Fig. 36-2. Although storyboard sheets can be purchased, the storyboard can be created on 3 in. x 5 in. cards.

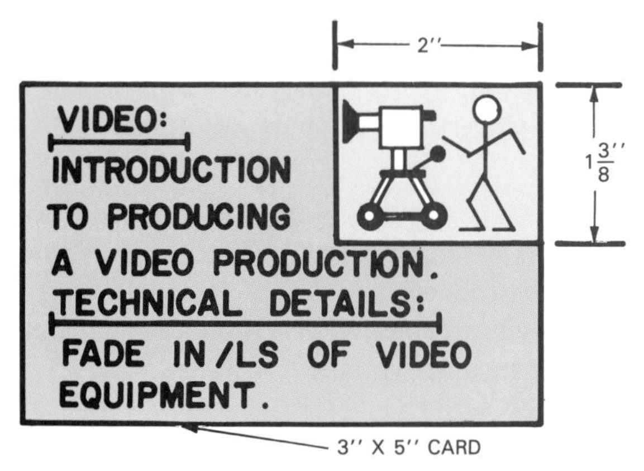

Fig. 36-2. A typical storyboard card looks like this.

To produce a storyboard, make quick, rough sketches on 3 in. x 5 in. cards to show scenes in the program. In addition to the picture, a written description of the scene and any technical details will be helpful. After the cards are completed, arrange them on a large board or table until a logical sequence is developed. As you arrange the cards, you may find it necessary to add or delete cards. The completed sequence is known as the storyboard.

SCRIPT

The **script** is the final step in designing the video production. It is based on the storyboard, but it provides more details.

As with storyboards, script forms can be purchased. However, you can make your own as shown in Fig. 36-3. The script usually provides the following details:

- Shot number and optional time.

EXAMPLE OF A SCRIPT

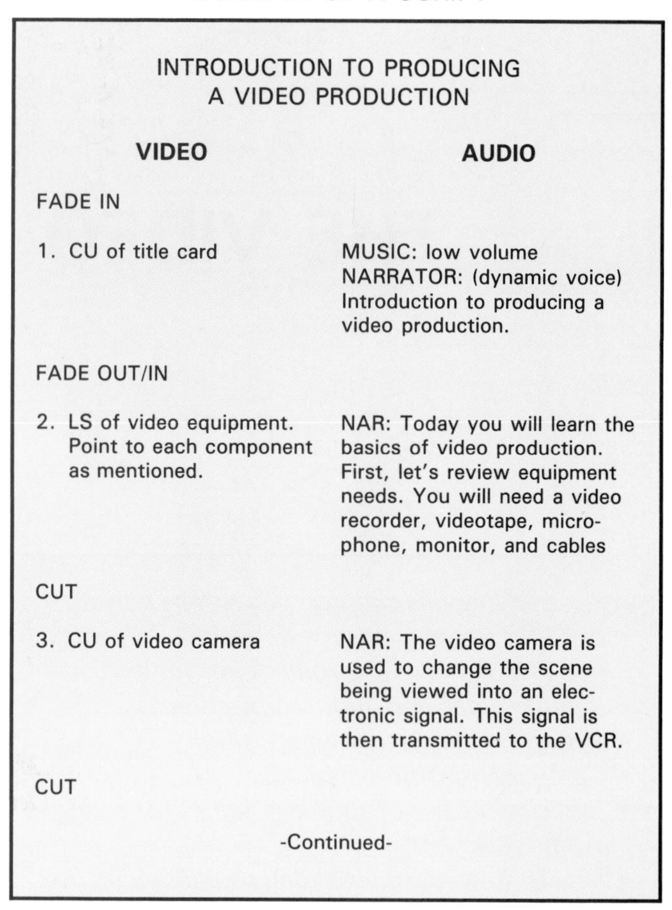

INTRODUCTION TO PRODUCING
A VIDEO PRODUCTION

VIDEO	AUDIO
FADE IN	
1. CU of title card	MUSIC: low volume NARRATOR: (dynamic voice) Introduction to producing a video production.
FADE OUT/IN	
2. LS of video equipment. Point to each component as mentioned.	NAR: Today you will learn the basics of video production. First, let's review equipment needs. You will need a video recorder, videotape, microphone, monitor, and cables
CUT	
3. CU of video camera	NAR: The video camera is used to change the scene being viewed into an electronic signal. This signal is then transmitted to the VCR.
CUT	

-Continued-

Fig. 36-3. The script provides details about the video production.

- Visual information about the scene, camera shots, graphics, and transition between scenes.
- Audio information such as narration, music, and sound effects.

Abbreviations are used frequently in script writing. For example, LS is used for a long camera shot. Some common abbreviations for camera shots were discussed in Chapter 35.

The planning steps presented are typical of those used by professional planners. However, there are instances when less planning might be acceptable. For example, some productions only progress to the storyboard step. You will have to make a decision with your instructor as to what planning steps are best for your production.

CONNECTIONS

Before connecting video equipment, it is important to know some terms. First, an **output** takes a signal out of a piece of equipment. An **input** brings a signal into equipment, Fig. 36-4. A **cable** is the channel for the signal. Cables have connectors known as **plugs** and

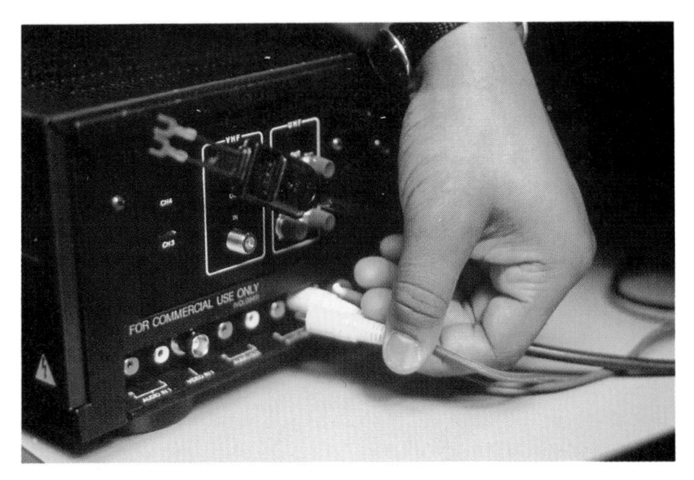

Fig. 36-4. This RCA plug is being connected to a jack. (Bowling Green State University, Visual Communication Technology Program)

sockets. The plug fits into a socket. Plugs are sometimes called "male" and sockets "female" or "jacks."

There are three types of plugs often used for audio signals. They are the phone plug, mini-phone plug, and the phono or RCA plug. These are shown in Fig. 36-5. More sophisticated audio cable uses a Cannon plug or three-pin connector for hook-ups, Fig. 36-6.

The standard connectors for video signals are the UHF, BNC, and F connectors. These are shown in Fig. 36-7. They are attached to coaxial cable. **Coaxial cable** is a round cable specially shielded against interference.

Several other types of connectors are used to carry multiple signals. One example is the **multiple-pin connector** that often carries both audio and video signals, Fig. 36-8. These should be connected carefully to avoid pin damage. When using a standard television receiver as a monitor, an **F connector** is used to send an RF signal that contains a mixed audio and video signal.

TYPICAL AUDIO SIGNAL PLUGS

MINI-PHONE
PHONE
RCA
THREE-PIN

Fig. 36-5. The typical plugs for audio signals are (from left to right) mini-phone, phone, RCA, and three-pin connector.

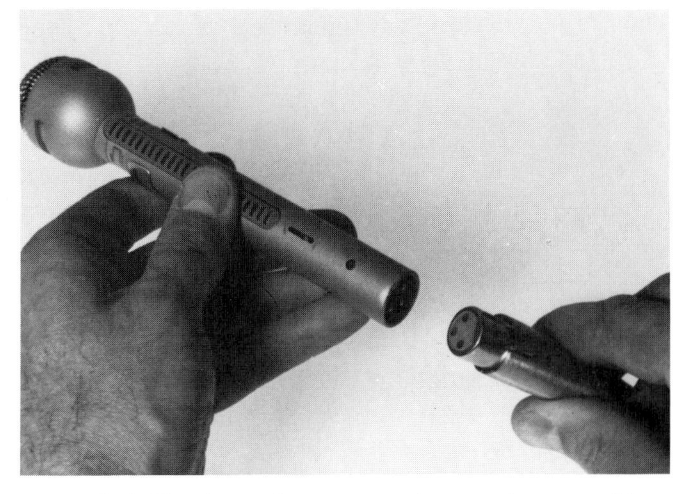

Fig. 36-6. The three-pin connector is sometimes used for microphone connections.

TYPICAL VIDEO CONNECTORS

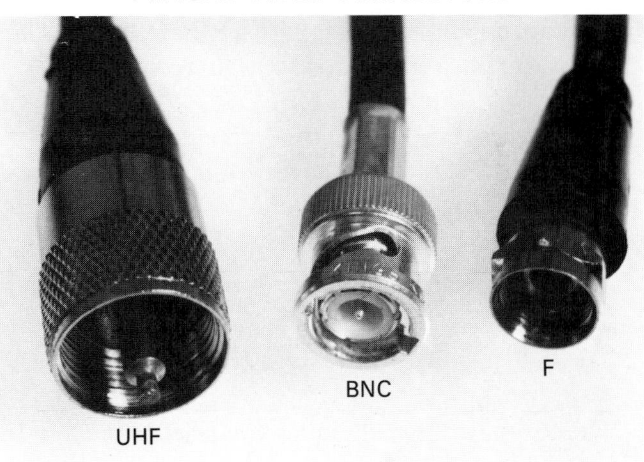

Fig. 36-7. Typical video connectors include the UHF, BNC, and F.

Fig. 36-8. This multiple-pin connector is used for some camera connections.

This is the same type of signal that is picked up by the television antenna in your home.

RECORD AND PLAYBACK

Once the equipment is connected properly, you are ready to begin recording. The following steps are followed when using most video recording equipment:

1. Check connections, Fig. 36-9. Then turn on the power switches.
2. Load and rewind the videotape. Set the counter at zero.
3. Adjust the camera for correct color and lighting conditions.
4. Set the desired audio level, if needed, for the sound equipment.
5. Focus on the subject, Fig. 36-10.
6. Depress PLAY and RECORD together and begin recording. For cameras with a RECORD/PAUSE button, push the button to begin recording.
7. To pause, depress the RECORD/PAUSE button or depress PAUSE on the video recorder.
8. Depress STOP when the recording is complete and record the counter number.
9. REWIND and then depress PLAY to review the production, Fig. 36-11.

COMPOSITION HINTS

The composition techniques you learned in Chapter 17 in relation to photography can be applied to producing a video. However, some additional composition hints are presented here.

- Use the zoom for an establishing long shot in the beginning. Zoom in for scene one. Use the zoom sparingly during the production.

Fig. 36-10. Some autofocus cameras have focusing zones that are the size of the focus area. A larger zone size is needed when the subject or camera is moving. (Panasonic)

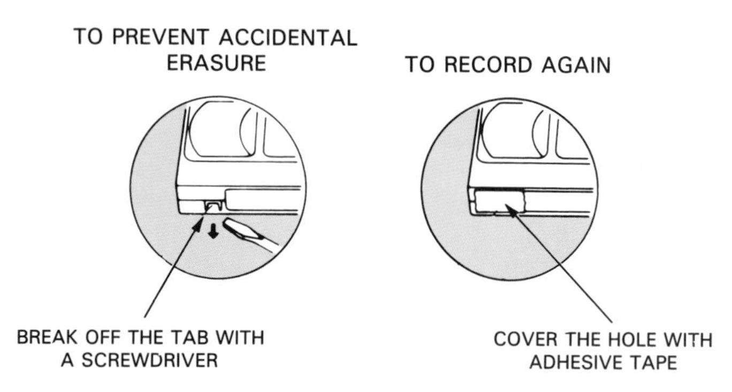

Fig. 36-11. Some cassettes have an erasure prevention tab that can be removed to prevent accidental erasure. (Panasonic)

- More light allows a higher camera f-stop. This provides greater depth of field that can be especially important in action shots.
- Too much camera movement or zoom movement is distracting to the viewer. Try to set up the scene so a minimum of movement is needed.
- A zoom lens in the telephoto setting gives very little depth of field.
- Your shots should be as close as possible, while still capturing the detail.
- Keep the action moving in the production.

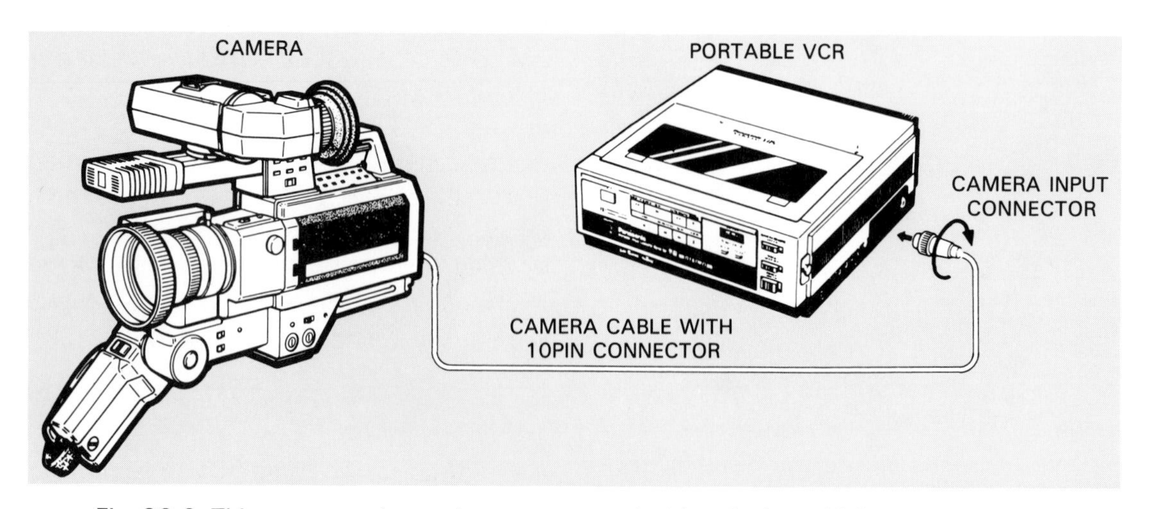

Fig. 36-9. This camera and recorder are connected with a single multiple-pin connector. (Panasonic)

TRANSITIONS

In the video industry, moving from one scene to another is called **transition**. This is often accomplished by using a **switching unit**, Fig. 36-12. Two or more cameras are attached to the switcher. The switcher technician can move from one camera to another very smoothly during the same recording. Electronic editing equipment is also used to splice together different scenes, Fig. 36-13. However, smooth transitions can also be made with a single camera unit using special techniques. These techniques include:

- Sharp cut.
- Wipe.
- Dissolve or fade transition.

SHARP CUT

A **sharp cut** is achieved by pausing the tape at the end of the first scene. The second scene is set up, and recording is started again. Avoid leaving the recorder in the PAUSE mode for over two minutes because this can cause tape wear. Use STANDBY instead. The sharp cut gives the viewer the feeling that no time has elapsed between scenes.

Fig. 36-12. A switching unit is helpful when a production requires the use of multiple cameras. (Bowling Green State University, Visual Communication Technology Program)

Fig. 36-13. This is video editing in progress. (Bowling Green State University, Visual Communication Technology Program)

WIPE

A **wipe** gives the appearance of "pushing" one scene out by another scene. It gives a feeling that time has elapsed between scenes. This can be accomplished in several ways. First, at the end of a scene, pan away from the scene very rapidly until the image blurs and then pause. After setting up the second scene, begin by panning rapidly into the scene from the opposite side.

The wipe can also be done by rapidly moving a sheet of cardboard or plastic over the lens from the side. Pause and set up the second scene. Begin the recording again by sliding the sheet rapidly off the lens onto the other side.

DISSOLVE OR FADE TRANSITION

Like the wipe, the **dissolve** or **fade transition** gives the feeling of time elapsing between scenes. Instead of a blurred image, however, the scene fades to black. This technique is accomplished by turning the camera aperture to the highest f-stop at the end of the scene. Then pause and change the scene. Begin the next scene by opening the aperture again. Refocus between scenes while not recording. Some cameras have an automatic fade-in/out button as shown in Fig. 36-14.

GRAPHICS

Graphics are words and illustrations that are added to the video production. These are usually used for titles and credits. The titles and credits on network television

Fig. 36-14. Some cameras provide automatic fade-in and fade-out by simply pressing a button. (Panasonic)

are produced using a **character generator**. This unit allows you to display electronic still images on the screen, or they may move across the screen. Some cameras have a small, built-in character generator, Fig. 36-15.

Graphics can be produced with very little equipment. First, you should know that the ratio of height to width on the television screen is 3:4. This is the ratio for the area where you will place your graphics. Second, use large lettering for the title so it will be visible on the relatively small television screen. The general rule to follow is 1/4 in. letter height for each 6 in. of screen height. Therefore, an 18 in. high screen requires 3/4 in. high lettering.

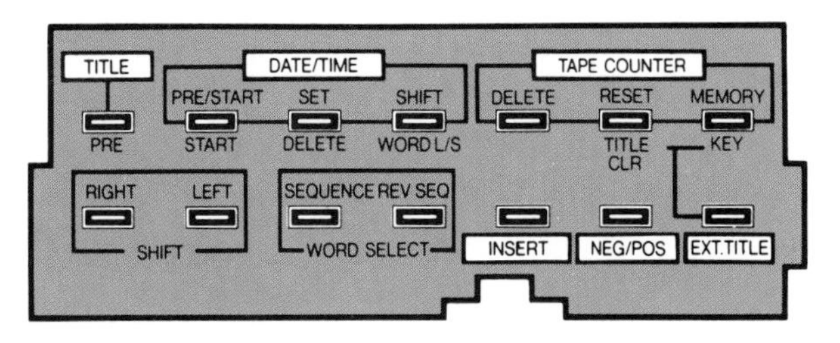

Fig. 36-15. These buttons on a video camera allow you to create an electronic title on the video tape. (Panasonic)

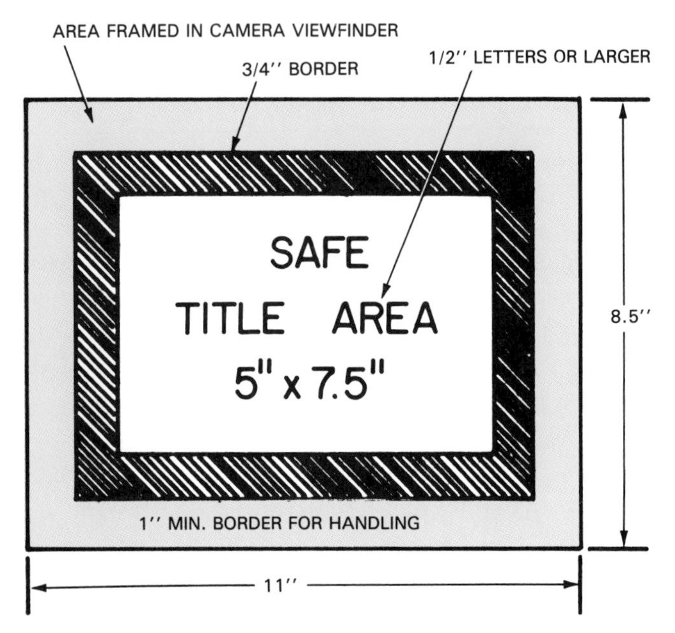

Fig. 36-16. This is a typical layout for a title card.

An example of a title card is shown in Fig. 36-16. The dead border or bleed area is for handling the card. The camera scan or picture area is the area that is seen in the camera viewfinder. The safe title area is where the graphics are placed. The safe title area is smaller than the scanned area. This is because television receivers often show less picture area than seen in the camera viewfinder.

A simple title card can be made from a stiff 8 1/2 in. x 11 in. paper, such as index paper or poster board. Draw a 6 in. x 8 in. rectangle in the center for the safe title area. Now draw a rectangle 3/4 in. larger on each side for the camera scan area. The space outside the scan area is the dead border area. Place the title on the card using 1/2 in. or larger lettering.

Lettering can be produced using a variety of techniques. Dry transfer or adhesive lettering is a simple technique to use. The title should be bold and simple without too much detail. Use an off-white or blue card for lettering. Also, lines that are drawn for margins and alignment should be drawn lightly. A non-photo blue pencil is good for this purpose.

When more than one graphic is shown in sequence, they should be of a uniform size so camera settings can remain the same. Graphics can be shown in sequence by placing them in a three-ring binder and flipping them down one after the other. You can also use the transition methods mentioned earlier.

If you want to produce a crawl, the title can be produced in a long strip and attached to a flat surface. Using the pedestal, the camera is simply moved down along the strip.

SOUND

Although many video cameras have built-in microphones, it is best to use an auxiliary or external microphone for quality sound. Both pick-up pattern and style of microphone should be considered for the production. If the talent will be moving around in the scene, small lavalier microphones are available that will clip onto clothing.

In some videos, it is not necessary for the talent to speak. A good example of this is a technical demonstration where only hands and equipment are seen. In this case, a second person can simply read the script into the microphone off camera. This way, the talent will not have to be concerned with lines. This technique also works well for adding music to the beginning and ending of a production.

If the talent will be speaking, cue cards can be made and held up off camera. Letters on the card should be large and legible from a distance. Sometimes, all that is needed on the cue card is an outline instead of the entire script.

Sound can be added after the video production is made by using the **audio-dub** found on some cameras and video recorders. This function places a new sound track on the tape. When audio-dubbing, you will need an external microphone and a monitor so you can view the video as you are adding the sound.

LIGHTING

The location of light is important in a video production. Under normal circumstances, you will probably want to avoid **backlighting**. This is light that is coming from behind the subject. Examples might include the sun, a lamp, or a window.

In some cases, it may be necessary to use auxiliary lighting for the video production. Inexpensive light fixtures with reflectors can be used for this purpose. The bulbs, however, should be rated at about 3200° Kelvin for the best results.

One, two, or three lights are used for lighting the scene, Fig. 36-17. When one or two lights are used, they are placed at an angle on either side of the camera. One is the **key light**. It is the brighter light. The other light is the **fill light**. It is used to fill in the shadows created by the key light. The fill light can be made less bright by simply moving it further away from the scene.

A backlight is often used in professional productions because it provides depth in the scene, while also reducing shadows. This is a light that is placed above and behind the subject.

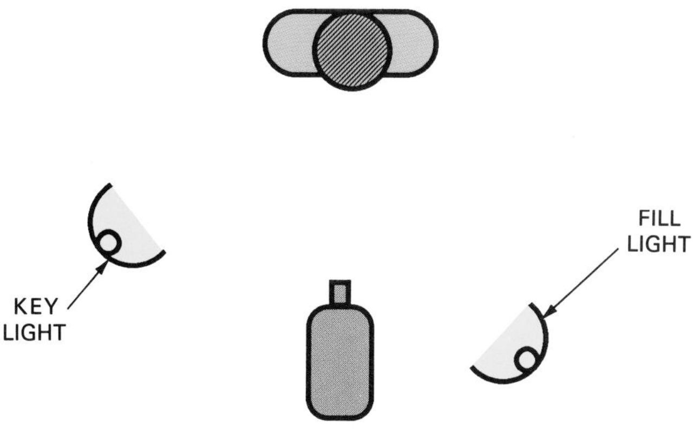

Fig. 36-17. This is a typical lighting set-up when using three lights.

SUMMARY

You can use a video production as an effective communication tool. This chapter has provided background on the techniques for producing a technically good video production without the use of expensive, extra equipment. With careful planning, a good production can be produced without editors, character generators, and switching units. Producing your own video can be a fun, creative, and exciting experience.

WORDS TO KNOW

All of the following words have been used in this chapter. Do you know their meanings?

audio-dub
backlighting
cable
character generator
coaxial cable
dissolve
F connector
fade transition
fill light
graphics
input
key light
multiple-pin connector
output
plugs
script
sharp cut
sockets
storyboard
switching unit
transition
wipe

REVIEWING YOUR KNOWLEDGE

Please do not write in this text. Write your answers on a separate sheet.

1. List the three steps in planning a video production.
2. The script for a video production usually provides what information?
   a. Shot number and optional time.
   b. Visual information about the scene, camera shots, graphics, and transition between scenes.
   c. Audio information such as narration, music, and sound effects.
   d. All of the above.

3. Which of the following is NOT a standard connection for video signals?
   a. UHF.
   b. Phono or RCA plug.
   c. BNC.
   d. F connector.
4. To record with most video recording equipment, you must press:
   a. RECORD only.
   b. Both RECORD and PLAY.
   c. Both RECORD and PAUSE.
   d. REWIND only.
5. True or false? When producing a video, a good composition technique would be to use the zoom frequently.
6. To give the viewer the feeling that no time has elapsed between scenes, a common transition technique that can be used is the:
   a. Sharp cut.
   b. Wipe.
   c. Dissolve or fade.
   d. None of the above.
7. On a title card, which area is where the graphics are placed?
   a. Bleed area.
   b. Camera scan area.
   c. Safe title area.
   d. None of the above.
8. True or false? When more than one graphic is shown in sequence, they should be of a uniform size so camera settings can remain the same.
9. Sound can be added after the video production is made by using the ________-________ found on some cameras and video recorders.
10. When lighting a scene, which light is the brighter light?
    a. The key light.
    b. The fill light.
    c. The backlight.
    d. They are all the same.

APPLYING YOUR KNOWLEDGE

1. Plan and produce a short video production that includes a title, transition, and lighting.
2. Make a display of the various connectors for audio and video signals.
3. Investigate the job opportunities in the television industry. Prepare a written report.
4. Prepare a video production on how to use video equipment.
5. Divide into teams and videotape one another presenting a topic. Review and critique your performance. Repeat the performance and try to correct flaws noted in your critique.

37

CHAPTER

LONG DISTANCE MESSAGES WITH WAVES

After studying this chapter, you will be able to:

- *Explain the relationship between electricity and magnetism.*
- *Describe the propagation of electromagnetic waves, their form, and modulation.*
- *Compare the broadcasting of AM radio, FM radio, and television.*
- *Discuss the use of microwaves in telecommunication technology.*
- *Explain how lasers are used in long distance communication.*

Many messages are sent long distances through the air, Fig. 37-1.

For instance, thanks to satellite technology, people all over the world can witness live events such as the Olympics. Electromagnetic waves are used for this purpose. In this chapter, you will explore how these waves are produced. You will also find out how they are used to send messages.

Fig. 37-1. Electromagnetic waves from a radio station are being received by this radio.

ELECTRICITY AND MAGNETISM

As was discussed in Chapter 34, when an electric current flows through a wire, a magnetic field surrounds the wire. If the wire is coiled, the field is "concentrated." An iron rod placed in the center will be magnetized. The opposite is also true. Magnetism can be used to produce an electric current. For example, if a magnet is moved within a coil of wire, an electric current will be induced. This relationship between electricity and magnetism is important for electromagnets as well as the reproduction of electromagnetic waves.

ELECTROMAGNETIC WAVES

During the mid 1800s, James Clerk Maxwell proposed that electromagnetic waves could be produced by rapidly changing current direction in an electrical circuit. In 1886, Heinrich Rudolph Hertz proved this to be true by transmitting these waves with an electrical device known as an **oscillator**.

To create an electromagnetic wave, electric current direction must change very rapidly or oscillate. When this is rapid enough, the magnetic field will "break off" into space. This is the beginning of an electromagnetic wave, Fig. 37-2.

As you may recall, a magnet can induce a current in a wire conductor. This same process can take place with atoms in the air instead of atoms in a wire. The magnetic field, which is "pushed" into space, induces an electric field in the surrounding atoms. This electric field, in turn, induces a magnetic field. This continual action is an **electromagnetic wave**.

An electric current must change directions very rapidly to begin emitting waves. An oscillator is used in transmitting equipment for this purpose. Once a wave is emitted, it travels at 186,000 miles per second.

Wave form

An electromagnetic wave can be shown by the letter "S" on its side. This is known as a **sine wave**. The S is the electric field. An S is used because the magnetic field alternates and pushes the electric field first up and then down. One complete up and down cycle is known as one **wavelength**. Refer again to Fig. 37-2.

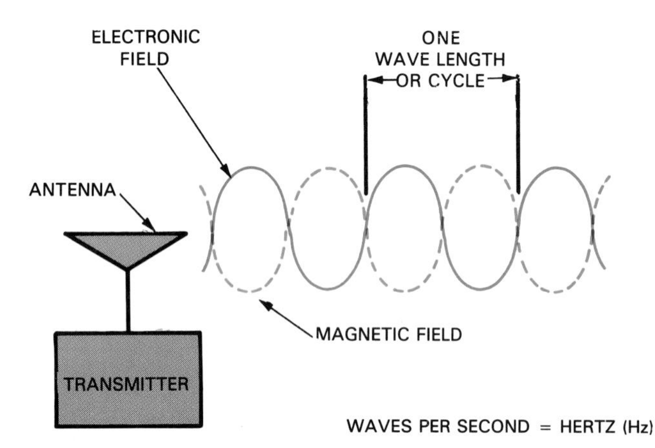

Fig. 37-2. An electromagnetic wave consists of an electric field and a magnetic field.

Wave frequency

Wave frequency is a measure of how many waves are produced in one second. Instead of "waves per second" we use the term **hertz (Hz)**. This is in honor of the man that first transmitted these waves, Heinrich Hertz.

Electromagnetic waves travel 186,000 miles per second. If two waves are produced in one second, this is known as 2 Hz. Each of these waves would be 93,000 miles long. As the wave frequency becomes higher, wavelength becomes shorter.

The electromagnetic spectrum chart, shown in Fig. 37-3, shows wavelengths and how they are used. Waves at the left end can be many miles long. In comparison, you can fit thousands of waves of visible light in one inch.

Very large numbers of waves can be produced each second. For this reason, it is helpful to use abbreviations. Refer again to Fig. 37-3. For example, one kilohertz (kHz) is the same as 1,000 Hz. Also, one megahertz is the same as 1,000,000 Hz.

Modulation

Waves can be used by placing messages on them. This is known as **modulation**. If the message is placed on the wave by changing the height, this is known as **amplitude modulation (AM)**. If the frequency is varied, this is known as **frequency modulation (FM)**. Some messages, such as television signals, use both AM and FM to carry the signal.

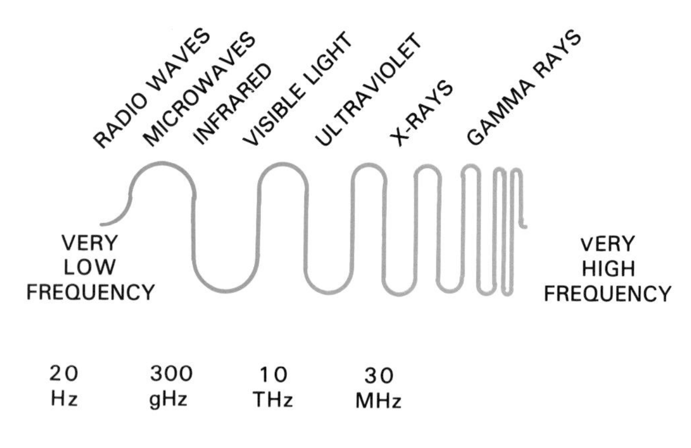

20 Hz   300 gHz   10 THz   30 MHz

1 KILOHERTZ (kHz) = 1 THOUSAND Hz

1 MEGAHERTZ (MHz) = 1 MILLION Hz

1 GIGAHERTZ (gHz) = 1 BILLION Hz

1 TERAHERTZ (tHz) = 1 TRILLION Hz

Fig. 37-3. This is the electromagnetic spectrum and some typical abbreviations.

DIVIDING THE ELECTROMAGNETIC SPECTRUM

Since the first electromagnetic waves were generated by Heinrich Hertz in the 1800s, there has been increasing competition for using the spectrum for communication. Since the portion of the spectrum most often used for broadcasting is limited, it has been necessary to assign frequencies for different uses, Fig. 37-4. The **Federal Communications Commission (FCC)**, a government agency, has this responsibility in the United States, Fig. 37-5.

The FCC assigns frequencies based on the bandwidth requirements of the different broadcasting technologies. A **bandwidth** is the difference between the highest and lowest frequencies needed for a signal and is measured in hertz (cycles per second). For example, a television signal requires a bandwidth of about 6 MHz (6 million hertz).

Based on these different requirements, the spectrum is divided into larger bandwidths by the FCC. The band from 535 to 1605 kHz is reserved for AM radio broadcasts. The band from 54 to 216 MHz is allocated to the VHF television channels (2-13). These larger bandwidths are then further divided into smaller bandwidths or channels such as a radio or television station.

Frequency Range (Bandwidth)	Frequency Designation	Use Examples
535-1605 kHz	MF (Medium Frequency)	AM Radio
27 MHz	HF (High Frequency)	CB Radio
30-50 MHz	VHF (Very High Frequency)	Police and Fire
50-54 MHz	VHF	Amateur Radio
54-216 MHz	VHF	TV Channels 2-13
88-108 MHz	VHF	FM Radio
470-890 MHz	UHF (Ultra High Frequency)	TV Channels 14-83
1.3-1.6 gHz	UHF	Radar

Fig. 37-4. Frequencies are assigned for different uses.

Based on signal ranges, more than one user can occupy the same bandwidth. For example, the channel 12 station on your television will most likely be different than channel 12 in a city or state much further away.

AM RADIO

AM is considered to be low frequency radio broadcasting. The signal is transmitted from the radio station as ground waves and sky waves, Fig. 37-6. Ground waves are attracted to the earth and travel near the surface. Sky waves are transmitted up to the ionosphere, which bounces them back to earth. Sky waves can greatly increase the range of an AM radio station. At night, the cool ionosphere is especially good at reflecting these waves. This is one reason distant stations can be received better in the evening.

Fig. 37-5. The FCC assigns the frequency that this handheld scanner and terminal must use for transmitting radio signals. The transmitting antenna is shown on the top of the terminal. (Norand Data Systems)

FM RADIO

FM radio is transmitted in the VHF part of the spectrum. At this frequency, the waves travel in a straight line rather than follow the earth's surface. Refer again to Fig. 37-6. This is known as a direct wave. For this reason, 70 miles is about the maximum range for FM stations.

TELEVISION

Television signals are in the VHF and UHF range. Channels 2-13 on the television dial are in the VHF range. UHF signals are received on channels 14-70.

Television signals combine both AM and FM waves. Amplitude is changed to carry the picture, and frequency is changed to carry the sound. Since there is more interference on AM, it is common to lose the picture and still receive the sound.

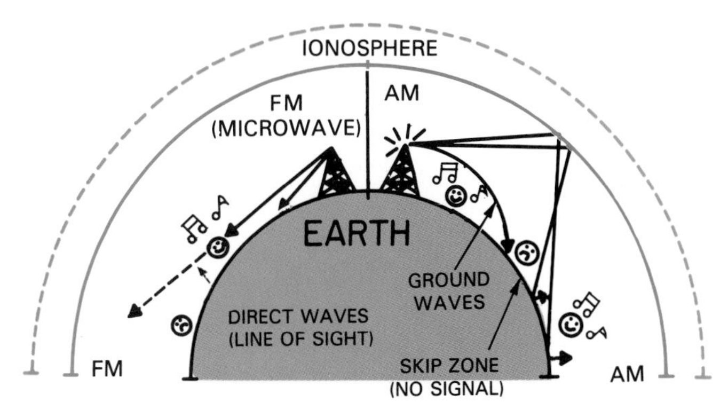

Fig. 37-6. AM and FM radio waves are transmitted differently.

Television stations use direct waves to broadcast signals to homes. However, signals can also be sent from station to station by microwave or over the phone line.

MICROWAVES

Microwaves begin above UHF and go all the way to visible light. These waves travel by direct wave, or line-of-sight only.

Very short waves are easily reflected. For this reason, microwaves are useful for **radar** (**ra**dio **d**etection **a**nd **r**anging). Waves are transmitted, and a receiver picks up the returning wave that is reflected from objects. Radar is used for tracking planes and satellites. Radar has also been used to explore most of the planets in our solar system.

Microwaves are used for relaying signals across the country. Relay antennas are spaced about 40 miles apart and in line-of-sight. The incoming signal is received and transmitted to the next station. This system is used for telephone signals, television programs, and data communication, Fig. 37-7. For example, a television program can be sent to a station by microwave. The station then broadcasts the signal to homes.

Another use for microwaves is electronic news gathering (ENG). A specially equipped truck can be sent away from the radio or television station to record an event. After the news is recorded, a microwave signal is beamed from the truck to a relay antenna. The signal is passed on to the station where it is recorded for later use on the news.

Fig. 37-7. This dish antenna is used by the U.S. Navy to send and receive microwave signals. (U.S. Navy)

Microwaves and satellites

Microwaves travel great distances in a vacuum, such as outer space. For this reason, microwaves are used for relaying signals by satellite. An earth station sends the signal to a satellite. The satellite then transmits the signal to another location. Data, telephone, and broadcast communications are sent cross-country and around the world in this way, Fig. 37-8.

In 1945, Arthur C. Clarke, a science fiction writer best known for *2001: A Space Odyssey*, proposed that satellites could be placed in a stationary orbit. He predicted that if a satellite were placed high enough over the equator and traveled at a constant speed, it would appear to be stationary. This would allow it to be used 24 hours a day for relaying signals.

Scientists found that Clarke's prediction was correct. A satellite placed 22,300 miles over the equator will be synchronized with the earth's rotation and thus will appear to be stationary. This is called a **geostationary orbit** or **geosynchronous** orbit. This area above the equator is named the "Clarke Belt," in honor of Arthur Clarke.

The first geostationary communication satellite was the SYNCOM-I, launched by NASA in 1963, Fig. 37-9. Several years later, the Early Bird was placed in orbit by Intelsat, an organization created to provide international communication service by satellite. Intelsat now provides much of the telephone, television, and data communication between countries.

A signal is sent to a satellite from an earth station, Fig. 37-10. The signal is sent to a **transponder** that amplifies the weakened signal, changes the frequency, and transmits the signal back to earth. The incoming (uplink) signal must be a different frequency from the outgoing (downlink) signal so the two signals do not interfere with each other.

The downlink signal is transmitted back to earth over an area known as a "footprint." This is the area where the signal is focused. This signal is then received by television stations, cable stations, hotels, and other businesses.

COMMUNICATION USING LIGHT

Flashes of light have been used as a signaling method for many years. However, it has only been in the last 20 years that information has actually been placed on the

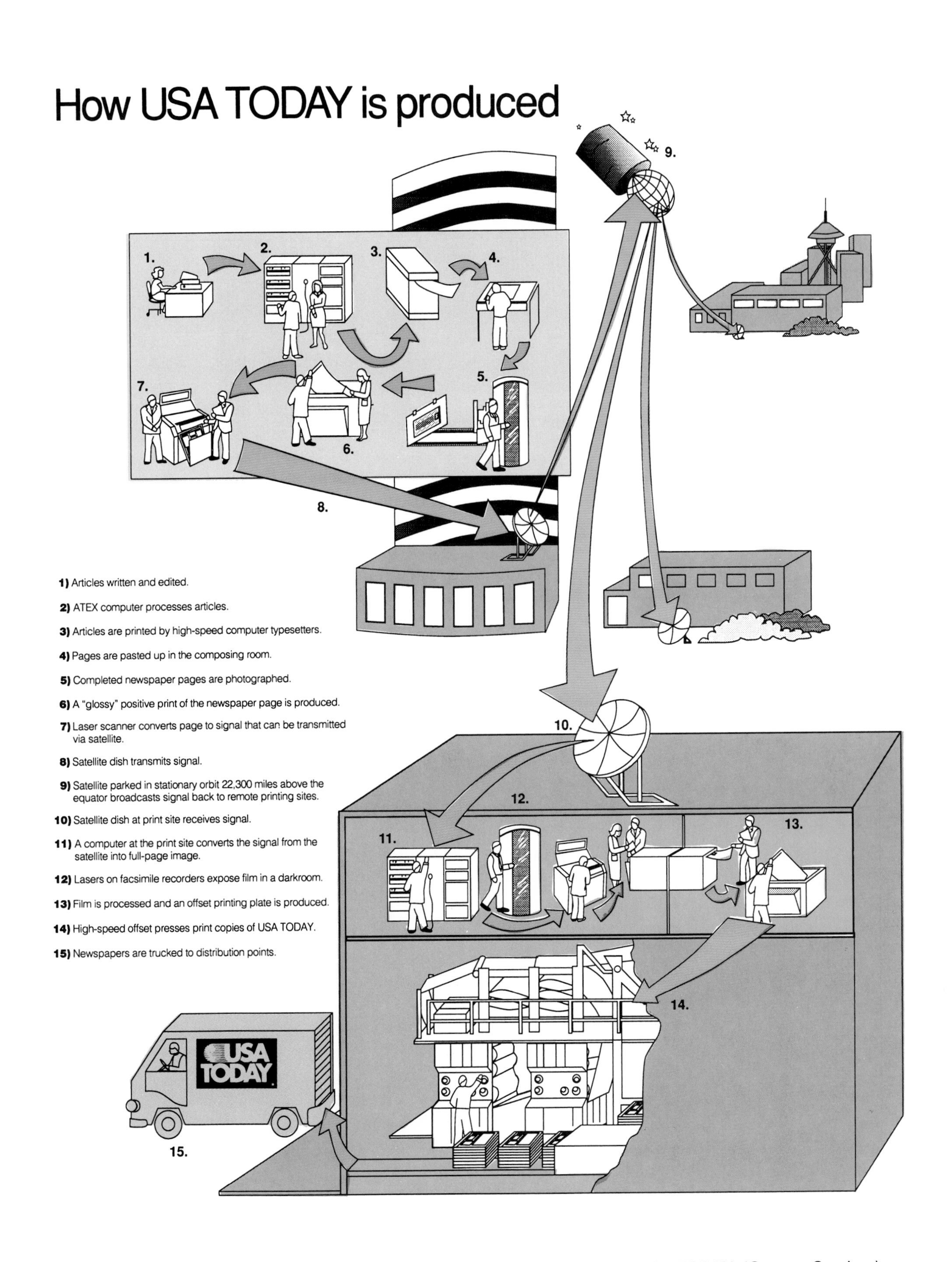

Fig. 37-8. This illustration shows how satellites are used in the production of USA TODAY. (Gannett Co., Inc.)

Fig. 37-9. This is an illustration of SYNCOM-I. (NASA)

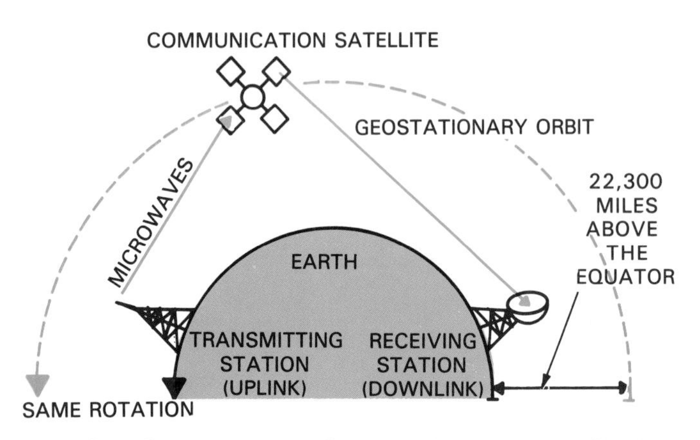

Fig. 37-10. The process of communicating by satellite is shown here.

visible light waves and used for communication purposes. Since light is of very high frequency and has a large information capacity, it is very desirable as a transmission channel.

Optical fibers are used for the transmission of light, Fig. 37-11. These are thin glass fibers that have a light transmitting core surrounded by another layer of glass (cladding). The cladding helps in keeping the light moving down the fiber, Fig. 37-12. The fibers are bundled together into larger fiber optic cables, Fig. 37-13.

The light source in fiber optic systems is a laser or LED (light emitting diode). The LED is used for shorter distances.

The word **laser** is an acronym that stands for **l**ight **a**mplification by **s**timulated **e**mission of **r**adiation. Lasers are used for long distance communication with fiber optics.

Most light sources are **incoherent light**, Fig. 37-14. This means that they emit a variety of wavelengths. The laser is **coherent light**. The waves are all one wavelength and oscillated together perfectly. Also, unlike incoherent light, the laser beam does not spread very much. These coherent light waves can be modulated to carry messages.

Fiber optics will be used more in the future for carrying many cross-country signals, such as telephone, radio, and television, Fig. 37-15. Fiber optics will primarily be used for digital transmission.

Fig. 37-11. These students are conducting a fiber optics experiment. Use of fiber optics has generated a revolution in communications. Fiber optics uses light to transmit sound, data, or video images over long distances more efficiently than transmission through wire. (ESE Instruments)

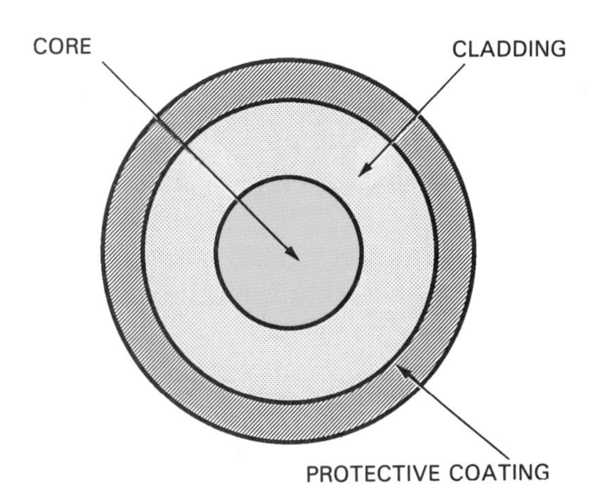

Fig. 37-12. A cross section of an optical fiber is shown here.

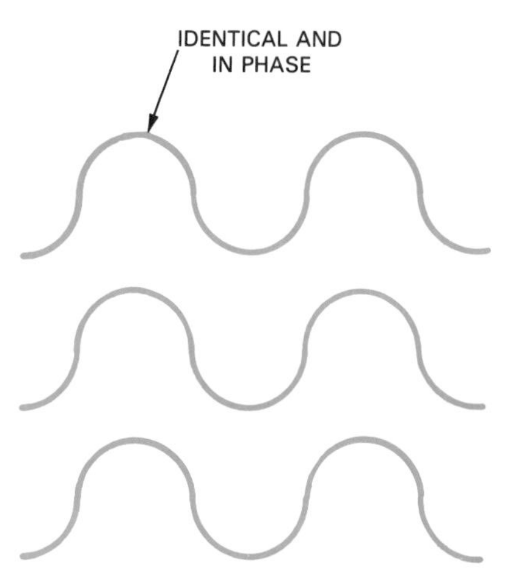

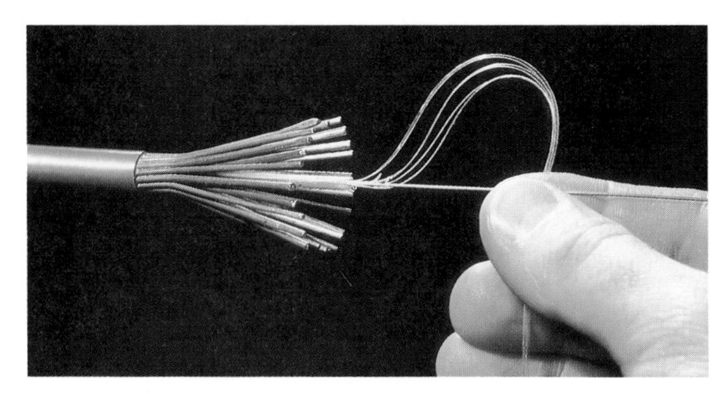

Fig. 37-13. This fiber optic cable will be placed under the sea for long distance communication between countries. (AT&T Bell Laboratories)

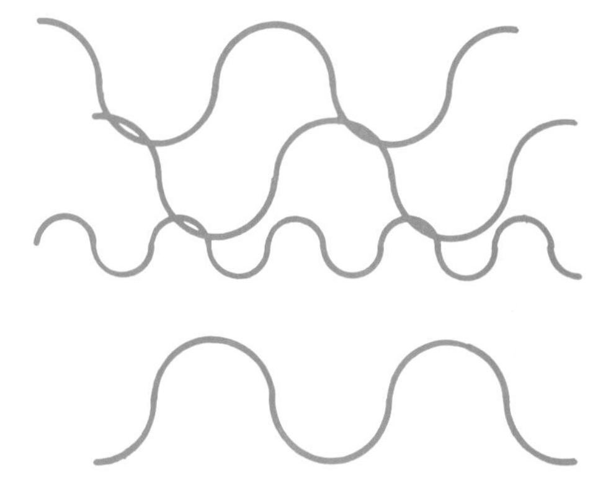

Fig. 37-14. A laser produces coherent light waves.

Fig. 37-15. A fiber optic cable is being placed underground. (United Telecom)

SUMMARY

In this chapter, you have explored how electromagnetic waves are produced. You have also seen how these waves can be used to carry information. Presently, the harnessing of electromagnetic waves has just begun. The future in this field is very exciting.

WORDS TO KNOW

All of the following words have been used in this chapter. Do you know their meanings?

amplitude modulation (AM)
bandwidth
coherent light
electromagnetic wave
Federal Communications Commission (FCC)
frequency modulation (FM)
geostationary orbit
geosynchronous
hertz (Hz)
incoherent light
laser
microwaves
modulation
optical fibers
oscillator
radar
sine wave
transponder
wave frequency
wavelength

REVIEWING YOUR KNOWLEDGE

Please do not write in this text. Write your answers on a separate sheet.

1. An electromagnetic wave consists of:
   a. An electric field.
   b. A magnetic field.
   c. Both an electric and a magnetic field.
   d. None of the above.
2. Hertz refers to:
   a. Waves per second.
   b. Waves per minute.
   c. Waves per hour.
   d. Waves per day.
3. True or false? The Federal Communications Commission has the responsibility of assigning frequencies for different uses.
4. True or false? FM radio waves follow the earth's surface.
5. ________ signals are in the VHF and UHF range.
6. Microwaves are commonly used for:
   a. Radar.
   b. Telephone signals.
   c. Electronic news gathering.
   d. All of the above.
7. True or false? A satellite placed 22,300 miles over the equator will be synchronized with the earth's rotation and thus will be in geostationary orbit.
8. True or false? A laser is an example of incoherent light.

APPLYING YOUR KNOWLEDGE

1. Investigate the accepted frequencies for various communication devices and proposed changes.
2. Demonstrate wave communication with a small, modulated laser. Be careful to follow all safety instructions provided.
3. Research the important technology breakthroughs leading to modern telecommunication. Prepare a written report.
4. Describe what you think the future of telecommunications will be like. Share your predictions with the class.
5. Invite a representative from a long distance telephone company to speak to the class about the field of telecommunications.

The armed forces often rely on communicating long distances with waves. Radar screens are monitored in the combat information center aboard the battleship USS Iowa. (U.S. Navy)

Section ACTIVITIES 6 ELECTRONIC COMMUNICATION

BASIC ACTIVITY: Radio Programming

Introduction:

Pretend that you are the station manager of a radio station. Several of your listeners are wondering why their favorite music is interrupted with news, weather, public service announcements, and especially commercials. In response to these comments, you decide to put together a two-minute explanation of why a radio station includes what it does in its program day. This explanation, addressing your listeners' concerns, will be broadcast on your station.

Guidelines:

- Individual or small group activity.
- Message should be two minutes in length.

Materials and equipment:

- Audio tape recorder.
- Literature related to radio programming.

Procedure:

1. Investigate radio programming by talking with local radio station personnel or by visiting the library. Find out about the programming elements that can be used. As part of this investigation, find out about a programming clock that is used for planning the elements that will be used in a certain time period.
2. Develop an outline and script for your radio spot. Be sure to keep your target audience in mind. You will want to explain why programming is set up as it is.
3. Record your script on an audio tape. Review and record again if needed. Use your best radio voice, and try to put energy into the message. Remember that you are talking to your listeners, who are necessary to stay in business.

Evaluation:

Considerations for evaluation are:

- Depth of investigation.
- Quality of script and completed tape.
- Individual or group effort.

ADVANCED ACTIVITY: Face the Issues

Introduction:

A video production is an excellent way to reach a mass audience. See Section Fig. 6-1. It is one of the most realistic and exciting forms of visual communication because it contains both moving images and audio. Through this activity, you will provide information to the public about an issue of importance to the community.

Section Fig. 6-1. A video production is a great way to reach a mass audience. (Bowling Green State University, Visual Communication Technology Program)

Guidelines:

- Small group activity.
- Video should be no longer than 5 minutes.
- A title and credits should be included.
- Video should contain at least two transitions, a zoom, and a pan.

Materials and equipment:

- Camcorder and external microphone, if available.
- Auxiliary lighting.
- Props as needed.

Procedure:

1. Decide on a topic of interest to the community or school for the video production and the intended

audience. Examples might include a school issue or recycling issue. Most issues are too complex for complete coverage in a 5 minute video, so you may want to deal with only part of the problem. For example, you could talk about how to recycle and what items are recyclable.

2. Research the topic and ways to present the topic. Watch local cable and news broadcasts for ideas. Then develop an outline and storyboard.
3. Divide the responsibilities among group members and proceed with planning. Write down job titles and decide who will be responsible for various tasks. Develop a script. In addition, prepare any titles, props, lighting, etc.
4. Practice the video several times before the final production. In the final production, try to make everything as professional as possible.
5. Shoot the final production and review it. If necessary, try it again.
6. Share the video with the rest of the class, and have them critique it.

Evaluation:

Considerations for evaluation are:

- Coverage of topic.
- Careful planning in evidence, including the outline, storyboard, and script.
- Quality of the completed video.

ABOVE AND BEYOND ACTIVITY: Comparing Long Distance Services

Introduction:

Since long distance service has been deregulated, many long distance carriers are now available to consumers. A good way to better understand the telephone industry is to compare the long distance services these companies offer.

Guidelines:

- Small group activity.

Materials and equipment:

- Available literature from your regional telephone company.
- Long distance carrier literature.
- Telephone directory.

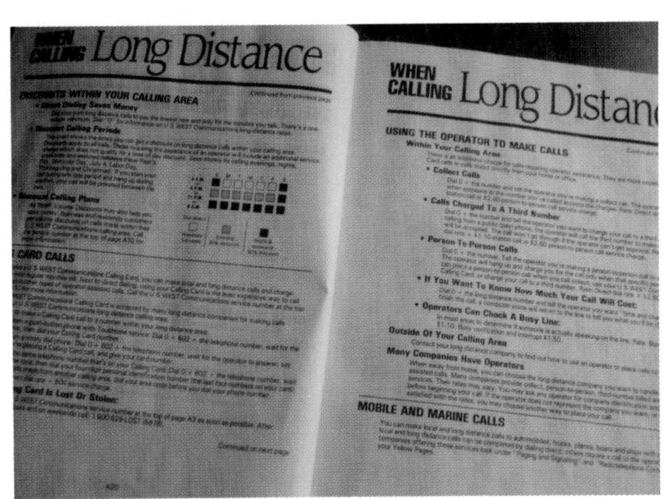

Section Fig. 6-2. A telephone directory will provide some of the information you need concerning local and long distance phone service.

Procedure

1. Before beginning your research on long distance options, determine the difference in local and long distance service. Check on the best times to make long distance calls. See Section Fig. 6-2.
2. Research the available long distance carriers. Make a list of the criteria that should be used in selecting a carrier. For example, what services does each carrier claim to provide?
3. Make a chart comparing long distance carriers on the criteria selected. In addition, make a chart showing how time of day and week affects long distance phone charges.
4. If time allows, present your findings to the class.

Evaluation:

Considerations for evaluation are:

- Depth of research.
- Content and design of charts.
- Group and individual effort.

NANCY WEHLAGE

SANTA FE PACIFIC RAILROAD

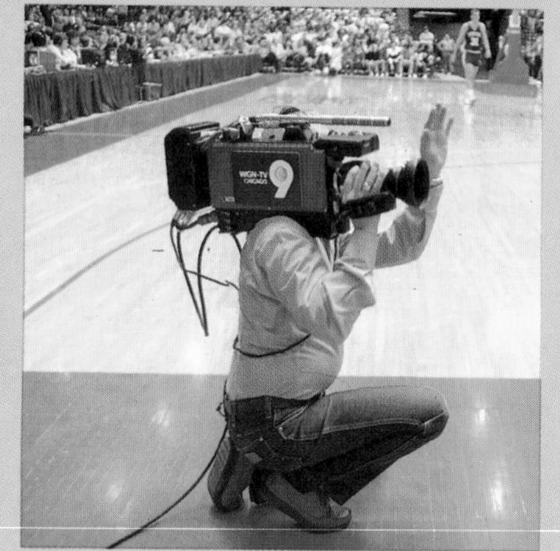

WGN–TV CHICAGO

IBM CORP.

SONOCO

HEIDELBERG EASTERN, INC.

section 7

Communication, Jobs, and You

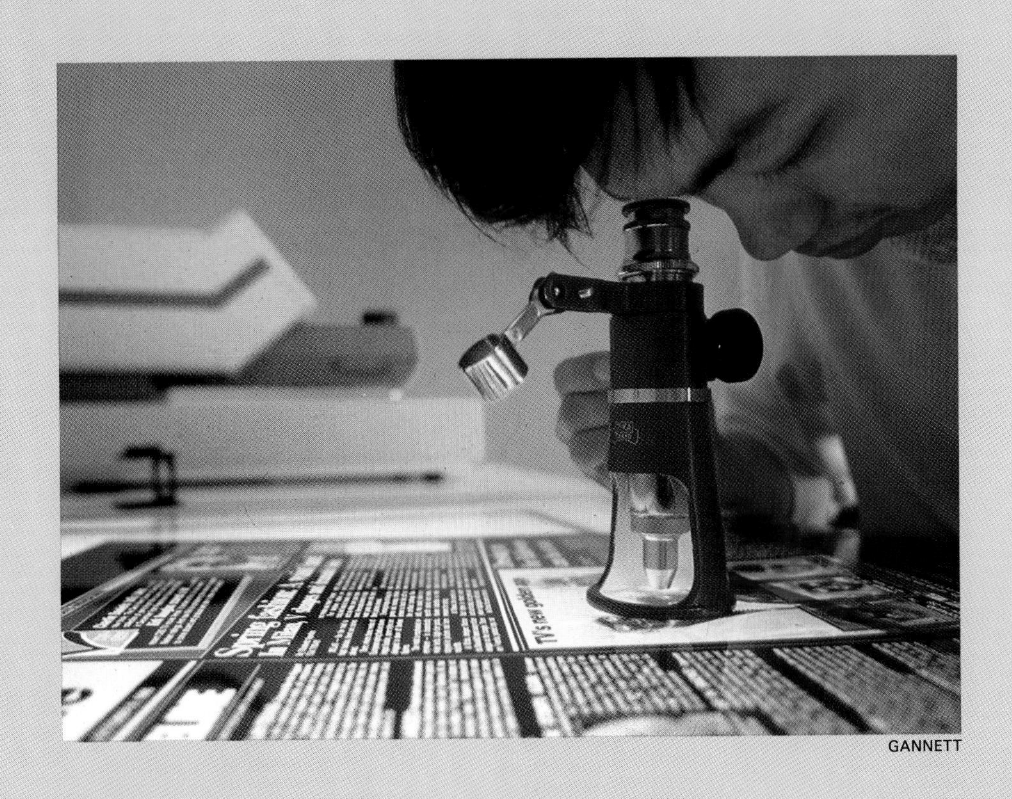

GANNETT

This freelance writer is an entrepreneur. Although being an entrepreneur can be rewarding, there are certain risks involved. (IBM Corp.)

38

CHAPTER

ORIENTATION TO BUSINESS

After studying this chapter, you will be able to:
- *Name and describe the three types of business ownership.*
- *Interpret an organizational chart.*
- *List important management functions.*
- *Identify the divisions of a company.*
- *Discuss the importance of communication in a business.*
- *Explain how to start a student enterprise.*

How do you apply the communication technology skills presented in this textbook? You can use them in practically all aspects of your daily life. However, you may also be interested in how you could apply them toward a career in this area.

There are two approaches that are available to you in finding a job in communications. You can start your own business or you can find an existing job in business and industry. People who start their own businesses are called **entrepreneurs**, Fig. 38-1.

Fig. 38-1. Some entrepreneurs have business cards printed to let potential customers know about the products or services they provide.

Whether you are interested in starting a business or finding an existing job, it is important to have some knowledge of business organization. This chapter will give you an overview of how businesses operate.

BUSINESS OWNERSHIP

There are three major forms of business ownership, Fig. 38-2. Each form has advantages and disadvantages. The three major forms of business ownership are:
- Single proprietorship.
- Partnership.
- Corporation.

SINGLE PROPRIETORSHIP

A **single proprietorship** exists when one person owns a business. A photography studio or a small printing shop might be examples of proprietorships. Because the

Fig. 38-2. Which form of business is most likely represented by this cartoon? (The correct answer is a partnership.)

owner has complete control of the company's operations, business decisions can be easier because they are made by one person. One problem with this type of business is that the owner is personally responsible for all company debts.

PARTNERSHIP

A **partnership** results when two or more owners start a business. Examples of partnerships may be an advertising agency or a family-run newspaper. An advantage of this type of business is that the partners share in the cost of starting the business. In addition, the owners have a variety of skills that help in making good decisions. A disadvantage is the necessity for joint decisions that can be more time-consuming. Also, as in the single proprietorship, the partners are liable for company debts.

CORPORATION

The **corporation** is a business formed and authorized by law with the rights and obligations of an individual. Examples of corporations include International Business Machines (IBM) and American Telephone and Telegraph (AT&T). The owners are persons who have purchased or obtained stock in the corporation. A **stock** is a share of ownership in a company. Many shares of stock can be sold to finance a large business.

Since there may be hundreds of stockholders in a corporation, a **board of directors** is elected to represent them. The board oversees the operation of the business.

A license is needed for operating a corporation. This is applied for through a government organization. In the case of a corporation, the license is known as a **charter**. The charter lists the company name and officers. In addition, general rules for the corporation are stated. These rules are listed in more detail in another document known as **corporate bylaws**. Bylaws include such details as officer election procedures, officer responsibilities, and board of directors meeting time and place.

The corporation overcomes some of the weaknesses of the other forms of business ownership. Unlike the other forms of ownership, stockholders have limited liability. This means that they are only responsible for debts up to the amount of money invested in the stock. In this case, an owner need not fear losing personal property, such as a home, because of business debts. If the corporation makes a profit, part of this is returned to the stockholders. These returned profits are known as **dividends**.

LINES OF RESPONSIBILITY

Lines or levels of responsibility exist in every company. In larger companies, these are usually represented as **organizational charts**, Fig. 38-3. People are

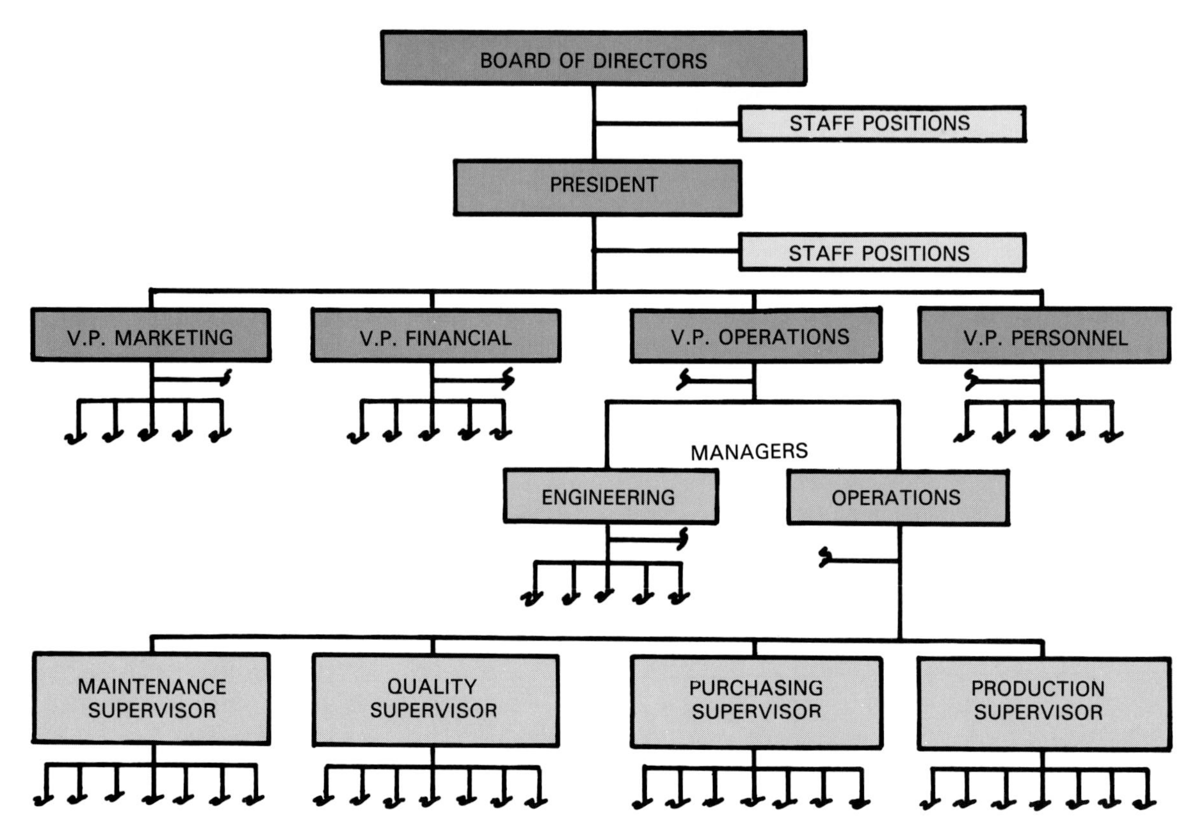

Fig. 38-3. This is an example of a typical organizational chart.

Fig. 38-4. Managers are responsible for planning, organizing, directing, and controlling activities. (IBM Corp.)

responsible to those above them on the chart. Normally, charts only show management positions. The charts provide a picture of the various responsibility levels within the company.

In the case of a large corporation, the board of directors appoints a president to operate the business. The president, in turn, appoints a vice president for each major division of the company. The vice presidents have additional levels of management as needed. This flow of authority, from president to supervisors, is generally known as **management**.

MANAGEMENT

Managers have the responsibility for coordinating activities that take place within levels that fall under them on the organizational chart. See Fig. 38-4. There are various levels of management in a company. However, at each level managers are responsible for:

- Planning activities.
- Organizing activities.
- Directing activities.
- Controlling activities.

Planning is the process of thinking ahead. For example, budgets are developed that project how much a product will cost. Planning is often done to reach a goal, such as increased sales and profits.

Organizing involves scheduling people and resources so they are used in the most efficient manner. For example, a production manager needs to set up production so enough skilled people and materials are available to produce a product.

Directing involves assigning people to specific activities. After assigning a job, the manager may need to provide further training and safety information.

Controlling is the act of checking to ensure that activities are taking place as expected. For example, quality control is used to check a product to ensure a high standard of quality.

COMPANY ORGANIZATION

A large company is usually divided into areas of specialization. Each area has a management team that directs and controls operations. Common divisions include:

- Research and development.
- Production.
- Marketing.
- Finance.
- Personnel.

RESEARCH AND DEVELOPMENT

When you are assigned to write a paper, where do you begin? The logical approach would be to research the topic and then to develop the paper. In business, **research and development** is conducted for new products or services. Based on this research, a new product or service is developed. Development begins with product designs. In many cases, a model or prototype is made to see how the product will look. The model is then tested to see if it performs as expected.

When the final design is chosen, final drawings are prepared. The research and development team uses the drawings to describe and explain the product to the production division of the company.

PRODUCTION

Production takes place after the design of the product. Production managers often use a **flowchart** to decide on the steps in producing the product, Fig. 38-5. Standard symbols are used on a flowchart for each step. For example, a square stands for inspection, which is important for quality control.

The flowchart helps managers decide on the resources needed. These might include natural resources, people, and equipment. It is the responsibility of production to make the best product possible with the least amount of resources at the lowest cost.

MARKETING

The **marketing** division is responsible for selling the product or service. Often, marketing surveys are done prior to production to make sure there is a demand for a product. Also, details about the product, such as design, can be obtained through this research.

Fig. 38-6. Advertisements, such as this one in the developmental stage, are an important part of marketing a product. (Autodesk, Inc.)

Advertising is also a part of marketing. This is the technique of providing information to consumers, to persuade them to buy a product. See Fig. 38-6. As a product is advertised, marketing is also involved in selling the product and sending it to the consumer.

FINANCE

The **finance** division of a company is responsible for money management. By handling and keeping records of all financial transactions, it is possible to ensure that bills are paid and hopefully, a profit is made.

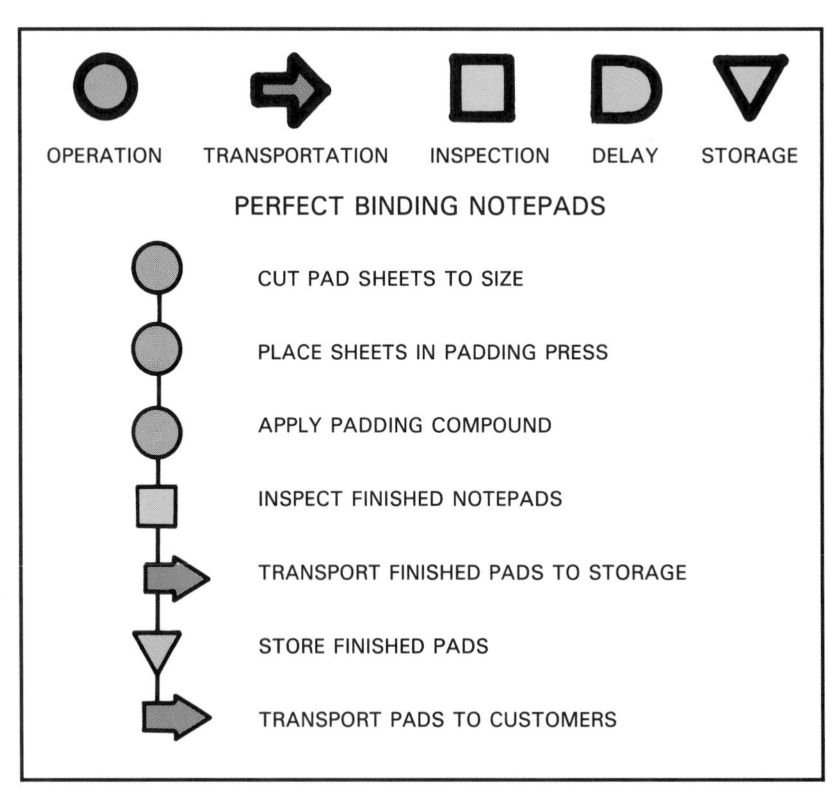

Fig. 38-5. This is an example of a product flowchart.

Just as a flowchart is important in production, the financial division uses a **budget** to predict how money will be used by the company. After production and sales, accountants provide financial reports on the actual costs and sales. If, for example, costs are higher than expected, it may be necessary to increase the price of the product in order to make a profit.

PERSONNEL

The **personnel** division is responsible for finding and keeping a qualified work force for the company. These responsibilities often include hiring, training, and safety education. Wages and benefits are also managed by this division. Benefits might include health insurance and a retirement program.

The personnel division is responsible for developing procedures for handling worker grievances. **Grievances** are complaints that an employee has about alleged unfair treatment in the workplace.

In some companies, **labor unions** are formed by employees to represent their interests, Fig. 38-7. This is decided by election. If employees are unionized, then collective bargaining is used to negotiate an agreement with management on pay and working conditions. Based on these negotiations, a contract is developed to be in effect for a certain length of time. The personnel division of the company is responsible for working with the labor union.

COMMUNICATION AND BUSINESS

Communication technology is utilized in all areas of business. In research and development, it is important to have technical drawing skills when designing products. Production may be involved with a communication product, Fig. 38-8. For example, a newspaper,

Fig. 38-7. This is the emblem for the Communication Workers of America, a labor union that represents many workers in the communications field.

Fig. 38-8. Many aspects of communication technology are used in producing a newspaper. (Gannett Co., Inc.)

magazine, and videotape are all communication products. In the marketing division, advertisements must be developed for print media, radio, and television, Fig. 38-9. Distribution of the product, such as television programs, might involve transmission. The personnel division is often involved with public relations, which is used to convey a certain company image to the public.

These examples are just a few of the ways communication technology skills are important in every business. In addition, many companies, such as a newspaper, are specifically involved with communication products or services.

A STUDENT ENTERPRISE

In order for you to better understand business operations, it is sometimes helpful to form a student enterprise within your classroom. Since a typical class has more students than would be needed for a single proprietorship or a partnership, it is a good idea to form a corporation, with each student becoming a stockholder.

ORGANIZING A STUDENT ENTERPRISE

A student enterprise can have divisions in much the same way as a large company. One difference is that it will probably be necessary for some people to be involved in more than one job.

The first step in forming a business is to decide on what product you want to produce or service you want to provide. (If there is no product or service, there is no need for the business.) A market survey can be developed for developing product ideas. A brainstorming session in class is also helpful, Fig. 38-10. Product ideas

Fig. 38-9. Developing an advertising campaign for a product is a responsibility of the marketing division of a company. (Colgate-Palmolive Co.)

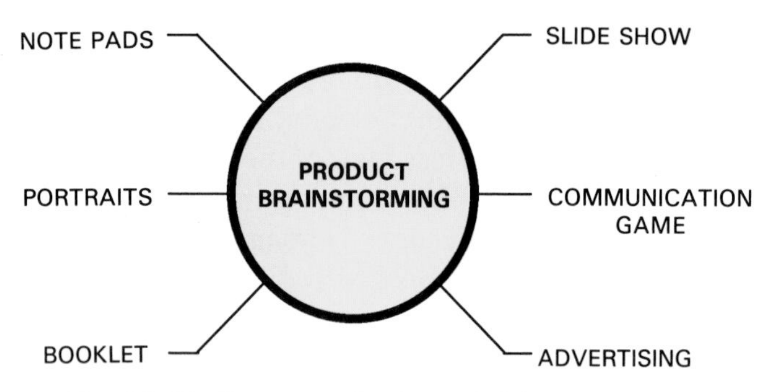

Fig. 38-10. Brainstorming is a helpful technique for deciding on a student enterprise product.

might include recycled paper notepads or slide/tape productions. Once a list of ideas is developed, it will be necessary to eliminate some options based on equipment limitations, time limitations, and low consumer demand. After a product is chosen, it is time to form the corporation.

In order to form a corporation an application must be developed for a charter. Since you are not applying for a real charter, the information can be less detailed. However, you will need to determine the company name, people involved, stock information, and activities in which the company will be engaged. This can be submitted to your instructor for approval.

The next step is to develop the corporate bylaws. These bylaws should list the corporate officers and their responsibilities, meeting times and dates, and a date indicating when the corporation will be dissolved.

MANAGING THE STUDENT ENTERPRISE

After the board of directors is elected, a vice-president for each division is elected or chosen. This person will be responsible for managing the operation for that portion of the company. These are some suggestions for activities in each division:

- *Research and Development*: Coordinate market research, design and refine product plans, and make a prototype; assist with mass production problems.
- *Production*: Develop a flowchart and actually produce the product based on the plans from research and development.
- *Marketing*: Advertise, sell, and distribute the products; provide input on the quantity needed.
- *Finance*: Develop a budget and keep financial records; assist with determining costs for the product; purchase supplies.
- *Personnel*: Help find qualified people for jobs, assist with training and safety, handle grievances; assign jobs as needed.

DISSOLVING THE STUDENT ENTERPRISE

The student enterprise is normally dissolved after a prescribed length of time. At that time, the financial division must be sure that all corporate debts are paid. Then any dividends can then be distributed among the stockholders.

After dissolving the corporation, it is helpful to discuss what you learned about business and how you might do things differently in a real business.

SUMMARY

A general knowledge of business is important in any company. This chapter has provided that general background along with information on creating a student business. The relationship of communication technology and business was also explored.

WORDS TO KNOW

All of the following words have been used in this chapter. Do you know their meanings?

board of directors
budget
charter
controlling
corporate bylaws
corporation
directing
dividends
entrepreneurs
finance
flowchart
grievances
labor unions
management
marketing
organizational chart
organizing
partnership
personnel
planning
production
research and development
single proprietorship
stock

REVIEWING YOUR KNOWLEDGE

Please do not write in this text. Write your answers on a separate sheet.

1. Name and describe the three major forms of business ownership.
2. Which of the following would represent a share of ownership in a company?
   a. Stock.
   b. Charter.
   c. Dividend.
   d. None of the above.
3. Name the four main responsibilities of managers.
4. Which of the following divisions would most likely use a flowchart?
   a. Personnel.
   b. Research and development.
   c. Production.
   d. Marketing.
5. True or false? The financial division is responsible for working with the labor union.
6. True or false? Communication technology is utilized in all areas of business.
7. The first step in organizing a student enterprise is:
   a. Apply for a charter.
   b. Develop the corporate bylaws.
   c. Decide on what product or service you want to provide.
   d. Elect a board of directors.
8. In managing a student enterprise, each of the following divisions should exist. Name at least one activity that would be the responsibility of each division.
   - Research and development.
   - Production.
   - Marketing.
   - Finance.
   - Personnel.

APPLYING YOUR KNOWLEDGE

1. Interview people involved in each of the three major forms of business ownership. Compare the advantages and disadvantages of each.
2. Develop an organizational chart for a local company. Describe the management responsibilities related to each level of the company structure.
3. Interview some local entrepreneurs with products related to the field of communication. Find out how they got started and about the rewards and challenges of being a entrepreneur.
4. Form a student enterprise and market a product. Discuss the similarities between the student enterprise and other businesses.

Careful research can be used in selecting a job that is right for you. (Jack Klasey)

39

CHAPTER

FINDING A JOB

After studying this chapter, you will be able to:
- *Discuss the relationship between student clubs and future employment.*
- *Identify three personal characteristics that should be considered when choosing a job.*
- *Name some valuable resources that can help you in a job search.*
- *Describe various training options for a job.*
- *Demonstrate job hunting skills.*
- *Describe correct procedures to follow in applying for a job.*
- *Recognize the importance of company benefits*
- *List ways in which you can succeed on the job.*

The information in this text can be used in a variety of ways. It can be used to start a new hobby, such as photography. It can also help you understand new developments in communication technology. Another important use for this information is to assist you in making career plans. You can begin making career plans while you are still in school. Try to take courses that will be an asset to you in your future career. By becoming involved in activities and student organizations related to your career interests, you can explore your options. (Information on careers related to the field of communications is presented in Chapter 40 of this text.)

STUDENT ORGANIZATIONS

Participation in student organizations is a good way to find out about careers, Fig. 39-1. Student organizations can help you develop the qualities necessary to be successful on the job. For instance, student organization activities can help teach responsibility and leadership skills. Being part of an organization is like being part of a company, and each person can play an important role in its success.

Fig. 39-1. Participating in a student organization, such as this one, can give you an opportunity to develop leadership skills and explore career options.

Student organizations are worthwhile because they help you to learn more about the community in which you live as well as employers. Field trips are often scheduled. Community leaders may also come to speak to club members. These and other club activities help you gain confidence in meeting and working with other people.

CHOOSING A JOB FOR YOU

There are thousands of types of jobs available today. In addition, jobs are constantly changing because of new technology and information How can you decide on the best job for you?

The *best* job for you is one that most closely matches your interests and abilities. By finding out more about yourself, you have a greater chance of finding the right kind of job, Fig. 39-2.

Fig. 39-2. The best job for you is one that closely matches your interests, abilities, aptitudes, and personality.

There are four major factors to consider when thinking about a job. They are:

- **Aptitudes**: These are the talents with which you were born that can be developed into abilities. For example, a musical aptitude can be developed into an ability to play an instrument.
- **Abilities**: These are developed through training. You will most likely find it easier to develop abilities in areas where you have a high aptitude.
- **Interests**: These are the things that you enjoy the most. For example, you may have an interest in fishing, sewing, or music. You may have many interests. Sometimes those interests can lead to a career choice. See Fig. 39-3. However, sometimes interests can change. For this reason, interest alone is not always the best way to make career choices. Jobs for which you have high aptitude *and* interest are usually better choices.
- **Personality:** This word is often used to refer to how well someone relates to others. Good personality means "likable." However, personality is more than this. Personality is the combination of behaviors that make each person unique. Different jobs require different personality traits. For example, a person in sales should be friendly, courteous, and assertive.

How can you find out about your aptitudes, abilities, interests, and personality? One way is to simply think about yourself and write down your thoughts. However, counselors and teachers often have inventories and forms that can help you to do this. This information can then be compared to categories of jobs. Certain job categories will most closely fit you.

Fig. 39-3. An interest in music might lead to a career as a disc jockey. (Westinghouse Electric Corp.)

Testing can provide a general idea of the types of jobs for which you are best suited. However, remember that abilities and interests may change over time. In 10 years, a job may appeal to you that is unappealing now.

What can you do if you want to make changes in yourself to match certain jobs? First, remember that humans are very flexible. You have the ability to make changes in yourself. For example, through education you can become more skilled in an area. You can cultivate new interests. In addition, you can change some aspects of your personality. For example, you can learn to be more responsible and dependable, which are traits needed on most jobs.

YOUR JOB SEARCH

After you have learned more about yourself, this information can be matched with available jobs. Jobs are sometimes placed in clusters to make it easier to group similar kinds of jobs.

One of the best listings of jobs is the **Dictionary of Occupational Titles (D.O.T.)** by the U.S. Department of Labor. Over 20,000 jobs and descriptions are placed in nine career clusters, Fig. 39-4.

In order to use the *D.O.T.*, you must understand the occupational codes, Fig. 39-5. The first three numbers of the code subdivide the nine occupational clusters into 603 occupational groups.

The second set of three numbers identify to what extent the job requires you to work with data, people, or things. A lower number, such as zero, means more need in that area. For example, a news photographer has a 062 code. The code means that it is very important

CAREER CLUSTERS

- Professional, Technical, and Managerial Occupations
- Clerical and Sales Occupations
- Service Occupations
- Agricultural, Fishery, Forestry, and Related Occupations
- Processing Occupations
- Machine Trade Occupations
- Bench Work Occupations
- Structural Work Occupations
- Miscellaneous Occupations

Fig. 39-4. These are the nine career clusters listed in the *Dictionary of Occupational Titles (D.O.T.).*

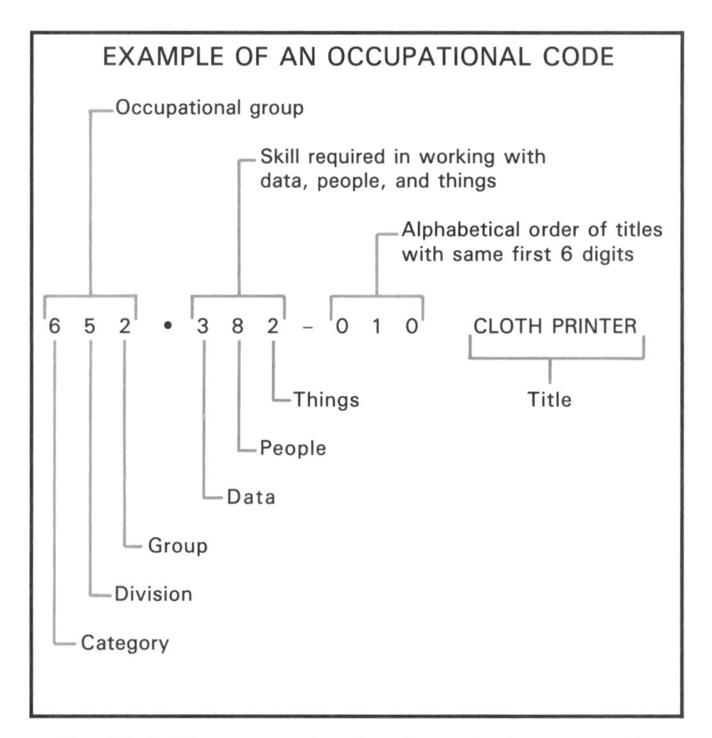

Fig. 39-5. The occupational code can be interpreted in this way.

that the person be able to work with data (0) and things (2). However, it is less important that the person is able to work with people (6).

The **Occupational Outlook Handbook** is published every two years by the Department of Labor. It provides updated information on present and future jobs. In combination, the *Dictionary of Occupational Titles* and the *Occupational Outlook Handbook* are valuable resources for a job search.

JOB TRAINING

Most jobs require at least some training. One method of obtaining the necessary skills is on-the-job training. Some on-the-job-training is required for nearly every job you can imagine.

Fig. 39-6. Some type of training is required for most jobs. (AM Multigraphics)

When technical skills are needed for a job, this training can be obtained in a variety of ways, Fig. 39-6. One method is a vocational program at a high school or technical institute. Another method is apprenticeship. An **apprenticeship** is a type of training in which a worker learns the skills for a trade while working on the job. The length of an apprenticeship varies from job to job. Written and performance tests are used to determine how well the apprentice has learned the trade.

Some jobs require a degree from a college or university. Many professions require this type of education. For instance, most high school teachers and engineers must have a college degree. In addition, it is sometimes necessary to obtain training beyond a college degree. For example, lawyers and medical doctors require advanced training.

The requirements of most jobs are constantly changing. Also, your interests will likely change over time. This is why it is important to keep in mind that training often takes place throughout your career, Fig. 39-7. For instance, you might receive training in a vocational program and find a job. After a few years, you may develop an interest in another job in the company. If the job requires additional training, the training might be available within the company or outside education may be required. In another instance, suppose your company acquires new equipment with which you are unfamiliar. Most likely, you will require additional training in order to do your job properly.

THE JOB HUNT

Where do you look for a job? The answer will vary depending on your qualifications and the type of job you desire. Usually, a combination of sources is your

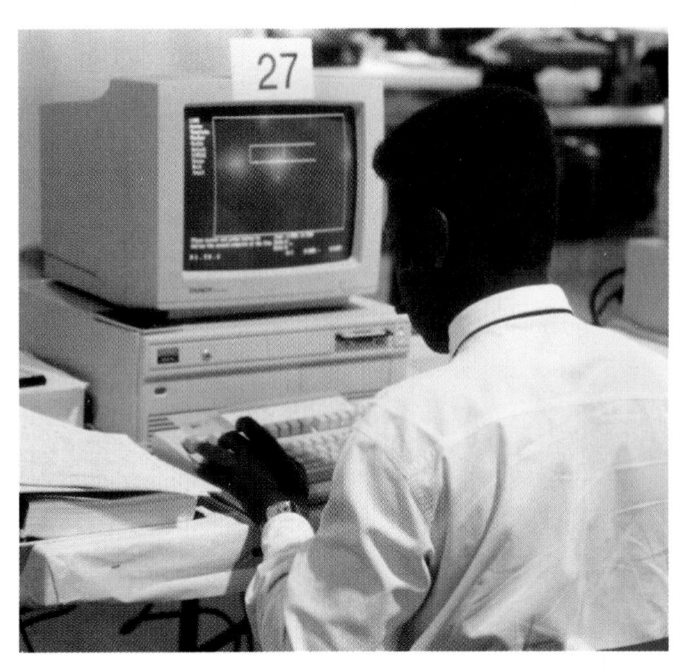

Fig. 39-7. Training often takes place throughout your career. As new developments take place, workers must keep up with the new technology. (Bud Smith)

best approach. Close acquaintances, such as friends or relatives, may be aware of available jobs. The classified section of the newspaper has a "Help Wanted" column that lists available jobs. The yellow pages of the telephone directory has lists of businesses within your area of interest. These can be contacted to see if any openings are available. You can also look for "Help Wanted" signs in windows of local businesses.

A variety of placement services may be used to help you find a job. Many schools have placement counselors to assist you. In addition, there are state employment agencies in every state. Private employment agencies charge a fee for helping you find a job. The fee is paid by the job seeker, employer, or a combination of both. Do not sign any contract with an employment agency until you read and understand all the terms of the agreement.

PREPARING INFORMATION ABOUT YOURSELF

Before you begin applying for jobs, you need to write down all the important facts about yourself. You can do this on a **personal information sheet**. The format can vary, but the following information should be included on your personal information sheet:

- *Name, address, telephone number.*
- *Social security number.*
- *Education and dates.* Begin with the most recent education. If you received good grades, you may want to mention your grade point average.
- *Employment record and dates.* List your work experience starting with the most recent employment. Don't forget to include any volunteer work.
- *Special honors and activities.* This section allows you to highlight your achievements and interests.
- *References.* These are persons the employer can contact about your character. Do not list relatives as references. Former employers, club advisors, teachers and religious leaders are better choices. Before listing someone as a reference, be sure to ask for her or his permission to do so.

Some potential employers may request that you send a **resume**, Fig. 39-8. This is similar to the personal information sheet, but it is carefully typed and copies are made. Many printing companies can assist you in preparing and printing resumes at a reasonable cost.

LETTER OF APPLICATION

If you are applying for a job through the mail, a letter of application should accompany your resume. Use a business letter format, and explain to the employer how your skills are right for the available job. A sample letter of application is shown in Fig. 39-9.

A good letter of application is brief and to the point. It should be sent to the person within a company who has the authority to hire you. If you do not know the name and title of the person to whom a letter of application should be written, call the company to get it. Be sure to ask about the correct spelling of the name.

JOB APPLICATION

Most employers will ask you to fill out a job application. In order to do this correctly, you can refer to your personal information sheet or resume.

Before you begin writing, read through the entire application form. Write as neatly as possible. Give complete and accurate information. Never lie. Respond to every question. If a question does not apply to you, draw a line through the space. If you simply leave it blank, the employer may think you carelessly skipped over the question.

THE JOB INTERVIEW

An interview is a way for the employer to learn more about you. It also gives you a chance to learn more about the job and the company.

Prior to the interview, learn more about the employer. List any questions you might have. Talk to people who work for the company or to people who know

MARSHA RICHARDSON
5103 Mansfield Lane
Shawnee, Kansas 66203
913/555-7463

JOB OBJECTIVE	Entry-level job as an editorial assistant.
EDUCATION	Shawnee Mission Southeast High School, Shawnee, Kansas Expected date of graduation: May 19xx Related courses: Communication Systems, Journalism I and II. Grade point average: 3.5 (A = 4.0)
WORK EXPERIENCE	Paper Carrier, Shawnee Sun Newspaper, from June 19xx to present. Publicity Assistant, Shawnee Community Theater Group, Summer, 19xx. As a volunteer, I produced posters and advertisements announcing productions. I also wrote and proofread program copy.
HONORS AND ACTIVITIES	First-place news story award in Kansas High School Publications Competition sponsored by Kansas State University. News editor and photographer for the school newspaper, The Herald. Member Quill and Scroll, served as vice president. Member of Student Council. Member of softball team.
INTERESTS	Photography, writing, swimming, sewing.
REFERENCES	Available upon request.

Fig. 39-8. A well-written resume can help you to make a good impression on an employer. A resume is your advertisement of yourself.

others who work there. The public library also contains valuable information about general types of business.

To make a good impression, you will want to look your best from head to toe. Dress appropriately for the interview. As a general rule, wear clothes that are equal to or a step above what you would wear on the job. Clothes should be clean and pressed. Hair should be clean and neatly styled.

Always go to an interview alone. However, there are several items you should take with you. These include a pen, a pencil, and a copy of your resume or personal information sheet. If you are applying for a job as an illustrator, writer, or photographer, you may want to bring along samples of your work.

Check the date, time, and place of the interview. Plan to arrive for the interview a few minutes ahead of time.

Greet the interviewer with a firm handshake. Do not chew gum or smoke. Try to look and act interested and alert. See Fig. 39-10.

5103 Mansfield Lane
Shawnee, KS 66203
May 3, 19xx

Sharon Skivers
Personnel Director
Cahill Publishing Company
5355 Wilson Road
Shawnee, KS 66203

Dear Ms. Skivers:

Through Mr. Donald Zimmer, Guidance Counselor at Shawnee Mission Southeast High School, I learned that your company plans to hire a part-time editorial assistant. I would like to apply for this position.

The communication technology and journalism courses I am taking in high school have fascinated me. I work on the school newspaper, The Herald, as the news editor and a photographer. With my education and experience on the school paper, I feel confident that I could perform well as an editorial assistant.

May I have an interview to discuss the job and my qualifications in greater detail? I can be reached after 4 p.m. at 555-7463. I will appreciate the opportunity to talk with you.

Sincerely,

Marsha Richardson

Marsha Richardson

Fig. 39-9. A letter of application should be written in business letter format, similar to this one. It should draw attention to your qualifications for the job for which you are applying.

Although interviews vary, the following suggestions may be helpful:

- Be pleasant, self-confident, polite, and honest during the interview. Look directly at the person asking questions.
- Listen carefully and speak clearly and slowly, using good grammar. Don't brag, but don't be bashful about your accomplishments.
- Don't criticize past employers during the interview. Also, do not mention personal or financial hardships to make the employer feel sorry for you. You might get sympathy but not the job.
- Do not be afraid to admit that you do not know something. However, let the employer know you are very willing to learn.
- When it is time for you to ask questions, ask questions that show you are interested in the entire company and being part of the organization. Ask about salary and fringe benefits. However, avoid asking questions concerning breaks and lunch hour. These

Fig. 39-10. If you were an employer, would you hire this job applicant? Why or why not?

types of questions may give the impression that you are most interested in finding out how little you have to do to "get by."

- Look for clues that the interview is over. Find out when you will be notified about the job. Thank the interviewer and leave.
- Write a brief thank-you note after the interview.
- If you do not hear from the company by a certain date, a follow-up visit or phone call can be made to find out how things are progressing. Never wait for only one job. Rather, continue searching during the waiting period.
- If you don't get the job, don't be too discouraged. Try to keep a positive attitude. Think of the interview as a learning experience.
- Following the interview, evaluate your performance during the interview. This will help you do even better in your next interview.

THE JOB OFFER

When you are offered a job, you should find out about working hours, the salary, and benefits. **Benefits** are extras that a company provides. Some examples are:

- Insurance. This includes health, dental, and/or life insurance.
- Retirement plan. This is a plan for providing you an income after retirement. Many plans require both the employer and employee to contribute together to the fund each pay period.
- Vacation and sick leave. This is time off work. Find out whether the company pays for vacation or sick days. Companies vary in their policies concerning vacation and sick days.
- Advancement. Find out what the opportunities are for promotions and salary increases.

After finding out about the job, make a decision to accept or reject the offer. If you want to consider the offer, ask if you can have a few days to consider it. If this is allowed, then let the interviewer know what day you will call. Make the call whether you decide to accept or reject the offer. Always be courteous.

SUCCEEDING ON THE JOB

The qualities that you project in an interview are also important in job success. A suitable appearance, good attendance, and responsibility are important in most jobs. Honesty, enthusiasm, and a positive attitude are also helpful.

Work hard during working hours. Cooperate with co-workers. You do not have to be best friends, but getting along on the job helps you and the company to be more productive. Avoid spreading gossip about co-workers. Those persons that delight in this activity are probably talking about you as well. In addition, gossip is often false and unkind information.

PAY INCREASES AND ADVANCEMENT

Pay increases and advancement are often the rewards for doing a job well for an extended period of time. You can increase your chances of promotion by finding out about other parts of the company and how you might fit in with your experience and skills. If additional education is needed, you will need to think about taking additional classes.

If you are interested in a supervisory job, you should be interested in working with people. You need to be able to give directions to others and to be fair, Fig. 39-11. Some management skills can be learned through

Fig. 39-11. People who advance to management positions are often those people who have demonstrated that they can perform the job well and get along with co-workers. (Geo. A. Hormel & Co.)

classes. However, companies may initially choose managers by observing how they get along with other employees on the job.

It may not always be in your best interest to accept an advancement. If you have job security and enjoy your present job, you may wish to consider remaining where you are, even if it means that your pay won't increase.

SUMMARY

This chapter has covered some helpful information in finding a job and succeeding on the job. However, it is not realistic for you to assume that you can make a career choice after reading one chapter. Right now, you may not know what career to choose, but you can develop the skills that will help you succeed in any career.

WORDS TO KNOW

All of the following words have been used in this chapter. Do you know their meanings?

abilities
apprenticeship
aptitudes
benefits
Dictionary of Occupational Titles (D.O.T.)
interests
Occupational Outlook Handbook
personal information sheet
personality
resume

REVIEWING YOUR KNOWLEDGE

Please do not write in this text. Write your answers on a separate sheet.

1. Student organizations are worthwhile because:
   a. They help you to learn more about the community in which you live as well as employers.
   b. They help you to develop responsibility and leadership skills.
   c. You really don't have to participate in order to be a member of a successful group.
   d. Both a and b above.
2. List four major factors to consider when thinking about what job would be best for you.
3. Name two sources of information, by the U.S. Department of Labor, that can aid you in a job search.
4. A(n) ________ is a type of training in which a worker learns the skills for a trade while working on the job.
5. True or false? On a personal information sheet or resume, education and employment should be listed starting with the most recent education or employment.
6. When addressing a letter of application, if you do not know the name and title of the person to whom the letter of application should be written:
   a. Guess.
   b. Call the company and get it.
   c. Address it to "Sir" or "Madam."
   d. None of the above.
7. When filling out a job application form:
   a. Write neatly.
   b. Give complete and accurate information.
   c. Respond to every question.
   d. All of the above.
8. When interviewing for a job:
   a. Take a friend along to keep you company.
   b. Chew gum to relax.
   c. Greet the interviewer with a firm handshake.
   d. Ask about breaks and lunch hours right away.
9. ________ are extras that a company provides such as insurance, a retirement plan, and vacation and sick leave.
10. True or false? Pay increases and advancement are often the rewards for doing a job well for an extended period of time.

APPLYING YOUR KNOWLEDGE

1. Join a student club and become involved. Encourage members to pursue club activities that will help in their future employment.
2. Learn to use the Dictionary of Occupational Titles. Look up requirements of jobs of interest to you and list them.
3. Prepare a resume and personal information sheet. If possible, do this on a word processor so it can be updated.
4. Perform mock interviews with a friend. Videotape the interviews. Then eveluate and review them citing areas for improvement.

40

CHAPTER

CAREERS IN COMMUNICATIONS

After studying this chapter, you will be able to:
- *List the six major career areas in the field of communications.*
- *Explain the effect of change on jobs.*
- *Identify examples of jobs in the field of communication technology.*
- *Describe major categories of jobs that often exist within companies.*

Rapid changes are taking place in technology. These changes and the transition from an industrial-based society to an information-based society will have an impact on your career choices. The importance of communication in all jobs is growing. It is reported that by the year 2000, about 80 percent of all jobs will deal primarily with information. If you have an understanding of communication technology, you should have a good chance of obtaining one of these jobs.

Communication is an important aspect of almost any job you can name. Obvious examples include teachers, secretaries, lawyers, stockbrokers, managers, and sales personnel. These people spend a great deal of time working with information, Fig. 40-1. However, people such as auto technicians and factory assembly workers also spend much of their time communicating. Service and parts manuals must be referenced, and various machines are used that have communication functions such as diagnostic equipment and computer-controlled machinery.

Fig. 40-1. Communication is an important part of a teacher's job. (Bowling Green State University, Visual Communication Technology Program)

Fig. 40-2. More jobs related to space will be available in the future. (NASA)

Jobs will continue to change in the future along with advancing technology, Fig. 40-2. This trend will continue, resulting in many new job titles in the future.

MAJOR CAREER AREAS IN COMMUNICATION

Reading a magazine, going to the movies, or talking on a car phone, are all made possible through the work of workers in communications careers.

Most careers in communications require training at a college, vocational technical institute, or training through an apprenticeship. Courses in art, graphic arts, photography, electronics, mathematics, speech, and writing can also help you to prepare for a career in communications.

As you have studied in this text, communications involves these major areas:

- Drafting and CAD.
- Photography.
- Graphic arts and desktop publishing.
- Audio communication.
- Moving images.
- Telecommunication.

Fig. 40-3. A career choice is not always an easy decision.

Making a career choice is like a matching game. Matching yourself with the career that is right for you is the object of the game, Fig. 40-3. In the following sections, you will explore various careers. The careers described are by no means a complete listing of the many careers available in communications. They are simply examples of the many careers available. You can find out further information about communications careers by visiting the library and talking with your guidance counselor.

DRAFTING AND CAD

Whether done on a drawing board or at a CAD station, drafting skills are necessary for many jobs within business and industry. See Fig. 40-4. **Drafters** utilize

Fig. 40-4. Drafting skills are being used in this job. (IBM Corp.)

drawing tools and computers to prepare drawings of products. **Industrial designers** help create the product as well as draw it. Engineers also design products and work with designers and drafters to draw them.

Technical illustrators prepare pictorial drawings for publication. Inking, airbrushing, and shading techniques help the illustrator produce realistic drawings.

PHOTOGRAPHY

Photography skills are important in many occupations. **Commercial photographers** often work with advertising agencies. **Photojournalists** work in the publishing industry. They produce photos for newspapers, books, and magazines, Fig. 40-5. Some photographers work for audio-visual services to produce slides and filmstrips for instructional presentations, Fig. 40-6. In addition, photography skills are important in the television and motion picture industries.

GRAPHIC ARTS AND DESKTOP PUBLISHING

The field of graphic arts and desktop publishing is involved with the high volume reproduction of images such as those that appear in books and magazines, Fig. 40-7. A variety of occupations is available in the writing, design, printing, and distribution of these products, Fig. 40-8. Writers, editors, electronic scanner operators, press operators, graphic designers, and technical sales representatives are a few examples, Fig. 40-9. In addition, suppliers or merchandisers furnish the resources for printing such as paper and ink.

Fig. 40-6. A person employed in an audio-visual career often uses slides to communicate ideas. (Polaroid Corp.)

Fig. 40-5. A photojournalist often covers major news events.

Fig. 40-7. This printer is about to inspect a printed sheet. (United Telecom)

Fig. 40-8. These people are designing a product that will be printed. (Xerox Corp.)

Fig. 40-10. Many people, such as engineers and technicians, are working behind the scenes so that this person can use his cellular phone. (Motorola, Inc.)

Fig. 40-9. Employees in this stripping department are preparing flats that will be used to make printing plates. (AM International, Inc.)

AUDIO COMMUNICATION

Many companies specialize in sound communication, Fig. 40-10. For example, radio stations need **audio technicians** to operate and maintain the equipment for broadcasting. Announcers, script writers, and music directors are also needed. The entire field of radio and television is known as the **broadcasting industry**.

Audio-visual personnel combine images and/or sound for presentation. These are sometimes called **media specialists**. For example, audio tapes are combined with slides and filmstrips for presentation.

MOVING IMAGES

Many positions are available that require a knowledge of moving images. For example, media specialists within a company prepare videotapes for training purposes. The television and motion picture industry also have a wide variety of available positions, including directors, camera operators, and video engineers, Fig. 40-11. Script writers, sound technicians, and lighting technicians are other examples, Fig. 40-12.

TELECOMMUNICATION

The field of telecommunication specializes in sending messages over long distances. People are needed for designing, installing, and maintaining telecommunication systems, Fig. 40-13. These systems not only transmit the human voice. Some are also involved in the transmission of television signals, radio signals, newspapers, and other data. Technicians with an electricity and electronics background are needed for many of the jobs in this field.

JOB CATEGORIES

Since specific jobs may change in the future, it is helpful to think about categories of jobs within a business or industry. Often, the major categories will remain the same, while specific jobs may change and merge.

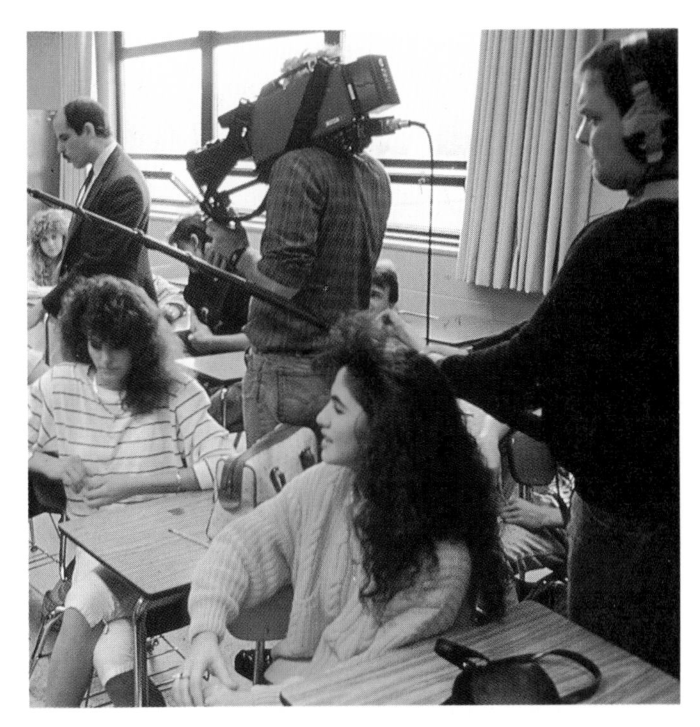

Fig. 40-11. A professional video production requires the work of several individuals.

Fig. 40-12. Script writing is an important job in the production of television programs and motion pictures. (Bowling Green State University, Visual Communication Technology Program)

Fig. 40-13. These workers are installing phone cables. (U.S. Sprint)

Fig. 40-14. This scientist is examining optical fibers that are being tested for undersea conditions. (AT&T)

Scientists and **engineers** within a company conduct research to develop better products and better ways of producing the products, Fig. 40-14. **Production personnel** include writers, technicians, and crafts persons, responsible for producing the product. **Merchandisers** provide the raw materials needed for the products. **Marketing directors** and **sales representatives** advertise, sell, and distribute the products. **Support personnel,** such as secretaries and bookkeepers, handle much of the company paperwork and data processing. Managers direct the people and activities in every part of a company, Fig. 40-15.

These job categories apply to industries involved in producing goods as well as industries that provide services. Of course, for smaller companies, some of these categories are combined since a smaller number of people must provide the human resources for the entire company.

Fig. 40-15. Managers provide leadership in a company. (Hewlett-Packard Corp.)

SUMMARY

This chapter has provided an overview of some of the career options available in communications. The career areas explored included drafting and CAD, photography, graphic arts and desktop publishing, audio communication, moving images, and telecommunication. As communication technology continues to change, the jobs will change. However, a broad understanding of the communication field should enable you to find it easier to change as the jobs change.

WORDS TO KNOW

All of the following words have been used in this chapter. Do you know their meanings?

audio technicians
broadcasting industry
commercial photographers
drafters
engineers
industrial designers
marketing directors
media specialists
merchandisers
photojournalists
production personnel
sales representatives
scientists
support personnel
technical illustrators

REVIEWING YOUR KNOWLEDGE

Please do not write in this text. Write your answers on a separate sheet.

1. It is reported that by the year 2000, about what percentage of all jobs will deal primarily with information?
   a. 20 percent.
   b. 40 percent.
   c. 60 percent.
   d. 80 percent.
2. True or false? Communication is an important aspect of almost any job you can name.
3. Name three courses that can help a person to prepare for a career in communications.
4. List the six major career areas of communications.
5. Which of the following utilize drawing tools and/or computers to prepare sketches and drawings of products?
   a. Audio technicians.
   b. Drafters.
   c. Media specialists.
   d. Press operators.
6. The entire field of radio and television is known as the ________ industry.
7. If you are interested in a career as a telecommunication technician, a background in which of the following would be best?
   a. Electricity and electronics.
   b. Writing.
   c. Graphic arts.
   d. Photography.
8. Within a company, which of the following often conduct research to develop better products and better ways of producing them?
   a. Support personnel.
   b. Merchandisers.
   c. Scientists and engineers.
   d. Production personnel.

APPLYING YOUR KNOWLEDGE

1. Investigate a communication career of particular interest to you. Find out more about the career through research as well as by talking with people in these jobs. Share your findings with the class in a written report.
2. If possible, try to spend a day "shadowing" someone employed in the communications field. Prepare a report about your experience.
3. Search for communication technology jobs in the classified ads in a newspaper. Try to categorize the available jobs into the major career areas of communication discussed in this text.
4. Predict what communication careers may be available 20 years from now. Describe what background you might need for these careers.

41
CHAPTER

PERSONAL AND GROUP SKILLS FOR SUCCESS

After studying this chapter, you will be able to:
- *Describe personal and group skills for success.*
- *Differentiate between leadership and management.*
- *Identify some common steps needed to plan a project.*
- *Describe the importance of values in being successful.*

There are a variety of skills that you can use to be successful in communication courses as well as in other parts of your life. These skills for success can be divided into two categories: personal skills and group skills. You can be successful in what you attempt as an individual and you can be successful in working with groups. Each of these will be explored in this chapter.

PERSONAL SKILLS FOR SUCCESS

Personal skills are those we develop as individuals. These are the skills that make us competent. They include skills such as personal leadership, personal management, listening, reading, writing, and presenting.

PERSONAL LEADERSHIP

Have you ever wanted to buy something very badly, sacrificing and saving so you could finally afford it? You had a goal and by working at it, finally reached the goal. This is how we create our future day by day, and this requires personal leadership.

One definition of a **leader** is a person who has a vision of the future, and is able to influence others to follow that vision. This is important for ensuring that everyone in a company is moving toward the same goal, so leaders are very important in the business world. However, leadership is important to us as individuals as well. We have to be leaders of ourselves. Personal leadership involves having a vision or goal for yourself, and then finding a way, step-by-step, to reach this goal. This is also called following your dreams.

One of the best ways to begin practicing personal leadership is to set goals for yourself. This is your vision for what you want to accomplish. Once the goals are made, then you will think of ways each day to help in accomplishing these goals. In essence, the goals act as a roadmap, helping to guide you through your daily activities. Fig. 41-1.

Remember that even with goals, you may occasionally fail. Most people that have great success have many failures as well, and this is just a fact of life. Being able to learn from these setbacks, and not being discouraged, is an important factor in success. Having a clear picture of what we want to accomplish (our goals) helps us to see beyond setbacks and continue striving for success.

PERSONAL MANAGEMENT

After setting goals (personal leadership) the next step is organizing your life to accomplish the goals you have made. Leaders have the vision, but **managers** figure out how to reach this vision through daily activities. So, personal leadership and management go hand in hand in reaching your goals.

Personal management involves planning and organization. Planning is your "to do" list. It might be on a pocket calendar, a 3″ × 5″ index card, or computer. Not only do we need the list, we need to decide on priority. Those things on your list that will help you the most in reaching your goals should have a high priority. Obviously, there may be those day-to-day things that must be done but are not directly related to your goals. Your list is a constant reminder of what is most important.

Personal management includes time management. **Time management** is simply a way to better organize your life to accomplish what needs to be accomplished. Your "to do" list is an important part of time management. Sticking to your priorities requires careful management of your time. Throughout a day you may have many opportunities to get off-course, including watching TV and playing computer games. In some cases, other people may monopolize all your time. A natural tendency is to try to please your friends, but if this distracts you from accomplishing your own goals, then others may be managing your life. Time management involves taking control of your own time and making decisions about how that time is spent.

Even though hard work is important to success, a balance is needed between work and leisure. So, it is a good idea for your goals to include both. Work is necessary for you to reach your goals, but relaxation helps you to get reenergized so you can be more productive.

Fig. 41-1. Goals are like a roadmap, helping to guide you through daily activities.

LISTENING

Habit 5 of Stephen Covey's *Seven Habits of Highly Effective People* is "seek first to understand, then to be understood." We often are thinking about what we want to say, and so do not truly hear what the other person is saying. We assume we know what the other person will say and answer accordingly.

Complete understanding of another person's point of view comes from careful listening, Fig. 41-2. When you take the time to fully understand another person, and truly care about their viewpoint, they will normally take time to better understand your point of view. This begins with good listening skills.

READING

We can certainly learn quite a bit about the world from television, but this still does not take the place of reading. Reading requires a bit more work, but the rewards are worth it. Your comprehension of reading materials goes up as you read more, you increase your vocabulary, and you stimulate your imagination.

Textbooks are a bit different than leisure reading material such as a fiction novel. Have you ever been reading a textbook and found yourself lost or confused? This can happen because textbooks are full of facts, some of which you may not think are important to you. Here is a strategy for dealing with this type of reading material:

1. Look through the entire book to find its various parts. Where is the Table of Contents and Index? Does the book have a glossary or special tables in the back?

Fig. 41-2. Careful listening helps you understand another person's point of view.

2. Think about how the subject matter in the book could be important in your life. There may even be a section in the book that talks about this.
3. Review the Table of Contents to find out what the units and chapters are about.
4. Look at a chapter and determine its make-up. Does it have an introduction and summary? Are there questions at the end? Also, look at the level of headings in the chapter. These are the headlines. First level headings might be centered whereas second level headings might be to the left. The first level heading will provide general information and the second level heading will provide more details. There can even be third or fourth level headings.
5. Glance through the chapter you have to read and get a sense of what the chapter is about.
6. Try to stay focused on your reading and take frequent breaks as needed.

WRITING

Writing begins with understanding your subject. This may require research in the library or on the Internet, or you may already have the needed knowledge. Once you understand the subject, then you are ready to determine the best way to communicate the information to your audience.

One of the most important lessons in writing is that even great writers often have poor first drafts. So, if you write your paper in one sitting, without ever writing a rough draft, chances are that the paper will not be your best effort since it is essentially still in rough draft form.

Another important lesson is that you can overcome inertia by freewriting. In physics, we learn that a body in motion tends to stay in motion and a body at rest tends to stay at rest. So, a force is needed to overcome the tendency to stay at rest. Freewriting helps to overcome the tendency to not begin the writing process. **Freewriting** is placing information on the paper or screen quickly, without regard for spelling or grammar. Begin by freewriting an entire page. This will help to get you in the writing mode, and will be the first part of your rough draft.

There are formulas for some types of writing, but they may not work in all cases. In the case of a report, you will typically need an introduction, body and summary. However, this may not be an effective formula when writing a fiction paper for an English class. In this case you can be much more creative, and the main concern will be writing an interesting and entertaining story.

In some cases, you can write much like you speak, by using contractions (e.g. won't), incomplete sentences and slang. An example might be: "Don't be a jerk!" This type of writing is considered informal, and typically reserved for fiction. When writing reports, you need to be more formal than this. Be sure you know what your instructor expects before beginning.

When writing, think about your audience and write accordingly. If you are writing technical information, and your audience has no previous knowledge on the subject, then technical terms will need to be defined in the paper.

Like most talents, there is certainly a wide range of abilities with regard to spelling. However, there is very little excuse for spelling errors in papers, since there are dictionaries and in the case of word processors, spellcheckers. See Fig. 41-3. Grammar checking is also available on most word processors.

Writing papers will often require **references** or a bibliography. A reference list includes only those sources (books, encyclopedias, etc.) that are used in the paper. A **bibliography** includes all sources you reviewed for the paper, some which you may not have actually used in writing it. Both references and bibliographies show the reader where you obtained information for the paper, including quotes. There are different formats for references and bibliographies, depending on the type of writing.

Proper use of references will ensure that you are not blamed for plagiarism. **Plagiarism** is taking the words or ideas of others and using them as your own. This includes *paraphrasing*, which is using different words to say the same thing as another person. It also includes using a quotation from another person as your own. Plagiarized writing is often noticeable because it will

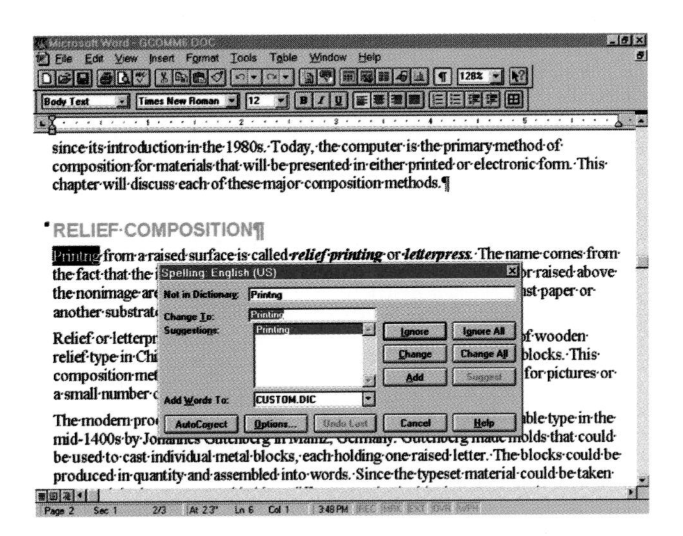

Fig. 41-3. Spellcheckers are a part of most word processing programs.

be different than the other writing in the paper. No grade is worth getting caught for plagiarism. Doing your own work helps you to learn more and you will be considered an honest individual.

PRESENTING INFORMATION

It is often necessary to make presentations, such as speeches and demonstrations. There are some simple techniques you can use to make a more effective presentation.

Prepare your presentation with your audience in mind. What are the main points you wish to get across? How can this best be done? Use language familiar to those to whom you are talking, but if you must use unfamiliar words, be sure to define them.

Visuals and other media can make your presentation more interesting, Fig. 41-4. However, text and illustrations should be large enough for everyone in the room to see them clearly.

Think about a *hook* for the beginning of your presentation. This is a technique for getting your audience interested in your topic. You might have a personal story that shows the importance of the topic. A current event from the newspaper or TV news could also be used. A less creative approach that also will work is to simply explain the importance of the topic.

It is very common to get nervous prior to speaking. However, remember that nervousness causes athletes to perform at their peak. Typically, the nervousness will cause you to be more alert, and will be a benefit in getting the presentation started.

Presenting is a bit like acting. You need to give the appearance of confidence, and maintain eye contact with your audience. You also need to dress appropriately for your audience. You must speak clearly and in a friendly manner. If your voice changes when you are nervous, make a conscious effort to use the same voice that you use in daily conversation, but speak a bit louder.

Fig. 41-4. Visual aids, like the one this teacher is using, make presentations more interesting.

GROUP SKILLS FOR SUCCESS

This section has information on how to do a better job in groups, either as a member or as a leader of the group. Being part of a group is a great way to build your skills in working with people. In school, you will often need to work in small groups to complete a project. You should also consider joining other groups, such as student clubs. The more you practice your group skills, the better you will be at working in groups.

BEING A TEAM PLAYER

Being a good team player simply means doing your fair share as part of the team. It also means doing what you can to make sure that all on the team are successful. Here are some ideas for being a successful member of a team:

- Be open to ideas other than your own. Carefully listen to other viewpoints and respect other's views.
- Get work done on time so that you are not a burden to the entire team. If you need to miss a meeting, be sure you let team members know ahead of time if possible. Spend extra time making up for work missed.
- Work equally well as a team member or leader. Be the kind of member you would like if you were a team leader, and the type of leader you would like if you were a team member. Experience on teams is helpful regardless of the position.

WIN-WIN

One of the most important concepts when working with others is the concept of **win-win**. This means that you try to find a solution that allows both people to be successful, Fig. 41-5. Some people do this more naturally than others, but everyone is capable of thinking this way.

Win-win actually begins prior to a meeting. You need to prepare yourself to believe in the possibility of solutions that include the thinking of others. So, good listening skills are important in finding win-win solutions.

Fig. 41-5. The concept of "win-win" is an important one for a leader to understand. It involves finding a solution that allows both sides to be successful.

GROUP LEADERSHIP

As mentioned before, we are all leaders and managers with our personal projects. However, there may also be instances where you will need to be a leader of a group. Actually, most leaders of small groups are both a leader and manager. What is the difference? Leaders can have a variety of job titles, such as manager, supervisor, or CEO (chief executive officer), or may not have a formal title at all! Regardless of title, they have the ability to inspire others to follow and get the job done. They help the group develop an overall vision or goal, and ensure that everyone has a role to play in accomplishing this goal. Essentially, they spend most of their time trying to help others succeed. Leaders normally have good interpersonal skills, they think win-win, they truly want others to be successful, and they show their appreciation for the work being done by others.

In a business, managers have the job title of manager or supervisor, and their job is to get everything and everyone organized to get the job done. They may or may not be able to inspire others to get things done (a leadership skill), but they have good organizational skills. Managers do not necessarily have to be leaders, although this is certainly helpful.

Let's look at a simple example of the difference in management and leadership. A family decides to go on a much needed vacation. The *leader* inspires everyone to take the trip, while the *manager* organizes the trip (plans where to stay each day, budget, etc.). Again, this may or may not be the same person.

You might think that there is only one type of personality that can be a manager or leader, but this is not the case. A wide variety of people serve in these positions. However, they may have different styles in leading. The important thing is not the style used, but the results.

Your goal as a group leader is to do the best that you can at being the leader and manager. What can you do to be successful? Most of these ideas have already been covered. Think win-win, help others be successful, show others your appreciation for a job well done, and keep everything and everyone organized and moving toward the goal. Also, be sure to delegate work, so you are not doing everything yourself.

LEADING A MEETING

Meetings can be formal or informal. They can be planned in advance or be done spur of the moment. Some use **parliamentary procedure**, which are formal rules for the meeting. If people in the meeting say "I make a motion to...", this is an indication parliamentary procedure is being followed.

Here are some ideas for more effective meetings. Pick those that you think are best for the type of meeting you are having.

- Reserve a room and send out an agenda in advance, being sure to include time, day of week, date, and location. It is also important to let everyone know the ending time of the meeting, so they can schedule other activities around the meeting.
- If brainstorming or problem-solving is needed during the meeting, learn about this prior to the meeting. You may wish to write down the steps on the board or provide this as a handout so everyone is aware of the procedure being used.
- Arrive at the meeting place early and set up the room so it is comfortable for everyone. For small groups, a circle arrangement works very well.
- Start the meeting by having everyone introduce themselves (if necessary) and going over what is on the agenda. Be open to changing the agenda if the group deems it to be necessary.
- Ensure that everyone that wishes to has an opportunity to speak and that people are polite to one another. When the conversation drifts from the agenda, your job will be to guide everyone back to the topic. Do not attempt to silence all conversation outside the topic, but keep it to a minimum. Some conversation on other topics actually helps members of the group to get to know one another, and makes meetings more fun.
- Think win-win as you lead the meeting, and be cautious about trying to promote your own ideas to the exclusion of others.

- Try to end the meeting at the scheduled time. Before leaving, schedule the next meeting so everyone can get it on their calendars.
- If the meeting requires parliamentary procedure, find a simple book on the topic at the library and read it before the meeting. Often, someone else at the meeting will help you with parliamentary procedure if you have not done it before.

PROJECT PLANNING AND MANAGEMENT

In order to complete projects correctly and on time, project management is required. This includes all the techniques used to take the project through to completion. There are special techniques and computer programs that can help with this task, but for most simple projects, this is not needed. Here are some steps that should be considered when starting a project.

1. **Establish a goal and objectives.** Write a general statement about what needs to be done (goal) and if needed, more specific statements about how to reach that goal (objectives). The goal might be to develop a 5 minute multimedia program, and one objective might be to write an outline for the program in a word processing program.
2. **Make a timeline.** List the steps needed to complete the project, including as much detail as needed. Indicate when each step needs to be completed in order to reach the deadline.
3. **Establish specifications.** The **specifications** tell you the level of quality expected for the project. Fig. 41-6. Examples include the quality of writing, size of item, workmanship, safety factors, etc. Specifications need to be carefully written, since success of the project is usually determined by whether it meets the specifications.
4. **List needed resources.**
5. **Decide on how you will evaluate the finished project.** Usually this will be based on your specifications.
6. **Divide up responsibilities.** It is helpful to have a project leader whose job is to make sure work is evenly divided and that all work is being accomplished.

SUCCESS AND VALUES

This chapter has been about being successful, as individuals and in groups, but success was never completely defined. Success can mean different things to different people. When we go to a movie and see a star on the screen who can command millions of dollars a picture, we often think of this as success. Or, when we hear about a CEO of a large corporation having her name placed on a new building, we may think of this as success. In both cases, we are defining success in terms of having a great career, money, and fame. See Fig. 41-7. Although these things are certainly nice, are they truly all we are after?

True success begins with having good values, and following these values in the way you live your life. What does this mean? Think about the question, what do I value? In other words, what is most important in my life? These are your **values**. We can choose to place value on money and fame alone, or choose to value honesty, courage and justice. People with good values ultimately triumph. They may or may not have fame and large sums of money, but they are living an admirable life.

An example may help. Betty is a great public speaker, and is hired by a company in South America that wants to improve its public image while clear-cutting thousands of acres of rainforest, with no plan for

Fig. 41-6. The quality of this space satellite development project will be determined by how it meets specifications. (NASA)

Fig. 41-7. Material possessions are nice to have, but true success is much more than just the things that money can buy. (Lexus)

reforestation. Betty uses her talents to convince others that the company is doing the right thing although she strongly believes the company is wrong. She makes a great living, and her life fits some definitions of success. But is this real success?

Betty's ability to convince others may cause valuable rainforest resources, such as future medicines, to be lost. She has professional success, but not personal success, because she is not true to her own values. She values the forest, but is willing to go against her own values to do her job. We might say she "sold out."

So, what is true success? True success begins with those things we often do not see, such as integrity and perseverance. It has to do with deciding what the best values are for living our life and following those values.

Should we place a value on making a good living and having nice things? Absolutely. This should be something we all value. However, personal success should involve obtaining these things in a manner that is consistent with other values, such as honesty and integrity.

Over thousands of years, great thinkers have given us information on values and their importance in our lives. Here are a few examples:

- Value yourself as an important and unique human being. If you do not value your own health and abilities, it is more difficult to help others.
- Have respect for your family and other loved ones.
- Have compassion for other human beings.
- Be honest and fair with others. Treating others as we would like to be treated builds relationships.
- Take responsibility for your own learning and success. We often have a tendency to blame others for not accomplishing a task, but normally we control what we accomplish.
- Care about the planet and leave a small footprint (minimum damage). Each of us creates pollution on Earth by using a variety of manufactured products. This impacts the natural environment as well as other people. Minimize the damage by being a careful consumer.
- Think win-win. Life is not always a competition. Look for solutions that create success for all.
- Be a good team player. This involves doing your share, and letting others know you appreciate their work.

SUMMARY

This chapter was a compilation of techniques that can help you to be successful, both as an individual and in groups. One of the most important steps you can take for success is to decide on your goals (personal leadership), and work to achieve them (personal management). It has been said that if we do not create our future someone else will create it for us. In striving to meet your own goals you are creating your own future.

There are techniques for working more effectively in groups and these have been described. You should seek out group opportunities since this will help you be a more effective group member and leader. The concept of win-win is extremely important when working in a group.

Project planning steps were described in this chapter. Planning is critical for a successful project. It ensures that you know the steps to complete the project and the time needed at each stage to meet the deadline. Planning should also include specifications for the project so you can evaluate the finished project based on the requirements.

Although there are special techniques for individual and group success, you need to be sure that this success is aligned with your personal values. If we have great public success that is not in alignment with our values, then we have not been personally successful.

WORDS TO KNOW

All of the following words have been used in this chapter. Do you know their meanings?

leader
manager
time management
freewriting
references
bibliography
plagiarism
win-win
parliamentary procedure
specifications
values

REVIEWING YOUR KNOWLEDGE

Please do not write in this text. Write your answers on a separate sheet.

1. True or false? Some leadership does not require other people.
2. True or false? Personal goals and vision are not the same thing.
3. True or false? A hook gets people interested in a presentation.
4. A person that can influence others to take action is known as a:
   a. Leader.
   b. Manager.
   c. Supervisor.
   d. CEO.

5. "Seek to understand, then to be understood" refers to what personal skill?
   a. Time management.
   b. Reading.
   c. Listening.
   d. Writing.
6. Which of the following is the best reason for using freewriting?
   a. Writing a rough draft.
   b. Overcoming inertia.
   c. Greater creativity.
   d. Disregard for spelling and grammar.
7. Finding a solution that is satisfactory to all is known as (best answer):
   a. Success.
   b. Leadership.
   c. Being a team player.
   d. Win-win.
8. In project planning, you write __________ to describe the level of quality expected in the finished project.
9. Describe two things a person can do to be a better time manager.
10. What is the difference in a reference list and bibliography?

APPLYING YOUR KNOWLEDGE

1. Discuss the effect of values on success. Give specific examples from current events that illustrate how personal values can be a detriment to success.
2. Divide into small groups. Decide on at least five personal goals that would help any person in the group be more successful.
3. Review a textbook, using the strategies outlined in this text. Discuss how this helped you better understand the textbook.
4. Divide into small groups and find several issues on which you disagree. Try to find a win-win solution.

42

CHAPTER

IMPACTS OF TECHNOLOGY

After studying this chapter, you will be able to:
- *Describe technology trade-offs.*
- *Identify types of technology impacts.*
- *Explain how technology assessment is done.*
- *Describe several technology forecasting techniques often used to predict future trends.*

Technology evolves from a need to extend human abilities. For instance, a telephone allows you to communicate all over the world, and a car allows you to travel long distances. You probably rarely think about the technology in your life and its impact. Perhaps, you may take these everyday technologies for granted because they have evolved over many generations and are commonplace in society. In fact, often only the newest technologies or technological problems capture attention, Fig. 42-1.

Fig. 42-1. It is usually the newer technologies that capture our attention. As space exploration continues, communication technology will play a vital role. (NASA)

Although most technologies have been developed for their positive benefits, negative consequences can also be a result. For example, cars help us travel over long distances, but they also pollute the air and require a nonrenewable energy source, gasoline, for operation. As this example reveals, there are often trade-offs in using technology. **Trade-offs** are things that are given up or compromised in return for others. It is important to understand these trade-offs in order to make wise decisions concerning technology.

The positive and negative effects of technology can be called the impacts of technology. Since technology is controlled by people, it is important to understand these impacts so that informed decisions can be made about the use of technology.

IMPACTS OF COMMUNICATION TECHNOLOGY

Think about how technology impacts the way you communicate. Examples surround you, such as books, magazines, radios, televisions, video players, computers, telephones, and even clocks. All of these have been made possible because of the evolution of technology. As with any technology, there are both advantages and disadvantages related to communication technology.

A good way to think about technological impacts is to move back in time and compare situations then and now. Suppose someone was injured in a home 150 years ago. Without a telephone, medical help was probably summoned by riding a horse into town. In bad weather conditions, a person probably was not able to travel at all, so treatment occurred at home. Because of this time delay, people died. In addition, treatment from doctors was much less advanced because of the lack of medical information at that time.

Now think about communication today and the reduced time needed for sending a message. If an injury occurs at home, a phone is available on which you can dial an emergency number for help. A computer at the telephone company assists in placing the call. If needed, the person who answers will immediately call the hospital for assistance. A medical team arrives and hooks up monitors to the patient to provide information about his or her condition. If needed, the team can use a two-way radio in their vehicle to obtain further instructions from a doctor. On the way to the hospital, a drug or procedure may be administered that is possible because scientists from different countries could communicate about medical research. As you can see, communication is very important through this entire process, Fig. 42-2.

The previous example shows how communication technology has had a positive impact on human lives. However, it is also necessary to be aware of negative impacts. For example, the term **information overload** has been used to describe the problem of bombarding people with more information than they can review and comprehend. Much of this information is in the form of advertisements that are carefully designed to encourage people to buy a product or service. It can be a problem to determine which of the information sources are most accurate and important. Another problem is the reduced amount of personal communication because of communication devices such as television, radio, and video players, Fig. 42-3.

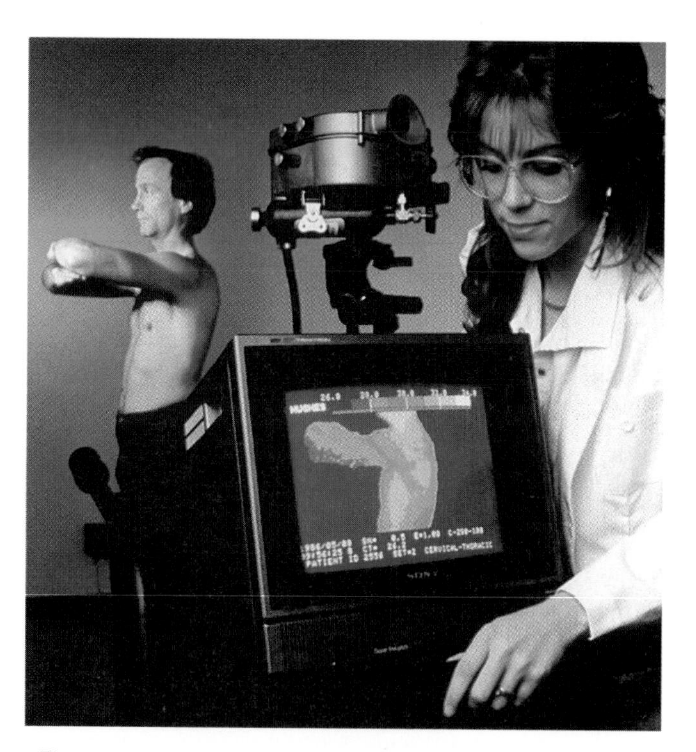

Fig. 42-2. Medical thermography, a spin-off from the space program, is used to identify temperature patterns in the body. Nerve function and location of pain can be identified in this manner. (NASA)

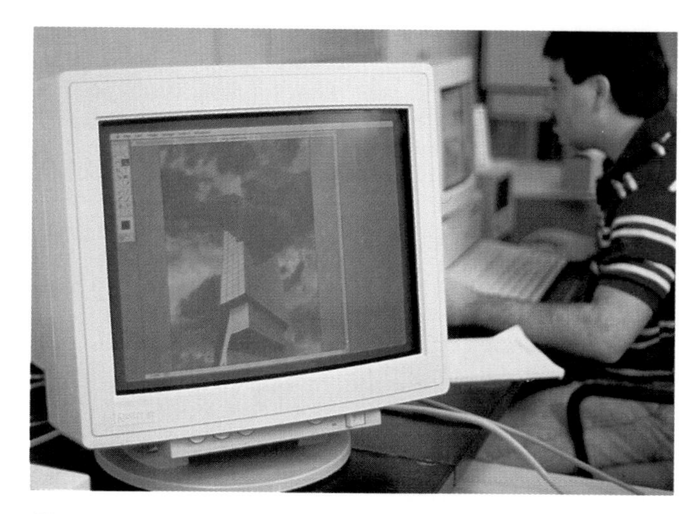

Fig. 42-3. One potential result of using some communication devices is reduced personal contact. However, utilization of some communication devices, such as computers, also helps us to work faster, thus possibly providing time for more personal contact. (Bowling Green State University, Visual Communication Technology Program)

TYPES OF IMPACTS ON COMMUNICATION TECHNOLOGY

Communication technology impacts almost every facet of your life. It is important to have some knowledge of these impacts. This is so you can make careful decisions about how communication technology is used. Some of the major categories of impacts include:

- Environmental.
- Social.
- Economic.
- Personal.

ENVIRONMENTAL IMPACTS

Almost daily, you can find news stories regarding environmental problems related to technology. Automobile exhaust contributes to unhealthy air in urban areas. Industrial waste is accidentally dumped in the ocean causing a problem with marine life. Forests are being destroyed for wood and paper products contributing to the greenhouse effect, Fig. 42-4. Acid rain, a result of pollution, causes additional problems for forests. These examples serve to demonstrate that there are obviously trade-offs for technological sophistication.

Communication technology has had many positive effects on the environment. People are much more aware of environmental problems because of the media. This, hopefully, results in decision making that reduces these negative impacts. These decisions might be in the form of voting for a candidate, recycling efforts, or carpooling, Fig. 42-5. In all of these cases,

Fig. 42-4. The manufacture of paper products can deplete our forests without safeguards, such as replanting and recycling. (Sonoco Products Co.)

communication technology helps us to make the best decisions by providing more information and reducing the time needed to receive the information. Therefore, even though there may be negative impacts with new communication methods, one beneficial aspect is the power of communication to provide information on a topic quickly.

Communication technology can help to reduce the use of nonrenewable resources. One example is the use of the telephone for conference calls. By using the telephone, the need for people to drive long distances for meetings can be minimized. Another example is the use of CAD and CAE (computer-aided engineering) systems to design and test more energy-efficient devices, such as cars and buildings, before they are actually constructed using actual materials.

Fig. 42-5. These products are made from recycled paper. (Sonoco Products Co.)

SOCIAL IMPACTS

Many aspects of society are impacted by technology. How people live, work, and play are all influenced, to some extent, by communication technology, Fig. 42-6.

There is a concern that radios and televisions are taking the place of conversation in families, thus weakening the family influence. In addition, there is concern that the stories of crime and violence on television may contribute to distrust among people. Another concern is the potential lack of privacy related to using electronic communication devices such as the telephone and credit cards.

Fig. 42-6. Communication technology is evolving to make communication with other people more convenient and efficient. (Motorola, Inc.)

Just as negative influences of communication technology exist, there are positive influences as well. People have become more aware of other societies through television that has, in effect, made the world an "electronic family." Increased awareness of other countries and their cultures may ultimately help in bringing about world peace and better living conditions. Television can be used for teaching courses on topics such as learning to read or a new hobby.

ECONOMIC IMPACTS

Economics, the study of finance, is another major area where technology plays an important role. Most new technologies are introduced so that a profit can be

Fig. 42-7. Consumer dollars are an important factor in the survival of many technologies.

made. This technology might be in the form of a new product for consumers or a device for making a product at a lower cost. If the technology fails to make a profit, it is often discontinued. This is why consumers are very important to the survival of many technologies.

Since businesses must compete for consumer dollars, there is pressure to produce low cost products and ones that appeal to buyers. The result is that there is a constant supply of new products and services at reasonable prices. See Fig. 42-7.

One of the major products in today's society is information. Information is passed on through the use of media broadcasts, newspapers, and advertisements to name just a few. Consumers are exposed to hundreds of ads each day. Each one has been carefully designed to entice consumers to buy. Since people normally have more choices available than money, decisions must be made. These decisions can be difficult, and some people may spend more than they have. An important skill for consumers is being able to learn how to distinguish between wants and needs. *Needs* might include clothes and food, while *wants* include television, eating out, and vacations. Wise consumers are careful about how they spend their money.

PERSONAL IMPACTS

Technology helps to bring about change. This change often requires adjustment. During the industrial revolution in the United States, people had to adjust to the crowded conditions in the cities and to jobs that often provided little personal satisfaction.

Today, as in the past, change continues to occur. In fact, the rate of change today is faster than ever before. Many of today's problems are often cited, at least in part, as evidence of people's inability to adjust to this change. Examples include information overload that is reported to lead to some psychological disorders and feelings of alienation.

Fortunately, as technological change occurs, we are learning more about helping people deal with this change. Employers are creating work environments that improve employee relations. In addition, a variety of education options is available for training and retraining, Fig. 42-8.

Fig. 42-8. These students are learning about communication technology. Training and retraining are an important part of living in a rapidly changing world. (Bowling Green State University, Visual Communication Technology Program)

Communication technology plays a major role in helping people adjust to change by spreading the message on ways to cope. People are amazingly adaptable, and most can learn strategies for dealing with rapid change.

CONSIDERATIONS FOR DESIGN

The potential for negative impacts of technology must be addressed when designing communication products and services. Local, state, national, and even international regulations can influence how some products are designed. In addition to style considerations, other issues include:

- Energy efficiency and protection of the environment.
- Safety and health.
- Security.
- Human factors engineering.

ENERGY EFFICIENCY AND PROTECTION OF THE ENVIRONMENT

Many design factors, with the goals of increasing energy efficiency and protection of the environment, must be considered. These include using recycled materials, energy-conserving devices, alternative energy systems, and control systems to ensure that damage to the environment does not occur. See Fig. 42-9.

A four-point plan, developed by the Environmental Protection Agency, outlines four waste management options that include, in order of priority:

1. Reducing the amount of waste produced.
2. Recycling, reusing, and composting.
3. Waste-to-energy incineration.
4. Landfilling.

Just as it is now, protection of the environment will be a concern for designers in the future. Not only do products and services need to meet government regulations

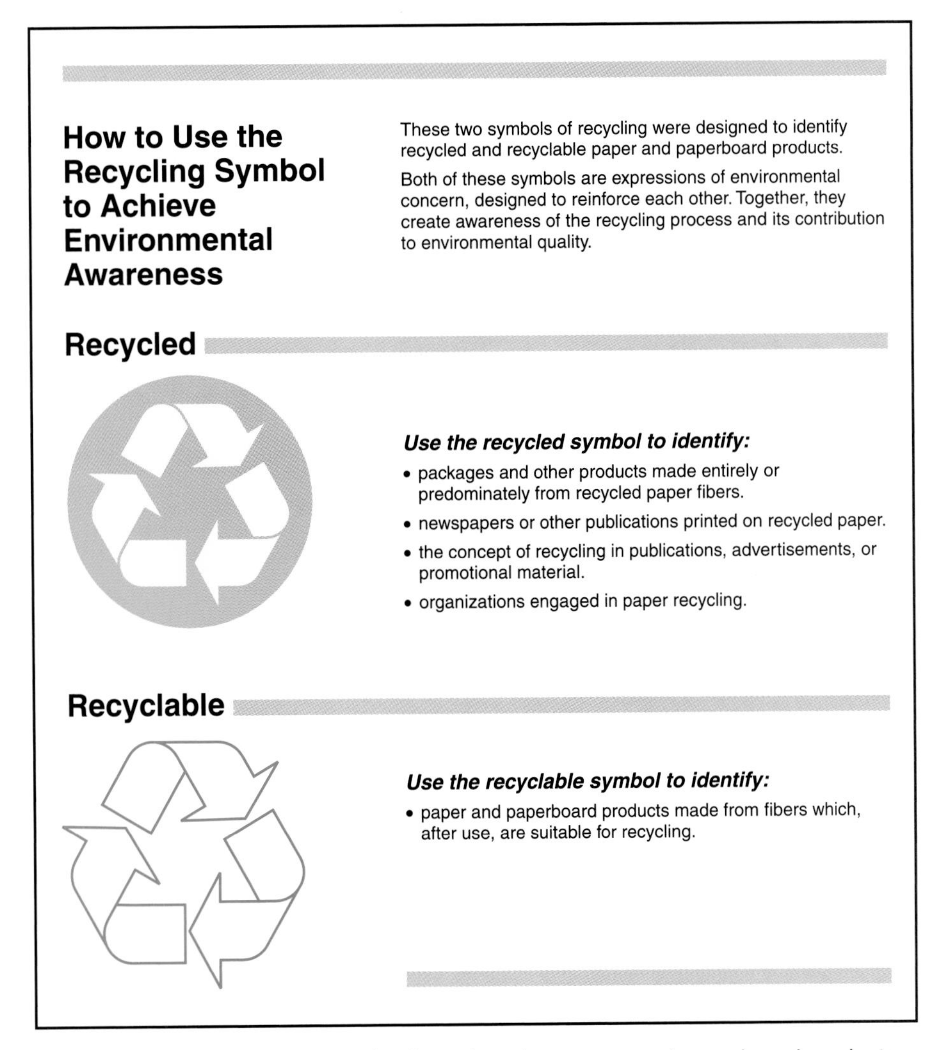

Fig. 42-9. The recycling symbol, when placed on paper and paperboard products, communicates the fact that the product is made predominantly from recycled materials and the product is recyclable.

related to the environment, but they must also be acceptable to consumers who are concerned about environmental issues. Recycling is one method that is being used to protect the environment.

Recycling

Recycling is a technology that involves four major steps. These include:

1. Collecting items that can be recycled.
2. Separating items.
3. Processing the items into new products.
4. Developing end-use markets.

As you can see, separation and collection alone do not make a recycling program. This process is not completed until a material is processed, purchased by a consumer, and used again. For recycling to be effective, it must be a joint effort among manufacturers, consumers, recyclers, and the government.

Many people feel that recycling is a hassle and not worth the time. Some believe that it is easier to simply throw garbage away and let it be hauled to a landfill. However, landfills are rapidly reaching capacity. See Fig. 42-10. New landfills are difficult to establish due to widespread public opposition.

Recycling is one way to ease the dependency on landfills. Items that are currently being recycled include newspapers, aluminum, steel cans, glass, motor oil, plastics, and organic wastes.

Fig. 42-10. Recycling is considered to be a better waste management alternative than creating more landfills such as this one. (Westinghouse Electric Corp.)

SAFETY AND HEALTH

Safety and health are other design considerations. Whether it is a design for a vehicle, telecommunication system, or computer, there are safety regulations that must be considered. Lawsuits related to safety and health problems can be quite expensive for businesses. Therefore, most companies want to make their products as safe as possible.

SECURITY

Security is a growing area of concern and must be considered by designers. Electronic files are available that can provide a person's credit rating, credit card numbers, and other information that must remain secure. Data communication provides convenience in our lives, but it also carries with it additional risks to privacy that need to be considered when designing and using such systems.

Other security problems face companies and individuals each day. Cable companies scramble premium channels because some people illegally attempt to receive the signals for free. Video and audio tapes, computer programs, and copyrighted written and graphic material are copied even though it is illegal to do so. Computer viruses are passed to different computers through infected disks and electronic services, destroying valuable data and programs.

COPYRIGHTS AND PATENTS

One protection for designers is the **copyright.** The U.S. Copyright Act of 1976 covers "original works of authorship." When the work is created and a copyright is obtained, whether it is a film, video, audio tape, computer program, or book, it is protected by copyright law. The owner of the copyright has exclusive rights to publish and sell the work, although the author can transfer these rights to others, such as publishing and recording companies. All copies of the work must contain a copyright notice as well as the date and the name of the copyright owner. For instance, this textbook has a copyright. The copyright notice appears on page 2 of this textbook.

The **patent** is another form of protection. A patent gives the inventor sole rights to his or her invention. Designers of original inventions should check on and apply for patents.

HUMAN FACTORS ENGINEERING

Another design consideration is **human factors engineering,** otherwise known as **ergonomics.** This is the field of study that deals with adapting machines and other technologies to humans. Comfort, less fatigue, safety, and reduced errors are all results of ergonomically correct technology, Fig. 42-11. Cars, computers, telephones, and assembly line equipment are just a few of the items that must be designed with people in mind.

Fig. 42-11. This filter reduces glare on a computer monitor, thus reducing eye fatigue. (NASA)

ASSESSING IMPACTS

Today, there is a growing awareness of the tremendous impact that technology can have on all aspects of your life. Since negative impacts can affect everyone, it is important to study these impacts to ensure that negative effects are reduced.

The study of technological impacts is known as **technology assessment.** This study can be for existing technologies or for future technologies that are being explored. Although it is very difficult to determine the exact impact a future technology will have, careful study can help lessen problems.

In 1973, the **Office of Technology Assessment (OTA)** was established as a government office with the task of assessing the impacts of new technologies before widespread application. This information is then passed on to Congress. Members of Congress must then decide if the benefits of the technology outweigh the risks.

The advantage of the OTA is that it is impartial. Any new technological development often has opponents and proponents. For example, a business might be in favor of a new device because it can make a profit by selling the device. On the other hand, an environmental group might be opposed to its manufacture because of a negative environmental impact. The OTA can look at all the issues and make a recommendation based on the benefits to all citizens, not just the special interest groups.

Assessing the impact of technology is never an exact science, and usually involves finding the opinions of experts. One technique that is used to solicit these opinions is the Delphi study developed by the Rand Corporation. The **Delphi study** is a way to solicit the opinions of experts while also providing them with additional information to consider on the topic.

The first step in a Delphi study is to find a group of experts on the topic of interest, Fig. 42-12. The first round of questions is then sent to them. Based on the responses, a second round of questions is then prepared. These questions are modified from the first round and reflect the overall opinions of the group for each question. Each expert is asked to reconsider his or her response based on the additional information from the other experts. It is not unusual to have five or more rounds of the questionnaire, with each being modified slightly. When complete, there should be greater agreement among the experts. Therefore, better conclusions can be drawn from their responses.

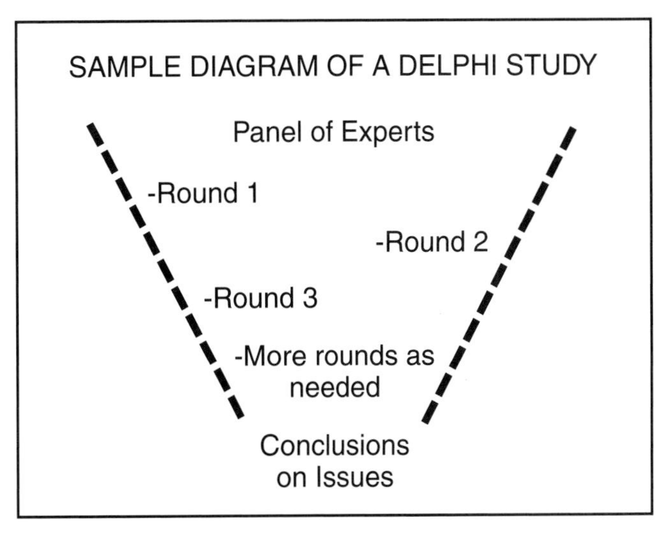

Fig. 42-12. A Delphi study involves sending rounds of questions to a panel of experts. When completed, there should be greater agreement among the experts on the topics of interest.

PREDICTING THE FUTURE

What does the future hold in store for us? Nobody knows for sure, but based on current trends, you can make an educated guess. A few examples will serve to illustrate this point.

A major impact on our lives in the future, as it is now, is information technology. Computers and telecommunications have helped to create what has been described as a "global village." We are now almost instantly aware of events as they happen around the world, and we have a greater understanding of other cultures through television, radio, and print media. Many businesses are now multinational, and must understand how to deal with workers and consumers from other countries. This merging of economies, cultures, and information will most likely bring the countries of the world more closely together. This may ultimately result in a single world community.

A variety of technologies, once experimental, are now becoming commonplace. For instance, the **Integrated Services Digital Network (ISDN)** provides a common standard for communication transmission worldwide. This is done by providing a single network for communication signals, such as voice, video, and data. This system is already available in many cities, and employs fiber optics that are capable of carrying the large information flow, Fig. 42-13.

Multimedia, which is the merger of separate media, such as print and video, will most likely play a larger role in education in the future. People will be able to explore a topic on their own, including video interview, graphics, photos, text, and sound. This system will make it easier to learn information by employing several of the senses simultaneously.

Virtual reality is another technology that is beginning to be used for business, education, and entertainment purposes, Fig. 42-14. This is a system that allows a person to enter another reality through a number of the senses. A helmet and gloves are worn that allow the user to see and move around in a computer-generated world. There are many uses being developed for this technology, such as allowing an architect to actually walk through a building being designed to verify the quality of the design before actual construction begins.

Fig. 42-13. Fiber optics will be the primary communication channel for ISDN. (United Telecom)

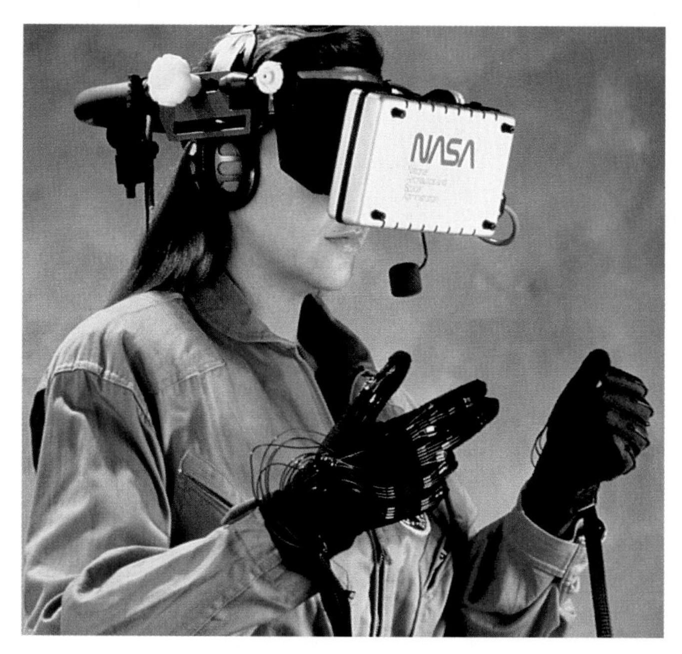

Fig. 42-14. A virtual reality system lets you see and move around in a computer-generated world. (NASA)

In addition to enhanced communication around the world, information technology will also continue to be important for communication in space. As space colonies become established, it will be important to be in constant contact, Fig. 42-15. In addition, future information technology will allow us to communicate with life beyond our solar system if it is discovered.

FORECASTING TECHNIQUES

Studying the future can help you to prepare for life in an ever changing world. When people think about and prepare for future possibilities, they are usually less fearful about changes as they occur. In addition, studying the future can help you to make wise decisions with your votes and dollars.

Several techniques are available for forecasting. One method is the Delphi study technique described earlier.

Trend extrapolation is a forecasting technique that is often used. A **trend** is an emerging pattern over time. It deals with past and current events. For example, you can safely say that increasing computer use is a trend in the United States as well as the world, Fig. 42-16.

Fig. 42-15. This is the interior of a 10,000 person space settlement of the future. The sphere would rotate at 1.9 rpm so gravity would be produced that would be much like on Earth. (NASA)

When you try to predict the continuation of a trend into the future, this is known as **trend extrapolation** or trend projection, a type of forecasting.

Trend extrapolations are done by first mapping the present trend on a graph. You can then make an educated guess as to what will happen in the future. The more expertise a person has in a field, the better the guess. This is why experts are often consulted for trend extrapolation.

Another forecasting technique, known as a **future wheel,** is a brainstorming technique. The wheel begins with a central issue or problem. Brainstorming then results in various possibilities that are placed like spokes on the wheel, Fig. 42-17. A possibility may result in other ideas that can be placed off the spoke. In fact, each spoke may become a separate issue and have its own set of spokes around it. For example, if the cost of complex virtual reality systems were below $100.00, what impact would that have on our society? Spokes might include positive impacts such as more manufacturers making the systems and many different applications in business and entertainment. Negative impacts might include people becoming addicted to the virtual reality

Fig. 42-16. Increasing computer use is a trend all over the world. (Canon USA, Inc.)

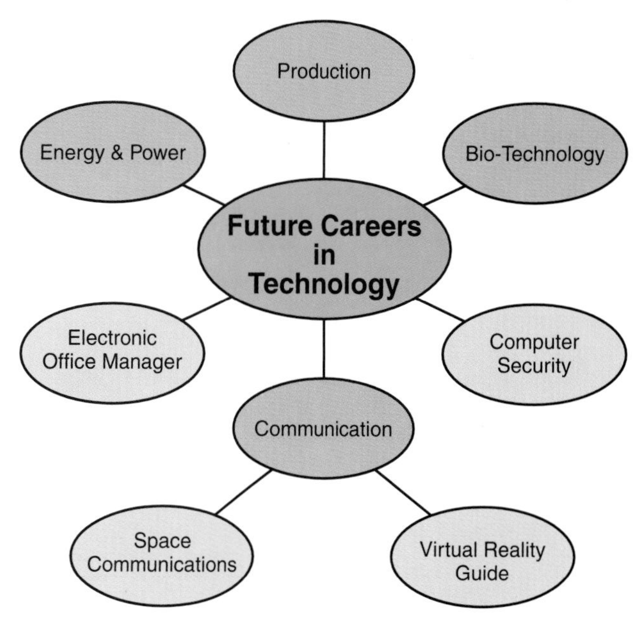

Fig. 42-17. This is the beginning of a future wheel. Any topic can have subtopics under it, shown by spokes and more circles.

environment or impaired functioning after exiting a virtual reality state. Other issues related to these impacts could be placed as spokes around each impact.

SUMMARY

This chapter has explored the advantages and trade-offs involved in new technologies. Various impacts were discussed, including environmental, social, personal, and economic impacts. Assessing technologies and techniques used for predicting how technological development will impact society were also explored. It is important for you to be aware of technological changes and their effects so that you are prepared to make wise decisions about the use of technology, now and in the future.

WORDS TO KNOW

All of the following words have been used in this chapter. Do you know their meanings?

copyright
Delphi study
economics
ergonomics
future wheel
human factors engineering
information overload
Integrated Services Digital Network (ISDN)
Office of Technology Assessment (OTA)
patent
recycling
technology assessment
trade-offs
trend
trend extrapolation
virtual reality

REVIEWING YOUR KNOWLEDGE

Please do not write in this text. Write your answers on a separate sheet.

1. The term ______ ______ has been used to describe the problem of bombarding people with more information than they can possibly review or comprehend.
2. The four major categories of impacts on communication technology are listed below. Give an example of each.
   - Environmental
   - Social
   - Economic
   - Personal
3. List the four waste management options developed by the Environmental Protection Agency, in order of priority.
4. An effective recycling program involves:
   a. Separation and collection.
   b. Processing items into new products and developing end-use items.
   c. Landfilling.
   d. Both a and b.
5. Name two forms of legal protection for authors and inventors.
6. True or false? Assessing the impact of technology is an exact science.
7. ______ ______ allows a person to see and move around in a computer-generated world.
8. Which forecasting technique is done by soliciting the opinions of experts while also providing them with additional information to consider on a topic, with the goal of achieving greater agreement among the experts?
   a. Trend extrapolation.
   b. Future wheel.
   c. Delphi study.
   d. None of the above.

APPLYING YOUR KNOWLEDGE

1. Choose a communication technology and discuss its positive and negative impacts.
2. Develop and administer a survey to determine the impacts of some communication technology. Summarize your findings.
3. Visit a business or industry to determine how technology has impacted their operation.
4. If your school does not already have one, implement a recycling program at your school. Discuss the environmental impacts of not having a recycling program versus having one.

This is a cutaway view of the interior of a proposed space habitat. In the future, living in space could be a reality. Communication technology will play a vital role. (NASA)

Section Activities 7 COMMUNICATION, JOBS, AND YOU

BASIC ACTIVITY
Convincing the Interviewer

Introduction:

A critical part of getting a job is the job interview. You may look great "on paper," but you also have to be able to express yourself well in an interview situation. This activity will give you a chance to practice your interviewing skills.

Guidelines:

- Work in groups of three to four students.
- Each interview should last no more than seven to ten minutes.

Materials and equipment:

- Video equipment.
- Props to simulate the office of a personnel director.

Procedure:

1. Begin by preparing a resume on the computer. Since you will not be sending it out, it does not have to be in final form, but should be well organized. Your resume should include information such as your name, address, phone number, social security number, education and dates, employment record and dates, honors, interests and hobbies, and personal references. You can refer to the resume in the job interview as well as use it in the future to help in filling out job applications.
2. Normally, a job application would be filled out, but we will assume that has been done and you are one of the top candidates for the job. Each top candidate has an opportunity to interview.
3. As a group, select a communication related job for which each of you will be interviewing. Prepare a list of questions for the personnel director. The personnel director is not necessarily limited to the list of questions.
4. Set up the video equipment and props. Decide on a rotation schedule so each person is able to serve as camera operator, personnel director, and interviewee during this activity.
5. Tape each interview twice. After all or part of the first interview, the group should critique the performance of the person being interviewed. Suggestions might be to maintain eye contact, use better posture, answer in a more forthright manner, etc. The objective is to have each person do his or her best. After the critique, make the final tape.
6. After the video is completed, discuss how each of the three jobs affected your perceptions of the interview. Do you think playing the different parts will help you in actual interview situations?

Evaluation:

Considerations for evaluation are:

- Complete and correct resume.
- Interview effort.
- Quality of the video.

ADVANCED ACTIVITY
The Job Search

Introduction:

A variety of jobs deal with communication technology, so there is probably an area that appeals to most everyone. See Section Fig. 7-1. This activity will involve you in investigating several of these jobs.

Guidelines:

- Report should be no longer than three pages, not including the title page.
- All note cards should contain source information. This includes, at minimum, the name of the publication, date published, and page number where the information was found.

Materials and equipment:

- Dictionary of Occupational Titles and Occupational Outlook Handbook.
- Other resources providing information about jobs now and in the future.
- Index cards.

Procedure:

1. Begin by considering your interests and abilities. Make a wish list of those aspects of a job you think you would like the best.
2. Decide on a category of communication technology in which you might like to be employed. Research

Section Fig. 7-1. There are probably many jobs that involve communication technology that fit your interests and aptitudes. (Bowling Green State University, Visual Communication Technology Program)

the specific jobs in that category and write a short description of your top two choices on index cards. Also, write down the source of the information. Include the requirements of the job, including training needed.

3. Investigate the future of the two jobs chosen. On two additional cards make notes on how the jobs are predicted to change in the future. Again, list the source of the information.
4. Use a computer to write a short report on your findings. Include information on how the two jobs chosen best match your interests and aptitudes, and which you think would be your best choice. Also include information about the future of the jobs.

Evaluation:

Considerations for evaluation are:

- Depth of job investigation.
- Quality of the report.

ABOVE AND BEYOND ACTIVITY
Future Wheel Predictions

Introduction:

The future wheel is a technique for brainstorming about the future. See Section Fig. 7-2. In this activity you will create a future wheel predicting the future of communication technology.

Guidelines:

- Small group activity.
- Note cards should have source information on them.

Materials and equipment:

- Publications and other resources containing information about the future.
- Science fiction books and movies.
- Poster board.

Procedure:

1. Without doing any investigation, construct a future wheel that projects communication technologies in the future. Take 25 minutes to complete this activity. Start with communication technology as the center and the categories in this book as spokes. From each of these spokes, make predictions. For example, what future technologies do you expect to see in the photography field? You can also place spokes off the technologies you list, and write the names of technologies within this category.
2. Investigate the future of communication technology through publications, the movies, or other means. On index cards, write the source investigated and thoughts that you may have for the future wheel.

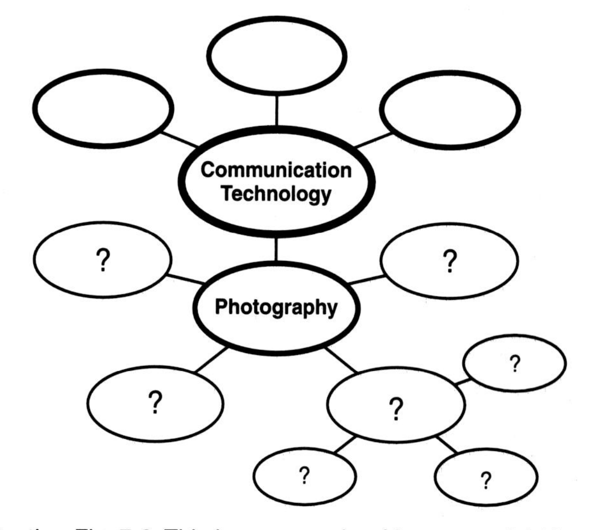

Section Fig. 7-2. This is an example of how you might begin the future wheel.

3. Spend another class period making a new future wheel. After it is complete, place the information on a poster board and display it.
4. Wrap up by discussing how this activity may have changed your thoughts about the future. If time permits, discuss some of the impacts of the technologies you predicted.

Evaluation:

Considerations for evaluation are:

- Effort in completing each future wheel.
- Depth of investigation about the future.
- Design and overall quality of the poster.

Glossary

A

Aberrations: lens defects that can affect the image recorded on film.

Acceptance sampling: a technique that involves inspecting a certain percentage of the goods or services as the basis for a decision on whether or not to accept or reject the entire lot.

Acoustic coupler: older type of modem that required placing the telephone handset directly on the modem to send and receive information.

Active focusing system: system that emits an ultrasonic or infrared beam that is reflected back to a camera sensor from the subject and used to set the distance for focusing.

Acute angles: those that are less than 90°.

Additive plates: presensitized litho plates that require chemicals to build up the image area.

Additive primaries: red, green, and blue wavelengths that combine to produce white light.

Adhesive binding: fastening pages or signatures together with an adhesive applied along the spine. It is used for paperback books and notepads. Another name for this type of binding is perfect binding.

Adjustable ruling pen: a pen that consists of two nibs that can be adjusted for a variety of line widths.

Advanced Photo System (APS): a new film format and camera system developed by a group of photographic equipment and film manufacturers. It uses a special size canister for film that is slightly smaller than 35mm.

Agate line: a printing measurement used for column depth. There are 14 agate lines in one inch.

Airbrush: a shading tool that uses compressed air to spray dots of ink on the drawing.

Algorithm: a written, step-by-step procedure for solving a problem.

Aligned system: one in which dimensions are placed to read from the bottom and right side.

Alphabet of Lines: a standard for lines used in drafting, developed by the American Society of Mechanical Engineers.

Alternating current: electric current that flows in alternate directions.

American National Standards Institute (ANSI): organization that establishes standards for materials and equipment.

American Point System: measurement system that is used in the printing and publishing fields. Measurements are made in units called points and picas.

American Standard Code for Information Interchange (ASCII): system developed to standardize the binary code for various characters and numbers.

Ampere: the unit of measure for electric current

Amplitude modulation (AM): a means of placing information on a carrier wave by varying the amplitude (height) of the wave, while the frequency of the wave stays the same.

Analog signal: any signal that continuously varies.

Analog-to-digital (A/D) converter: device that changes continually varying (analog) signals to on/off (digital) signals.

Angle: shape formed by two lines intersecting at a point.

Angular dimensioning: a method that specifies angles in degrees. The dimension line is drawn as an arc with the angle specified.

Angular perspective: see Two-point perspective.

Anilox roller: a roller used in flexographic printing. It has an engraved surface that holds a layer of liquid ink.

Animation: the addition of motion to drawn images or three-dimensional figures.

Aperture: the size of the hole behind the lens through which light passes.

Apprenticeship: a type of training in which a worker learns the skills for a trade while working on the job.

Aptitudes: talents a person was born with that can be developed into abilities.

Arc: a part of the circle circumference.

Architect's scale: a measuring device most often used for drawing buildings.

Arithmetic-logic section: portion of the computer's circuitry that performs mathematical functions as needed.

ASCII: American Standard Code for Information Interchange.

Assembly drawing: one that shows how parts of an object fit together.

Atoms: the smallest particle of a chemical element. An atom has a nucleus made up of protons and neutrons, and an outer shell of varying numbers of electrons.

Audio head: component that records the sound on videotape in a video recorder.

Auxiliary view: one that is used when a normal projection plane does not show a surface's true size and shape.

Averaging meter: meter that takes an average of all the light in the scene.

Axonometric drawings: pictorial drawings. There are several types, differing in the angles between the lines joining the front corner or axis.

B

Backlighting: lighting method in which the light is behind the subject.

Balance: refers to how elements are arranged horizontally or vertically in a design.

Bandwidth: the difference between the highest and lowest frequencies needed for a signal. Bandwidth is measured in hertz (cycles per second).

BASIC: a high level language for a computer program. BASIC is an acronym that stands for Beginner's All-Purpose Symbolic Instruction Code.

Basic density range (BDR): the maximum density range a halftone screen can record on film with a single exposure.

Baud rate: speed (in bits per second) at which a modem sends and receives data.

Benefits: extras that a company provides, such as insurance and paid vacations.

Binary code: a means of encoding information using on and off states of electricity.

Binding: the process of fastening pages together.

Bit-mapped graphics: see Raster display.

Bits: binary digits stored in computer memory as on (1) or off (0), related to whether the electrical circuit is on or off.

Blanket cylinder: on an offset lithographic press, the cylinder to which a rubberized cover called a "blanket" is attached. It contacts the printing plate and reverses the image.

Blind embossing: a type of embossing that has a raised surface, but does not involve printing on the embossed area.

Blind hole: a hole that does not pass through an object.

Blockout liquid: a thick liquid that can be used to repair pinholes in screen printing stencils.

Blowback: print of an enlarged drawing from microfilm.

Blueprints: copies of drawings that have a blue background and white image lines.

Board of directors: individuals elected to represent the shareholders of a business and oversee the operation of that business.

Body copy: in printing, small letter sizes used for text.

Bolt circle: a circular centerline used when holes are equally spaced in a circle.

Bone folder: creasing tool used for manual folding of paper.

Braille: a technology that enables blind people to read by printing words as raised dots. A person can "feel" the words by moving his or her fingers across a page.

Brainstorming: a special technique for group problem solving that involves generating as many ideas as possible in a short time.

Broadcast television: system in which programs are sent by atmospheric signals from one place to another, such as from a television station to your home.

Buckle folder: machine that uses rollers to push the paper against a stop. At this point, the paper is buckled and is caught between two more rollers that crease the fold.

Budget: document developed by a finance department to predict how money will be used by a company.

Bump exposure: an exposure made without the halftone screen for low-contrast copy. It enlarges the dots in the highlight area.

Burning-in: darkening an area of a print by giving it more exposure.

Bytes: a group of eight bits (binary digits). Using a binary code, these bytes can be used to symbolize numbers and letters.

C

Cabinet oblique: drawing that shows depth as half size.

Cable television: a system in which television signals are received by a large central antenna, then distributed over a coaxial cable network to subscribers who pay a fee for the service.

Callouts: notes that are used to supplement dimensions.

Camera body: the light-tight box which contains the film.

Canister film: film in a metal or plastic canister that serves as a light-tight protective package.

Cardioid microphone: one that picks up sound in a heart-shaped pattern.

Cartridge film: film in a closed container that is merely placed into the camera for automatic loading.

Cathode ray tube (CRT): most common type of monitor display. It works in a fashion similar to a television set.

Cavalier oblique: drawing with depth dimensions drawn full length.

Cel (celluloid) animation: a technique in which characters drawn on transparent celluloid sheets are placed on top of a background.

Cell: point where a row and column meet on an electronic spreadsheet.

Cell site: location that contains the transmitting and receiving equipment for communicating with cellular phones in that cell.

Center of interest (COI): the main subject of a photo.

Center-weighted meter: a meter that measures all the light in the picture but places more emphasis on the center of the picture.

Centerlines: lines used to show the center of round features on a multiview drawing.

Central processing unit (CPU): the microprocessor of the computer, where work is performed on the data.

Charge-coupled device (CCD): a light-sensitive microchip with characteristics similar to a camera tube, now used in many video cameras.

Chase: in relief printing, a metal frame that holds the image carrier.

Chemical etching: method of preparing a gravure cylinder, using an acid to etch the cells into the surface of the cylinder.

Chromogenic system: a type of film in which the emulsion layers have color added during processing.

Cinematography: the use of photographic techniques to make motion pictures.

Circle template: a tool used to trace specific circle diameters.

Civil engineer's scale: a measuring device divided in tenths of an inch, often used for drawings of highways and maps.

Clip art: copyright-free artist illustrations that are purchased in book or disk form for use in publications.

Clock speed: measure of how fast the computer processes data.

Closed loop system: a system with a feedback loop allows for monitoring the output and making corrections if needed.

Closed-circuit television: see Nonbroadcast television.

Coaxial cable: a cable that is specially shielded to avoid interference.

Coherent light: light waves that are all one wavelength and oscillate together perfectly.

Cold composition: methods for generating type that do not use hot metal.

Collating: the process of placing individual sheets in order for binding.

Color electrostatic process: often used in color copiers, it is similar to xerography. Cyan, magenta, yellow, and black toners are applied to the paper and fused with heat.

Color negative film: is used for color prints. The colors are recorded on the film as opposite or complementary colors. When a print is made from the negative, the colors will be correct.

Color positive film: also known as reversal film, is used for slides. Light is projected through color positive film for viewing on a screen.

Color separation: the process of converting a color original to negative form for reproduction. The color original is photographed (or electronically scanned) four times using color filters and a halftone screen. This produces one set of color separation negatives, each with a dot pattern, that can be used to make printing plates for the process colors (cyan, yellow, magenta, and black).

Color systems: systems that assign a number or code to each color as an accurate way to describe colors.

Color temperature: the sum of the wavelengths in a particular color of light, measured in degrees (Kelvin).

Color wheel: a graphic treatment that shows how colors are related to one another.

Communication network: a number of computers or terminals linked together, allowing them to share data, software, and peripherals.

Communication technology: a variety of technical means used to transfer information.

Compact disc-read only memory (CD-ROM): large capacity storage device that can only be read (information cannot be added, changed, or deleted). This type of storage device is handy for storage of large databases, such as a library's card catalog.

Compact disc-rewritable (CD-RW): a compact disc that can be written to (recorded upon) as well as read.

Compass: a tool for drawing circles and arcs.

Complementary harmony: the use of colors directly across from each other on the color wheel.

Composition: in photography, the process of visually arranging the elements in the scene for the best photo. Also used to describe the resulting arrangement.

Compound lens: a lens made up of individual elements, used to correct lens defects.

Computer: an electronic device that processes information under the control of a program.

Computer numerical control (CNC): a method for using computer-operated machines to perform drilling, machining, welding, and cutting.

Computer software: program containing the specific instructions for directing the operation of the computer.

Computer-aided drafting and design (CADD): the process of using special software and a computer system to produce designs and technical drawings.

Computer-aided manufacturing (CAM): using a computer to operate machines in industry.

Condenser microphone: in this device, batteries supply power to a metal plate attached to the diaphragm. As the charged plate vibrates, the charge is transfered to another metal plate, creating the electrical signal.

Conductors: materials that allow the free movement of electrons in their atoms.

Contact print: a positive photo image made by placing a negative on top of printing paper and exposing it to light. Often called a proof sheet.

Continuous light source: a source, such as a floodlamp, that gives constant lighting.

Continuous tone image: photograph or similar illustration that exhibits a variety of tones (for example, different grays in a black-and-white photo).

Continuous tone shading: technique that provides the most realistic appearance; patterns are varied to show the surface gradually becoming shaded.

Convergent lens: a lens that bends light inward.

Cool colors: blue, violet, and green; often called receding colors because they seem to move away or stay in the background.

Coordinate system: numerical locations used in CAD to precisely place or locate entities (objects) on a drawing.

Copy: CAD command used to make a duplicate of an entity for placement elsewhere on a drawing. In journalistic and printing use, the written words making up the text of a magazine article, advertisement, or similar item.

Copy density range: shadow density minus the highlight density.

Copyfitting: the process of determining the amount of copy that will fit into a specific amount of space.

Copyright: legal protection for "original works of authorship." The owner of a copyright has exclusive rights to publish and sell the work.

Corporation: a business formed and authorized by law with the rights and obligations of an individual. The owners are persons who have purchased or obtained stock in the corporation.

Crash: hard disk failure in which the read/write head bounces and damages the platter.

Cropping: a method of improving a photograph by selecting only part of the negative to print.

Crystal microphone: one in which a crystal vibrates, causing it to produce an electrical signal.

Cutting plane line: line used to show where the section view is located.

Cybernetics: a form of machine-to-machine communication in which one machine signals or controls another.

Cycolor process: a color copier process that uses special film with dye capsules embedded in it. After exposure, the film is pressed against a chemically treated paper and the dye is transferred to the paper wherever the capsules have not been exposed to light.

D

Daisy wheel printer: impact printer with characters arranged on a plastic or metal wheel.

Data communication software: program used, along with a modem, to communicate computer data over distances via telephone lines.

Database software: program used to file important information in computer memory.

Debugging: term that describes the process of finding and solving problems in computer software.

Decibels: units used in measuring the loudness of a sound.

Dedicated flash: a unit designed to work only with a particular camera.

Demodulating: separating the carrier wave from the information that is being carried.

Demon letters: characters that can be confused when reversed as relief-printing type, such as b and d.

Density: a measurement that shows the "darkness" of an image; a measure of the light-absorbing ability of an object.

Design: a process that has the function of creating and refining ideas.

Desktop publishing system: system that uses a personal computer, specialized software, and a laser printer as a system for page layout.

Developer: chemical that reacts with film emulsion to change exposed silver halide crystals into black metallic silver. It turns a latent image into a visible image.

Dialog boxes: display screens that allow operators to make selections or enter new information.

Diazo prints: see Whiteprints.

Die-cutting: finishing operation done by pressing a sharp steel rule into the paper surface.

Diffusion transfer system: a system for printmaking in which a negative is exposed and placed in contact with a receiver sheet. After being through a processor, the sheets are separated. The negative image transfers as a positive image onto the receiver.

Digital camera: a camera that records a picture as electronic signals on a special cartridge or disk, instead of film. The pictures can be transferred to a computer and printed out on a laser or ink jet printer.

Digital signal: one that has only *on* and *off* states.

Digital-to-analog (D/A) converter: device that changes on/off (digital) signals to continually varying (analog) signals.

Digitizers: devices that allow existing drawings to be input into a computer system by tracing over them.

Dimensioning: size and location information on a drawing.

Dimetric drawings: axonometric drawings in which two of the angles between the lines joining the front corner or axis are equal.

Direct connect modem: modem that attaches directly to the phone line for sending information. The modem can be built into the computer, or can be a separate unit.

Direct current: electric current that flows in one direction.

Direct image plate: lithographic plate upon which the image is manually placed by drawing or typing.

Direct thermal copying: process that uses heat to transfer an image to the paper from an original that has a carbon-based image.

Direct thermal printer: one that uses a chemically coated paper. Heat causes an image to form on the paper.

Direct transfer: a process in which a light-sensitive photopolymer emulsion is applied to a gravure cylinder and serves as a mask while the desired image is chemically etched into the cylinder.

Directional microphone: one that picks up sound from directly in front of its location.

Disk operating system (DOS): system software that manages the computer, peripheral devices, and software so they all work together correctly.

Display type: large letter sizes that are used to attract attention.

Divergent lens: a lens that spreads light rays.

Divider: a tool that looks like a compass, but both legs have a steel point at the ends.

Dodging: lighting an area of a photographic print by giving it less exposure.

Dolly: moving a video or motion picture camera toward or away from the subject.

Dot matrix printer: output device that forms letters and graphic images as a series of small dots.

Downloadable fonts: type fonts that must be sent to the printer from the computer.

Drafters: workers who use manual drawing tools and computers to prepare drawings of products.

Drafting film: a plastic sheet with one side roughened to accept ink and pencil lines.

Drafting machine: device attached to the drawing table that takes the place of the T-square, triangle, protractor, and scale.

Drilling: a process used to put round holes in paper.

Dry mounting: a method of mounting prints that involves using a tissue that has an adhesive on both sides. The tissue is placed between the board and print, and heat is applied with a press.

Ductor roller: on an offset press, roller that transfers ink and dampening fluid from the fountain rollers to the form rollers.

Dummy: a quickly made representation of a finished printed product.

Duplicators: sheet-fed offset lithography presses that have a small plate size and fewer ink and dampening rollers than larger presses.

Dynamic microphone: microphone that has a diaphragm attached to a coil of wire that is between the poles of a magnet. As sound waves make the diaphragm and coil vibrate, an electric signal is induced in the wire.

E

e-mail (electronic mail): a system that allows computer users to send and receive messages. Individuals must use passwords or codes to retrieve their messages from electronic storage.

Editing: changing something that has been previously drawn on a CAD system.

Electric circuit: a complete path for current.

Electric current: a flow of electrons through a conductor.

Electrical signal: the modulated current in telephone wires that varies in flow in relation to the sound vibrations.

Electromagnetic induction: the scientific principle that an electric current can be caused *(induced)* in an electric wire by a moving magnetic field. It is the principle used to generate most electrical power.

Electromotive force (emf): see Voltage.

Electronic bulletin board: data communication service that allows the user to post messages electronically via computer.

Electronic publishing: broad term for the sophisticated page layout systems used in the graphic arts industry.

Electrostatic copies: copies of drawings made with a powder that is fused to the paper by heat to form the image.

Ellipse: a geometric shape that is an elongated circle.

Em quad: in hot-metal composition, a space the same width as the type size being set.

Embossing: a method of raising the paper surface by pressing it between a convex (raised surface) die and a concave (sunken surface) die.

Engineers: professionals within a company who conduct research to develop better products and better ways of producing the products.

Entities: the elements (points, lines, arcs, and circles) that are combined to make up a drawing. These are sometimes called primitives.

Entrepreneurs: people who start their own businesses.

Ergonomics: the field of study that deals with adapting machines and other technologies to human physical and psychological needs.

Extension lines: lines used to move dimensions off a drawing so it is not cluttered.

F

Fade: visual transition that gives the feeling of time elapsing between scenes. The scene fades to black, then from black to the next scene.

Fax machines: devices that can electronically transmit or receive pictures or text.

Federal Communications Commission (FCC): a government agency responsible for assigning broadcast frequencies.

Feeding system: method used to introduce paper in sheet form (sheet-fed) or roll form (web-fed) to a printing press.

Fiber optics: transmitting system using glass fibers that have a light-transmitting core surrounded by another layer of glass (cladding).

Fiber-base paper: photographic printing medium with a paper (rather than plastic) base.

File server: computer that is used to store the software in a communication network.

Fill light: light used to fill in the shadows created by the key light.

Fillet: a small arc found on inside corners of parts.

Film speed: a rating of the sensitivity of the film to light.

Finance: division of a company responsible for money management.

Finishing operations: additional operations performed on printed items, such as scoring, folding, perforating, die-cutting, drilling, and foil stamping.

Fixer: film processing chemical that removes the unexposed emulsion and makes the image permanent.

Flash: a device, usually electronic, that provides a short burst of light for photography.

Flash exposure: in graphic arts photography, the exposure that places detail in the shadow area.

Flat: a special masking sheet with negatives attached to it, used to make the lithographic plate.

Flatbed plotter: output device for drawings, designed so that the substrate remains stationary while the pen moves over the surface horizontally (x axis) and vertically (y axis).

Flexography: a popular form of relief printing using an image carrier (plate) made of flexible rubber or plastic.

Floppy diskette: a 3.5″ disk used on most computers for data storage. It has a hard case to protect the thin plastic disk from damage.

Flowchart: diagram showing all the steps in producing a product, designed to help managers decide on the resources needed.

Fogging: picture defect that results from accidentally exposing the film to light.

Font: originally used to mean a specific typeface, style, and size. In electronic publishing, however, font sometimes means the same thing as typeface.

Form rollers: in offset lithography, rollers that deliver the ink and dampening fluid to the plate when printing.

Format: term that refers to a certain size of film and kind of packaging.

Fountain roller: roller used to transfer ink or dampening fluid from fountains to ductor rollers on an offset press.

Frequency: the number of waves (sound or electromagnetic) that pass a given point in one second.

Frequency modulation (FM): a means of placing information on a carrier wave by varying the frequency of the wave, while the amplitude (height) of the wave stays the same.

Friction binding: a loose-leaf binding method that uses a plastic U-shaped channel that slips over the binding edge.

Frontlighting: light coming from behind photographer and striking the front of the subject.

f-stop: the lens aperture or size of the hole where light enters the camera body.

Full duplex transmission: data communication system that allows sending and receiving to take place at the same time.

G

Gathering: stacking signatures for binding. Thick products, such as books, are usually assembled by gathering.

General oblique: drawing with a depth dimension between half and full length.

Geostationary orbit: satellite orbit synchronized with the earth's rotation so it appears to be stationary. Also called a "geosynchronous orbit."

Ghosting: faded image area that normally follows an area that requires a lot of ink.

Graph paper: drawing material that has a grid that aids in keeping lines straight, parallel, or of equal length.

Graphic arts: the communication system involved with mass production of graphic images.

Graphic arts photography: the process of changing flat images into photographic images. Also known as image conversion.

Graphic tablet: input device that has a flat surface on which to draw, with an electrical grid underneath.

Graphical user interface (GUI): see Windows.

Graphics software: programs used to create pictures, such as charts in reports or technical drawings.

Gravure printing: printing from a recessed image.

Gray scale: see Sensitivity guide.

Grid: a set of reference dots on the screen in rows and columns, used as a drawing aid.

Grievances: formal complaints by workers of alleged unfair treatment.

Gripper margin: area of a masking sheet next to the lead edge, where no image can be placed. This is where the press grippers pull the paper into the press for printing.

H

Halftone: a continuous tone image that has been broken into different sizes of dots to give a line image with the appearance of a continuous tone.

Hard drive: device for magnetic storage of large amounts of data. A hard drive has one or more metal platters in a sealed case, along with the read/write head.

Hardcopy devices: output devices that provide information in readable printed form.

Hertz (Hz): unit of measure for electrical frequency, in cycles per second.

Hexagons: six-sided polygons.

Hickies: foreign particles on the plate or blanket that can cause light or dark spots on the printed product.

Hidden lines: lines used to show edges that are hidden from view inside an object.

Hot stamping: relief printing process used primarily for decoration. Instead of ink, the type form presses a metallic foil on the surface to be printed.

Hot-type composition: setting type which is cast as needed from a molten mixture of lead, tin, and antimony.

Hue: the name given to a color.

I

Icons: simple pictures on a computer screen that represent a command.

Image conversion: see Graphic arts photography.

Image generation: a creative process, the actual production of images.

Impact composition: a method of creating an image by striking a substrate such as paper. Typewritten copy is an example.

Impact printers: output devices that create an image by pressing a carbon ribbon against a substrate.

Imposition: placing pages in the correct position for printing.

Impression cylinder: offset lithography press cylinder that has grippers which pull the paper between it and the blanket. The paper receives the image from the blanket.

Incident meters: light meters that measure the light at the subject location.

Indirect stencil: stencil with a light-sensitive emulsion coated on a plastic backing sheet. After exposure, the stencil can be adhered to a screen, and the plastic backing peeled away.

Industrial designers: people who help create the product as well as draw it.

Ink and dampening system: rollers that apply ink and dampening fluid to the plate.

Ink jet printer: output device that forms an image by spraying drops of ink on the page through small nozzles or jets. Images are made up of tiny dots.

Ink set-off: printing problem that results when ink from one sheet is transferred to the back of the next sheet.

Input/output (I/O) devices: those that may be used to either input or output data (such as modems or disk drives).

Inserting: placing signatures inside one another for binding. Thin booklets and magazines are usually assembled in this manner.

Instant picture camera: one that allows quick viewing of pictures. A special film "develops itself" in minutes.

Instrument drawing: a manually produced drawing that provides the detailed size and shape information necessary for production.

Insulators: materials that have atoms with tightly bound electrons.

Integrated circuits: often called microchips, they are very small and made up of many transistors and other solid-state devices.

Integrated software: software that combines several programs, such as word processing, database management, and an electronic spreadsheet.

Interactive video: a method of using the computer with a video player to access information. Videodiscs are often used for this because they provide random access.

Internet: a "network of networks," allowing computer users to connect to other computers worldwide for information or entertainment.

Irregular curve: a curved line with no single center.

ISO (ASA) number: rating scale that gives a measure of film speed. As ISO numbers become larger, the film speed increases. ISO stands for International Standards Organization.

Isometric drawings: axonometric drawings in which the angles between the lines joining the front corner or axis angles are equal.

J

Joystick: input device that can be used in a manner similar to a mouse.

Justify: adding spaces between letters and words to make line lengths even, resulting in straight margins.

K

Key light: the brighter light used on a scene that is being photographed.

Key: master negative made from the base paste-up when using the overlay method of mechanical color separation for printing.

Keyboard: an input device used with most computer systems for entering text.

Knife folder: machine that uses a steel knife that pushes the paper between two rollers for folding.

Knife-cut stencil: screen-printing stencil on which a knife is used to cut a thin film. The parts that will print are then peeled away from the backing sheet.

L

Labor unions: organizations formed by employees to conduct collective bargaining on pay and working conditions.

Laser printer: a common output device used in desktop publishing systems.

Layering: CAD features that allows drawings to overlay one another on the screen. Layers can be viewed or plotted together or separately.

Leading: in typography, the total line spacing including the type size and extra space.

Lens: the part of the camera that ensures the image on the film is in focus.

Letter fold: an example of a parallel fold.

Lettering: the accurate drawing of letters and numbers on a drawing.

Letterpress printing: the oldest relief printing method, which uses cast metal type or plates as the image carrier.

Letterspacing: adjusting the space between letters in a word so they appear equal.

Light pen: input device used by touching it to the monitor. A sensor in the pen determines the screen location and sends this information to the computer.

Line copy: illustration or type that has a uniformly dark, or dense image.

Line gauge: a special ruler used to make type measurements.

Liquid crystal display (LCD): display in which each pixel consists of a crystal that changes position when given an electrical charge. This new position allows light to pass through the crystal, thus forming the image.

Lithography: printing done from a smooth surface.

Local area network (LAN): a network that resides in a small area, such as one or several buildings.

Loose-leaf binding: binding methods that allow the easy addition and removal of sheets.

M

Machine language: a program consisting entirely of binary codes.

Magnetic sound recording: a method of recording in which sound waves are first transduced into an electrical signal by a microphone. An electromagnetic recording head converts the signal into variations in magnetism that is transferred to tape with a metallic coating.

Main exposure: the detail exposure, made with the halftone screen over the film.

Mark-up information: instructions needed to complete the printed product, such as paper size and finish, width of margins, ink colors, folding requirements, and related information.

Material Safety Data Sheet (MSDS): a form that an employer must keep for each hazardous substance used in the workplace. This form contains the chemical name, safe exposure levels, explosion and health hazard information, and precautions for safe use.

Math co-processor: a second processor installed in some computers to handle the large number of mathematical calculations needed for computer graphics.

Mechanical: see Paste-up.

Memory: areas where information is stored for use by the computer.

Menu-driven software: programs with menus displaying options from which the user can choose.

Mesh count: the designation, in threads per inch, of how closely the fibers of screen fabric are woven together.

Microfilming: a method for reducing a drawing on film for storage.

Microphone: device that changes or transduces sound waves into an electrical signal.

Microprocessor: a very small integrated circuit (IC) that is a major component in computers.

Microwaves: electromagnetic waves that begin above UHF and go all the way to visible light. They travel by direct wave, or line-of-sight, only.

Million instructions per second (MIPS): an accurate measure of computer processor capability.

Mimeograph duplication: stencil process used for fairly short runs of duplication.

Modem: input/output device for sending and receiving computer information, typically over telephone lines.

Monitor: screen of the computer.

Mouse: one of the most common devices used for input. Movement of the mouse results in a corresponding movement of the cursor on the screen.

Multidirectional microphone: a microphone that picks up sound from all directions.

Multimedia software: programs that allow the computer to be used to mix different media, such as sound and video.

Multiplexing: a system in which more than one signal is simultaneously transmitted through a single channel, such as a wire.

Multiview drawing: one that provides a combination of views (usually top, front, and side) of an object.

N

Node: in a communication network, an end point, such as a personal computer.

Nonbroadcast television: a category of television that is not sent by a broadcast signal. It may be played from a VCR to a single receiver, or on various monitors throughout a building. When played throughout a building, it is sometimes referred to as "closed-circuit" television.

Nonimage area: in graphic arts, the background, which is not printed.

Nonverbal communication: conveying information using factors other than words.

O

Oblique drawing: one with a front face that is shown in its true size and shape, while the top and side faces recede back from the front.

Obtuse angles: those that are greater than 90°.

Occupational Outlook Handbook: reference published every two years by the U.S. Department of Labor. It provides updated information on present and future jobs.

Occupational Safety and Health Administration (OHSA): the federal agency that is responsible for administering workplace safety regulations.

Octagons: eight-sided polygons.

Offset lithography: type of lithographic printing in which the image is "offset," or transferred to the blanket before printing occurs.

Ohm's law: scientific principle that describes the relationships among current, voltage, and resistance.

One-point perspective: view in which the object's front face is true shape and is against or parallel to the projection plane. It is also known as parallel perspective.

Opaque inks: inks that cover what is underneath when printed, and stay the same color.

Optical center: a visually pleasing center point slightly above the mathematical center.

Optical character recognition (OCR): capability of specialized software, when used with a scanner, to automatically input text matter.

Organizational charts: graphic representation of the various responsibility levels within a company.

Orthochromatic film: the high contrast film used in graphic arts.

Orthographic projection: multiview drawing in which various sides of the object are drawn to accurately communicate what the object looks like and how it is made. These views are in a specific place in relation to one another.

Oscillation: rapid change in direction.

Output devices: plotters, printers, display screens, and other devices used to present the results of CAD and other computer applications.

P

Packaging: a container or a wrapper for a product.

Pan: pivoting a video or motion picture camera left or right.

Panchromatic film: one that is sensitive to about the same range of color as the human eye.

Parallax error: a problem encountered with viewfinder cameras, in which the image seen through the viewing lens is slightly different from the image seen by the camera lens.

Parallel circuit: electrical circuit with branches, with each branch leading to a device needing current.

Parallel fold: one made by aligning the folds parallel with one another.

Parallel perspective: see One-point perspective.

Parallel processing: computer communication method in which data is sent in parallel streams.

Parity bit: in data communication, an extra bit often used at the end of a character code to check for errors.

Partnership: ownership of a business by two or more persons.

Passive focusing system: system that uses the normal light reflected from the subject to determine correct focus.

Paste-up: assembling images by placing them on a board made of heavy white paper. The resulting layout is referred to as a mechanical.

Patent: a form of protection that gives the inventor sole rights to his or her invention.

Percentage indicator: device used to determine the dot percentage in different areas of a halftone negative.

Perfect binding: see Adhesive binding.

Perforating: cutting small slits into paper so it can be separated.

Persistence of vision: the physical principle that states a person retains an image for a split second after the image leaves the field of view. This allows still images to be shown at a high rate of speed and appear as motion.

Personal computer: common name for the desktop computers widely used in homes, schools, and offices.

Perspective drawing: pictorial drawing in which dimensions become smaller as they move away from the viewer. A perspective drawing is the most realistic pictorial.

Photo direct plate: lithographic plate made directly from a paste-up and usually used for short runs.

Photographic stencil: a light-sensitive screen printing stencil on which the nonimage areas are hardened by exposure to light. The image areas, which are not hardened, can be washed away when the stencil is developed.

Photography: the art or process of using reflected light to produce pictures on light-sensitive film.

Photopolymers: light-sensitive plastics that harden when exposed to ultraviolet light.

Phototypesetters: obsolete composition machines that produce type by photographic means.

Picas: measuring units used in printing. A pica equals 12 points, or 1/6 inch.

Piezoelectric effect: the ability of some crystals to convert physical pressure into an electrical signal.

Pinhole camera: the simplest type of camera, simply a light-tight container with a pinhole at one end to admit light.

Pinholes: small, unwanted image areas on the screen printing stencil.

Pixels: dots on the computer's display screen. Pixels are turned on or off to create the image.

Plate cylinder: press cylinder that holds the printing Quoins: wedge-like metal locking devices used to firmly plate.

Platen press: named for the area (platen) where the item to be printed is placed, this press uses a flat image carrier for printing.

Plotter: output device that uses pens to make a "hard copy" of a drawing on paper.

Points: measuring units used in printing. A point equals 1/72 inch.

Polygon: a shape made from straight lines. At least three lines are required to make a polygon.

Primary light sources: those that emit light.

Primitives: see Entities.

Printed circuit board: assembly with printed copper paths connecting the integrated circuits and other components.

Printing: the mass production of graphic images by placing ink on paper. Also, the act of reproducing a photograph on light-sensitive paper from a film negative.

Process camera: large camera used in graphic arts for photographing flat copy.

Process color printing: printing that uses the colors yellow, magenta, cyan, and black to reproduce colored, continuous tone illustrations.

Profile view: view of an object from the right side in a multiview drawing.

Proof copy: text and headlines ready to be checked for errors.

Proofreading: the process of checking typeset material for errors.

Protractor: a drawing tool, marked off in degrees, that is used to draw inclined lines and angles.

Pull-down menu: a listing of command options that appear in a column on the screen when an item is selected from the main menu.

Punching: a process that involves pressing a die through the paper to form a hole. Any hole shape can be made with this technique.

Q

Quadrilaterals: four-sided polygons.

Quoins: wedge-like metal locking devices used to firmly secure type forms in a chase.

R

Radar (radio detection and ranging): a system used for tracking planes and satellites. A signal is sent out and a receiver picks up the returning wave that is reflected from objects.

Radius dimensioning: used for dimensioning round features such as arcs.

Random access memory (RAM): the "working memory" of a computer, where information is stored while being processed.

Rangefinder camera: an adjustable-focus type of viewfinder camera.

Raster display: image created by a series of pixels (dots) on the screen.

Read-only memory (ROM): installed when the computer is assembled, this memory permanently stores instructions necessary to perform tasks.

Recycling: a technology that involves the major steps of collecting items that can be recycled, separating items, processing the items into new products, and developing end-use markets.

Reflection densitometer: instrument used to make precise readings of copy density by reflecting a spot of light from the area being read.

Refraction: bending of light rays by a lens.

Relief printing: printing from a raised surface.

Reprography: term used to describe the entire category of duplicating and copying processes.

Resin-coated paper: photographic printing paper with a plastic coating on either side of the paper base.

Resistance: opposition to the flow of electric current.

Resume: brief statement of education, work experience, and other important facts compiled by a job seeker.

Reverse: a printing technique in which the image is an area that is not printed.

Revision block: a record of changes to the original drawing, placed above the title block.

Rhythm: in design, the feeling of movement achieved by repeating elements.

Roll film: film that must be manually threaded onto a take-up spool inside the camera.

Rotary press: one that uses a curved image carrier so the plate can revolve. The substrate is printed as it passes between the revolving plate and impression cylinder.

Rotate: CAD command used to move an entity at an angle around a point.

Rotogravure: gravure process in which the cylinder rotates for printing.

Rule of thirds: a composition technique in which the scene to be photographed is divided into thirds like a tic-tac-toe board. The subject is placed at one of the four intersection points.

S

Saddle binding: a method using wire staples inserted in the spine, often used to bind signatures that are inserted into one another. Magazines are often saddle bound.

Safelight: a light that can be used in the darkroom without exposing the film emulsion.

Sans serif type: a style of type without serifs. (The word "sans" is French for "without.")

Scaling: enlarging or reducing copy proportionately, usually to fit a layout.

Scanner: device that converts illustrations and text into digital form so they can be imported into pages for desktop publishing.

Scoring: crushing paper along the fold line in preparation for folding across the grain.

Screen fabric: mesh made of woven threads and used to support the stencil in screen printing.

Screen printing: printing done with a stencil on a screen.

Script: the final step in designing the video production, containing all necessary audio and video information.

Script type: a style that resembles handwriting.

Scumming: ink that prints in the nonimage areas.

Secondary light sources: those that reflect light emitted by a primary source.

Section view: one that is sometimes used to show internal features of an object.

Semiconductor: see Solid-state devices.

Sensitivity guide: an important graphic arts tool consisting of a scale of gray values. It is photographed along with copy and used to measure density as film is developed.

Serial processing: computer communication method in which data is sent in single file.

Series circuit: electrical circuit with devices that require current arranged in a line, so the current passes through each device in sequence.

Serif type: Type that has thin strokes, called serifs, usually at the ends of letters. Also known as Roman.

Sewn binding: the most durable form of binding. Thread is used for binding the sheets together.

Sheet film: film cut into individual sheets and loaded into a light-tight filmholder for use in a view camera.

Sheet-fed: press feeding system in which sheets are removed from the feed tray for printing with a friction wheel or vacuum-type sucker feet.

Shutter release: button that is pressed to open the shutter and allow light to strike the film.

Shutter speed: the length of time the shutter is open to expose the film to light.

SI Metric System: "International System of Units," a measuring system that has been adopted by most countries in the world. See U.S. Customary Measurement System.

Side wire binding: often used for gathered signatures, with staples placed along the binding edge.

Sidelighting: lighting method in which the light is to the side of the subject.

Siderography: process of producing multiple images from a master die.

Signature: multiple pages printed on both sides of a single sheet and then folded.

Single lens reflex camera (SLR): a type of camera that allows the photographer to look directly through the picture-taking lens, eliminating parallax error.

Slitting: cutting paper on a folder or press equipped with slitting wheels.

Solid model: the most sophisticated modeling technique, in which the model is stored in the computer as an object with mass or volume.

Solid-state devices: electronic components, such as transistors, that are made from a semiconductor (a crystal material, such as silicon, that can be made to conduct electricity under certain conditions).

Sonar (sound navigation and ranging): a means of determining water depth by sending out ultrasonic waves and timing how long it takes for the echo to return.

Spirit duplication: a system involving a spirit master (image carrier) containing an aniline dye. Alcohol dampened paper picks up the dye from the master.

Spoilage allowance: adjustment to printing paper quantities for sheets that are used in achieving ink/dampening fluid balance.

Spot color: a second ink color used for some of the image on a page.

Spot meter: a reflective meter that measures the reflected light in a very small area in the picture.

Spotfacing: finishing method that levels the area around a hole.

Spreadsheet: computer program typically used for financial record keeping.

Staple method: a common method, using metal staples, of attaching fabric to a screen printing frame.

Statistical process control: a mathematical system for checking to be sure that processes are performed to a certain standard.

Stencil: the image carrier for screen printing.

Stereophonic: recording and playback method that uses multiple microphones to record sound, then plays back the resulting "sound mix" over separate speakers.

Still video camera: a camera that electronically records still images on a diskette.

Stitcher: machine that makes staples from a roll of wire and uses them to saddle-bind or side-wire-bind printed items.

Storyboard: series of sketches used to show the type and sequence of scenes in the production.

Stripping: the process of placing litho and halftone negatives on a masking sheet to form a flat.

Subtractive plates: presensitized litho plates that already have a coating to build up the image area. Processing removes this coating from the nonimage areas.

Subtractive primaries: yellow, magenta, or cyan, the colors that are used to subtract colors from white light.

Surprint: the technique of printing over an area that is already printed.

Synchronous transmission: data communication method in which a transmitter clock sends a timing signal to the receiving modem to ensure that they are synchronized.

T

Tangent point: the exact point where a line or arc joins another arc.

Technical fountain pen: a drawing pen that comes in a variety of point sizes according to the line width needed.

Technology: a constantly evolving body of knowledge that deals with the technical way in which we change the world to meet our needs and wants; the practical application of scientific knowledge.

Telecommunications: the process of sending information over long distances by electronic means.

Template: a device used to help in drawing shapes and symbols.

Thermal printer: output device that uses heat to form the image.

Thermography: a method for raising a printed surface by placing a resin powder on an image with wet ink. When heated, the powder expands and fuses with the ink.

Time and temperature method: film processing technique in which developing time remains constant. Changes in image density are produced by changing the exposure time.

Tint: a special printing effect produced when the image is broken into dots of equal size.

Title block: a rectangle in the lower right corner of a drawing that contains the title, drafter's name, scale, and date.

Tolerance: the amount a size can vary.

Topology: the configuration of nodes and links in a communication network.

Tracing paper: thin, untreated paper that can be used for pencil and ink drafting.

Track ball: input device with a ball on top, used to control movement of the cursor.

Transceiver: the combined transmitter and receiver unit that is used by a mobile cellular telephone.

Transducer: a device that changes energy from one form to another, such as electrical energy to sound energy (as in a loudspeaker).

Transistor: see Solid-state devices.

Translucent paper: paper that transmits light and can be used for most copying methods.

Transmitting: the process of placing a message on a channel so that it can be sent to the receiver.

Transparent inks: ink colors that combine with the color underneath when printed.

Transponder: device used on a satellite that amplifies the weakened signal, changes the frequency, and transmits the signal back to earth.

Trend extrapolation: attempting to predict the continuation of a trend into the future.

Triadic harmony: achieved by using colors that divide the color wheel into equal thirds.

Trimetric drawings: axonometric drawings in which the angles between the lines joining the front corner or axis are unequal.

Tripod: a folding three-legged stand used to steady a camera.

Truck: moving a video or motion picture camera left or right.

Truncated: word used to describe a solid with the end removed so the resulting face is not parallel to the base.

T-square: a drawing tool that consists of a head attached at a 90° angle to a blade.

Twin lens reflex camera (TLR): camera with two lenses, one above the other, that are focused together. The top lens is the viewing lens; the bottom, the picture taking lens.

Two-point perspective: perspective with two vanishing points. It is at an angle to the projection plane. Two-point perspective is also known as angular perspective.

Typeface: a specific type design.

Typography: the art of effectively using type to convey a message.

U

Ultrasound: sound above the frequency range humans can hear.

Unidirectional system: one in which all dimensions are placed to read from the bottom of the drawing.

Units: a measure of distance on a drawing.

Unity: the goal of design, achieved when all parts of a design look as if they belong together.

U.S. Customary Measurement System: the form of linear measurement used in the United States. Distance is measured in inches, feet, yards, and miles. See SI Metric System.

Utilities: programs that carry out useful functions such as formatting disks, copying disks, deleting files, and viewing directories.

V

Vanishing point: in a perspective drawing, the point where receding dimensions eventually meet.

Vector graphics: also known as "object-oriented" graphics. The software defines objects mathematically so they print or plot very precisely.

Vellum: treated tracing paper that contains oils and waxes that improve its quality.

Vertex: the point where two lines intersect to form an angle.

Video head: component that records the picture on videotape in a video recorder.

Video monitor: receiver designed to accept separate video and audio signals.

Videodisc: an optical storage medium for video information. Images are stored digitally and played back with a laser beam.

Videotape: a thin ribbon of material with a metallic coating that stores pictures and sound magnetically.

View camera: large camera used by professional photographers to eliminate distortion in a picture.

Viewfinder camera: a camera with a viewing lens through which the photographer looks to compose a picture. A separate camera lens is used to focus the scene on the film.

Virtual reality: a technology that allows a person to enter another reality through a number of the senses. A helmet and gloves are worn that permit the user to see and move around in a computer-generated world.

Visible line shading: technique used to do shading by varying the thickness of the visible lines.

Visual densitometer: a card with holes in the center of density patches, used to find extreme highlight and shadow values.

Voltage: the electromotive force (emf) needed to achieve a flow of electrons.

W

Warm colors: red, orange, and yellow; often called advancing colors because they seem to move forward and suggest activity.

Watt: the measure of electrical power (the rate of doing work or using energy).

Wave frequency: measure of how many waves are produced in one second.

Wave guide: a channel (basically a hollow metal tube) used for transmitting electromagnetic waves that are of a lower frequency than visible light.

Wavelength: equal to a complete up and down cycle, or alternation, of an electromagnetic wave.

Web-fed: press feeding system that uses a roll of paper.

Welding: a binding method that involves exposing the binding edge to ultrasonic waves. This fuses the sheets together.

Wetting agent: chemical used to prevent water spots on developed film.

Whiteprints: copies of drawings that have a white background with blue image lines. Also called diazo prints.

Widows: in typographic terms, short lines at the end of a paragraph or bottom of the column.

Windows: a user interface that is widely used on IBM-compatible personal computers. It allows the display of information on one or more rectangular portions of the screen while the program screen is still underneath. (Also called a graphical user interface or windowing system).

Wipe: visual transition that gives the appearance of pushing one scene out by another scene.

Wireframe model: three-dimensional CAD drawing in which all edges of the object, including edges that would normally be out of sight to the observer, are shown as lines.

Word processor: software that simplifies writing and rewriting.

Word spacing: the amount of space between words, which should be equal to the height of the letters.

World Wide Web: one aspect of the Internet, consisting of pages of information and graphics that can be downloaded and displayed on a computer.

X

Xerography: widely used photocopying process employing an electrostatic charge, powdered toner, and heat.

Z

Zoom: CAD command that allows the user to enlarge or reduce a portion of the drawing on the screen.

Index/Glossary Reference

USING THE INDEX/GLOSSARY REFERENCE

The Index/Glossary Reference serves two purposes. It can be used as an index for finding topics in the text. It also provides an easy method of locating definitions in the text.

A **bold typeface** is used to give page numbers of definitions. If you need the definition of a technical term, simply turn to the page number printed in bold type. The defined word is also printed in bold type.

The Index/Glossary Reference has several advantages over a conventional glossary because it allows you to obtain more information about the terms you want to define. You can read the definition in the context of the book and also refer to the illustrations relating to the new terms.

A

*For the definitions of technical terms, turn to the page numbers printed in **bold type**.*

D

E

F

I

J

K

L

M

P

*For the definitions of technical terms, turn to the page numbers printed in **bold type**.*

Q

R

S

T